D0805374

Topics in Photosynthesis — Volume 1

THE INTACT CHLOROPLAST

TOPICS IN PHOTOSYNTHESIS — VOLUME 1

The Intact Chloroplast

Edited by

J. Barber

Imperial College of Science and Technology,
Department of Botany, Prince Consort Road,
London SW7 2BB, United Kingdom

ELSEVIER SCIENTIFIC PUBLISHING COMPANY

AMSTERDAM — NEW YORK — OXFORD — 1976

Published by:

ELSEVIER/NORTH-HOLLAND BIOMEDICAL PRESS
335 Jan van Galenstraat
P.O. Box 211, Amsterdam, The Netherlands

Distributors for the United States and Canada:

ELSEVIER/NORTH-HOLLAND INC.
52, Vanderbilt Avenue
New York, N.Y. 10017

Library of Congress Cataloging in Publication Data
Main entry under title:

The Intact chloroplast.

(Topics in photosynthesis ; v. 1)
Includes index.
1. Chloroplasts. I. Barber, James. 1940-
QK882.I53 581.8'733 76-23261
ISBN 0-444-41451-7

Printed in The Netherlands

Foreword

When Joseph Priestley had discovered that vegetation was the agent responsible for reversing the effects of animal respiration Benjamin Franklin in 1773 commented "that the vegetable creation should restore the air which is spoiled by the animal part of it looks like a rational system". At the beginning of the 19th century the system had come to be described in chemical terms. Robert Mayer in 1845 showed that it was indeed rational: the radiant energy received from the sun was continually restoring the energy being degraded by animal life. The process of energy conversion in a green plant seems far more efficient than any artificial system that modern technology can devise. Hence the attraction to a study of photosynthesis both for a practical standpoint and for the endeavour to find aesthetic appreciation in science. The secret of this efficient natural machinery seems to lie hidden in the chloroplast.

Von Sachs had observed the rapid formation of starch in chloroplasts when leaves were illuminated. He regarded this as being a property of the chloroplasts which was independent from the rest of the cell. The whole process, then, would be the transformation of the atmospheric carbon dioxide plus water into starch plus oxygen. To prove this posed the problem of removing chloroplasts from the cell and preserving their activity. But it was well established that even relatively slight damage to the plant tissue would completely obliterate all observable photosynthesis. However, in few cases, when the tissue of a green leaf was carefully cut under hypertonic sugar solution apparently undamaged chloroplasts could be seen under a microscope to have been freed from the cells. These chloroplasts would show a residue of their photosynthetic activity by producing oxygen which could be detected with motile bacteria.

Some years ago in a practical class in biochemistry for plant-physiological students, one was faced with the question "why do we always have to use pestles and mortars". This would seem to have emphasized the apparent hopelessness for the success of any effort to obtain preparations of undamaged chloroplasts in bulk. However that might be, the possibility of success with green leaves really seemed to emerge at Dytchleys, Essex, under the auspices of Queen Mary College in London when C. P. Whittingham was Professor of Botany. This depended upon two conditions, skillful craftmanship and, as Martin Gibbs pointed out later, the addition of a substance to abolish the latent period which otherwise would mask the activity. Thus, since 1964 the method, developed by one of the contributors to this present volume (D. A. Walker), has led, with repeated modifications, to the general accessibility of whole chloroplasts for investigations in photosynthesis.

Dr. Barber is to be congratulated on his weaving together these comparative studies of chloroplasts. They give expert accounts of structure, biochemistry and development representing the several individual approaches of his contributors.

Robin Hill, F.R.S.

Biochemistry Department
University of Cambridge
Cambridge, U.K., 1976

Preface

During recent years research in photosynthesis has considerably expanded and it is often difficult for a worker in one particular area to keep abreast with progress and development of new concepts in other parts of the field. This is a pity and can have serious implications as it may lead to fragmentation of research which would benefit by being coordinated. The aim of this new series on photosynthesis is to produce volumes which will attempt to develop a particular theme making them useful both to specialists and non-specialists.

This volume, the first of the series, deals with the structural, biophysical, biochemical and physiological aspects of intact chloroplasts. That is, the chloroplast is dealt with as a complete organelle retaining both its intrathylakoid and stromal compartments. The stroma is not only considered in terms of its biochemical activity but also as a physiological medium for the thylakoid membranes. It is now clear that the photochemical and electron transfer events of photosynthesis are to some extent under the control of the biochemical activity and ionic composition of the stroma and the importance of these interrelationships is discussed in the earlier chapters of this book. The later chapters deal more with the biochemical activity of chloroplasts and like the earlier chapters emphasise the role of the outer chloroplast membranes or envelope as a biochemically active diffusion barrier. They also raise the question of the autonomy of the organelle within the intact cell. Various biochemical interactions between the chloroplast and cytoplasm are dealt with and the last chapter gives an overall picture of the collaborative integration of chloroplast and cytoplasmic functioning.

Contributors have written within the theme of the title by developing their own individual approach but where necessary have provided sufficient background concepts to make the book valuable to students and non-specialists. I have tried to maintain continuity throughout and there is extensive cross referencing.

In chapter 1 Jim Coombs and Denis Greenwood have given a comprehensive account of higher plant chloroplast structure and have also discussed in some detail the structural characteristics of the photosynthetic apparatus of algae. The diversity of algal chloroplast structure raises some important points not only about the origin of higher plant chloroplasts but also about the relationships between chloroplast structure and function. The next two chapters by Willem Vredenberg and myself respectively describe how changes in the ionic composition of the chloroplast's compartments can affect primary photochemical events giving rise to phenomena such as electrical potential differences, energy and electron transfer processes and changes in chlorophyll fluorescence yield. Chapter 4 by David Hall and chapter 5 by Heinrich Krause and Ulrich Heber essentially deal with the mechanisms of light induced

ATP and NADPH production and discuss how the relative concentrations of these two substances change in the chloroplast under different conditions. The role of the outer membranes in controlling movements of metabolites in and out of the chloroplast is also discussed in chapter 4 but is considered in greater detail in chapter 6 by Hans Heldt. In chapters 7 and 8 David Walker and Jim Coombs respectively have presented up-to-date accounts of photosynthetic CO_2 fixation in both C3 and C4 metabolism. David Walker has discussed photosynthetic induction in terms of the autocatalytic nature of the Calvin cycle while Jim Coombs has dealt with C4 metabolism on a broad basis emphasising the problems of the complexity of conventional schemes. The ability of chloroplasts to produce reducing power for reduction processes other than CO_2 fixation is emphasised in chapter 9 by Jens Schwenn and Achim Trebst on the photoassimilation of SO_4^{2-} and in chapter 11. In chapter 11, Rachel Leech and Denis Murphy deal with NO_3^- reduction and amino acid synthesis as well as discussing the role of chloroplasts in lipid synthesis. These and earlier chapters clearly indicate that the intact chloroplast is not an autonomous organelle and this is emphasised by John Ellis in chapter 10 in which he outlines the limitations of chloroplasts in producing their own proteins and nucleic acids. The final chapter by John Raven calls on much of the knowledge presented in the preceding chapters and discusses the sharing of reduced carbon, ATP and photoreductant between the chloroplast and cytoplasm in intact cells.

Overall the book emphasises that in order to obtain a full description of photosynthesis it will be necessary to identify all those processes which take place within chloroplasts and define the relationships of these processes with each other and with cytoplasmic functioning. Much of the knowledge presented in the book has been gained by using isolated chloroplasts having functional envelopes and classified as Type A by David Hall (see chapter 4). These are ideal preparations for bridging the gap between studies with conventional broken chloroplasts and intact cells and have the distinct experimental advantage in that simple removal of the outer membrane by osmotic shock can be compensated by "reconstitution" of the suspending media to a condition which mimics the stroma. Such a technique has played a particularly important part in our present day understanding of CO_2 fixation as emphasised by David Walker in chapter 7 and I am sure will represent an important approach for future research into various aspects of the photosynthetic processes.

It has given me considerable pleasure to edit this volume and I thank the contributors for presenting their specialised knowledge in a clear and comprehensive manner. I also thank my colleagues, Alison Telfer, Mike Hipkins, John Mills and Geoff Searle for their help and in particular I would like to thank my wife, Lyn, and Jennifer Nicolson who bore the brunt of the tedious jobs which befall both a contributor and an editor. Finally, it is with great respect that I thank Robin Hill for writing the foreword for this book and acknowledge his outstanding contributions to our present day knowledge of photosynthesis.

London, 1976 J. Barber

List of Contributors

J. Barber, Department of Botany, Imperial College of Science and Technology, Prince Consort Road, London SW7 2BB, United Kingdom
J. Coombs, Tate and Lyle Ltd., Group Research and Development, P.O. Box 68, Reading, United Kingdom
R. J. Ellis, Department of Biological Sciences, University of Warwick, Coventry CV4 7AL, United Kingdom
A. D. Greenwood, Department of Botany, Imperial College of Science and Technology, Prince Consort Road, London SW7 2BB, United Kingdom
D. O. Hall, University of London, King's College, 68 Half Moon Lane, London SE24 9JF, United Kingdom
U. Heber, Botanisches Institut der Universität Düsseldorf, 4 Düsseldorf, German Federal Republic
H. W. Heldt, Institut für Physiologische Chemie und Physikalische Biochemie der Universität München, 8 München 2, Goethestrasse 33, German Federal Republic
G. H. Krause, Botanisches Institut der Universität Düsseldorf, 4 Düsseldorf, German Federal Republic
R. M. Leech, Department of Biology, University of York, York, United Kingdom
D. J. Murphy, Department of Biology, University of York, York, United Kingdom
J. A. Raven, Department of Biological Sciences, University of Dundee, Dundee DD1 4HN, Scotland
J. D. Schwenn, Department of Biology, Ruhr University, Bochum, German Federal Republic
A. Trebst, Department of Biology, Ruhr University, Bochum, German Federal Republic
W. J. Vredenberg, Center for Agrobiological Research, P.O. Box 14, Wageningen, The Netherlands
D. A. Walker, Department of Botany, The University of Sheffield, Sheffield, United Kingdom

Contents

For a detailed list of contents the reader is referred to the first page of each chapter

The Intact Chloroplast — edited by J. Barber

Chapter 1

Compartmentation of the Photosynthetic Apparatus

J. COOMBS and A. D. GREENWOOD*

*Tate and Lyle Ltd., Group Research and Development, P.O. Box 68, Reading, and *Department of Botany, Imperial College of Science, Prince Consort Road, London (Great Britain)*

CONTENTS

Abbrevations used in plates:
(1) Structure: A, A space; Bp, broken plastid; BS, bundle sheath; Ch, chloroplast; Cr, chloroplast endoplasmic reticulum; CW, cell wall; Cy, cytoplasm; D, DNA region; Di, disc; En, envelope; ER, endoplasmic reticulum; F, fret; G, grana; Gb, girdle band; Ge, genophore; Go, Golgi; Ip, intact plastid; L, lamellae; Lo, locullus; Lt, limiting thylakoid; M, margin; m, matrix of the starch compartment; Mb, micro-bodies; Mc, mesophyll cell; Mi, mitochondria; N, nucleus; Nm, nuclear membrane; P, partition; Pc, perichloroplastic space; Ph, phycobilisomes; Pl, plastoglobuli; Pm, plasma membrane; Pr, peripheral reticulum; Py, pyrenoid; Ri, ribosomes; Sr, stroma; St, starch; T, tonoplast; TB, thylakoid band; Th, thylakoid; V, vacuole; Ve, vesicle; Vt, vascular tissue.
(2) Fixatives and stains: Glut/Os, 2—3% aqueous glutaldehyde followed by Os in same buffer; Os, fixed in 1% aqueous osmium tetroxide buffered at about pH 7.0; Pb, sections stained with Reynold's lead citrate solution; Pm, fixed in unbuffered 2% aqueous potassium permanganate; U, sections unstained; Ua, sections stained with 2% aqueous uranyl acetate.

1.1. INTRODUCTION

In higher plants the process of photosynthesis occurs within specific cytoplasmic compartments, the chloroplasts. The structure of such organelles is, of course, well documented and has been discussed in detail in a number of recent reviews and monographs. These include descriptions of the structure of mature chloroplasts [138], relationships between this structure and photosynthetic function [2,109] and aspects of chloroplast development [72,73]. In particular an excellent comprehensive review has recently been published by Gunning and Steer [53a].

The present volume is concerned mainly with the concept of the higher plant chloroplast as a separate cellular compartment, with a specific and restricted biochemical function.

This chapter will consider to what extent the higher chloroplast represents an optimal structural development, modified from more primitive ancestral forms, to carry out the complex physical and biochemical reactions which are described in more detail in other chapters.

With the possible exception of the C4 plants (see Chapter 8) the chloroplasts of higher plants are remarkably constant in basic features of composition, enzymology and function. The main purpose of recent work has been to correlate the known biophysical and biochemical reactions with specific structural components.

The actual observation of fine details of chloroplast ultrastructure has only been possible since the advent of the electron microscope — a period of 25 years or so. Furthermore, difficulties arise during sample preparation when the plant material is processed through a variety of fixation, dehydration and embedding procedures followed by staining if necessary with electron opaque substances. The possibilities that artifacts may arise during such treatment makes interpretation of electron micrographs subjective and has resulted in a literature in which development of new techniques leads rapidly to new suggestions concerning the detailed structure of the chloroplast. However, in spite of such problems and notably by employing a limited range of special procedures for fixation and embedment a consistent pattern of chloroplast structure may now be recognised in sections of intact cells and in more varied preparations of the isolated organelles. The interpretation of many features in micrographs of sections, or in whole mounts of chloroplast fragments, have been largely substantiated and additional details observed by using the independent and more recently developed technique of freeze-etching [97,131]. This technique involves the replication of surfaces exposed by fracture of a rapidly frozen specimen of fresh or minimally processed tissue or cell homogenate with or without a further process of etching to controlled depths by sublimation of ice. In particular it has provided information on the substructure of the internal membranes, adding to previous information obtained by examination of metal-shadowed or negatively stained preparations of burst chloroplasts.

1.2. GENERAL CHARACTERISTICS OF HIGHER PLANT CHLOROPLASTS

1.2.1. Size and shape

Most published electron micrographs of ultrathin sections of higher plant leaves show the characteristic lens shape of the chloroplast, in sectional profile, with a long axis of 5 to 10 μm (Plate II). This view is selected since, in general, it reveals the stacking of the internal membranes to the best advantage. However, as shown in Plate Ia, when sectioned in a plane parallel to the main surfaces of the plastid both the chloroplast and the more conspicuous of the internal membranous layers appear discoid. Such disc-like internal layers are also revealed in metal-shadowed preparations of fragments of isolated chloroplasts (Plate Ic). These types of images, which were typical of those obtained in early studies of chloroplast structure [28,51,132], suggested that the chloroplast consists of a bag of disc-like membranous lamellae of uniform diameter, stacked like "piles of pennies" to form the cylindrical *grana* which had been previously observed in the light microscope.

These early studies indicated the three major structural regions of the chloroplast; i.e. the pair of outer membranes or *envelope* (Plate Ib), the amorphous *stroma* and the highly organised internal lamellar structure (Plate Id). It is now clear that in general the light-dependent biophysical reactions (see Chapters 2, 3 and 4) of photosynthesis are associated with the internal membrane system whereas the dark biochemical reactions of CO_2 assimilation are dependent on the soluble proteins of the stroma (see Chapters 7 and 8). It is also clear that the outer envelope represents a selectively permeable barrier which may regulate the movement of fixed-carbon, reducing power and chemical energy into and out of the chloroplast (see particularly Chapters 5 and 6).

Early studies were carried out using plant material which had been fixed with compounds such as potassium permanganate [143] (Plate Ib) or osmium tetroxide [132,133] (Plate Ia), which are electron opaque. Although these fixatives preserve the membrane components or replace them by opaque residue, they do not enable much detail to be resolved in the stroma which is often partially lost. Improved fixation techniques using compounds such as glutaraldehyde [118], by which some otherwise labile protein components are cross-linked, and additional staining with lead citrate [113] or uranyl acetate not only preserved much of the organisation in the stroma (Plate II) but enabled the formulation of more detailed concepts of structure for the chloroplast membranes.

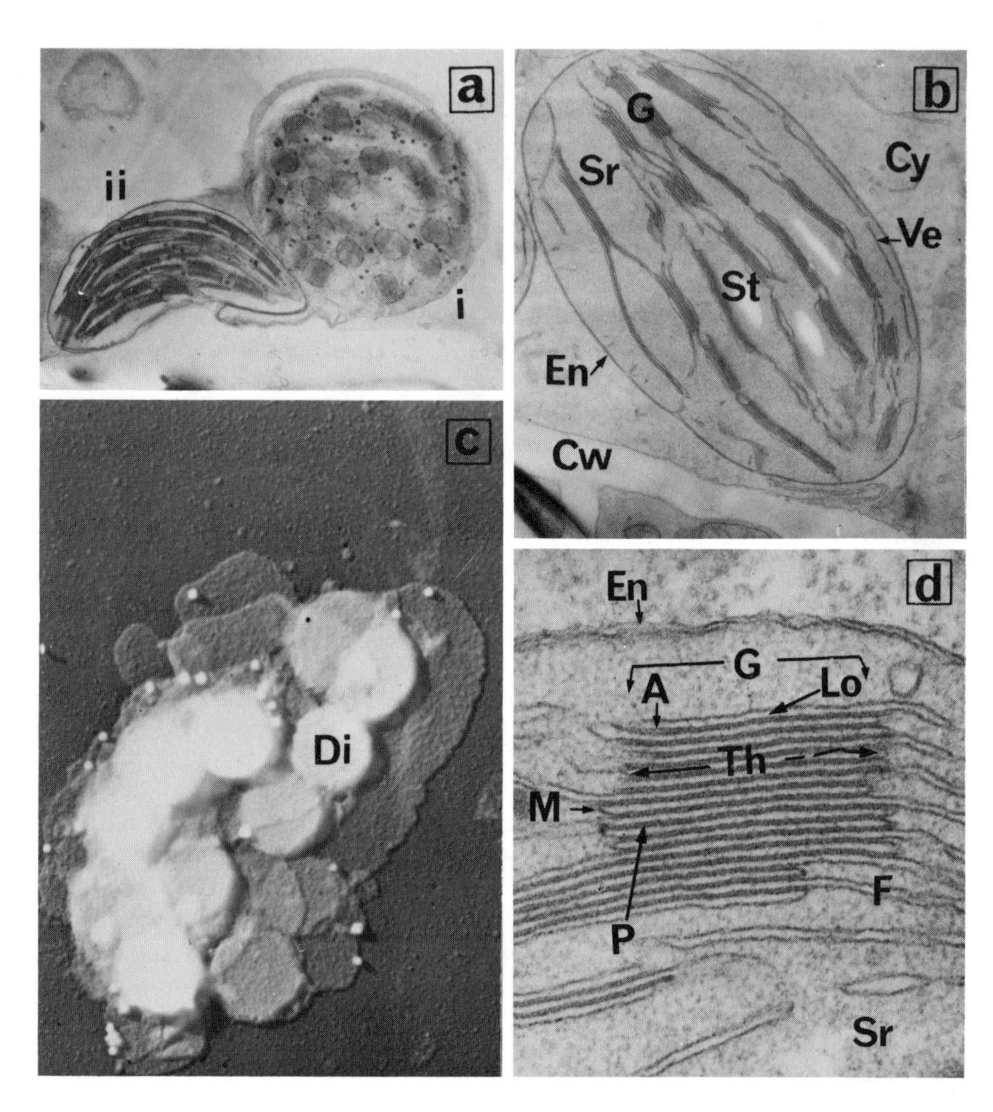

Plate I. Electron micrographs showing aspects of organisation in higher plant chloroplasts. (a) Two chloroplasts in a cell of *Vicia faba* sectioned in a plane parallel (i) and at right angles (ii) to the major plane of the internal lamellae. Os; × 10 000. (b) Chloroplast in a cell of *V. faba* showing vesicles (Ve) immediately within the envelope (En) with some connections to the inner membrane of the latter. The stromal contents have been destroyed by Pm fixation allowing clearer differentation of the envelope and internal lamellae. Pm, U; × 20 000. (c) Electron micrograph of a fragment of the lamellae system from a preparation of chloroplasts of *V. faba* isolated in 0.3 M glucose showing granal discs (Di) and stroma lamellae in face view. Dried and washed preparation shadowed with Au/Pd; × 36 000. (d) Section of a granum (G) of a chloroplast from *Spinacia oleracea* showing arrangement of thylakoids (Th) to form partitions (P) with central space (A). The partitions and margins (M) bound the electron-translucent loculus (Lo). The grana are linked by the stromal lamellae or frets (F) passing through the stroma (Sr). Note also the limiting membranes of the envelope (En). Glut/Os, Ua, Pb; × 90 000.

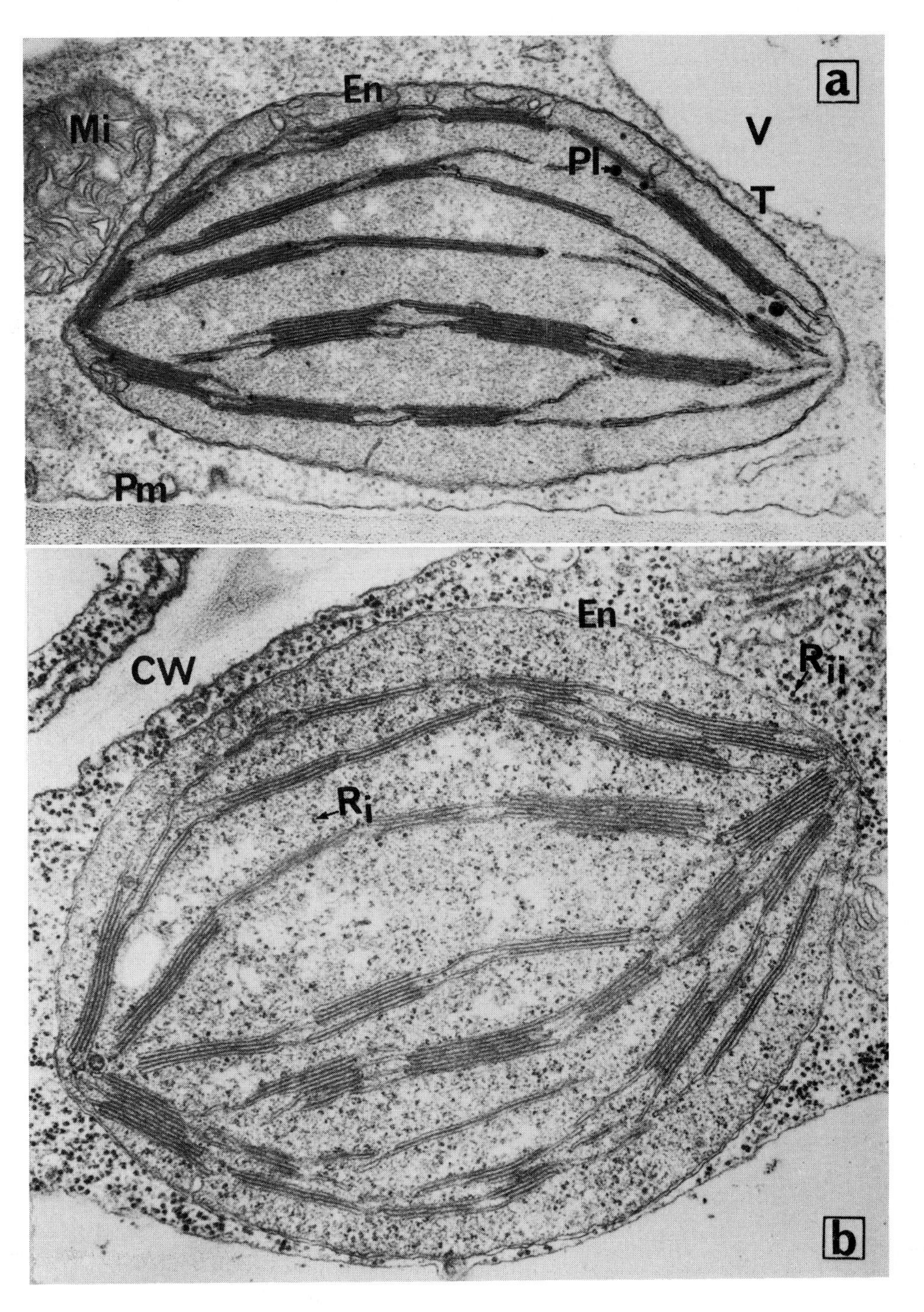

Plate II. Sections of chloroplasts in cells of *S. oleracea* cut from the same embedded specimen showing the effect of staining the sections in (a) lead citrate (Glut/Os, Pb; × 22 000) or (b) uranyl acetate followed by lead citrate (Glut/Os, Ua, Pb; × 32 000). Note the accentuation of membranes and plastoglobuli (Pl) in (a) and ribosomes (R_i and R_{ii}) in (b). Also plate (b) shows that the chloroplastic ribosomes (R_i) are smaller than those found in the cytoplasm (R_{ii}).

1.2.2. Outer membranes

In the higher plants the outer envelope invariably consists of two membranes separated by an electron-translucent space of about 10 nm.

In general three types of image have been recorded for the membranes of the chloroplast envelope, reflecting the various staining and/or fixation techniques used. These are (a) the typical unit membrane [116] structure consisting of two electron-dense regions about 2 nm thick with a central translucent region of comparable width [58,143]; (b) a single electron-opaque layer of up to about 10 nm deep; (c) a membrane composed of globular subunits [145,146]. The unit membrane type of image is usually associated with permanganate or osmium fixation (Plate I) and the presence of globular sub-units with glutaraldehyde fixation and post-staining with lead. Not uncommonly the inner of the two membranes appears thinner than the outer membrane.

Although the structure of the outer envelope has not been investigated in detail some attention has been paid to the small invaginations which can be seen arising from the inner membrane (Plate IIa). It has been suggested that these may represent cross-sectional views of plate-like structures arising from the inner membrane [124]. In the C4 plants (Plates VIII, X, XI] the extent of the vesicular structures lying below the envelope is much greater. This ramifying association of anastomosing vesicles, which has been termed the *peripheral reticulum* [75], is considered in more detail in section 1.6.

Studies on chloroplast development (reviewed in refs. 73, 95) have indicated that invaginations of, or vesicles formed from, the inner membrane of the envelope may contribute to the development of the granal system. However, such observations have to be correlated with reports that the chemical composition, and in particular the lipid composition, of the isolated chloroplast envelopes differs significantly from that of the internal granal membranes [91].

1.2.3. Stroma and inclusions

The envelope encloses the granular matrix or *stroma*, in which lies the internal system of lamellae. The dry matter of the stroma is largely protein. In C3 plants the major component of this protein is known as fraction I [71] the main catalytic activity of which is that associated with ribulose-*bis*-phosphate carboxylase (see Chapter 7). The degree of structure which can be resolved in the stroma depends to a large extent on both the fixatives and the electron-opaque materials used in the preparation of the specimens. However, even with improved techniques using glutaraldehyde, lead citrate and uranyl acetate (Plate II), little if any detail of the enzyme components can be resolved. However, protein aggregates, presumably composed of fraction I protein, have been described in chloroplasts of *Avena* following fixation in OsO_4/glutaraldehyde. These so called *stroma centres* [53,54] consist of an ordered or paracrystalline arrangement of fibres of about 8.5 nm in diameter and up to 200 nm in length arranged in bundles with hexagonal close packing. Similar crystalline bodies have been reported in normal preparations of leaves from certain other higher plants [112], however, they can also arise as artifacts of fixation [126].

Chloroplasts contain also a number of discrete particles embedded in the stroma, which in general represent stored photosynthate. The most common of these are starch grains, which may be up to 2 μm in length, and plastoglobuli [4]. The starch grains do not usually take up electron-opaque materials and thus appear as electron-transparent areas (Plate Ib), although in other preparations electron-opaque regions are observed (Plates IVa,VIIIa,Xc). Although little work has been done to positively identify these bodies as starch grains under the electron microscope or explain the reason for the electron transparent and opaque regions extrapolation from histochemical observations under the light microscope indicate their true identity. In higher plants the starch grains appear to lie freely within the stroma and are not associated with a specific membrane system.

Following fixation with OsO_4 many small round electron-opaque bodies, which have been termed *plastoglobuli*, are seen in almost all chloroplasts [4,45]. These plastoglobuli again lie free in the stroma and apparently function as storage of lipid material, in particular plastoquinone and tocophoryl-quinones [84,87]. In aged leaves, or during transition of chloroplasts to chromoplasts, the composition of these globules may change and carotenoid-type pigments accumulate within them [86].

Electron-opaque material, similar in appearance to the substance ferritin observed in animal tissues, has also been observed in plant tissue [67]. Such material, which has been termed phytoferritin, is always associated with plastids, often in senescing leaves [8]. The individual particles of phytoferritin, which are about 10 nm in diameter, may be arranged in a variety of regular patterns ranging from rows to paracrystalline aggregates arranged in an octahedral manner with a subunit at each of the six vertices [115].

Ribosomes are seen in varying abundance in the stroma of most higher plant chloroplasts [25,99] becoming intensely stained when the sections are treated with uranyl acetate and lead citrate (Plate IIb) but are less conspicuously differentiated with lead staining alone (Plate IIa). These are less obvious or not seen when the plant material is fixed with permanganate or osmium tetroxide alone (Plate Ia,b). The individual ribosomes are about 17 nm in diameter and thus smaller than the cytoplasmic ribosomes which have a diameter of about 25 nm. This difference in size is borne out by separation of ribosomes on density gradients or following gel electrophoresis. Using the appropriate cytological, autoradiographical and biochemical techniques it has been established that these particles, which under some conditions appear as polysomes attached to the chloroplast membranes

[41], do in fact contain RNA and play an important role in the biosynthesis of part of the fraction I protein (see Chapter 10). Similar techniques have established the presence of DNA in the stroma of a wide range of plants including algae, Bryophyta [90], Pteridophyta, Gymnospermae and Angiospermae [152]. This DNA which appears as a mesh of 2.5 nm fibrils, removed by treatment with DNAase, can be distinguished from the nuclear DNA by its base composition, buoyant density and other characteristics. It has been possible to isolate the individual DNA strands from chloroplasts. These have been shown to consist of closed loops, or circular molecules of about $9 \cdot 10^7$ daltons [136].

1.2.4. Internal lamellae system

The structure of the internal lamella system is extremely complex. When chloroplasts are burst, and the resulting fragments shadowed with heavy metal (Plate Ic) the appearance is that of a number of prominent discs of more or less uniform diameter accompanied by more extensive sheets of membrane. The disc-like appearance of the internal membranes can also be seen in the tangential profile shown on the right hand side of Plate Ia.

The conventional view is that shown in Plate Ib. In such sections two distinct features of the membranous structure can be recognised. The first, closely packed, regions (Plate Id) correspond to the *grana* seen under the light microscope. The second comprises the less dense but extensive interconnecting stromal lamellae.

On the basis of early information of the type shown in Plate I it was suggested that the grana consisted of a series of separate sac-like discs between 0.2 and 0.4 μm in diameter, stacked to form the cylindrical granal structures [28,51,132,133]. It was further suggested that the grana were discrete structures connected one with another by similar but more extensive discs comprising the intergranal lamellae [69,94,95,128]. Subsequent studies in which serial sectioning and negative staining were used led to a rapid understanding of the true complexity of the inner membrane system. The progressive stages of elucidation leading to the present concepts are shown diagrammatically in Fig. 1.1. The major advances were the realisations that the intergranal lamellae were perforated so forming an interconnecting fretwork system [58,143] the layers of which could in fact be connected to lamellae at several levels within the same granum [150]. Further studies indicated that the membranes were probably connected in such a way that individual *frets* followed an oblique course in association with each successive layer of an individual granum [58,59,141]. Extensive investigations by Paolillo et al. [103–105] indicated that the multiple connections with a single fret appear to form a right-handed helix in attachment to consecutive levels at the margin of a granum and that groups of several parallel frets may be similarly associated with a single granum

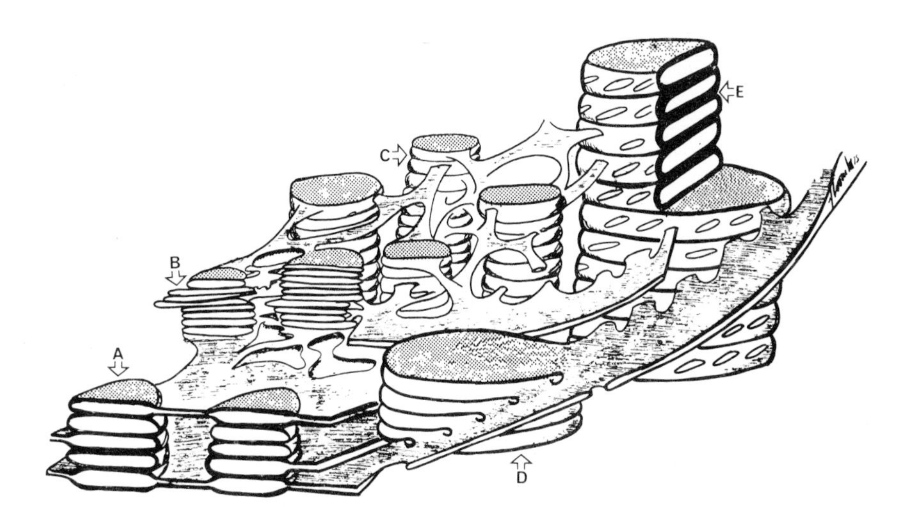

Fig. 1.1. Schematic representation showing evolution of the concept of membrane stacking in the grana of chloroplasts from higher plants: A. Early suggestions showing grana formed of individual discs (small thylakoids) linked by more extensive stromal discs (large thylakoids); B. Perforated stromal lamellae (frets) linking grana at several levels; C. Frets connecting with several thylakoids within the same grana; D. Spiral formation of grana membranes; E. Spiral fretwork arranged around and interconnecting the individual grana. (Based on the suggestions of Menke, Weier, Wehrmeyer, Heslop-Harrison, Paolillo, Thomson and others as indicated in the text.)

although the interconnecting frets are typically fewer (often about half the number) than the layers in a granum. Comparable results consistent with such a structure have been observed in a wide range of higher plants including those species most often used in previous investigations: *Zea mays*, *Phaseolus vulgaris*, *Cannabis sativa*, *Elodea canadensis*, *Nicotiana rustica*, *Pisum sativum* and *Spinacia oleracea*. Hence, it would now appear that the complex internal system of lamellae is derived by the folds and connections of a single continuous sheet of membrane that divides the volume within the envelope into two compartments each forming a separate continuum [59]. One compartment is that of the stroma proper, the other represents the intracisternal volume enclosed within the membranes of the grana and frets.

If it is accepted that the internal membrane system is indeed comprised of a single folded and perforated sheet the terms *disc* and *lamellae* lose much of their original meaning. Strictly "disc" should not be used [49] since it implies a view of the general shape and isolation of units of structure in the membrane system that is no longer tenable. "Lamella" retains its usefulness in reference to the uniformly flattened form of the predominant parts of the system for which the convenient term *thylakoid* is often virtually a synonym although coined [95] to express the concept of the lamellae as a specific type of cisterna. Both terms are frequently employed in the plural form, for example in describing sectional profiles where the integral nature of the membrane system is not self-evident or in doubt and where only separate or sparsely interconnected parts are evident. Additional terminology developed mainly by Weier and his associates [137,142,143,147,148,149] retains its usefulness for particular purposes of description or allusion to special functional or structural attributes of the parts of the system to which it applies. This includes the term *fret*, already discussed, to described the interconnecting parts that may otherwise be referred to as stromal or *intergranal* lamellae or thylakoids. The areas of paired membranes constituted by the close contact or adhesion of the surfaces of the adjacent thylakoid layers within the granum are termed *partitions*. The membranes exposed to the stroma at the edge of the granal thylakoids — which may preferentially accumulate heavy metals [119] — are termed *margins*. The partitions plus the margins enclose the electron-translucent *loculus*. In some preparations the partitions may be resolved as two dark layers separated by a second translucent region, the A space [58].

1.2.5. Membrane structure

At a higher level of resolution the appearance of the thylakoid membranes depends to a large extent on the fixation procedure and the electron-opaque materials used in sample preparation. In general terms the various images produced are similar to those recorded for the envelope membranes, ranging from the tri-partite (unit) membrane to membranes apparently comprised of globular subunits [64,100,146]. A highly detailed membrane structure, based on the arrangement of globular proteins has been proposed [10,144] which suggests that the partitions consist of two layers of subunits and the frets of one layer of subunits.

The concept of a globular or particulate substructure is supported by the more recent results obtained with the use of freeze fracture and etching. However, there has been some difference of opinion in the interpretation of these recent results with respect to earlier studies of this type on chloroplasts [98]. Problems of interpretation arose since it was not self-evident whether a fracture had occurred along an interface at the surface of a membrane or in a hydrophobic region within it [21] and varied conclusions were drawn as to the superficial or internal location in the membrane of the particles thus exposed. However, these differences now appear to be substantially resolved [109]. Different authors have used different terminology to describe the fracture faces; recently attempts have been made to produce a generally accepted terminology to be used in freeze-etch studies [22,53a].

It appears that the layered substructure of the membranes allows them to split more or less readily along several internal planes of weakness [23,107]. In contrast the external surfaces of membranes in contact with the stroma are hydrophilic and when frozen rarely, if ever, become planes of fracture.

However, these surfaces have been observed in preparations of isolated chloroplast lamellae washed free of stroma and then subjected to freeze fracture followed by deep etching to remove a layer of ice covering the thylakoids. Characteristic particles, about 10 to 11 nm in size, are attached to the external surfaces (A′ faces) of the frets but these can be removed [3,107a] by washing the lamellae with solutions containing EDTA prior to freeze-etching which shows little remaining surface relief. The structure of the membranes is asymmetric and fracture occurs most easily along a preferred plane to leave exposed one or other of two non-identical faces (B and C) that are readily recognised by the different types and numbers of particles distributed upon them. These particles in the interior of the membrane are not susceptible to removal by prior treatment of the lamellae with EDTA. The characteristic particles of the B face are larger than any seen in other types of biological membrane and estimations of their size vary mainly between about 13—17 nm in diameter and about 8—9 nm in height. The B particles are few in number and perhaps sometimes absent in the frets but are numerous in the grana, lying in a layer close to the locular surface of the membrane. The opposed face of fracture (C) exposes a layer of smaller particles, about 9—12 nm in size that occur in large numbers in both grana and frets; they are situated adjacent to the A′ surfaces and in corresponding positions in the partitions of the grana. The membranes of the partitions are occasionally split apart by freeze fracture and then reveal smooth faces (A) lacking the superficial particles carried on the A′ surfaces. The other surface of the membrane bounding the internal compartment of the thylakoids, is not uncommonly exposed by deep etching and presents a further type of face (D) characterised by "bulges" measuring about 18.5 × 15.5 nm which are regarded as caused by the larger B particles in this side of the membrane. These results indicate that at least two types of particles exist on the thylakoid membranes. The larger particles occur mainly in the partitions whereas the small particles occur in both stromal and granal lamellae.

The concept of individual particles or units associated with photosynthesis is not new. The photosynthetic unit was defined in 1932 [39] and subsequently the term *quantasome* was used for particles in ordered array observed on the surfaces of dried metal-shadowed or negatively stained fragments of isolated grana [106,108]. It was suggested that these quantasomes, which appeared to be comprised of four subunits about 6.5 nm in diameter, could be equivalent to the photosynthetic unit. This view has been questioned [66] on the basis that the particles could represent proteins of stromal origin with enzymic functions, such as fraction 1 protein or an ATPase. Attempts made to relate these observations and concepts to those arrived at with the aid of the freeze-etching techniques have proved less successful than might have been anticipated. Thus the presence of the particles on the A′ face after the stroma has been washed away and their removal by treatment with

EDTA can be correlated with ATPase activity. In contrast to this quantasomes may remain after treatment with EDTA [107]. The displays of quantasomes as originally seen in dried preparations resemble closely the views of the D face after freeze etching and it has been proposed that the characteristic bulges of the latter together with the underlying large particles can be better equated with quantasomes [109]. The size of such a structured unit and the arrangement of the large particles in paracrystalline order [106] when they are present in large numbers on B faces support this suggestion. However, the fret membranes in which large particles are scarce or absent contain chlorophyll and mediate photochemical reactions. For these and other reasons the concept of a quantasome as the structural counterpart of a functional photosynthetic unit in the full classical sense of this term is in doubt.

Techniques of sonication, use of detergents and other disruptive procedures have been developed by which it is possible to produce membrane fractions enriched in both large and small subunits from isolated chloroplasts [2,3,121]. Comparison of the functional competence of such isolated fractions with that of intact membranes in which the grana and frets remain connected suggests that the smaller particles may be associated with photosystem I and the larger particles with photosystem II. The combined investigations of ultrastructure and function suggest that the morphological differentiation of the thylakoid in the form of grana and frets facilitates a corresponding pattern of partial functions that is essential to the characteristic modes of photosynthesis in higher plants. It seems also that a more than casual relationship exists between the observable substructure and the location of specific membrane-bound activities the further elucidation of which will remain a fruitful aim of research.

1.3. ORIGIN OF HIGHER PLANT CHLOROPLASTS

A full explanation in molecular terms of photosynthesis in higher plants can be obtained only by the intensive investigation of selected systems, aspects of which are the main concern in this volume. The foreseeable outcome of continued progress along such lines will surely come not far short of a complete account of the flux of small molecules and the energy transformations mediated by the photosynthetic equipment of the cell as well as the specification of its macromolecular framework with the steps by which it is put together and maintained under genetic and physiological control. Yet what is true for one type of organism does not hold for another and our understanding of higher plants will be incomplete, if not defective, without some concepts of how the particular pathways of metabolism and the accompanying idiosyncrasies of cell structure in extant species have come about.

We know that they can only have been derived by modification in descent of patterns in progenitors that were not identically equipped and could not have lived in the range of habits now occupied by higher plants.

With the exception of the C4 plants the structure, composition, enzymology and function of chloroplasts from higher plants are remarkably uniform. This uniformity does not reflect strictures imposed by either the plastid or the nuclear genome since in developing organs, such as root, stem, fruit or leaves similar proplastids are capable of developing into a variety of specialised organelles including chloroplasts, leucoplasts, chromoplasts, amyloplasts or etioplasts. Furthermore, in many instances these diverse organelles are interconvertable. Hence it might be concluded that the specialised chloroplasts of higher plants have evolved (in adaption to terrestrial conditions) an optimum arrangement for photosynthesis, variable only within limits yet capable of adjustment to function in the wide variety of environment accessible to contemporary species. One of the more obvious special attributes that may contribute to such a quality of flexible efficiency is the granal type of membrane system, constant in its fundamentals but variable in features that may modulate its function, such as the total number and arrangement of grana in a chloroplast or the number of thylakoid layers and frets associated with individual grana. In this context it may be noted that the chloroplast dimorphism evident in some C4 plants and the possession generally of numerous types of specialised non-photosynthetic plastids in different cells is possible only in a complex organism with a highly developed transport system nourishing a variety of tissues, some of which are not themselves photosynthetic. In contrast unicellular organisms normally carry at any one time only a single type of plastid, either a chloroplast or a leucoplast and are correspondingly limited in their modes of nutrition.

Unicellular species of plants are found only in the so-called lower groups which evolved primarily in aquatic environments to which they have mostly remained restricted. When these forms (including bacteria, blue-green algae and eukaryotic algae) are examined in detail, a wide variety of structures, pigmentation and organisation is seen apparently carrying out photosynthesis with a similar degree of efficiency and matching in many respects that of higher plants. The basis of diversity in these groups is however of a different order to that in higher forms. The living algae belong to more than a dozen separate major groups, each given the taxonomic status of a class (see Table I) and distinguished by differences in the basic biochemistry and organisation of the cells, whereas the major subdivisions of the higher plants have a unity at the cellular level corresponding basically with only one of the distinctive types of cell in the algae. Increased knowledge of the ultrastructure has considerably clarified the identity of the natural groups of algae and consolidated the concept of an algal class as conforming to a uniformity of cell type and chemistry. Most of the groups listed in Table I are now generally recognised as distinct classes but with the important qualification that the

TABLE I

THE MAJOR GROUPS OF ALGAE CLASSIFIED INTO DIVISIONS AND MAJOR SECTIONS

Based on the recommendations of Christensen [27b] but attributing to the Euglenophyceae a greater affinity to the Chromophyta than to the Chlorophyta and with some alterations in the classes to conform with those in generally accepted current lists.

	Divisions	Classes	
PROKARYOTA			
	Cyanophyta	Cyanophyceae	
EUKARYOTA			
ACONTA	Rhodophyta	Rhodophyceae	
	Chromophyta	Cryptophyceae	Raphidophyceae
		Dinophyceae	Chrysophyceae [a]
		Haptophyceae	Bacillariophyceae [a]
			Phaeophyceae [a]
		Eustigmatophyceae	Xanthophyceae [a]
CONTOPHORA	(Euglenophyta)	Euglenophyceae	
	Chlorophyta	Chlorophyceae	Prasinophyceae
		Charophyceae	

[a]These 4 classes closely resemble each other in chemical and cytological characteristics whilst being morphologically quite distinct.

subdivisions of the Chlorophyta are currently in debate. These classes vary greatly in size and apparent relationship, some resembling each other more closely than others, and there remain wide differences of opinion in allocating them to divisions or larger categories. Almost every class listed in Table I has been accorded the level of a separate division in one or another recent system of classification, according to differing views on their phylogeny. In detail these remain matters of argument and for present purposes the classes have been allotted to a small number of divisions in accordance with the system advocated by Christensen [27b]. This recognises the isolated positions of the Cyanophyceae and Rhodophyceae and the association of the remainder into 2 main assemblages characterised generally by the presence of either chlorophyll *b* or *c*. This scheme expresses the fundamental nature of the traditional gross classification of the algae into major phyla on the basis of the photosynthetic pigments and is supported by other criteria such as the nature of the reserve products, type of flagella, structure and composition of cell wall, etc. Most authors consider the Cyanophyceae should be given divisional rank since they form the only prokaryotic group and are probably nearest in attributes to a prototypical alga. The others are all eukaryotic but the Rhodophyceae cannot be accorded close affinity with any other particular class and are also usually given a separate division. The others show a general resemblance in possessing motile cells with the complex eukaryotic type of flagella or in being clearly derived from such forms and have other common features suggesting relationships. The Chromophyta include 4 groups that resemble each other closely in cytology and chemistry and 5 which are individually more distinct although possessing general features, notably of cell compartmentation and chloroplast structure, common to the division. The Chlorophyceae are a cytologically uniform assemblage having the same basic cell characteristics as higher plants and without close resemblance to any of the Chromophyta. Until recently the Chlorophyta were often lumped together as a single class but with a large number of different orders. Electron microscopy has disclosed a greater diversity in cell details than was supposed but the limits of the classes Prasinophyceae and Chlorophyceae and the allocation of species between them remain as yet incompletely determined. The relationships of the 3 classes listed are uncertain and it cannot yet be said which of these, if any, may be regarded as fundamentally in closest relationship to higher plants. Nevertheless, it is now widely accepted that the higher plants may have evolved from an ancestral stock in common with the Chlorophyta probably by the diversification on land of an originally single line that acquired early in its progress the typical cyclic life-history and the development stage recalled in the term Embryophyta that embraces all known living and fossil forms. A number of different hypotheses have been advanced to account for the cellular diversity in the other groups of oxygenic phototrophes that comprise the algae. These hypotheses involve two alternative and conflicting views of the origins and subsequent evolution of chloroplasts. One view assumes that the chloroplast arose as it were in situ in the cells of a primaeval ancestor [27b,27], having been formed by the differentiation of internal membranes that previously served also other functions. On this basis the diversification of chloroplasts has occurred by changes in the successive progeny of a single ancestor evolving along branching pathways represented now only by the different groups of algae that became isolated by the extinction of intermediate forms. The higher plants would thus be regarded as ultimate branches of a main limb represented by the Chlorophyta. This concept requires only one prokaryotic source (perhaps a blue-green alga) for all eukaryotic species of organisms. However, the total complexity of the simplest type of eukaryotic cell is so much greater than that of any prokarotic one as to have prompted the revival [114] of earlier speculations that chloroplasts originated from photosynthetic cells that had become symbiotic in cells of another type. From this point of view the subsequent evolution of chloroplasts was, at least initially, by a reduction in the functions and structure of a previously fully autonomous organism rather than solely by the development of greater complexity in an originally simpler part of the equipment of a cell. This theory also necessarily invokes the assimilation of at least two different endosymbionts, although not simultaneously, to provide mitochondria as well as chloroplasts. The extreme elaborations of this idea [93a] have become incredible in their full motley which may implicate as many as 5 separate acts of endosymbiosis to elaborate the various eukaryotic organelles. In relation to chloroplasts it has been pointed out that pathways of basic metabolism that are stereotyped in higher plants and animals are often a main topic of variation in prokaryotic groups. This lends special support [112a,130] to extensions of the endosymbiotic theory that attempt to account for the several photochemically variant types of algal chloroplasts by supposing them to have originated separately from a like number of correspondingly pigmented different prokaryotes. It must then be assumed that free living forms of these have become extinct, excepting the Cyanophyta.

Any proposed phylogenetic tree is of course speculative, this is particularly so for the algae where fossil evidence is either not available, or has not been used. However, studies of fossils do suggest that organisms similar to the blue-green algae appeared in the pre-cambrian era prior to the emergence of eukaryotic forms of life. Again it is assumed that these arose after the photosynthetic bacteria, and gave rise to other algal forms. In attempting to distinguish between these various possibilities one can consider differences in the pigmentation, membrane organisation and photosynthetic storage products of the varied extant algal groups. Table II sets out the algal classes in relation to the organisation of the cells and features of the photosynthetic systems and storage products. The plan is based on the pigment groups and a progression from left to right of classes with increasing complexity of cell

TABLE II

THE MAIN CLASSES OF ALGAE
Cell compartments, classification and photosynthesis

Compartments	PROKARYOTIC Thylakoids in the cytoplasm	EUKARYOTIC Thylakoids in chloroplasts			
Enclosing membranes	No enclosing membranes	Chloroplasts in the cytoplasm Envelope of 2 membranes		Chloroplasts in sub-compartments of the cytoplasm Envelope + ER cisterna = 4 membranes (only 3 membranes in some, see* below)	
		ACONTA	CONTOPHORA (with flagella)		
Secondary pigments	Phycobilisomes *on* membranes		Chlorophylls b or c and carotenoids *in* membranes		
Thylakoid arrangement	No stacks (Thylakoids form simple lamellae)		Stacks, compound lamellae, grana	Compound lamellae or bands of 2 or 3 thylakoids usually. Often very regular 3-thylakoid bands { } = girdle lamella.	
Classification	CYANOPHYTA	RHODOPHYTA	CHLOROPHYTA	CHROMOPHYTA (and EUGLENOPHYCEAE)	
Secondary pigment groups (A) Phycobilins (B) Chlorophyll b (C) Chlorophyll c			(B) Charophyceae Prasinophyceae Chlorophyceae		Euglenophyceae*
				(C) Dinophyceae*	Haptophyceae Eustigmatophyceae Raphidophyceae Chrysophyceae, Phaeophyceae Bacillariophyceae, Xanthophyceae
	(A) Cyanophyceae (and some genera of uncertain affinities)	Rhodophyceae		Cryptophyceae	
Typical carbohydrate reserves	"Glycogen"	Amylopectin	Amylopectins and Amyloses		Laminarin, paramylon, etc.
	α — (1,4)glucans "starches"				β — (1,3)glucans no starch

compartmentation. This arrangement brings out parallel trends in photosynthesis, progressing for example from water-soluble to wholly lipidic accessory pigments and from simple to more elaborate and specialised lamellar systems with systematic differences in the nature and location of stored carbohydrates. The Cyanophyta, Rhodophyta and Chromophyta are placed in line because of overlap in several features of pigmentation and structure and the classes in the latter show a lateral spread across the diagram in accordance with their considerable diversity. The uniformity of the Chlorophyta and lack of correspondence in special features with the other divisions places them in a separate but intermediate position of cell complexity. The trends shown in the diagram may be interpreted as indicating similar directions of change in photosynthetic systems that have been selected independently along several evolutionary pathways; they may therefore point to the emergence of certain optimal functional and structural patterns and suggest the nature of underlying common factors that modulate the efficiency of photosynthetic processes. It also indicates the remoteness of the higher plants in levels of cell organisation from the more advanced Chromophyta and the Euglenophyceae. The arrangement emphasises a central gap in the continuity of its overall pattern. It might be supposed that this represents the extinction of intermediate forms in an essentially monofilial evolution from a Rhodophycean stock along the divergent limbs of the Chlorophyta and Chromophyta or alternatively that the various divisions of the eukaryotic algae were of independent origin from separate prokaryotic groups of which the Cyanophyceae may have been one. However, the systematic resemblances among the Chromophyta and Euglenophyceae suggest a monophyletic origin for all of these rather than several polyphyletic sources as recently argued [93a,112a]. Moreover, the detailed comparisons, referred to in later sections, allow rather precise specification of intermediate and supposedly extinct types of photosynthetic organisms that would be required to complete the hypothetical framework of a unitary scheme to explain the evolution of all eukaryotic organisms from a single source as proposed by several authors [27,27b].

Beyond the algae, the situation is much clearer, largely substantiated by fossil evidence, the primary assumption being that the higher plants evolved from a multicellular filamentous stock in the Chlorophyta that gave rise to parenchymatous plants evolving further mainly in terrestrial habitats.

1.4. BACTERIA AS PRIMITIVE FORMS?

Most of the discussion will relate to the oxygenic photo-autotrophs, that is plants in which water is the ultimate electron donor for photosynthesis, oxygen being produced as a by-product. This contrasts with the situation in the photosynthetic bacteria which are never capable of using water as electron donor and which do not evolve O_2. In general the photosynthetic bacteria are capable of assimilating CO_2 into organic compounds, using ribulose-*bis*-phosphate carboxylase in the light under anaerobic conditions where molecular hydrogen, reduced sulphur compounds or inorganic substances serve as electron donors.

The light reactions of photosynthesis in these bacteria are associated with species of chlorophyll that do not occur in oxygen-producing phototrophes [29]. In the green bacteria (Chlorobacteriaceae) the chlorophyll, as found in *Chlorobium spp.* [29,30] is located in a special type of pigment-filled vesicle of simple form that is not limited by a membrane of standard type but by a thin (2—3 nm) layer of different construction and composition. However, chlorophyll in purple bacteria (Thiorhodaceae and Athiorhodaceae) is membrane-bound as it is in higher plants but the photosynthetic membranes, like those in all other prokaryotes, are not confined by a separate limiting envelope. In these bacteria the chlorophyll is incorporated in the cell membrane and invaginations of it that may be vesicular, tubular or flat in shape according to species or to the nutritional state of the cells. In some species the pigmented structures are thylakoid-like and may show a degree of "stacking" [29,60,65,139]. Some purple bacteria are able to respire aerobically and membrane-bound components of the respiratory mechanism are carried in the same membrane as the chlorophyll. In these organisms therefore the membrane that serves as the limiting boundary of the protoplast and mediates the exchange of materials with the environment functions additionally in photosynthesis and respiration. It is of interest that the ability to carry out photosynthesis and to produce chlorophyll is lost in the presence of oxygen in the purple bacteria where the respiratory or photosynthetic function of the membranes may be selected by environmental conditions. The dark reactions in the fixation of CO_2 and the reduction of its products in bacteria proceed in the general cytoplasm along with other processes of metabolism. In other words compartmentation of function is at a minimum.

In view of the simplicity of their structure the photosynthetic bacteria appear to represent primitive photosynthetic forms. Although this may be the case, consideration of pigmentation in this group of organisms does not suggest a close relationship with the oxygenic photoautotrophes since a variety of distinct chlorophyll forms are observed. Members of the purple bacteria possess either bacteriochlorophyll *a* or *b*, whereas the green bacteria contain bacteriochlorophylls *c* and *d* (chlorobium chlorophylls) in which the tetrapyrrole head is esterified with farnesol in contrast to other chlorophylls which contain phytol although the ring oxidation levels and consequently the spectra are very similar to chlorophylls of higher plants. Hence, although the bacterial phototrophes may have ancestors in common with the oxygenic phototrophes they have diversified independently and no clear relationship can be seen.

1.5. OXYGENIC PHOTOTROPHES: ALGAE

1.5.1. Pigmentation

All known photosynthetic organisms other than the bacteria produce oxygen. Oxygenic photosynthesis is the only mode found in the eukaryotes; it is also found in one group of prokaryotes — the blue-green algae (Cyanophyta). The fact that both prokaryotes and eukaryotes possess the ability to use water as the electron donor indicates that formation of oxygen does not depend on the segregation of the photosynthetic apparatus into a true chloroplast. The most important common factor appears to be the possession of chlorophyll *a*, and of an ancillary pigmentation system. In the blue-green algae, red algae and cryptomonads the secondary pigments are the water-soluble linear tetrapyrroles phycocyanin (often more abundant in Cyanophyta) and phycoerythrin (often more abundant in Rhodophyta). In general these are associated with specific proteins. The phycobiliproteins of the Rhodophyta and the Cyanophyta have very close resemblances chemically and are located as granular aggregates on the surfaces of thylakoids, those of the Cryptophyceae are rather different in both respects. The further evolution of oxygenic photosynthetic systems appears to have involved the loss of these water-soluble chromoproteins [130] and the adoption of wholly lipid-soluble complements of ancillary pigments including some additional chlorophylls. Chlorophyll *b* and chlorophyll *c* are not found in Cyanophyta or Rhodophyta but one or other is now known to occur systematically in all [11] the remaining eukaryotic groups. These constitute the groups which have flagella or are considered on good evidence to have had them ancestrally. The pigmentation, except in one class, is in accordance with the taxonomic criteria for dividing the classes into two natural assemblages [27b]. The majority, including the Cryptophyceae, contain chlorophyll *c* and comprise the Chromophyta. Chlorophyll *b* is present in all the rest. The Prasinophyceae, Chlorophyceae, and Charophyceae are classified as the Chlorophyta and resemble higher plants (Embryophyta) chemically and cytologically. The Euglenophyceae are difficult to place in either of the two main divisions because they possess chlorophyll *b* but in many respects resemble the Chromophyta. Two types of chlorophyll *c* have been identified [68a] and it is of interest that c_1 is absent from Cryptophyceae and Dinophyceae whereas both forms, c_1 and c_2, occur together in other classes of Chromophyta so far examined [93b].

1.5.2. Structure in relationship to cell compartments

The various structural peculiarities of the major algal groups will be briefly considered, more detailed treatment can be found in references 11, 35,49,52,80,83,92.

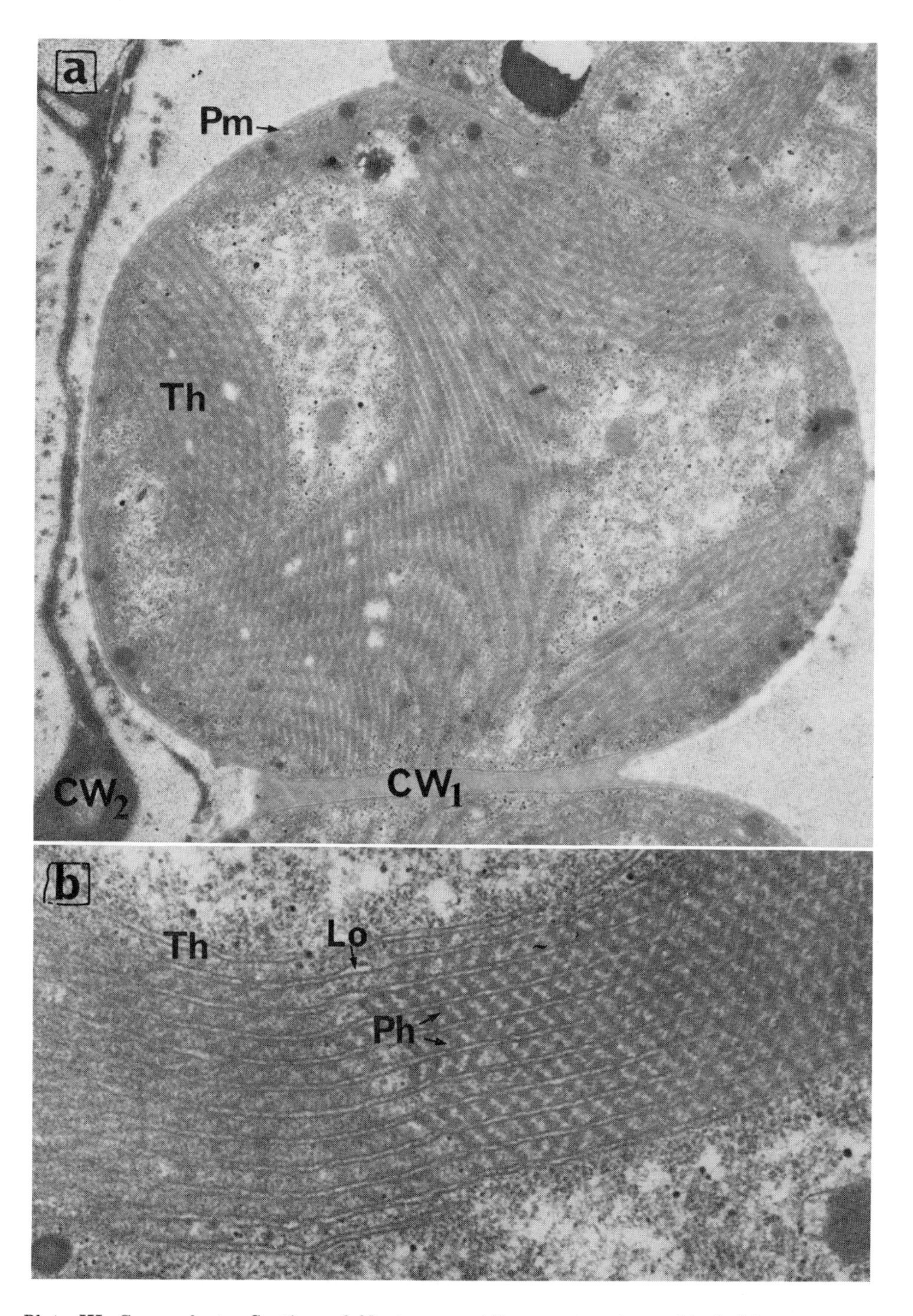

Plate III. Cyanophyta. Section of *Nostoc punctiforme* an endosymbiotic blue-green alga in cortical cells of the rhizome of *Gunnera manicata*. (a) Part of a trichome in longitudinal section including an entire cell bounded by the cell wall (CW_1) and plasma membrane (Pm) showing groups of parallel thylakoids (Th) in the general cytoplasm made evident by the regular arrangement of attached rod-shaped phycobilisomes. Glut/Os, Ua, Pb, × 30 000. (b) Small region of the cell at higher magnification (× 60 000) showing arrangement of phycobilisomes (Ph), sectioned transversely, on the membranes of the thylakoids (Th).

1.5.2.1. Cyanophyta

The simplest compartmentation is found amongst the blue-green algae [80] (Plate III). In common with some of the purple bacteria the membranes carrying the photoreceptor pigments are associated with respiratory and photosynthetic function [15] but in the blue-green algae the processes are not mutually exclusive and both may continue in the light. All the internal membranes of the cell appear to be involved and constitute a cisternal system, typically in the form of extensive lamellae. By analogy with the correspondingly pigmented cisternae of chloroplasts the term thylakoids is applied also to the internal membranes of the blue-green algal cell [94] although in the absence of an enclosing envelope the thylakoids are surrounded by a matrix of cytoplasm [43,102] rather than the specialised stroma of a chloroplast. The thylakoids in most species of blue-green algae tend to be peripheral rather than central in location, often parallel to cell surfaces but with variations in orientation, shape and distribution according to species and physiological conditions of growth [80]. The soluble phycobiliprotein pigments, after immobilisation by the use of glutaraldehyde as a fixative, have recently been shown [29,43] by electron microscopy to be located in loose attachment to the external surfaces of the thylakoids. In a number of instances large aggregates, phycobilisomes, have been observed as particles (35 mm) or parallel rods (Plate III) resembling those previously found [42] in a corresponding position on the thylakoids of rhodophycean chloroplasts. However, the phycobilisomes in most species of blue-green algae are small [44,130], tend to be closely packed, and cannot always be observed. The visible structure is somewhat variable in the cells of most species and often absent from some areas of the thylakoids. It may be assumed that the aggregated forms of the pigments and their distribution on the membranes depend upon the composition and amounts of the chromoproteins present and the physiological conditions of the cells. Evidence of attached phycobiliproteins is never seen on the cell membrane. In common with most photosynthetic bacteria some blue-green algae can fix nitrogen [1] by a mechanism dependent upon membrane-bound components carried with those of aerobic respiration in the pigmented membranes of the thylakoids. However, the reductive enzyme systems of nitrogen fixation are sensitive to oxygen and do not generally operate in the oxygenic photosynthetic cells of blue-green algae. Nitrogen fixation normally occurs only in certain multicellular species in which a proportion of the vegetative cells can become modified into specialised heterocysts adapted to this function [41a,133a], although their activity remains dependent upon attachment to unaltered cells in the same trichome. The cellular transformation is under environmental control, being suppressed for example by the availability of combined nitrogen as ammonium salts. The thylakoids become reorganised into a sinuous reticular system tending to become concentrated towards the ends of the heterocysts but more widely distributed than the typical peripheral arrangement of the photosynthetic lamellae. The phycobiliproteins disappear from the cells but the membranes retain chlorophyll *a*. Respiratory activity is enhanced rather than diminished in these cells and the membranes still support System I photochemical reactions although these are anoxygenic and linked to the reduction of nitrogen rather than products of CO_2 uptake which do not appear to be available in mature heterocysts. The cytoplasm retains ribosomes but typical granular reserves such as the cyanophycin, polyphosphate and polyglucan particles of blue-green algae disappear during maturation of heterocysts. The peculiar polyhedral bodies that are universally present in the photosynthetic cells also vanish. Similar polyhedral inclusions recognised by their shape and uniform finely granular contents retained by a distinctive thin pellicle (approx. 3 nm), are found in some other prokaryotic autotrophes but not in any eukaryotic organisms. It is of great interest that the contents of the polyhedral bodies in *Thiobacillus neapolitanus* (a non-photosynthetic sulphur bacterium) have recently been found [124a] to consist largely of the enzyme ribulose *bis*-phosphate carboxylase, essential for the uptake of CO_2 in autotrophes, and the name carboxysomes has accordingly been proposed for the particles. If, as seems probable, the polyhedral bodies in the blue-green algae are also of this nature their absence from heterocysts is consistent with the altered metabolism of these cells.

In comparison with other photosynthetic prokaryotes the blue-green algae show a greater localisation of function and specialisation of the membranes. The gross compartmentation of the cells somewhat resembles that in purple bacteria and the occasional connection of thylakoids with the cell membrane leaves open the possibility of common ancestry with these bacteria, involving however in subsequent evolution the separation of the photochemical and respiratory apparatus from the cell membrane itself and the acquisition by the thylakoid membranes of the unique ability to support photosynthetic processes concomitantly with those of aerobic respiration. The production of oxygen is associated with possession of forms of chlorophyll *a* in the photoreactive centres but the efficient operation of the System II component of the photochemistry appears also to be dependent upon the phycobiliproteins although these pigments function in an apparently secondary role as light-gathering molecules accessory to the reactive forms of chlorophyll *a*. The ability of some multicellular species of blue-green algae to differentiate another type of cell is of interest for several reasons. It is parallel at the prokaryotic level to the development in multicellular higher plants of cells with varied types of plastids of non-photosynthetic function that are dependent upon chlorophyll containing cells in other tissues. In the blue-green algae as in higher plants the processes of cell differentiation are under a degree of environmental as well as endogenous control with the additional comparison that the genetic competence of heterocysts is fully retained since they have been observed in isolation to regenerate a vegetative cell that may give rise to a new trichome [2].

1.5.2.2. Rhodophyta

The morphologically simpler orders of the Rhodophyceae have relatively small and slightly vacuolated cells containing typically a single, axile chloroplast, usually deeply lobed (stellate) and having a central pyrenoid as shown for the unicellular *Porphyridium* in Plate IVa. The thylakoids in the peripheral zones of the chloroplasts tend to lie, evenly spaced, in groups parallel to the main surfaces of the organelle and extending into the lobes [42,127,101]. Characteristically such groups terminate more or less abruptly at the ends of the lobes where the edges of lamellae come almost into contact with the inner membrane of the envelope. Phycobilisomes are carried on the stromal surfaces of the thylakoids but not on the membranes of the envelope which appear to be unconnected with those of the lamellae. Connections between adjacent lamellae are not infrequently seen and it is probable that the membranes are all part of a single thylakoid. The lamellae do not become closely apposed to form paired or multilayered structures such as occur in the chloroplasts of all other algae. In *Porphyridium* the lamellae become more sinuous and less regularly arranged towards the centre of the chloroplast where individual lamellae enter the pyrenoid at several points. Within the pyrenoid the lamellae interconnect more frequently to form a reticulum. Patterns vary and in some species, e.g. *Rhodella maculata* [40] thylakoids do not enter the pyrenoid and may terminate at its surface. The matrix of the pyrenoid, as in many algae, has a uniform, finely granular texture contrasting with the stroma and lacking its characteristic inclusions, such as lipid droplets although not limited by a membrane or other visible boundary structure. The more advanced Rhodophyceae are placed by some authors in a separate class or sub-class (Florideae). They are all filamentous plants with special septal pores and other intercellular connections. The cells may attain a large size and become multinucleate in some species. The cells have large central vacuoles with peripheral cytoplasm that contains variable numbers of lens-shaped chloroplasts. Pyrenoids are found in only a few of the simpler Florideae and the lamellae, seen in section, may run in parallel order from margin to margin [13,18,24,47] as shown for *Griffithsia floscuslosa* in Plate IV. It is typical that edges of the internal thylakoids do not reach the chloroplast boundary but terminate within or occasionally connect with a single thylakoid that surrounds them, lying close to the envelope and following its contour. This distinctive feature of Florideae was named the "inner limiting disc" [18] following earlier terminology, but now would be better referred to simply as the *limiting* or *peripheral* lamellae or thylakoid. The term "girdle lamellae" or "girdle thylakoid" has sometimes been applied to it but should be reserved for a superficially similar feature of organisation in certain Chromophyta for which it was named [44a] because it cannot be assumed that the two structures are homologous.

Plate IV also includes two views of the thylakoids at higher magnification which show the phycobilisomes attached to the outer surface of the mem-

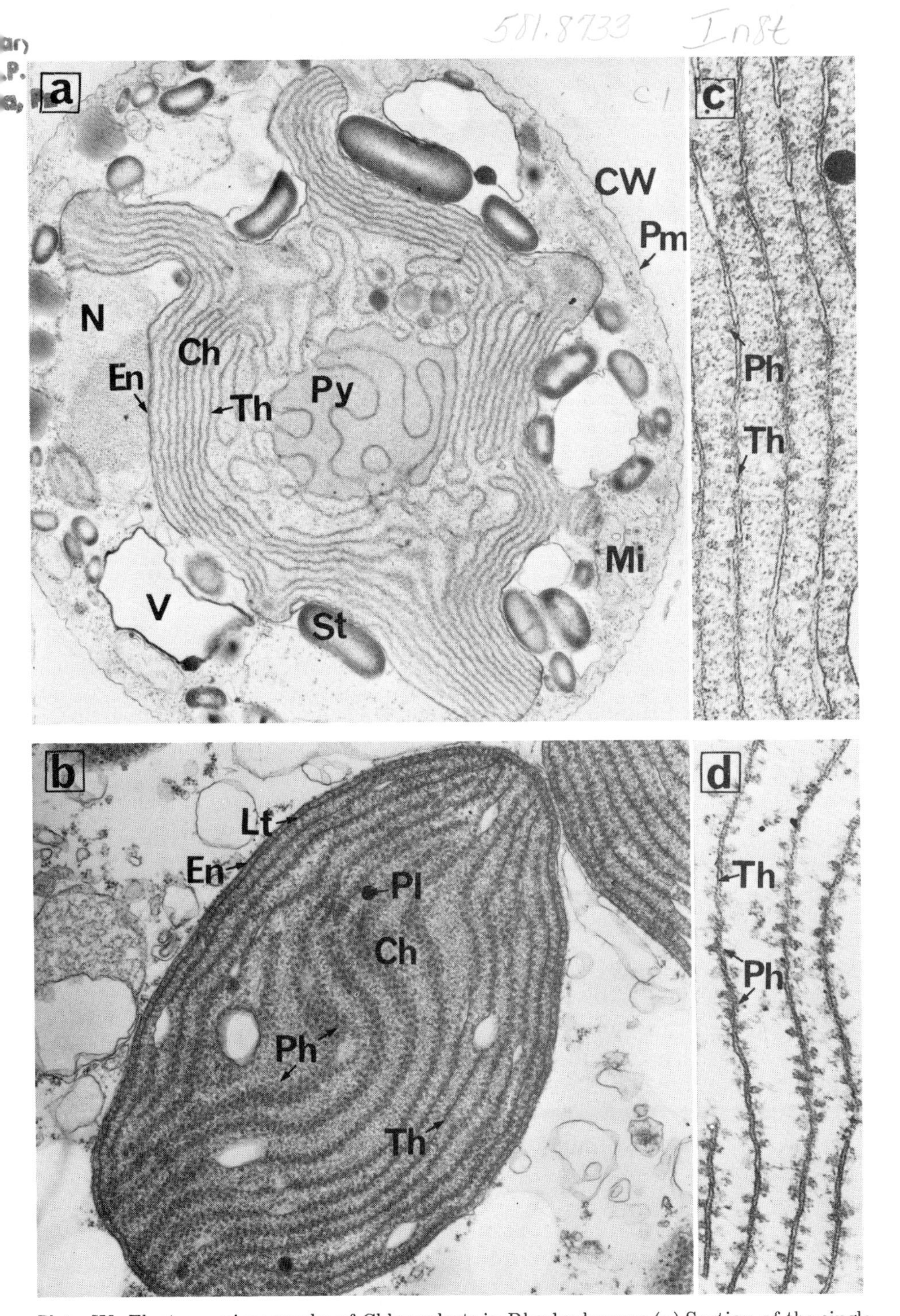

Plate IV. Electron micrographs of Chloroplasts in Rhodophyceae (a) Section of the single axile chloroplast (Ch) in a cell of *Porphyridium cruentum* showing parallel thylakoids (Th), separated by tracts of stroma, arranged peripherally around the central pyrenoid (Py) and extending into the characteristic projecting lobes of the plastid. Note the typically sinuous course of thylakoids which penetrate the pyrenoid and that starch (St) is deposited exclusively outside the chloroplast in the general cytoplasm. A portion of the nucleus (N) can be seen to be in intimate contact with the chloroplast envelope (En), Glut/Os, Pb; × 15 000. (b) Section of a discoid plastid in a cell of *Griffithsia flosculosa* showing the parallel arrangement of single thylakoids (Th) enclosed by an outer limiting thylakoid (Lt) positioned close to the envelope (En). This characteristic limiting thylakoid of the advanced members of the Rhodophyceae is absent in *Porphyridium* (see above). Glut/Os, Pb; × 22 000 (c) and (d) More highly magnified sections (× 40 000) of thylakoids (Th) of *G. flosculosa* showing phycobilisomes (Ph) under conditions of fixation where the stroma is retained (c) or lost (d).

branes, where they may remain when the stroma is washed out (cf. c and d). It should be noted that although the phycobilisomes are of about the same size as those observed in the cyanophycean species illustrated (35 nm) they do not show an identically ordered pattern. The particular substructure of thylakoids from red algae has been studied using freeze-etch techniques [12,101] where only two fracture faces are indicated in the unstacked thylakoids.

Starch grains are deposited in the cytoplasm, never in the chloroplast. In a few species where depressions between the chloroplast lobes penetrate deeply enough to invaginate the pyrenoid, or the latter is superficially rather than centrally placed, grains of starch may be formed in close opposition to the envelope and fitting its shape immediately overlying the pyrenoidal matrix. The nuclei are also characteristically positioned between the chloroplast lobes with their membranes in contact with the chloroplast envelope and in *Rhodella maculata* [40] the nucleus invaginates the pyrenoid resembling conditions met with regularly in some Chlorophyta. The Florideae resemble higher plants in lacking pyrenoids and also in showing normally no close structural involvement of other organelles or cytoplasmic membranes with the chloroplasts. The morphologically most advanced species have elaborate tissue differentiation and include examples with juvenile cells in the growing points that contain only proplastids with rudimentary thylakoids. The further development of the thylakoid system is from the peripheral lamella, which is always present, and does not involve the inner membrane of the envelope as occurs in higher plants.

In conclusion it may be commented that the structure, organisation and chemistry of the thylakoids in Rhodophyceae carries conviction of a primitive homology with the lamellae of the Cyanophyceae but the segregation of these components within a chloroplast, however it may have been derived, also involves a changed localisation of function that reflects the more elaborate compartmentation of the cell. Respiration is not directly associated with the photosynthetic apparatus and the storage of photosynthate is carried out in the cytoplasm externally to the chloroplast. The compartmentation of Rhodophycean cells is comparable in structural organisation with that of Chlorophyta and higher plants but the distribution of the photosynthetic functions is different and in the latter may include the temporary storage of insoluble reserves of photosynthate or the varied partitioning of partial processes of carbon dioxide uptake and assimilation between chloroplast and cytoplasm. In general the superficial resemblance in some specific details of structure of advanced Rhodophyceae to higher plants may be regarded as the consequences of convergent evolution rather than of more direct close relationship by descent.

A number of unicellular algae fall into consideration here because they have nuclei but possess chromatophores with pigments of Cyanophycean type; phycoerthrins are absent as are chlorophylls *b* and *c* and starch forms in the cytoplasm. These attributes and other common features of the photosynthetic system place these organisms collectively in a position in Table II that centres with the Rhodophyta. However, the uniformity of the photosynthetic equipment is at variance with the diversity of cell types found among them and they are difficult to place in the taxonomic classification of algae. The species *Cyanidium caldarum* has recently been included in the Rhodophyceae or regarded as a monotypic representation of a separate class in the Rhodophyta. Other genera would be attributed to various different classes in the Chlorophyta and Chromophyta were it not for their anomalous pigmentation and these have come to be generally regarded as probably endosymbiotic associations of blue-green algae (cyanelles) contained in host cells of other eukaryotic phototrophes that have lost their original pigments and chloroplasts. In most instances the hosts and cyanelles have not been isolated in separated culture and unequivocal proof of this hypothesis is difficult to obtain. It will be sufficient to draw attention here to two of these, *Glaucocystis* and *Cyanophora* that have been investigated with the electron microscope. The chromatophores in both genera have only a limiting envelope of two membranes to separate them from the cytoplasm and contain a thylakoid system of unassociated lamellae bearing external phycobilisomes. The granular matrix includes DNA and ribosomes of prokaryotic type but lacks any of the specialised inclusions of the cyanophycean cell although possessing lipid droplets and a distinctive region of more finely granular texture and greater density. The cyanelles have therefore few of the attributes of a free living alga unlike the phycobionts illustrated here (Plate III) within the cortical cells of *Gunnera* which retain cell walls and capsules and an essentially normal internal ultrastructure. The cells of *Glaucocystis* have a cell wall and the superficial morphology of *Oocystis* but the ultrastructure shows rudimentary flagella, not found in this genus and features of cell organisation not typical of the Chlorophyceae. *Cyanophora* was regarded as a cryptomonad but the ultrastructure of the flagella and periplast are unique with only general affinity as much to the Dinophyceae as the Cryptophyceae [34a]. Concerning the systematic position of the cyanelles they have been considered so unlike any other Cyanophyceae as to be placed in a separate family without known free living forms. All in all the ultrastructure increases the difficulty of assigning either the hosts or the putative endobiont to a place among other algae and leaves their affinities uncertain even as to class relationships. On the other hand there is nothing in the ultrastructure of the chromatophores that might not be found in a chloroplast comparable with or perhaps rather simpler than the Rhodophyceae. The dense inclusions may for example be pyrenoidal in nature. More intensive study may resolve the nature of these and other similar organisms. In the present state of knowledge they seem as likely to be single species as endosymbiotic systems of recent origin. If this proves to be true they would represent additional groups of algae with phycobilin accessory pigments and simple chloroplasts resembling the Rhodophyta but with other cell characters found otherwise only in the Chlorophyta and Chromophyta. Whilst it seems unlikely any of those at present known could bridge the gap between the cyanophycean and the eukaryotic types of photosynthetic system they include cells with other characteristics of the flagellate groups and it is not impossible that some may be found with a photosynthetic apparatus of intermediate character that could elucidate further the relationships and evolution of the varied pigment and membrane systems in eukaryotic phototrophes.

1.5.2.3. Chromophyta

The Chromophyta are more diverse than either the Rhodophyta or the Chlorophyta and are listed in Table I and set out in relations to the characteristics of the photosynthetic systems in Table II. Each class shows a distinctive and sometimes very large range of morphology and on these grounds alone some classes stand in considerable isolation and have been classified into more than one division by different authors. Knowledge of the ultrastructure has in general supported the identity of the main classes adopted previously. The Haptophyceae [92,93] and Eustigmatophyceae [61] have been added and the removal of these algae respectively from the Chrysophyceae and Xanthophyceae has further consolidated the matter. An important result of the more substantial basis of cytological and chemical detail recently provided has been to reveal an underlying unity in this large assemblage of algae which is particularly evident in common features of organisation in the photosynthetic systems that distinguish them from all other organisms. This is apparent at the cellular level as well as in the nature of the lamellae and other components.

In all classes of the Chromophyta the chloroplasts are segregated from the cytoplasmic continuum of the cell by two additional membranes (sometimes apparently reduced to one). It may therefore be said that the cells have an additional subcompartment of a type not found in any other eukaryotic organisms. This will be discussed in greater detail below.

The arrangement of the thylakoids varies systematically from simpler systems in some classes to a more complex arrangement, basically of the same type, that can be interpreted as a more advanced form which having emerged became stabilised and is found now in a number of classes. This structural pattern is as distinctive as the granum in the Chlorophyta and higher plants and in its own context may likewise be perhaps an optimum configuration derived independently from a rudimentary level of thylakoid apposition or "stacking". The range of thylakoid elaboration is comparatively small and centres around a common pattern of organisation into extensive *compound lamellae* [44a] or *bands* [49] that traverse the stroma parallel to the main surfaces of the chloroplast and are joined together by local interconnections passing from one to the next parallel lamellar plane. The internal membranes forming the lamellae do not appear to connect with the chloroplast envelope. Each band is composed usually of 2 or 3 thylakoids (occasionally 4 and rarely up to 6). The majority of species except the Cryptophyceae have regular systems of very uniform bands of 3 thylakoids. The Dinophyceae show the greatest variations with some species having a high proportion of paired or even single thylakoids. The lamellae in Haptophyceae tend to be more dissected by mutual interconnections. The further specialisation of a *girdle lamella* [44a] or *girdle band* [49] is found in the 5 classes enclosed in square brackets in Table II. This feature is illustrated in Plate VI in one of the Phaeophyceae and in Plate VIIb in a flagellate of the Raphidophyceae. Some of the parallel bands terminate towards the margins of the plastid but two of them are always connected by a curved link (to which the term girdle lamella Gb refers) that encloses the ends of several (Plate VIIb) or all of the remainder (Plate VIb). This detail of construction is related to the regular links which occur between the bands and are seen in Plate VIIb to involve a local separation of the constituent thylakoids to form a junction complex in which they terminate and rejoin one another. It may be inferred from these connections that the membranes are continuous with each other forming a single three-dimensional network as in higher plants. However, the membranes do not ever appear invaginated to form the partitions of a stack [49,85] and the resemblance to granal and fret connections is probably merely superficial; there is no strict analogue to the stroma lamellae in higher plants. The links in the girdled types of chloroplast involve the divergence of a segment from one band into continuity with another immediately adjacent to it, where it takes the place of a comparable segment that has passed into the next lamella and so on in like order across successive planes of the system, leaving eventually one free lamellar edge in the outermost band of the set. The lamella connections occur therefore in related series, aspects of which can be observed in Plate VIIb. The lamellae interconnect more frequently near the margins of chloroplasts. The connections with the girdle band become therefore repeatedly transferred from one lamella to another. At different points on the perimeter of the organelle, as found from serial sections (Greenwood, unpublished), the girdle band thus enfolds the ends of varying numbers of the internal bands. This idiosyncrasy of organisation is thus intimately related to the architecture of the membrane system and can hardly be regarded as likely to have originated in this precise form other than once in some common ancestry to these classes. The groups with girdled chloroplasts include Chrysophyceae [47], Bacillariophyceae [32,37,38], Phaeophyceae [19,20,31,40,47] and Xanthophyceae [44a] (Table II), that for other reasons have been long regarded as sharing a close relationship at the cellular level. These all possess tubular mastigonemes on one flagellum [34a], another key feature which cannot easily be imagined to have evolved on more than one occasion. The absence of a girdle band in some Xanthophyceae may perhaps be assumed to have been from secondary loss. Further evidence of the advanced specialisation of girdled plastids is the localisation of the DNA genophore of the plastid in a position immediately within the curved girdle link and it appears, at least in some, to have the form of a continuous ring [11,13]. The stacking of the thylakoids in surface contact in the lamellae of Chromophyta is widely regarded as corresponding to that in the grana of higher plants. However, the degree of apposition is more varied than in higher plants, differing with different staining reactions in electron micrographs, hence doubts have been recently expressed of any homology with higher plants [11]. If it is accepted that the apposition of lamellae has some general functional advantages it is conceivable that this condition has arisen more than once in chloroplasts and may not be identical in its molecular basis in all groups of algae.

The Cryptophyceae (Plate V) are unique in the pigmentation, structure and organisation of the thylakoids. They carry phycobilin pigments as well as chlorophyll c_2. The phycobilins are chemically different from those of Cyanophyceae and Rhodophyceae and have not been observed to form phycobilisomes on the exterior of the thylakoid membranes. However, the thylakoids are characterisitically rather wider and more irregular than in other organisms and particularly so in some genera, including *Hemiselmis* (Plate Vb and d) and *Chroomonas* [34] in both of which they also contain conspicuous dense contents (Plate Vd). Recent studies [44] have shown that the dense contents are removed when the phycobiliproteins are dissolved out and with other evidence leave little doubt that these pigments are located in the interior of the thylakoids rather than on the surface, where presumably they could preclude interfacial contact such as occurs in stacked arrangements. The thylakoids are associated in various ways [34,44,48,49,89] but preponderantly only in pairs. In some instances the thylakoids appear to be very loosely associated (Plate Vd) but in others (Plate Vc) a dark line can be observed that occupies the equivalent position to the A space in a granal partition, suggesting a more intimate union perhaps involving some specialised component that appears structurally as a bonding layer.

The Cryptophyceae are in many respects an isolated and specialised group. They have a supporting periplast within the plasmalemma as do the Dinophyceae, Euglenophyceae and the anomalous Cyanophora but all other classes of algae have external scales or cell walls. It is therefore tempting to suppose they are a specialised terminal remnant of an otherwise extinct branch of evolution and retain a transitional condition of the photosynthetic apparatus in which the ancestral biliproteins, which by their external position precluded the advantage of membrane apposition, have in part given their functions at the membrane surface to lipidic secondary pigments. It is of interest that the Cryptophyceae possess the tubular mastigonemes found also in the more advanced classes with girdled chloroplasts. The Dinophyceae and Euglenophyceae bear different types of mastigoneme and the Haptophyceae seem to lack them. The thylakoid organisation can be

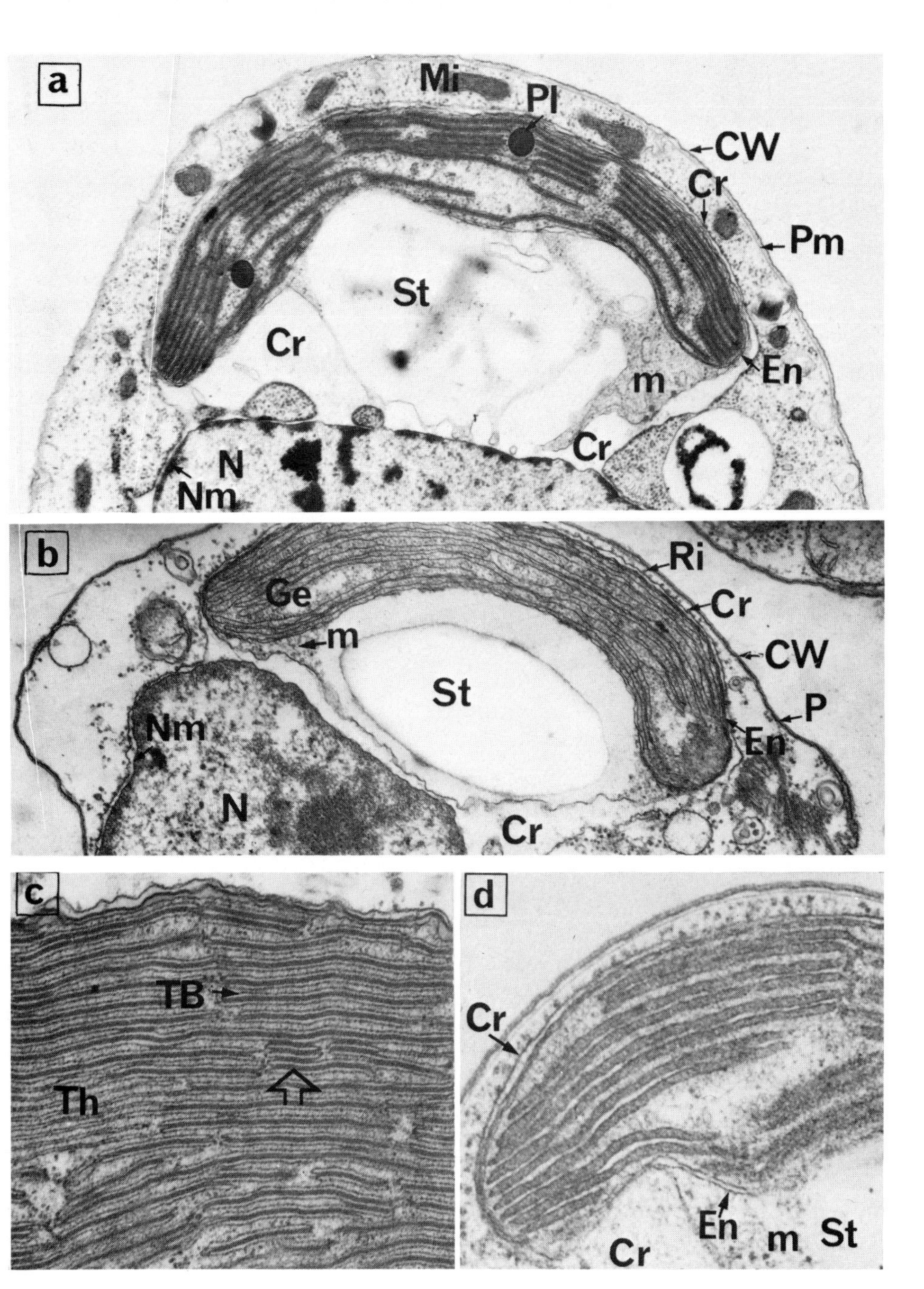

Plate V. Sections of the chloroplasts and associated cell components in Cryptophyceae. (a) *Cryptomonas sp.* Glut/Os, Ua, Pb; × 20 000. (b) *Hemiselmis rufescens* Glut/Os, Ua, Pb; × 27 000. (c) *Cryptomonas sp.* Os, Ua, Pb; × 42 000. (d) *Hemiselmis rufescens* Glut/Os, Ua, Pb; × 50 000. Note in (a) and (b) the wide zone of confluence between the chloroplast ER (Cr) and the envelope of the adjacent nucleus. The two membranes of the chloroplast ER, seen in (a), (b) and (d), converge towards the chloroplast envelope (En) to form a tight complex of 4 membranes bounding the outer side of the chloroplast; on the inner side the space between the chloroplast envelope and the ER is expanded to form a large compartment containing the starch grains and a granular matrix (m) with included ribosome-like particles. The lamellae of the chloroplast are in general formed of paired thylakoid bands (TB) although regions of unassociated thylakoids (Th) and of greater stacking (arrow in C) can be seen. Bands are often arranged in parallel order towards the outer side of the chloroplast (a) and there are no girdle bands (d). The thylakoids in *Hemiselmis* are wider and more irregular than in *Cryptomonas* with obvious dense contents and are more loosely associated in pairs (compare (c) and (d)).

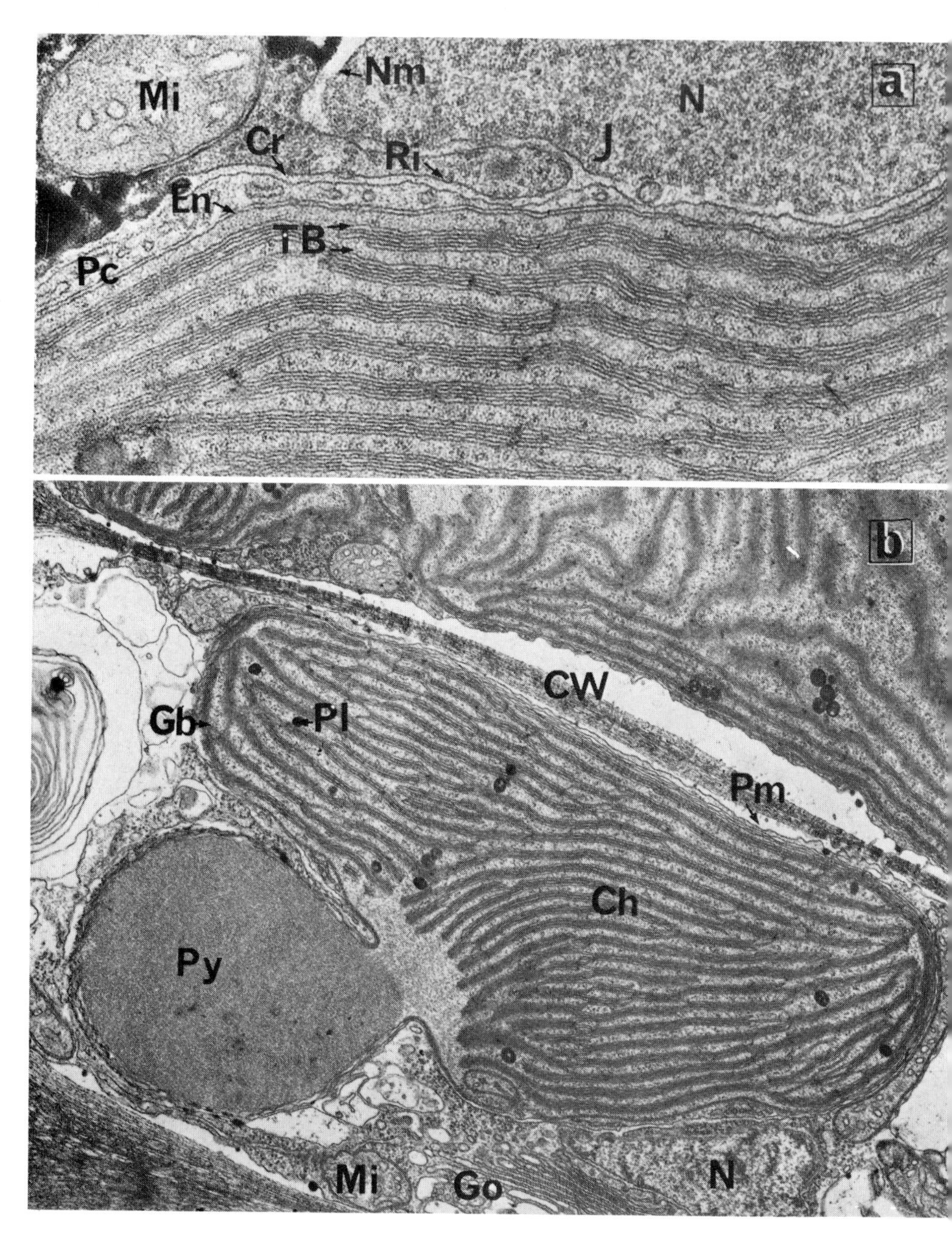

Plate VI. Sections of cells of *Scytosiphon lomentaria* showing details of the chloroplast and its association with the nucleus. (a) Showing the membranes of the nucleus and the adjacent chloroplast. The nuclear envelope, limited internally by the membrane (Nm), can be followed to its point of confluence J with the cisterna of the chloroplast endoplasmic reticulum (Cr). The outer membrane of the nuclear envelope is continuous with that of the chloroplast ER. Note the small vesicular profiles in the compartment Pc between the chloroplast ER and the chloroplast envelope (En). Ribosomes (Ri) are present in the general cytoplasm. Os, Ua, Pb; × 55 000. (b) Showing the regular compound lamellae each of 3 thylakoids and the girdle lamella, which encloses the lamellae at both ends of the profile. Parts of the chloroplast ER can be seen around the perimeter of the plastid including the stalked pyrenoid (Py), which is not penetrated by lamellae, and part of a nucleus in association with the Golgi body (Go) and the chloroplast. Os, Ua, Pb; × 22 500.

interpreted as showing a trend from the less ordered and more variable simpler systems in Dinophyceae and Cryptophyceae to the elaborate and structurally stereotyped girdled systems of the classes traditionally centred around the Chrysophyceae.

The salient features of the cell compartmentation that are unique to the Chromophyta are most obvious in species with one or two relatively large chloroplasts that stand immediately adjacent to a nucleus, as in most of the flagellate groups or in the smaller cells of many of the brown seaweeds. In the cell of *Scytosiphon lomentaria* shown in Plate VIa the two closely placed membranes of the chloroplast envelope (En) do not come into contact with the general cytoplasm of the cell but are separated from it by another pair of membranes (Cr), with a rather wider space between them that is in continuity with the perinuclear space. The outer of these two membranes is in contact with the cytoplasm and can be traced as an extension of the outer membrane of the nuclear envelope. The pair of membranes (Cr) have therefore the normal relationships to the nucleus of a cisterna of the ER and have been termed the chloroplast ER [19,49]. The lumen of the chloroplast ER may be confluent for a considerable distance with that of the nuclear envelope. The chloroplast ER appears to be continuous around the plastid but not necessarily in immediate contact with the chloroplast envelope. In Plate VI a space Pc can be observed between the outer membrane of the chloroplast ER and has been referred to as the *perichloroplastic space* [11] but will sometimes be spoken of here as the chloroplast cell compartment. The membranes come together over other parts of the plastid, particularly those remote from the nucleus, the chloroplast compartment usually becomes greatly reduced and a tight complex of four membranes may then constitute the boundary of the chloroplast. The relationships of the nuclear membrane system to the chloroplast are shown in the diagram (Fig. 1.2) representing an imaginary flagellate cell from which most other components have been omitted. The perichloroplastic space contains but a few empty-looking vesicles or membranes that appear often to be merely invaginations of the chloroplast ER. This condition is found with only small variations in all classes except the Cryptophyceae and Dinophyceae. In some large cells the plastids become remote from nuclei but still appear surrounded by 4 membranes, no doubt by detachment from the nuclear envelope as shown in the lower part of Fig. 1.2. In other examples, four membranes cannot be clearly resolved but usually three can be identified. This is the case in Vacuolaria. The Dinophyceae [33] also typically show three membranes and a nuclear connection has not been reported. These instances may be explained as a congenital and complete fusion of one or another pair of the membranes. Small areas of casual fusion of membranes derived in various ways as shown at 1 or at 2 in Fig. 1.2 are commonly found in electron micrographs. The relationships shown in Fig 1.2 indicate that the small spaces Pc correspond to the lumen of a sub-compartment isolated by the chloroplast ER from the

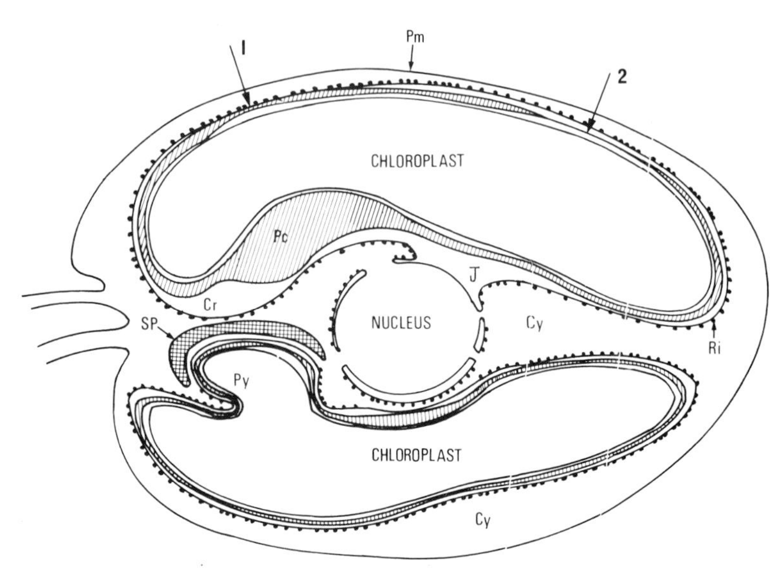

Fig. 1.2. Diagram showing the cell compartments in Chromophyta. Only certain generalised essentials of structure are drawn in an imaginary cell shown as a flagellate. Chloroplasts are surrounded by the limiting membranes of the chloroplast ER (Cr), shown characteristically in confluence with the nuclear envelope at J. The result is to confine the chloroplasts within subcompartments (Pc) in isolation from the general cytoplasmic continuum (Cy) of the cell. Parts of the ER(1) or the subcompartment (2) may be obliterated by the apposition or fusion of membranes. A capping grain or vesicle (SP) of reserve carbohydrate may form in apposition to the membranes covering the projecting pyrenoids (Py) of some species. The lower chloroplast is shown detached from the nucleus as occurs casually or regularly in some forms.

main cytoplasmic space of the cell, but containing the chloroplast. The absence of any other contents apart from a few empty vesicles attests to the complete separation of this compartment from the general cytoplasm. The corresponding condition in Cryptophyceae is remarkably different. In Plate Va and b the origin of the chloroplast ER from the nucleus is clearly seen and the convergence of its membranes into apposition with the chloroplast envelope at the ends of the profile. Thus on the outer side of the chloroplast there is a four-membrane complex of the usual type. On the inner side adjacent to the nucleus the perichloroplastic space is greatly inflated. It contains the starch grains which are the characteristic reserve product in all Cryptophyceae. In addition, however, the starch grains are embedded in a granular matrix (m in Plate Va and b). Ribosome-like particles, often numerous [48] can be found in this matrix. Although it is difficult to prove that there are no fenestrae in the chloroplast ER through which ribosomes from the cytoplasm might pass into the starch compartment such openings cannot be large since other cell components are not present. It must therefore be considered possible that in the Cryptophyceae the formation of the chloroplast ER has sequestered from the general cell contents a ribosome-containing matrix and perhaps the machinery for starch deposition in addition to the chloroplasts.

If this relationship represents a primitive condition the ancestral forms prior to the evolution of the chloroplast ER would presumably have had cytoplasmic starch. This is the standard location of starch in the Rhodophyceae and the anomalous *Cyanophora* which both also have biliprotein pigments. The Dinophyceae are the only other class of Chromophyta that produce starch and in these also it is formed in the cytoplasm. If the Dinophyceae possessed ancestrally a complete chloroplast ER which contained starch, the membranes must be presumed to have become detached from the nucleus in a way that allowed renewed communication of the starch compartment with the general cytoplasm. The absence of starch and contents from the perichloroplastic space in other Chromophyta could be more readily explained simply as following the loss of the ribosome population, since this may be essential for the production of dependent enzymic proteins to allow starch metabolism to occur. This hypothesis has the merit that it would thus explain why starch is not formed at all in the brown seaweeds, the diatoms and so many other of the chlorophyll *c* containing plants. It also has the advantage that the absence of starch in Euglenophyceae could be accounted for in the same way. Apart from the presence of chlorophyll *b* the euglenoids have the general attributes of Chromophyta [49]. The chloroplast lamellae (Plate VIII) vary greatly [9,83] but in a number of species they have the form of bands, not as regular as those of advanced Chromophyta but often of three thylakoids. They often have a very clear set of three membranes surrounding the plastids (Plate VIIa) and connections of the outer membrane and the nucleus have been observed [46,83]. Whilst the affinities of the euglenoids are with the Chromophyta rather than the Chlorophyta [49] they cannot be accorded a close relationship with any existing class but like the Cryptophyceae and Dinophyceae their attributes may be in part those of a common ancestral stock. Perhaps the most obvious possibility is that they diverged at a stage in the diversification of secondary pigments before the Chromophyta had acquired the specialised chlorophyll *c*. In all this speculation it is perhaps most difficult to conceive how ribosomes could be maintained in a compartment closed off by 2 membranes from the main nucleus of the cell and by the chloroplast envelope from the stromal nucleoid. The recent observation of a nucleus-like structure in the starch compartment of the Crytophyceae [44b] may provide a solution to this problem. If this structure proves to have genetic functions its loss in evolution might then have carried with it the inevitable loss also of other dependent components such as the ribosomes and the starch-forming machinery.

1.5.2.4. Chlorophyta

The Chlorophyta comprise another large assemblage of algae which appear to be derived from a flagellate ancestry. They range widely in morphology and can be classified into many more or less natural sub-groups but unlike the Chromophyta they show great uniformity in basic cell organisation and chemistry. Until recently, before Prasinophyceae were distinguished, they could be related to only one type of flagellated cell. In contrast to this constancy in basic cell type the chloroplasts show a much greater diversity of structure than is found in the Chromophyta. The chloroplast envelope is a structural constant in eukaryotic photosynthetic cells separating the apparatus for energy trapping and transfer from the cytoplasm. The localisation of other partial processes of photosynthesis and immediately dependent functions is not so fixed. In the Chlorophyta, as in higher plants, the machinery for the temporary storage of photosynthate as starch occurs within the stroma surrounding the lamellae but not in the cytoplasm as in other algae. The chloroplasts are not consistently associated with membranes of the nucleus or the ER. The Chlorophyta resemble the Rhodophyta in the general compartmentation of the cells (Plate VIId). The arrangement of the lamellae ranges in various patterns of stacking to that of a granal system. Invaginations of the inner membrane of the envelope, rare in other algae, are a common feature, particularly in developmental stages. Invaginations of the envelope usually become less numerous or absent at maturity, and the formation of a system of peripheral vesicles is unusual. The envelope may connect with thylakoids [85] during development but such connections are also rare at maturity. Although the term "stack" is in common usage for all forms of lamella apposition a distinction has been drawn [34a,49] between the type of firm bonding and the invagination of one thylakoid by another that characterise the formation of the partitions of true grana and the less firm association without such involvement of adjacent thylakoids that characterise the "bands" of the Chromophyta. All Chromophyta [26,47,50, 81,85,120,145] show a granal type of stacking in the above sense although it is by no means certain that other types of stacking may not also occur.

As to the systematic occurrence of different grades of lamella system in Chlorophyta, variable arrangements of 2—6 thylakoid stacks are common in most of the natural groups including the volvocales (flagellates) and the simpler filamentous orders. The multilayered formations may be extensive, appearing as bands, or be smaller in area individually and connected by single or paired thylakoids, approaching the appearance of grana with connecting frets. The Conjugales and the Charophyceae regularly possess a more typically granal like system, the former of many small grana, the latter closely resembling higher plants. The Chlorococcales, including *Chlorella spp.*, may have extensive regular parallel bands each of 3 to 4 thylakoids or show numerous interconnections and changes in thickness (Plate VII). The Prasinophyceae [92] also have varied systems of interconnected bands or shorter stacks. It appears that the affinities of some of the filamentous algae traditionally regarded as Chlorophyceae may be closer to the Prasinophyceae [110], which so far as yet determined are mainly flagellates. The division of the cells in some of the forms in question has features regarded [110] as resembling higher plants more than other morphologically similar

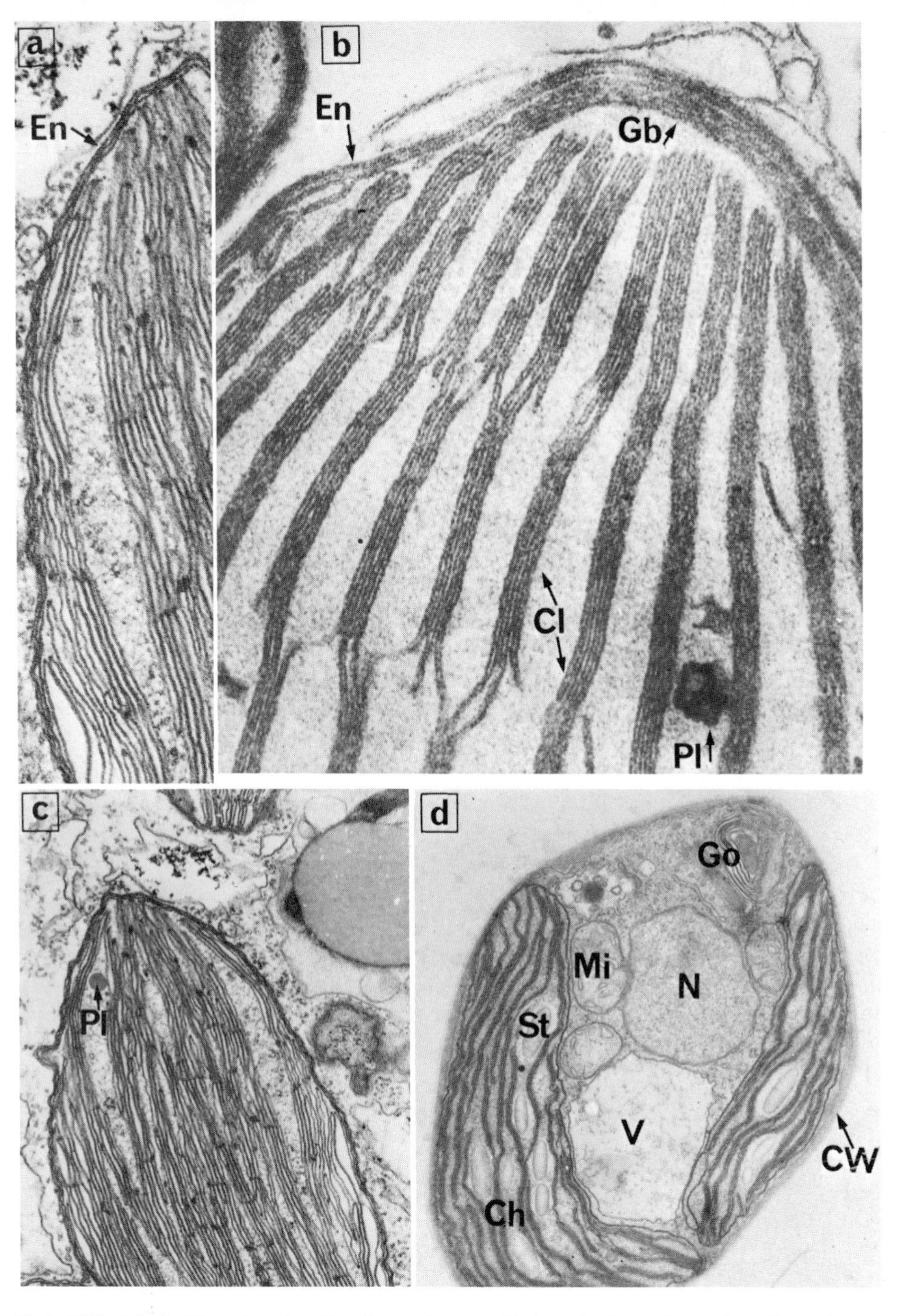

Plate VII. (a) Section showing the 3 membranes (En) at the boundary of a chloroplast of *Euglena spirogyra*. Os, Ua, Pb; × 75 000. (b) Section of part of a chloroplast of *Vacuolaria virescens* (Raphidophyceae) showing details of the envelope (En) with additional membranes, the uniform compound lamellae or bands (Cl) and the girdle lamella or band (Gb). Note the series of related interconnections between adjacent bands where individual thylakoids become separate to form a junction complex in which they terminate and rejoin. Pm; × 80 000. (c) Section of part of a chloroplast of *Euglena spirogyra*, showing irregularly stacked thylakoids, the absence of a girdle lamella in the margin and the boundary complex of 3 membranes. Os, Ua, Pb; × 40 000. (d) Section of cell of *Coccomyxa sp.* (Chlorophyceae) showing the chloroplast (Ch) and main organelles. The internal membranes of the chloroplast form a system of compound lamellae, often of 3 thylakoids each and superficially like those in Plate c above but not so regular in structure or interconnection and with terminations in the plastid margin in absence of a girdle lamella. Glut/Os, Pb; × 15 500.

filamentous green algae. It is therefore possible that stages by which the cellular organisation typical of higher plants, including perhaps the details of chloroplast structure, may have been approached in algal forms will be further elucidated. However, variations of structure between bands and grana are widely distributed in different natural orders and may vary in relation to developmental stages and physiological conditions. The question arises to what extent, if any, these variations in pattern affect the kinetics or efficiency of photosynthesis.

Pyrenoids [70,92] occur in the Prasinophyceae and widely, though not in all species, in Chlorophyceae. They are usually internal in position, or sometimes superficial but never greatly projecting or stalked, as they frequently are in Chromophyta where reserve products of photosynthesis are formed in the cytoplasm as insoluble grains or in containing vesicles. In the Chlorophyta large starch grains form in close apposition to the curved surfaces of the pyrenoids but usually occur also elsewhere in the stroma between the lamellae, as they do in absence of pyrenoids (Plate VII). The apposed grains may form an almost continuous starch shell but are interrupted where thylakoids enter the pyrenoid. Pyrenoids may be penetrated by many or few thylakoids and sometimes retain a degree of stacking but are more usually reduced to one lamella or occasionally to a tubular form [34a,92]. Thylakoids are absent from the pyrenoids of some species but when present these extensions of the membrane system may perhaps have a conductive function [92]. In some Prasinophyceae [92] the pyrenoids are superficially placed and may be deeply invaginated by tongues of cytoplasm but still remain covered by the 2 membranes of the chloroplast envelope or by the nucleus, when 4 membranes can be observed between the nucleoplasm and the matrix of the pyrenoid. Recently the structure of pyrenoids has been extensively surveyed [34a] and the relationships of pyrenoids to other cell components has been cogently discussed [92]. The matrix probably consists largely of enzymes with synthetic function but it may also probably serve for storage [17]. The composition of pyrenoids has been little investigated but a detailed analysis of the pyrenoid of *Eremosphera viridis* [64a] has shown that it consists mainly of protein; 90% of which comprises 2 components resembling fraction I protein and having its enzymic properties; other Calvin cycle enzymes were also identified. This finding is particularly significant in view of the close association of starch grains, or other oligosaccharide stores with the surfaces or limiting membranes of pyrenoids because phosphoglyceric acid, a product of the Calvin cycle, is known [53a] to stimulate the activity of some starch-forming enzymes. Capping grains, or sometimes vesicles (Phaeophyceae) of cytoplasmic storage products in contact with the covering membranes of pyrenoids occur in organisms as diverse as the Rhodophyta and the varied classes of the Chromophyta.

1.6. THE C4 PLANT

It is possible to imagine an evolution of the higher plant chloroplast from a cyanophyte-like ancestor, through formation of the limiting membrane and loss of phycobilisomes leading to the chloroplast with stacked lamellae. The capacity to store starch within the chloroplast must also have evolved. At the same time a progression from single cell to thallus occurred, followed by differentiation of cell types into tissues, which led to the higher plants through grades of organisation which in part can be discerned today in features of extant forms from the Bryophytes, Pteridophytes, Gymnosperms to the Angiosperms. However, during this evolution the essential structure and function of the chloroplast remained unaltered [96] with the possible exception of the C4 plants.

There is of course variation in the degree of stacking seen in chloroplasts of higher plants, if the spinach chloroplast as shown in plate VIIIa is regarded as representative variations can arise from mutations [122,123,125] which result in a reduction in stacking, or conversely in the formation of large grana [135]. In addition nutritional status [111] and other environmental factors such as light and temperature [5] can affect the observed structure. However, it is only within the bundle sheath chloroplasts of advanced C4 plants (Panicoid grasses) that a consistent variation occurs which is independent of environmental factors etc.

The structural aspects of C4 plants have been reviewed in some detail on a number of occasions [75—77]. The most prominent characteristic of the C4 plants is the arrangement of the chlorenchymatous tissue in concentric layers around the vascular tissue (Kranz-type anatomy). In most C4 species there are in fact two layers, the outer mesophyll cells and the inner bundle sheath layer. Early studies were carried out using species of the panicoid grasses of commercial interest such as maize, sorghum and sugar cane [6,62, 63,78,79,117,144] in which the bundle sheath chloroplasts differ conspicuously from those of the mesophyll cells, in that grana are much reduced or absent (Plate XIb). A similar anatomical arrangement was subsequently established in all C4 species, many of which — including the dicotyledons *Amaranthus* and *Atriplex* [16,36,74] — possess grana in both types of chloroplasts. Starch is always preferentially accumulated in the bundle sheath chloroplasts; however, it may also occur within the mesophyll cell plastids.

On the basis of a wide range of studies of C4 plants it has been suggested that they can be divided into three distinct groups [55,57]. These groups are distinguished primarily on the basis of the most active enzyme capable of decarboxylating the C4 organic acids which occurs in a given species (see Chapter 8). However, the three groups of plants may also differ at the morphological and ultrastructural level. Panicoid grasses and other C4 plants in which malate is preferentially formed during photosynthesis (malate formers)

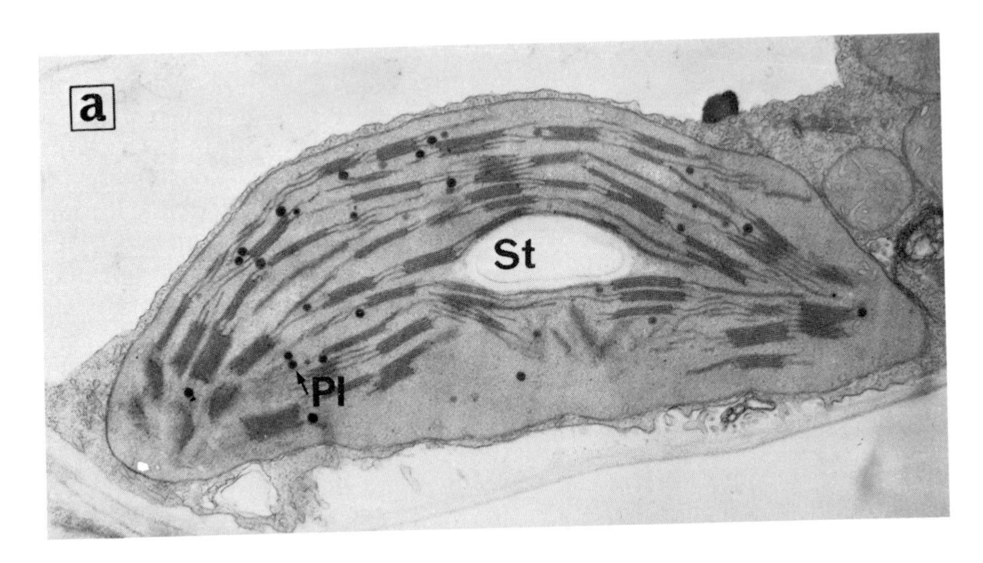

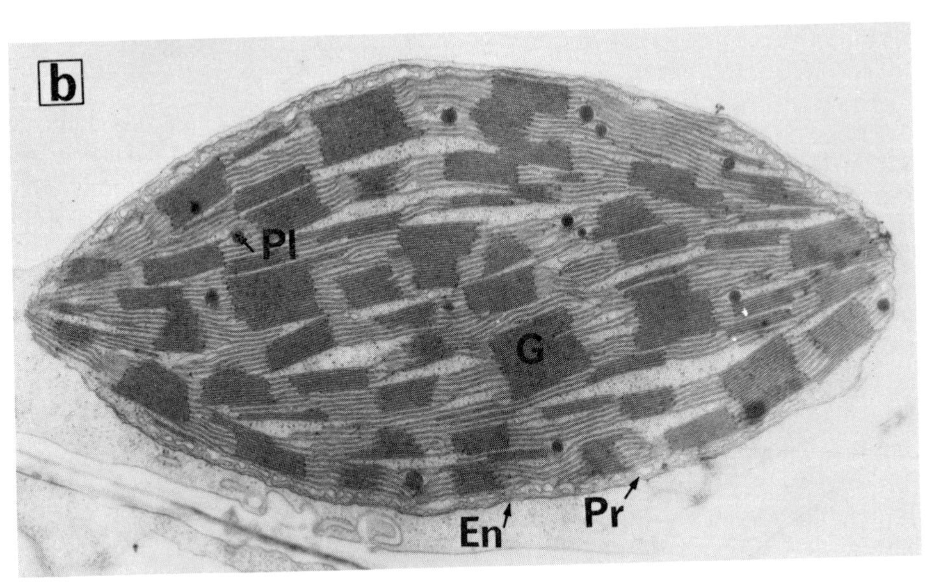

Plate VIII. (a) Typical C3 chloroplast from *Spinacia oleracea*. Glut/Os, Pb; × 18 000). (b) Grana-containing chloroplast from mesophyll cell of *Zea mays*. Glut/Os, Pb; × 20 000).

generally have chloroplasts with reduced grana in the bundle sheath cells. Those species in which aspartate is of greater importance (mainly Eragrostoideae) can be further divided on the basis of the position of the chloroplasts within the bundle sheath cells (Plate IX). As illustrated here in both fresh-cut (freezing microtome) and fixed and embedded material it is easy to distinguish the location of the chloroplasts as centripetal (a,c) or centrifugal (b,d). The similarity in position of chloroplasts in material from various sources and sectioned in various ways suggests that this is not an artifact.

Studies of enzyme levels [55,57] have indicated that those species with the centrifugal arrangement of chloroplasts may have higher levels of PEP carboxykinase, whereas those with a centripetal arrangement have higher levels of NAD-specific malic enzyme. Further differences can be observed between the two types of plants at the ultrastructural level (Plate X), since the chloroplasts of the centripetal type are closely associated with mitochondrion-like bodies (c). These differ in both structure and frequency from the mitochondria found in mesophyll cells of the same species and within leaf cells in general of other C4 plants.

The relationship between the mitochondria and peripheral reticulum has been studied in some detail for the asparatate type dicotyledons *Amaranthus edulis* and *Atriplex spongiosa* [27a]. Both mitochondria and protrusions or buds from the peripheral reticulum are superficially similar in ultrastructural characteristics. However, variations are found in the location of cytochrome *c* oxidase detected using 3′,3-diaminobenzidine and in changes in configuration of mitochondria after incubation with inhibitors of oxidative phosphorylation which enable the two structures to be distinguished. In particular the peripheral reticulum was found not to possess cytochrome *c* oxidase. Futhermore, developmental studies indicate that the mitochondria are present in the bundle sheath prior to the appearance of the peripheral reticulum. Hence it was concluded that the mitochondrion-like bodies were of two types, mitochondria or protrusions of the peripheral reticulum. It is possible that the latter represent the site of the NAD-malic enzyme.

All chloroplasts from leaves of C4 plants can be distinguished from those of other higher plants in that they possess a more pronounced system of vesicles, the peripheral reticulum, situated immediately within the envelope. This could be associated with the rapid transport of material into and out of the chloroplasts.

Otherwise the mesophyll chloroplasts of malate formers, and both types of chloroplasts in aspartate formers appear similar in structure to those of C3 plants. Plastoglobuli, ribosomes, starch grains and DNA-fibril-like regions can all be recognised. This is of particular significance since it has been suggested that the mesophyll chloroplasts fulfill a very restricted function (see chapter 8) catalysing the conversion of pyruvate to phosphoenol pyruvate in most species, and the reduction of oxaloacetate to malate in malate formers, but are not capable of the fixation of CO_2 or the formation

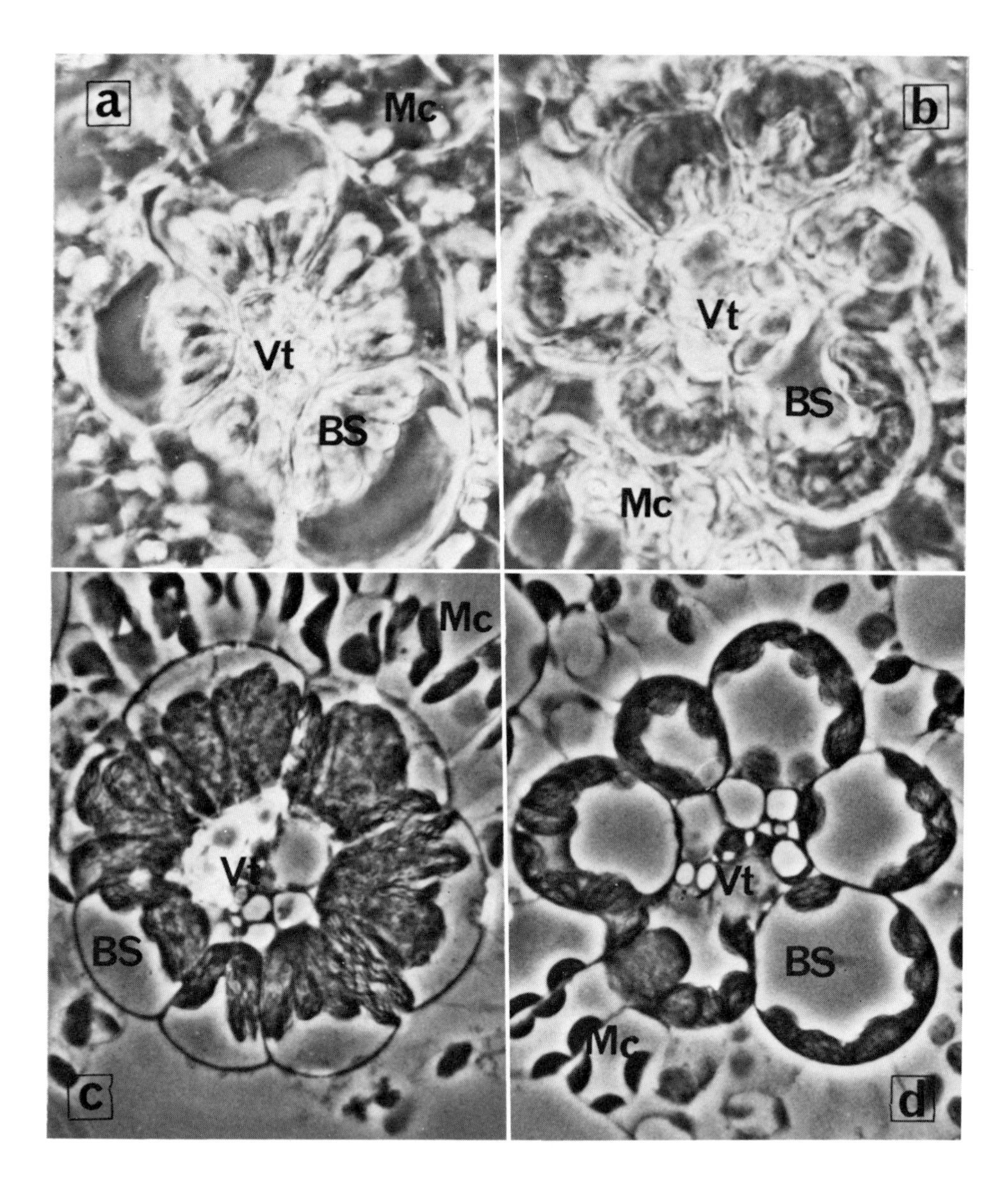

Plate IX. Photomicrographs of fresh-frozen (a,b) or fixed-embedded (c,d) and sectioned leaves of *Sporobolus spp*. Showing centripetal *(S.giganteus,* a,c) or centrifugal *(S.aeroides* b,d) arrangement of bundle sheath chloroplasts (× 2000).

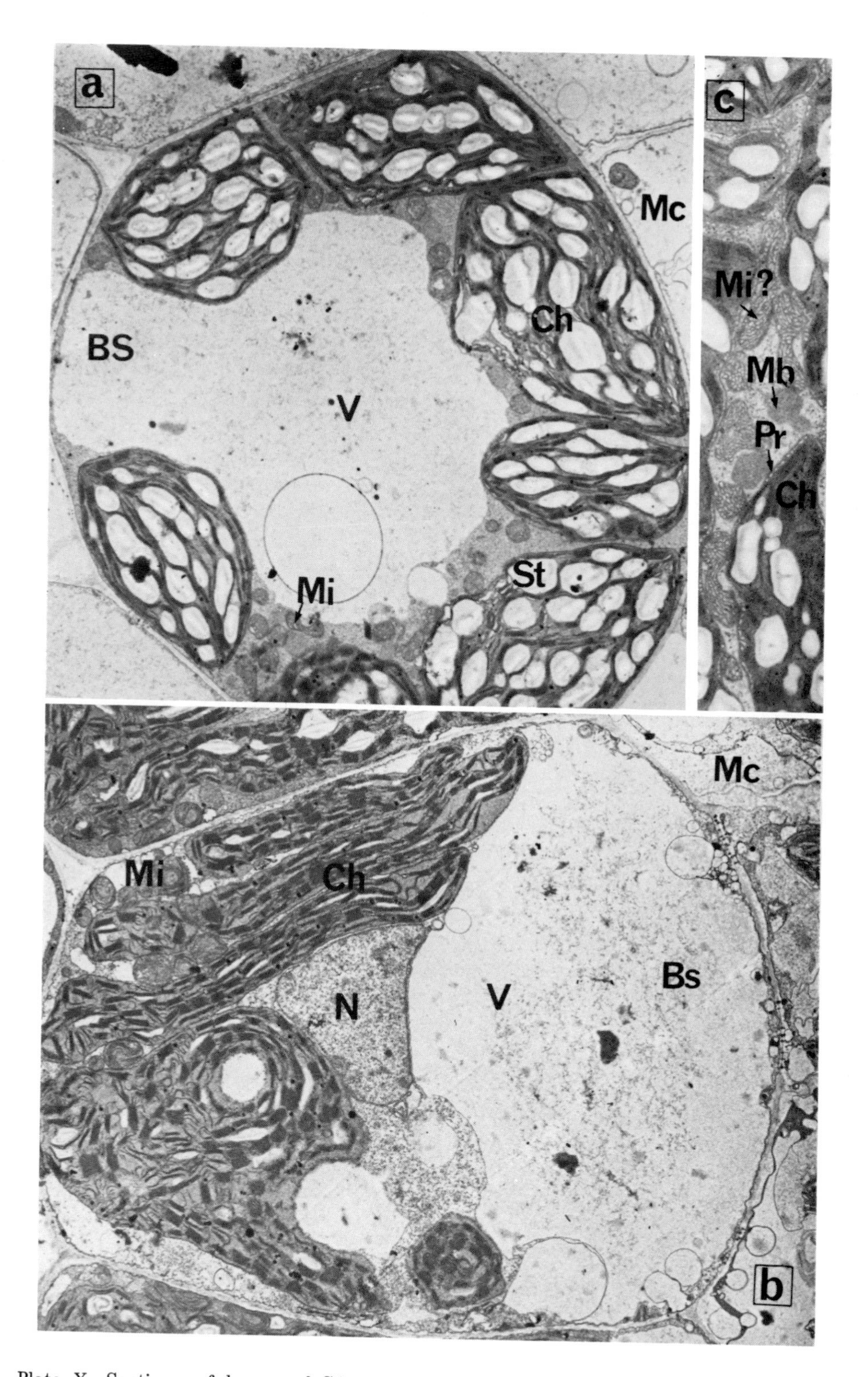

Plate X. Sections of leaves of C4 species showing (a) centrifugal *(S. aeroides)* or (b) centripetal arrangement of bundle sheath chloroplasts *(S.giganteus)*. Chloroplasts of both mesophyll and bundle sheath cells possess grana; an increased frequency of mitochondrial like bodies (Mi) is seen in species with centripetal chloroplasts (c). Glut/Os, Ua, Pb; (a) and (b) × 5000; (c) × 15 000.

of hexose. In contrast the reduced chloroplasts of the bundle-sheath of malate formers (Plate XIb) apparently possess normal carboxylation reactions but are restricted in photosystem II activity [151]. As a result there is apparently a spatial separation of the light-dependent and the primary carboxylation of ribulose-*bis*-phosphate reactions of photosynthesis between the two cell layers in these plants. Furthermore, in the aspartate formers with high levels of NAD-malic enzyme it is suggested that the mitochondria play an integral role in the process of photosynthetic carbon metabolism.

According to many current views the C4 plants represent a comparatively recent evolutionary development in land plants, with the establishment of two specific cell layers which have been further differentiated by the development of chloroplasts with a restricted biochemical function. However, apart from some variation in size, and the extent to which the plastids accumulate starch, there is little structural variation in the chloroplasts of mesophyll or bundle sheath cells of many C4 species.

1.7. ISOLATION OF INTACT CHLOROPLASTS

The problem of equating structure and function is accentuated by the difficulties which arise in separating the photosynthetic apparatus of most plants from within the rigid cellulose cell walls. In the last decade or so techniques have been developed to do this. Initial suggestions [82,129,140] indicated that two distinct types of chloroplast could be identified under phase contrast microscopy. These were the intact membrane-bound, class I, chloroplasts which appeared as refringent particles and within which the grana could not be seen (Plate XIIa), and the larger broken, envelope-free class II plastids. However, both aqueous [82,129,140] and non-aqueous [134] methods produce chloroplasts which vary greatly in structural integrity (Plate XIIb). A further classification of the types of isolated chloroplasts prepared, based on isolation techniques, has been proposed [56] (see Chapter 4). Plastids of these types can be recognised in the mixed population produced by aqueous methods. These range from the intact envelope-bound chloroplast (Plate XIIc) with an intact outer membrane and dense staining stroma, to chloroplasts which appear intact (Plate XIId) but in which the stroma is depleted and the thylakoids slightly swollen. It is possible that these represent plastids in which the outer envelope has broken, part of the contents have been lost and the envelope has re-fused [88]. Loss of the outer envelope and stroma (Plate XIIe) produces plastids which may still be distinguished from those in which the internal membrane system has started to "unfold" as found in washed broken chloroplasts suspended in low metal cation-containing media [68] (see Chapter 3). It is obvious that any detailed study of biochemical aspects of chloroplast function requires an equally detailed investigation of structural aspects.

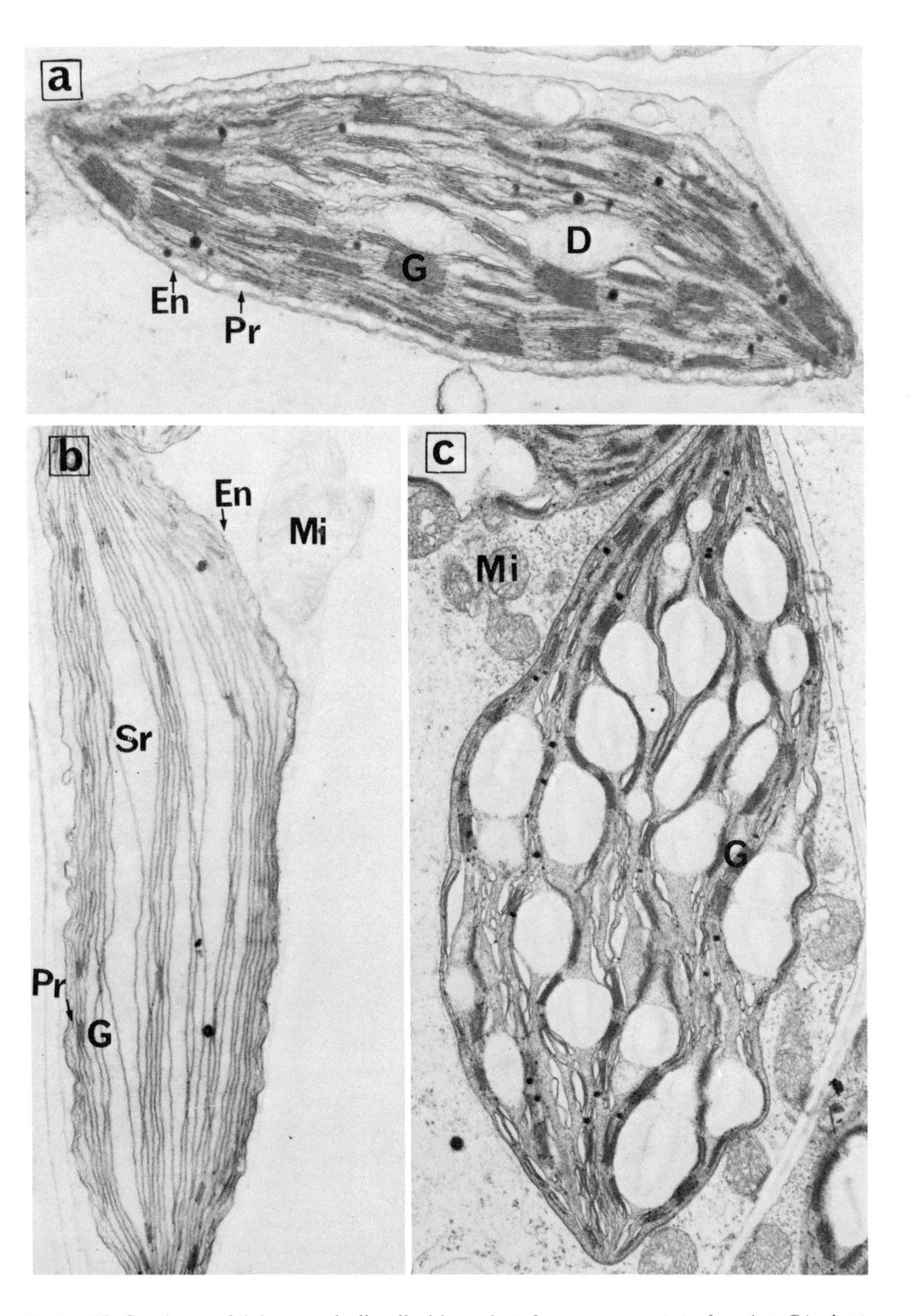

Plate XI. Sections of (a) mesophyll cell chloroplast from an aspartate-forming C4 plant *(S.aeroides)*; (b) bundle sheath cell chloroplast from *Z. mays*, malate-former, showing reduced grana; (c) Bundle sheath cell chloroplast from aspartate former *(S.aeroides)* showing pronounced grana. Note the peripheral reticulum (Pr) present in chloroplast from both cell types. Glut/Os, Ua, Pb. (a) × 18 000; (b) × 15 000; (c) × 10 000.

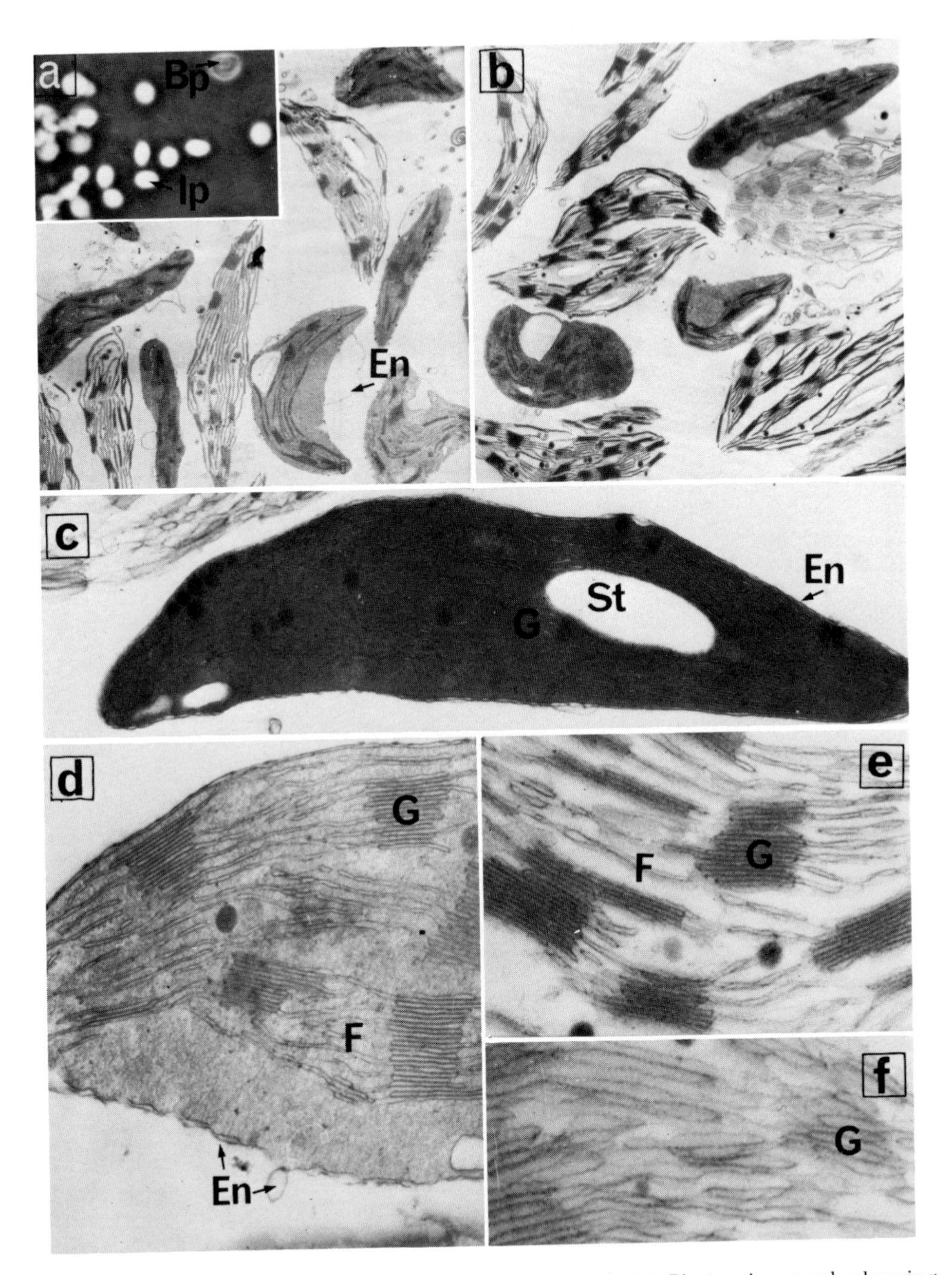

Plate XII. Isolated chloroplasts from *Spinacia oleracea.* (a) Photomicrograph showing intact (Ip) and broken (Bp) plastids. (b) Field view showing mixed chloroplast types (× 5000). (c) Intact chloroplast with dense contents. (d) Apparently intact chloroplast with swollen thylakoids and envelope. (e) Envelope-free chloroplast with grana intact. (f) Envelope-free chloroplast with disrupted granal structure. Glut/Os, Ua, Pb; (c) to (d) × 20 000.

1.8. SUMMARY

A detailed description of the ultrastructure of the higher plant chloroplast as typified by that found in C3 species is now possible and the various partial reactions of photosynthesis and carbon metabolism can be assigned to, or associated with specific structural entities. Departures from this pattern occur in two directions. First, in the algae alternate arrangements are seen in both the internal organisation of the lamellae system and in the number and arrangement of limiting membranes. Secondly, in the C4 species of higher plants variations of biochemistry occur; however, in many of such species this is reflected in specific differences in anatomical form rather than in variations of chloroplast ultrastructure.

At the highest level of resolution a uniform pattern of thylakoid substructure of globular subunits is seen. Variations in degree of thylakoid aggregation appear to reflect the presence or absence of specific ancillary photosynthetic pigments rather than a specific photochemical event. In the same way variations in the arrangement of limiting membranes may be associated with the nature and location of stored photosynthate. The chloroplast ER, and the varied relationships of pyrenoids to the cytoplasm and nuclear envelope found in algae, and the peripheral reticulum of the C4 plants may be associated with the rapid transfer of organic material to and from the plastids.

Although it is possible to postulate an evolution of the higher plant chloroplast through the algae it would appear likely that the present forms represent contemporary modifications of a more immediate ancestor to fill specific ecological or environmental niches. In the same way it is not possible to draw final conclusions on the origin of higher plant chloroplasts as endosymbiotic blue-green algae on the basis of structural considerations alone, since evidence for such an origin includes comparative studies of the nucleic acids and proteins of chloroplasts, mitochondria and prokaryotic organisms.

At the biochemical level the greatest divergence from a central pattern has been suggested on the basis of observations on C4 plants. With the exception of the mesophyll chloroplasts of such species all photosynthetic systems, whether prokaryote or eukaryote, contain fraction I protein and show ribulose-*bis*-phosphate carboxylase activity. The apparent absence of this protein from the structurally normal plastids found in such mesophyll cells is a problem which requires further investigation.

REFERENCES

1 Albertsson, P.A. and Leyon, H. (1954) Exptl. Cell Res., [7], 288–290.

2 Arntzen, C.J. and Briantais, J.M. (1975) in Bioenergetics of Photosynthesis (Govindjee, ed), pp. 51–94, Academic Press, New York.

3 Arntzen, C.J., Dilley, R.A. and Crane, F.L. (1969) J. Cell Biol. [43], 16—31.
4 Bailey, J.L. and Whyborn, A.G. (1963) Biochim. Biophys. Acta, [78], 163—174.
5 Ballantine, J.E. and Forde, B.J. (1970) Am. J. Bot., [57], 1150—1159.
6 Bangeon, J. (1973) J. Microsc., [16], 233—242.
7 Bartels, P.G. and Howshaw, R.W. (1968) Planta, [82], 293—298.
8 Barton, R. (1970) Planta, [94], 73—77.
9 Benshaul, U., Schiff, J.A. and Epstein, H.T. (1964) Plant Physiol., [39], 231—240.
10 Benson, A.A. (1966) J. Am. Oil Chem. Soc., [48], 265—270.
11 Bisulputra, T. (1974) in Algal Physiology and Biochemistry (Stewart, W.D., ed), pp. 124—160, Blackwell, Oxford.
12 Bisulputra, T. and Bailey, A. (1973) Protoplasma, [76], 443—454.
13 Bisulputra, T. and Bisulputra, A.A. (1967) J. Ultrastruct. Res., [17], 14—22.
14 Bisulputra, T. and Bisulputra, A.A. (1969) J. Ultrastruct. Res., [29], 151—170.
15 Bisulputra, T., Brown, D.L. and Weier, T.E. (1969) J. Ultrastruct. Res., [27], 182—190.
16 Bisulputra, T., Downton, W.J. and Tregunna, E.B. (1969) Can. J. Bot. [47], 15—21.
17 Bisulputra, T. and Weier, T.E. (1964) Am. J. Bot., [51], 881—892.
18 Bouck, G.B. (1962) J. Cell Biol., [12], 553—564.
19 Bouck, G.B. (1965) J. Cell Biol., [26], 523—537.
20 Bourne, V.L. and Cole, K. (1968) Can. J. Bot., [46], 1369—1375.
21 Branton, D. (1969) Ann. Rev. Plant Physiol., [20], 209—238.
22 Branton, D., Bullivant, S., Gilula, N.B., Karnosvsky, M.J., Moor, H., Muhlethaler, K., Northcote, D.H., Packer, L., Satir, B., Satir, P., Speth, V., Staehelin, L.A., Steere, R.L. and Wenstein, R.S. (1975) Science, [190], 54—56.
23 Branton, D. and Park, R.B. (1967) J. Ultrastruct. Res., [19], 283—303.
24 Brown, D.L. and Weier, T.E. (1966) J. Physiol., [4], 199—206.
25 Brown, F.A.M. and Gunning, B.E.S. (1966) in Biochemistry of Chloroplasts (Goodwin, T.W., ed), pp. 365—373, Academic Press, New York.
26 Burr, F.A. and West, J.A. (1970) Phycologia, [9], 17—37.
27 Cavalier-Smith, T. (1975) Nature, [256], 463—468.
27a Chapman, E.A., Bain, J.M., Gove, D.W. (1975) Aust. J. Plant Physiol.
27b Christensen, T. (1964) in Algae and Man (Jackson, D.F., ed), pp. 59—64, Plenum, New York.
28 Cohen, M. and Bowler, E. (1953) Protoplasma, [42], 414—416.
29 Cohen-Bazire, G. (1971) in Biological Ultrastructure (Harris, P., ed), pp. 65—90, Oregon State University Press, Corvallis.
30 Cohen-Bazire, G., Pfennig, N. and Kunisawa, R. (1964) J. Cell Biol., [22], 207—225.
31 Cole, K. (1970) Phycologia, [9], 275—283.
32 Coombs, J., Lauritis, J.A., Darley, W.M. and Volcani, B.E. (1968) Z. Pflanzenphysiol., [59], 124—152.
33 Dodge, J.D. (1968) J. Cell. Science, [3], 41—48.
34 Dodge, J.D. (1969) Arch. Mikrobiol., [69], 266—280.
34a Dodge, J.D. (1973) The Fine Structure of Algal Cells. Academic Press, London.
35 Dodge, J.D. (1974) Sci. Prog. (Oxford), [61], 257—274.
36 Downton, W.J.S., Bisulputra, T. and Tregunna, E.B. (1969) Can. J. Bot., [47], 915—919.
37 Drum, R.W. (1963) J. Cell Biol., [18], 429—440.
38 Drum, R.W. and Pankratz, H.S. (1964) Am. J. Bot., [51], 405—418.
39 Emerson, R. and Arnold, W.J. (1932) J. Gen. Physiol., [16], 191—198.
40 Evans, L.V. (1970) Br. Phycol. J., [5], 1—13.
41 Falk, H. (1969) J. Cell Biol., [42], 582.
41a Fay, P. (1973) in The Biology of Blue-green Algae (Carr, N.S. and Whitton, B.A.,

eds), pp. 238—259, Blackwell, Oxford.
42 Gantt, E. and Conti, S.F. (1965) J. Cell Biol., 26, 365—381.
43 Gantt, E. and Conti, S.F. (1969) J. Bacteriol., [97], 1486—1493.
44 Gantt, E., Edwards, M.R. and Provosoli, L. (1971) J. Cell Biol., [48], 280—290.
44a Greenwood, A.D., Abstr. Xth Int. Congr. Bot., Edinburgh, pp. 212—213.
44b Greenwood, A.D., Abstr. VIIIth Int. Congr. Electron Microscopy, Canberra, Vol. 2, pp. 566—567.
45 Greenwood, A.D., Leech, R.M. and Williams, J.P. (1963) Biochim. biophys. Acta, [78], 148—162.
46 Gibbs, S.P. (1960) J. Ultrastruct. Res., [4], 127—148.
47 Gibbs, S.P. (1962) J. Ultrastruct. Res., [7], 418—435.
48 Gibbs, S.P. (1962) J. Cell Biol., [14], 433—444.
49 Gibbs, S.P. (1970). Ann. N.Y. Acad. Sci., [175], 454—473.
50 Goodenough, U.W. and Staehelin, L.A. (1971) J. Cell Biol., [48], 594—619.
51 Granick, S. and Porter, K.R. (1947) Am. J. Bot., [34], 545—550.
52 Griffith, D.J. (1970) Bot. Rev., [36], 29—58.
53 Gunning, B.E.S. (1965) J. Cell Biol., [24], 79—93.
53a Gunning, B.E.S. and Steer, M.W. (1975) Ultrastructure and the Biology of Plant Cells, Arnold, London.
54 Gunning, B.E.S., Steer, M.W. and Cochrane, M.P. (1968) J. Cell Sci., [3], 445—456.
55 Gutierrez, M., Gracen, V.E. and Edwards, G.E. (1974) Planta, [119], 279—300.
56 Hall, D.O. (1972) Nature New Biol., [235], 125—126.
57 Hatch, M.D., Kagawa, T. and Craig, S. (1975) Aust. J. Plant Physiol., [2], 111—118.
58 Heslop-Harrison, J. (1963) Planta, [60], 243—260.
59 Heslop-Harrison, J. (1966) Sci. Prog. (Oxford), [54], 519—541.
60 Hickman, D.D. and Frenkel, A.W. (1965) J. Cell Biol., [25], 261—278.
61 Hibberd, D.G. and Leedale, G.F. (1970) Nature, [225], 758—759.
62 Hilliard, J.H. and West, S.H. (1970) Science, [168], 494—496.
63 Hodge, A.J., Mclean, J.D. and Mercer, F.V. (1955) J. Biophys. Biochem. Cytol., [1], 605—614.
64 Hohl, A.R. and Hepton, A. (1965) J. Ultrastruct. Res. [12], 542—546.
64a Holdsworth, R.H. (1971) J. Cell Biol., [52], 499—513.
65 Holt, S.C., Conti, S.F. and Fuller, R.C. (1966) J. Bacteriol. [91], 311—323.
66 Howell, S.H. and Moudrianakis, E.N. (1967) J. Mol. Biol., [27], 323—333.
67 Hyde, B.B., Hodge, A.J., Kahn, A. and Birnstiel, M.L. (1963) J. Ultrastruct. Res., [9], 248—258.
68 Izawa, S. and Good, N.E. (1966) Plant Physiol., [41], 544—552.
68a Jeffrey, S.W. (1969) Biochim. Biophys. Acta, [177], 456—467.
69 Kahn, A. and von Wettstein, D. (1961) J. Ultrastruct. Res., [5], 557—574.
70 Kowallik, K. (1969) J. Cell Sci., [5], 251—269.
71 Kawashima, N. and Wildman, S.G. (1970) An. Rev. Plant Physiol., [21], 325—358.
72 Kirk, J.T.O. (1971) An. Rev. Biochem., [40], 161—196.
73 Kirk, J.T.O. and Tilney-Bassett, R.A.E. (1967) The Plastids: their Chemistry Structure, Growth and Inheritance. Freeman, London.
74 Laetsch, W.M. (1968) Am. J. Bot., [55], 875—883.
75 Laetsch, W.M. (1969) Sci. Prog. (Oxford), [57], 323—351.
76 Laetsch, W.M. (1971 in Photosynthesis and Photorespiration (Hatch, M.D., Osmond, C.B. and Satyer, R.O., eds), pp. 323—349, Wiley/Interscience, New York.
77 Laetsch, W.M. (1974) An. Rev. Plant Physiol., [25], 27—52.
78 Laetsch, W.M. and Price, I. (1969) Am. J. Bot., [56], 77—87.
79 Laetsch, W.M., Stetler, D.A. and Vlitos, A.J. (1965) Z. Pflanzenphysiol., [54], 472—474.
80 Lang, N.J. (1968) An. Rev. Microbiol., [22], 15—46.

81 Lang, N.J. (1963) Am. J. Bot., [50], 280—300.
82 Leech, R.M. (1964) Biochim. Biophys. Acta, [79], 637—639.
83 Leedale, G.F. (1967) Euglenoid Flagellates. Prentice Hall, Englewood Cliffs.
84 Leggett-Bailey, J. and Whyborn, A.G. (1963) Biochim. Biophys. Acta, [78], 163—174.
85 Lembi, C.A. and Lang, N.J. (1965) Am. J. Bot. [52], 464—477.
86 Lichtenthaler, H.K. (1968) Endeavour, [27], 144—149.
87 Lichtenthaler, H.K. (1969) Protoplasma, [68], 65—77.
88 Lilley, R. McC., Fitzgerald, M.P., Rienits, K.G. and Walker, D.A. (1975) New Phytol., [75], 1—10.
89 Lucas, I.A.N. (1970) J. Phycol., [6], 30—38.
90 Luttge, U. and Krapf, G. (1968) Planta, [81], 132—138.
91 Mackender, R.O. and Leech, R.M. (1970) Nature, [228], 1347—1349.
92 Manton, I. (1966) In Biochemistry of Chloroplasts (Goodwin, T.W., ed), pp. 23—47, Academic Press, London.
93 Manton, I. (1966) J. Cell. Sci., [1], 187—192.
93a Margulis, L. (1970) The Origin of Eukaryotic Cells, Yale University, Press, New Haven.
93b Meeks, J.C. (1974) in Algal Physiology and Biochemistry (Stewart, N.D., ed), pp. 161—175, Blackwell, Oxford.
94 Menke, W. (1960) Experientia, [16], 537—539.
95 Menke, W. (1962) An. Rev. Plant Physiol., [13], 27—44.
96 Menke, W. (1966) in Biochemistry of Chloroplasts (Goodwin, T.W., ed), pp. 3—18, Academic Press, London.
97 Moor, H., Muhlethaler, K., Waldner, H. and Frey-Wyssling, A. (1961) J. Biophys. Biochem. Cytol., [10], 1—13.
98 Muhlethaler, K., Moor, H. and Szarkowski, J.W. (1965) Planta, [67], 305—323.
99 Murakami, S. (1963). Exptl. Cell Res., [32], 398—400.
100 Murakami, S. (1964) J. Electron Micros., [13], 234—236.
101 Neushal, M. (1970) Am. J. Bot., [57], 1231—1239.
102 Pankratz, H.S. and Bowen, C.C. (1963) Am. J. Bot., [50], 387—399.
103 Paolillo, D.J. (1970) J. Cell Sci., [6], 243—255.
104 Paolillo, D.J. and Falk, R.H. (1966) Am. J. Bot., [53], 173—180.
105 Paolillo, D.J., Mackay, N.C. and Griffus, J.P. (1969) Am. J. Bot. [56], 344—347.
106 Park, R.B. and Biggins, J. (1964) Science, [144], 1009—1011.
107 Park, R.B. and Pheifhofer, A.A. (1968) Proc. Natl. Acad. Sci. USA, [60], 337—343.
107a Park, R.B. and Pheifhofer, A.A. (1969) J. Cell Sci., [5], 313—319.
108 Park, R. and Pon, N. (1961) J. Mol. Biol., [3], 1—11.
109 Park, R.B. and Sane, P.V. (1971) An. Rev. Plant Physiol., [22], 395—430.
110 Pickett-Heaps, J.D. (1975) Green Algae. Sinauer Associates, Sunderland, Mass.
111 Possingham, J.V., Vesk, M. and Mercer, F.V. (1964) J. Ultrastruct Res., [11], 68—83.
112 Price, J.L. and Thomson, W.W. (1967) Nature, [214], 1148—1149.
112a Raven, P.H. (1970) Science, [169], 641—646.
113 Reynolds, E.S. (1963) J. Cell Biol., [17], 208—212.
114 Ris, H. and Plaut, W. (1962) J. Cell Biol., [13], 383—391.
115 Robards, A.W. and Robinson, C.L. (1968) Planta, [82], 179—188.
116 Robertson, J.D. (1959) Biochem. Soc. Symposium, [16], 30—43.
117 Rosado-Alberio, J., Weier, T.E. and Stocking, C.R. (1968) Plant Physiol., [43], 1325—1331.
118 Sabatini, D.D., Bensch, K. and Barrnett, R.J. (1963) J. Cell Biol., [17], 19—58.
119 Sabinis, D.D., Gordon, M. and Galson, A.W. (1969) Plant Physiol., [44], 1355—1363.

120 Sager, R. and Palade, G.E. (1957) J. Biophys. Biochem. Cytol., [3], 463—488.
121 Sane, P.V., Goodchild, V.S. and Park, R.B. (1970) Biochim. Biophys. Acta, [216], 162—178.
122 Schmid, G.H. and Gaffron, H. (1967) J. Gen. Physiol., [50], 563—582.
123 Schmid, G.H., Price, M. and Gaffron, H. (1966) J. Microsc., [5], 205—209.
124 Schotz, F. and Diers, L. (1967) in Le Chloroplaste (Sironval, C., ed), pp. 21—29, Masson, Paris.
124a Shively, J.M., Ball, F.L. and Kline, B.W. (1973) J. Bacteriol.[116], 1405—1411.
125 Shumway, L.K. and Weier, T.E. (1967) Am.J.Bot. [54], 773—780.
126 Shumway, L.K., Weier, T.E. and Stocking, R.C. (1967) Planta, [76], 182—189.
127 Speer, H.L., Dougherty, W. and Jones, R.T. (1964) J.Ultrastruct. Res., [11], 84—96.
128 Spencer, D. and Wildman, S.G. (1962) Aust.J.Biol.Sci., [15], 599—610.
129 Spencer, D. and Unt, H. (1965) Aust. J. Biol. Sci., [18], 197—210.
130 Stanier, R.Y. (1974) in 24th Symp. Soc. Gen. Microbiol., pp.00—00.
131 Steere, R. (1957) J. Biophys. Biochem. Cytol., [3], 45—52.
132 Steinman, E. (1952) Exptl. Cell Res., [3], 367—372.
133 Steinman, E. and Sjostrand, F.S. (1955) Exptl. Cell Res., [8], 15—23.
133a Steward, W.F.D. (1973) in The Biology of Blue-green Algae (Carr, N.S. and Whitton, B.A., eds), pp. 260—278, Blackwell, Oxford.
134 Stocking, C.R., Shumway, L.K., Weier, T.E. and Greenwood, D. (1968). J. Cell. Biol., [36], 270—275.
135 Susalla, A.A. and Mahlberg, P.G. (1975) Am. J. Bot. [62], 878—883.
136 Thomas, R.R., Kolodier, R. and Tewari, K.K. (1975) in Genetic Aspects of Photosynthesis (Nasyrov, Yu.E. and Sestak, Z., eds), pp. 9—30, Junk, The Hague.
137 Thomson, W.W. (1968) J. Exptl. Bot., [16], 169—176.
138 Thomson, W.W. (1974) in Dynamic Aspects of Plant Ultrastructure (Robards, A.W., ed), pp. 138—177, McGraw-Hill, London.
139 Valle, A.E., and Wolfe, R.S. (1958) J. Bacteriol., [75], 480—488.
140 Walker, D.A. (1965) Plant Physiol., [40], 1157—1161.
141 Wehrmeyer, W. (1964) Planta, [62], 272—293.
142 Weier, T.E. (1961) Am. J. Bot., [51], 615—630.
143 Weier, T.E. and Thomson, W.W. (1962) J. Cell. Biol., [13], 89—108.
144 Weier, T.E. and Benson, A.A. (1967) Am. J. Bot., [54], 389—402.
145 Weier, T.E., Bisulputra, T. and Harrison, A. (1966) J. Ultrastruct. Res., [15], 38—56.
146 Weier, T.E., Engelbrecht, A.H.P., Harrison, A. and Risley, E.B. (1965) J. Ultrastruct. Res., [13], 92—111.
147 Weier, T.E., Shumway, K.L. and Stocking, C.R. (1968) Protoplasma, [66], 339—355.
148 Weier, T.E., Stocking, C.R. and Shumway, L.K. (1966) Brookhaven Symp. Biol., [19], 353—374.
149 Weier, T.E., Stocking, C.R., Bracher, C. and Risley, E. (1965) Am. J. Bot., [52], 339—343.
150 Weier, T.E., Stocking, C.R., Thomson, W.W. and Drever, H. (1963) J. Ultrastruct. Res., [8], 122—143.
151 Woo, K.C., Pyliotis, N.A. and Downton, W.J.S. (1971) Z. Pflanzenphysiol., [64], 400—413.
152 Yokomura, T. (1967) Cytologia, [32], 361—377.

The Intact Chloroplast — edited by J. Barber

Chapter 2

Electrical Interactions and Gradients between Chloroplast Compartments and Cytoplasm

W. J. VREDENBERG

Center for Agrobiological Research, P.O. Box 14, Wageningen (The Netherlands)

CONTENTS

Abbreviations: CCCP, Carbonyl cyanide-*m*-chlorophenylhydrazone; DCMU, 3-(3,4-dichlorophenyl)-1,1-dimethylurea; DCPIP, 2,6-dichlorophenol-indophenol; PS1(2), photosystem one(two).

2.1. INTRODUCTION

Growth and development of living organisms are determined by a variety of physiological processes. The physico-chemical and biochemical reactions by which these processes are driven, can hardly be expected to proceed with an optimal rate and functional efficiency if they did not occur in differentiated cellular and sub-cellular compartments with special structural architecture of their interior and limiting membranes. Each living cell is surrounded by a membrane which, depending on the composition and interaction of its constituting lipids and proteins, acts as a selective and specific barrier [56,76, 77]. One of the primary functions of the membrane is to screen within its enclosure a quantity of different enzymes and substrates, which as a whole represents an integral set of one or more metabolic reactions that occur within the enclosed compartment. A second important function of a membrane is to exert a control on the ionic composition of the phases at both sides of the barrier. Several enzymatic reactions are known to depend on the relative activity of specific ions in the reaction phase. The capacity of the membrane for exerting these control functions is due to, and dependent on, limitations for passive diffusion of charged and non-charged plasma components through the barrier, set by its selective permeability characteristics [28]. In other chapters in this book various examples are given to show the regulatory functions of the chloroplast-enclosing membrane in maintaining optimal conditions for light-triggered and enzyme-regulated photosynthetic reactions, as well as in controlling the translocation of substrates and reaction intermediates or products.

This chapter will deal mainly with phenomena which are closely associated with a direct action of the chloroplast membranes on the energy regulation of metabolic reactions in the phases separated by the membrane. In general such action apparently is elaborated in cases where so-called (energy dependent) active ion pumps are operating in a membrane. In addition to controlling passive diffusion, the membrane is structurally involved as a functional site at which energy is converted into a form which triggers the rate of metabolic reactions at one or more distant sites. As far as ion pumps are concerned, these pumps are considered as active membrane components because of their capability of maintaining an electrochemical potential gradient of certain ions across the membrane at the expense of metabolic energy [81]. The electrochemical potential gradient in its turn may serve as an energy source for the passive movement of other ions and thus either directly or indirectly exerts control on other metabolic reactions.

2.1.1. Structure and function of chloroplasts

The structural aspects of the chloroplast and the chloroplast compartments have been discussed in detail in Chapter 1. Chloroplasts are surrounded by two membranes (inner and outer) which enclose the lamellar inner structures (thylakoids) and the stroma. Usually a major part of thylakoids is oriented in an appressed form to constitute granum stacks (grana lamellae). Single thylakoids, usually connecting granum stacks, are called stroma lamellae. The specialized and unique function of chloroplasts in providing and organizing the machinery of converting light energy into useful biological energy is associated with the thylakoid membranes [57,64]. The photochemical conversion of light energy into chemical energy takes place in a light trapping process and accompanying electron transfer through the photosynthetic electron transport chain with a coupled phosphorylation system (cf. recent reviews [2,15,48,63,84,93,94] and proceedings of recent photosynthesis meetings [8,30]). This process is performed by the photochemical systems which are located in the lamellar membranes of the chloroplasts (cf. refs. 15, 64). The most widely accepted scheme for this part of the photosynthetic energy conservation mechanism is the so-called Z-scheme, which postulates two photochemical systems, system one (PS1) and system two (PS2), operating in series. Light energy is absorbed by antenna pigments (chlorophyll *a* and accessory pigments) and transferred to the photoactive chlorophyll molecules P700 and P680 of the reaction centers of PS1 and PS2, respectively. Upon excitation of the reaction center an electron is transferred from the oxidizable donor molecules P700 and P680 to the primary acceptor molecules X (system one) or Q (system two), which become reduced. Oxidized P680 and P700 are reduced by the primary donor ZH and cyt *f* or plastocyanin, respectively. Reduced Q (QH) is oxidized by the oxidized photoproducts of system one via a sequence of redox reactions through carriers in the electron transport chain connecting both systems. A simplified scheme, in which cyclic electron transport in system one and pathways for and coupling sites of photophosphorylation have been omitted, is depicted in Fig. 2.1. The scheme as it is presented here only intends to indicate the probable position of carriers in the transport chain. It suffices for the material to be discussed in this chapter.

By means of the two light-trapping processes in PS1 and PS2 electrons are transferred from H_2O to $NADP^+$ with concomitant synthesis of ATP (not shown) and oxygen evolution. NADPH and ATP are necessary for the reduction of CO_2 into carbohydrates.

The requirements set by the chemiosmotic hypothesis of Mitchell [58] on the coupling between ATP production and photosynthetic electron transport, combined with the insights into the structure of chloroplasts and into the flow of energy through the photosynthetic electron transport chain, have suggested an essential relationship between the structure and function of the chloroplast membranes, in particular with respect to the energy transduction mechanism. Ample evidence has accumulated that the thylakoid membrane is actively involved in the regulatory energetic processes associated with the primary photosynthetic reactions [55,78,79,94].

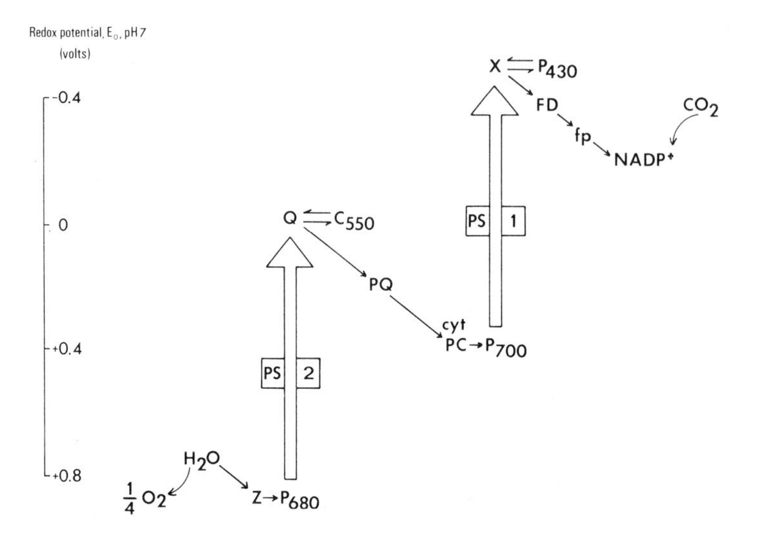

Fig. 2.1. Z-scheme of photosynthetic electron transfer from water to NADP+ (simplified form, i.e. coupling sites of photophosphorylation not shown, reaction sites of cyt *b*-559 and of cyt b_6 not indicated). P680 and P700: trapping centers of (PS2) and (PS1), respectively; X and Q: primary electron acceptors of PS1 and PS2, respectively. C550 and P430 are compounds absorbing at 550 and 430 nm, respectively. PQ, plastoquinone; PC, plastocyanin; FD ferredoxin; fp, flavoprotein.

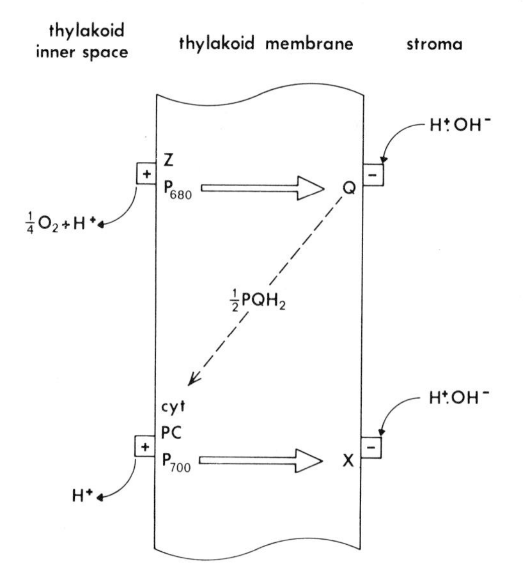

Fig. 2.2. Photosynthetic electron flow through the vectorially oriented electron transport chain and coupled proton translocation across the thylakoid membrane. The proton binding and proton releasing site at the outside and inside of the membrane, respectively, are indicated.

2.1.2. *Energy coupling in chloroplast lamellar membranes*

As mentioned already, the inner lamellar membranes of chloroplasts are the seat of the pigments, carriers and enzymes of the photosynthetic apparatus which participate in the generation of biological energy (NADPH and ATP). If the thylakoid membrane would only function as an organizer which is responsible for an efficient assembly of carriers and enzymes that catalyze the redox reactions and the generation of ATP, it could be considered primarily as a passive element in the energy transduction mechanism. Such function would be compatible with the chemical hypothesis on energy coupling, formulated first by Slater [80]. As an alternative Mitchell [58] has proposed a chemiosmotic energy coupling mechanism in which some of the free energy of redox reactions in the electron transport system is converted into an electrochemical potential gradient of protons (proton motive force) across the membrane. The chemiosmotic hypothesis requires the necessary fulfilment of essential conditions which are listed below. The tests of these requirements have been the target of numerous investigations and will be discussed here only briefly and selectively with relevance to the material discussed in this chapter. For more extensive discussions the reader is referred to the original papers and the reviews dealing with these topics [37,44,58, 65,79].

(i) A reversible ATP-ase should be anisotropically oriented in the membrane as to couple the passive translocation of protons at the enzyme site across the membrane to the flow of anhydro bond equivalents between water and the couple ATP/ADP + P_i. In their classical work [43,44], Jagendorf and his co-workers have shown indeed that ATP synthesis occurs in the dark in isolated chloroplasts at the energetic expense of a pH-gradient across the membrane, induced by an acid-base transition of the suspending medium. Recent work carried out in other laboratories has shown in addition that the imposition of a transient electric potential (inside positive) stimulates ATP synthesis substantially [72,86].

(ii) The reaction centers and redox carriers of the electron transport chain should be vectorially arranged across the coupling membrane. Energy trapping in the reaction centers and subsequent electron transfer from the photoactive pigments P700 and P680 to the primary acceptors X and Q, respectively, would then be associated with a charge separation, giving rise to a localised vectorial electric field (Fig. 2.2). Subsequent protonation and de-protonation of the active sites, and flow of reducing equivalents such as hydrogen groups and electron pairs between spatially oriented redox carriers of the reaction chain would result in the translocation of protons across the membrane. It has been shown indeed that a decrease in pH of the suspending medium occurs upon illumination of isolated chloroplasts [40,60,73]. This was interpreted as evidence, and experiments of Rumberg and associates have confirmed [68,69] that protons are translocated into the thylakoid inner

space concomitantly with electron transport through the photosynthetic reaction chain. Witt and co-workers [46,74,94] have been the first to suggest evidence that the light-induced absorption changes with a maximum at 515 nm, first discovered by Duysens [27], is an indicator of the formation of an electric field across the thylakoid membrane, due to the primary charge separation in the light-activated reaction centers. These changes were interpreted in terms of a wavelength shift of the absorption band of absorbing photosynthetic pigments (cf. refs. 3 and 4). Such electrochromic shifts have been confirmed to occur in response to the generation of a vectorial electrical field across artificial model membranes in which absorbing pigments were embedded [29,70]. We will discuss the impact of this phenomenon in more detail in a subsequent section (section 2.2.1.).

Trebst [84] and Trebst and Hauska [85] have recently reviewed the evidence, obtained by various methods (action of selective and specific antibodies against enzymes and redox carriers; action of hydrophilic and lipophilic donors and acceptors), for the sidedness of the reaction chain in the thylakoid membrane. This evidence points to a vectorial orientation of the electron transport chain in the membrane with the donor molecules (P700 and P680) of both systems at the inner facing side and the acceptor molecules at the stroma facing boundary of the membrane.

(iii) Proton translocation across the closed thylakoid membrane should be accompanied by charge neutralizing passive influx of anions and/or efflux of cations. The function of these systems is to regulate the pH and osmotic differential across the membrane with respect to the active charging of the membrane capacity and the energy dissipation through the passive ion conducting channels, including proton conduction through the ATPase system. Movement of anions and cations across the thylakoid membrane in association with light-stimulated proton influx has been observed under a variety of conditions in isolated chloroplasts [24,26,41,63] as well as in chloroplasts in situ [61]. However, there is as yet little agreement on the identity of the ionic species which are mainly involved in this process. This seems to me not too surprising because, as will be discussed in the next section, the density of the accompanying ion fluxes will strongly depend on the chemical composition of the chloroplast environment and on the permeability characteristics of the membrane.

(iv) The energy coupling capacity of the thylakoid membrane, reflected by the magnitude of the electrochemical gradient of the hydrogen ion, will be dependent on the membrane permeability coefficients and on the concentration of the permeating ions, including protons. Uncoupling is expected to occur in the presence of antibiotics which modify the permeability characteristics of the membrane, provided that protons are present at non-limiting concentrations, under low pH conditions. Coupling is expected to be weak at low pH values at which the electrochemical gradient is small due to the fact that the passive proton fluxes can compete with the fluxes of

other permeating ions. These phenomena have received ample attention [6,7], but certain aspects will be considered in more detail in a next section.

2.1.3. Electrogenesis and coupled ion movements in chloroplasts

Although knowledge about the energy coupling at the thylakoid membrane is still incomplete in its very details, there is reasonably good evidence, that the coupling is achieved by a light-triggered electrogenic hydrogen ion pump which is responsible for the generation of an electrochemical potential gradient of protons across the membrane. This evidence has been discussed in detail with respect to experimental data on spectroscopic changes of membrane pigments and artificial probes [22,51,52,94], on changes in delayed light emission [12,22,50,98], as well as on formation of a pH-gradient and (associated) ion movements across the thylakoid membrane [24,67,69,73].

The following simplified model scheme (Fig. 2.3) accounts for most of the evidence obtained so far. The pump tends to drive protons inwards which results in an electrogenic potential E_e, positive inside, across the membrane. This potential induces a discharging passive efflux of cations and (or) an influx of anions, with a concomitant active influx of hydrogen ions, ϕ^a_H, into the thylakoid interior. ϕ^a_H, will be equal, but opposite in sign, to the sum of the passive currents carried by the mobile ions. The relative contribution of an ionic species j to the passive fluxes is determined by the actual membrane conductance g_p, which in a first approximation is proportional to $P_j \cdot c_j$, in which P_j is the permeability coefficient and c_j the actual concentration of the ion j. The steady state of the system is characterized by a membrane potential E_m at which the mobile ions are passively distributed in equilibrium with this potential, and by a proton concentration gradient maintained at a steady-state level by the active and passive proton fluxes,

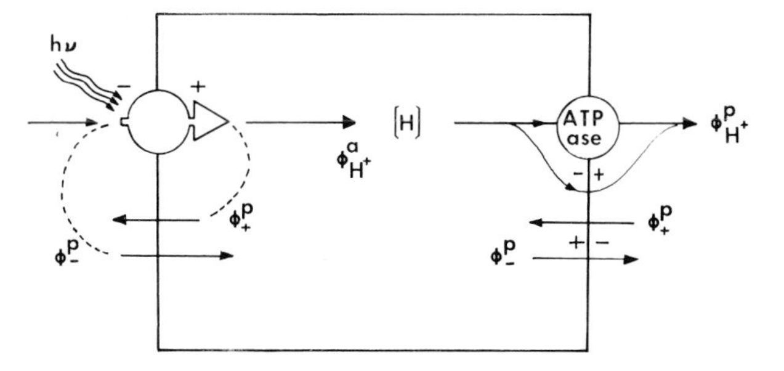

Fig. 2.3. Model scheme of the electron transport-coupled electrogenic hydrogen ion transport and of the passive ion transport across the thylakoid membrane (see details in the text).

ϕ_H^a and ϕ_H^p, respectively. In the dark the electrogenic components of the membrane potential becomes zero, and the initial dark potential will be equal to the diffusion potential set by the gradients of the ions transported during the operation of the pump. Re-equilibration of these ions to the dark-equilibrium state will occur by passive efflux of hydrogen ions in exchange for the other permeated ions.

The model can be represented by the following (simplified) equivalent electric circuit diagram of the membrane (Fig. 2.4.). The electrogenic pump acts in the membrane as an electromotive force E_e in series with an associated conductance g_p. At the onset of illumination the membrane capacitance C_m is charged at an initial potential $[^LE_m]_i = E_e$. The initial charging flux, which is equal to $E_e \cdot g_p$, is determined by the initial rate of charge separation in the photosynthetic reaction centers. Under steady-state light conditions this rate is equal to the (lower) steady-state electron transfer rate through the electron transport chain. Consequently the charging flux is smaller under these conditions. We will formally describe these variations in terms of variations in g_p, at constant E_e. Under steady-state light conditions the charging flux through g_p is balanced by an equal flux through the passive membrane conductance g. Both fluxes will be responsible for the creation of a diffusion potential set by the induced concentration gradients of the passively moving ions (through g) and of the actively transported hydrogen ions (through g_p). This diffusion potential contributes to the membrane potential and is represented by an ionic capacitor which under steady-state conditions of the pump, acts as a battery with an e.m.f. equal to E_{diff}.

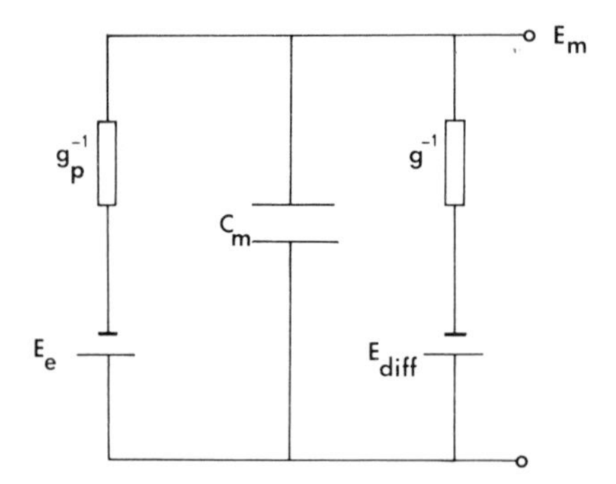

Fig. 2.4. Simplified equivalent electric circuit of the thylakoid membrane. E_e and E_{diff} are e.m.f.'s of the electrogenic H^+-pump and of the ion gradients, respectively; g_p and g are the conductances associated with the active and passive channels, respectively. C_m is the capacitance, and E_m the potential of the membrane.

Under steady-state conditions the membrane potential will be

$$[^{L}E_m]_{ss} = \frac{g_p}{g_p + g} E_e + \frac{g}{g_p + g} E_{diff} \tag{1}$$

According to Eqn. 1, the electrogenic (first) term decreases and the diffusion (second) term increases when g is increased with respect to g_p. Such higher passive conduction is achieved for instance by membrane modifying agents, like CCCP and valinomycin, which are known to increase the membrane conductance of protons (g_H) and potassium (g_K), respectively. $[^{L}E_m]_{ss} \simeq E_e$, when $g << g_p$, and $[^{L}E_m]_{ss} \simeq E_{diff}$ when $g >> g_p$, according to Eqn. 1. The latter situation is likely to be induced after switching off the light. In this case $E_e = 0$, and the active channel is suggested to be converted into a low conducting pathway, $g_p \rightarrow 0$.

The passive behavior of the membrane in the dark after the pump has been switched off is determined by the concentration gradients of the ions j across the membrane, associated with Nernst potentials E_j^N, and by the membrane conductances g_j of the ions. When it is assumed that the membrane conductance is mainly due to potassium, chloride and hydrogen ions, i.e. $g \simeq g_K + g_{Cl} + g_H$, then the initial membrane potential in the dark $[^{D}E_m]_i$ is given by:

$$[^{D}E_m]_i \;(= E_{diff}) = \frac{g_k}{g} E_K^N + \frac{g_{Cl}}{g} E_{Cl}^N + \frac{g_H}{g} E_H^N \tag{2}$$

$[^{D}E_m]_i \simeq E_H^N$ under conditions at which $g_H > g_K + g_{Cl}$, and is about equal to $0.5(E_K + E_{Cl})$ when $g_{Cl} \simeq g_K >> g_H$. The initial membrane potential in the dark will decay to zero potential at which all ions are at passive equilibrium in the dark, i.e. when inside and outside concentration of all ions are equal. The rate of this decay is determined by the rate of charge equilibration through the passive conductances.

2.2. ELECTRICAL POTENTIALS — ORIGIN AND ENERGY RELATIONSHIPS

The ability of compartmentalized living organisms to preserve the necessary conditions for the maintenance of the vital life processes, is closely related to the integrity and functioning of the limiting membranes. The proper membrane functioning is reflected by the behavior of the membrane to act as a site at which an electrical potential gradient is maintained, or even generated. Knowledge about the origin and energy dependency of the membrane potential and the possibility to measure its magnitude is of considerable importance for an understanding of the energetic behavior of the organism and of the functional cellular (sub-)units.

Membrane potentials can arise from three different origins: *(i) electrogenesis* by charge separation across the membrane, driven by electron and/or coupled ion transfer through a (photo-)chemically active enzyme system, oriented in the membrane; *(ii) diffusion* of unequally distributed anions and cations with differing mobilities in the membrane; and *(iii) non-uniform distribution (Gibbs—Donnan equilibrium)* of electrolytes between a phase containing non-diffusible fixed charges and the electrolyte solution. In contrast to the first two potentials, the Gibbs—Donnan potential is an equilibrium situation from which no useful energy may be derived. Since we will discuss the electrical potential (changes) of chloroplast membrane mainly in relation to the formation of ion distribution and the energy conservation mechanism (i.e. those associated with the changes in free energy), the Gibbs—Donnan potential will not be discussed further in this chapter. For a general discussion of Donnan equilibria, especially with respect to ion distribution in chloroplasts, the reader is referred to the literature [16,25].

2.2.1. Electrogenesis at the thylakoid membrane

Electrogenic potential generation is expected to occur in association with a charge separation across the membrane. Junge and Witt [46] have been the first to suggest that such a charge separation occurs across the thylakoid membrane in conjunction with the excitation of the photosynthetic reaction centers embedded in this membrane. If ΔQ charges (electrons) are involved in the charge-separating act, the electrical membrane capacitance C_m will be charged at a potential E_e, which equals:

$$E_e = \frac{\Delta Q}{C_m} \tag{3}$$

The capacitance of the membrane, considered as a parallel plate condenser, is given by $C_m = \epsilon\epsilon_0 A/l$, in which ϵ is the effective dielectric constant, and l the thickness of the non-aqueous insulating lipid layer in the membrane. A is the surface area of the membrane, covered by one complete electron transport chain (system one + system two), and ϵ_0 is the dielectric constant in vacuum. With $\epsilon_0 = 8.84 \cdot 10^{-12}$ (in SI-units), $\epsilon = 2$, $l = 3$ nm and $A = 30 \times 30$ nm^2, and considering that 2 electrons are transferred in one single unit (system one + system two), i.e. $\Delta Q = 2 \cdot 1.6 \cdot 10^{-19}$ Coulomb, it was approximated that $E_e \sim 50$ mV [46,74].

The magnitude of E_e is determined by the number of reaction chains per unit area of thylakoid surface, and by the electrical capacity of the membrane. A capacity of 0.6 μF/cm^2, as approximated by Junge and Witt [46] is within the range usually found for the capacitance of biological membranes (~1 μF/cm^2). The X-ray scattering results, indicating that 1 complete photosynthetic unit occupies a membrane surface area of about 10^{-11} cm^2 [53] are consistent with the results of other estimates on the surface area of the photosynthetic unit. Assuming an average chlorophyll concentration of 25 mM in chloroplasts (cf. ref. 62), and considering that about 500 chlorophyll molecules are associated with a photosynthetic unit, Vredenberg [88, 91] calculated that 1 unit is associated with a membrane surface area of about $4 \cdot 10^{-11}$ cm^2. Thus there seems to be reasonable good evidence that the initial charge separation in the photosynthetic reaction centers leads to the generation of a potential across the thylakoid membrane of 10 to 50 mV. Variations in this magnitude can be expected to occur in association with variations in the number of photosynthetic units per unit area of membrane surface, and with variations in the lipid composition of the membrane, i.e. with variations in the membrane capacitance.

Various methods have been applied to demonstrate the primary electrogenesis at the thylakoid membrane upon light excitation:

(i) The rapid light-induced rise in absorbance at 515 nm, denoted as the absorbance change of "pigment" P 515, first discovered by Duysens [27] and characterized by a light minus dark difference spectrum with a minimum at about 490 nm and a maximum around 515 nm, has been argued to be due to a shift in the absorption bands of the native pigments embedded in the membrane occurring in response to an electric field (electrochromic band shifts) [46]. The rise time of this change was found to be less than 20 ns [97]. The extent of the change has been interpreted to reflect the magnitude of the membrane potential associated with the electric field [94]. Millivolt calibration of the 515 nm change has been made with reference to the magnitude of the potential change E_e in a single turn-over saturating light flash, as calculated above [74,94]. Similar spectral shifts of carotenoids observed in photosynthetic bacteria [4,42], have been shown to occur in response to a membrane potential, induced by KCl additions to valinomycin-treated chromatophore preparations. Millivolt calibration of these light-induced spectral shifts was obtained with references to the salt-induced changes [22,42].

(ii) Charge polarization was measured by external electrodes in a layer of stripped chloroplasts (chloroplast thylakoids) which was illuminated by a light beam perpendicularly on the layer [31,96]. The results indicated that the outside surface of the chloroplast particles becomes negatively charged at the onset of illumination. The charge density gradient across the layer was shown to be associated with the light intensity gradient in the layer [32]. The actual rise time of the generation of charge unbalance was found to be less than 1 μs [32,96].

(iii) It has been shown [10,11] that ms delayed light emission can be used as an indirect method of estimating the magnitude of electrical potentials across the thylakoid membrane. Magnitude and kinetics of salt-induced increases in ms delayed light emission were found to be consistent with the

suggestion [12,9,8] that the intensity of delayed light emission, L, is an exponential function of the membrane potential, ΔE, i.e. $L \alpha \exp(\Delta E)$. Millivolt calibration of ms delayed light emission was made by measuring the initial intensity of this emission in response to diffusion potentials induced by addition of KCl to the chloroplast suspension in the presence of valinomycin. It has been shown [10] that in low salt chloroplasts, illuminated by 1.5 ms light flashes at a dark interval of 2.5 ms, the electrical potential across the thylakoid membrane, measured in the dark 1 ms after the flash, is in the range between 50 and 80 mV (inside positive).

(iv) Bulychev et al. [18,19] have been the first to show that light-induced potential changes occur across the membranes of intact chloroplasts, as probed by micro-capillary glass electrodes inserted into a single chloroplast of mesophyll cells of leaves of the plant *Peperomia metallica*. The characteristics of the initial potential rise (rise time approx. 0.5 ms, insensitivity to temperature and to the electron transfer inhibitor DCMU) indicated that these potential changes are associated with primary effects in the photosynthetic machinery, and suggested that these were reflections of photoelectrogenic events at the thylakoid membrane [89,90].

The method listed under *(i)* has been applied in experimental approaches in which the magnitude of the electric potential across the membrane during continuous illumination (energization) was estimated [36]. However, the application of this method needs justifications with respect to probable limitations set by the method, due to possible interferences of the determinant membrane signals with other processes occurring in the phases faced by the membrane. Amongst these are changes in concentration gradients of ions transported either actively or passively in conjunction with the electrogenic potential, which might cause deviations from the assumed linear potential relationship. Method *(iv)* would not be limited by these effects as it measures directly the actual potential across the membrane. However, the relevance of this method needs justification with respect to the position of the microelectrode in the inner chloroplast structures and, if localized in a thylakoid granum stack, to the induction of possible artificial ionic leaks in the membrane.

The relative magnitude of the electrogenic potential and its contribution to the steady-state membrane potential in the light cannot be predicted a priori. However, one can make an approximation on the magnitude of the electrogenic charge flux across the membrane under steady-state light conditions. According to the approximations referred to before, one photosynthetic unit is covering a membrane area of about $(1-5) \cdot 10^{-11}$ cm^2. As the transfer time of an electron through the electron transport chain connecting both systems is about 10 ms (cf. ref. 94), the steady-state flux across the two charge-separating sites in saturating light will be 100 electrons. s^{-1}. This means that under steady-state conditions the charging electrogenic flux across the membrane is $(0.4-2) \cdot 10^{13}$ ions. cm^{-2}.s^{-1}, or $(7-35) \cdot 10^{-12}$ equiv.

$cm^{-2}.s^{-1}$. The passive, potential-dependent ion fluxes will ultimately determine the magnitude of the steady-state electrical potential across the membrane. Their magnitude is equal, except for the sign, to the charging electrogenic flux under these conditions.

2.2.2. *Passive ion fluxes and diffusion potentials*

Spontaneous, or passive movement of a charged species j across a membrane will occur due to a difference in the electrochemical potential of the species across the membrane. The electrochemical potential of species j, under conditions at which its chemical activity is equal to the concentration, and pressure and gravitational terms can be neglected, is defined as

$$\bar{\mu}_j = \bar{\mu}_j^o + RT \ln c_j + z_j FE \tag{4}$$

where $\bar{\mu}_j^o$ is the reference standard potential (c_j = 1 M and E = 0), RT = 582 cal/mole at 20°, z_j is the valence of the species, F is the Faraday constant (= 96 489 Coulomb/mole) and E is the membrane potential. The rate of movement of species j across the membrane, ϕ_j, is determined by the electrochemical potential gradient, by the mobility u_j, and concentration c_j within the membrane, and is expressed by the flux equation

$$\phi_j = u_j \cdot c_j \cdot \left(-\frac{\partial \bar{\mu}_j}{\partial x}\right) \tag{5}$$

Substitution of Eqn. 4 gives

$$\phi_j = -u_j \frac{RT}{F} \frac{\partial c_j}{\partial x} - z_j u_j c_j \frac{\partial E}{\partial x} \tag{6}$$

With the assumption that the electrical potential difference across the membrane varies linearly with distance, e.g. $\frac{\partial E}{\partial x} \equiv \frac{\Delta E}{\Delta x}$ in which ΔE is the potential difference across and Δx the thickness of the membrane, the Goldman constant field equation [35] is derived by integration of Eqn. 6:

$$\phi_j = \frac{z_j F \Delta E}{RT} P_j \frac{[c_j^o - c_j^i \exp(z_j F \Delta E/RT)]}{1 - \exp(z_j F \Delta E/RT)} \tag{7}$$

where suffix o and i refer to the opposite (outside and inside) phases separated by the membrane. $P_j = u_j.RT.K_j/\Delta x$ is the membrane permeability coefficient, in which K_j is the partition coefficient of the ion within the membrane, characterizing the lipophilicity of the ion. For the situation at which only monovalent ions are assumed to be involved and no net charge is transported by the algebraic sum of all fluxes, i.e. when $z_j = \pm 1$ and $\Sigma \phi_j = 0$, the constant field (Goldman) equation for the membrane diffusion potential is obtained after summation of the individual fluxes (Eqn. 7).

Solving for ΔE then yields

$$\Delta E = \frac{RT}{F} \ln \left[\frac{(\Sigma z_j P_j c_j^o)_{cation} + (\Sigma z_j P_j c_j^i)_{anion}}{(\Sigma z_j P_j c_j^i)_{cation} + (\Sigma z_j P_j c_j^o)_{anion}} \right] \quad (8)$$

Let us consider the situation which is assumed to be likely one for chloroplasts in situ (cf. refs. 24,41), that H^+, K^+ and Cl^- are the main ions involved in the transport processes across the membrane.
Then

$$\Delta E = \frac{RT}{F} \ln \left[\frac{P_H c_H^o + P_K c_K^o + P_{Cl} c_{Cl}^i}{P_H c_H^i + P_K c_K^i + P_{Cl} c_{Cl}^o} \right] \quad (8a)$$

According to Eqn. $\Delta E = 0$, when $c_j^o = c_j^i$ and $\phi_j = 0$ for all ions j. We presume that this is the steady-state condition in non-energized thylakoids (chloroplasts). During energization the electrogenic process causes an adjustment of the passive fluxes across the membrane. According to Eqn. 7 the initial fluxes after the initiation of an electrogenic potential E_e is equal to

$$\phi_j(0) = -z_j \frac{F}{RT} E_e P_j c_j \quad (7a)$$

With E_e positive inside, H^+ and K^+ will be passively transported out and Cl^- transported in, concomitantly with an active influx of protons into the thylakoids (e.g. Fig. 2.3). As Eqn. 7a shows, the highest passive flux will be of the ionic species having the largest P x concentration product. So far P-values and concentrations of K and Cl have been estimated for the thylakoids (chloroplasts).

2.2.3. Permeability and ion conductivity

A sudden KCl pulse given to a Cl-free suspension of broken chloroplasts containing 2 mM K^+ will cause the generation of a membrane diffusion potential which, according to Eqn. 8a is equal to $\Delta E = (RT/F) \ln [c_K^o/(2 + \beta c_{Cl}^o)]$, where $\beta = P_{Cl}/P_K$. Barber [10,11] has shown that the generation and decay of this potential can be followed by analyzing the changes in intensity of delayed light emission. Intensity of delayed light emission is assumed to be exponentially dependent on the membrane potential. He estimated that $\beta = 0.47$ in the absence and 0.04 in the presence of valinomycin. The decay kinetics of the potential (light emission) after the pulse give a measure of salt entry into the thylakoid. Because $\beta < 1$, this entry is controlled by the rate of Cl^- entry. Fig. 2.5 shows, for a typical experiment [10], that following a 50 mM KCl pulse the initial Cl^- entry is equal to $3.75 \cdot 10^6$ equiv. $ml^{-1}.s^{-1}$. With an assumed surface area of $27.6 \cdot 10^{-10}$ cm^2 and a volume of $25 \cdot 10^{-16}$ cm^3 of a thylakoid, the flux across the thylakoid membrane is calculated to be $\phi_{Cl} = 3.4 \cdot 10^{-12}$ equiv. $cm^{-2}.s^{-1}$. Substitution of this value into Eqn. 7 yields that $P_{Cl} \simeq 1.8 \cdot 10^{-8}$ $cm.s^{-1}$. With $\beta = 0.47$, P_K then is equal to $3.6 \cdot 10^{-8}$ $cm.s^{-1}$. Similarly $P_K = 4.5 \cdot 10^{-7}$ $cm.s^{-1}$ in the presence of valinomycin ($\beta = 0.04$). It should be noted that, due to an apparent error in the calculation the values of fluxes and permeability coefficients calculated in the original paper [10] were incorrect (see also ref. 91). Schuldiner and Avron [71] have evidenced that Cl is one of the most permeable anions through the thylakoid membrane. The P_K value for

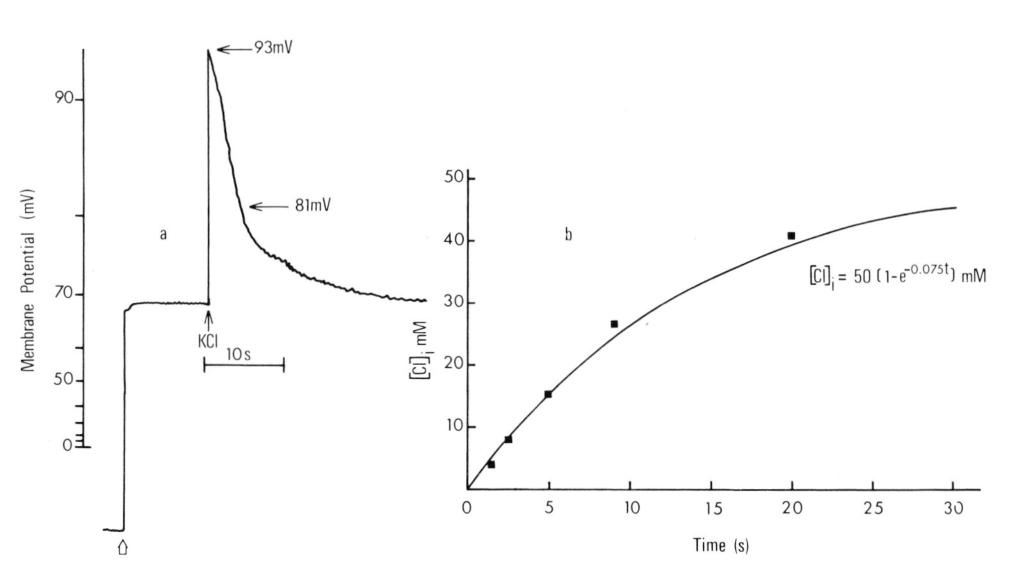

Fig. 2.5. (a) Kinetics of a 50 mM KCl-induced change in ms-delayed light emission, measured in broken chloroplasts, with mV-scale of associated membrane potential across thylakoid membrane. (b) Influx of Cl^- estimated from the decay of the signal in (a). The closed squares are experimental points and the curve has been drawn according to the equation shown, where t is the time (in seconds) and $(Cl)_i$ is the concentration of Cl^- (in mequiv./l) in the inner thylakoid space. (Reprinted from ref. 10).

the thylakoid membrane is of the same order of magnitude as the one estimated for cytoplasmic membranes of algal cells which has been reported to be between $0.2 \cdot 10^{-8}$ cm.s^{-1} [14] and $4 \cdot 10^{-8}$ cm.s^{-1} [49]. The P_H value of the thylakoid membrane is unknown as yet. It has been concluded [17,49] that the P_H value of the plasmalemma membrane of Nitella cells is of the order of 10^{-3} cm.s^{-1}. However, these conclusions have been criticized [87, 92]. Estimates on the ionic content of chloroplasts of different plant species [54,61] have indicated that potassium and chlorides are the most abundant cation and anion, respectively, present at a concentration of 100 to 300 mM. These concentrations are about the same as those reported for the cytoplasm of plant cells [54,61]. An approximate equal distribution of K^+ (and Cl^-) in the cytoplasm and the chloroplast stroma, might point to the importance of using high salt media during isolation of intact chloroplasts to preserve their ionic content and membrane integrity. The relatively low concentration of K^+ in chloroplasts recently reported by Gimmler et al. [34], might be due to the fact that these chloroplasts were isolated in K^+-free media, or in media containing 0.5 mM K^+. According to results reported by Heldt [39], the proton concentration inside a thylakoid in the dark is about 10^{-7} to 10^{-8} M (pH = 7.5 to 7.9).

Although with some reservations, due to the uncertainties about the actual ion concentrations (see Chapter 3 of this volume), it seems reasonable to assume that for chloroplasts in situ the initial passive ion fluxes across the thylakoid membrane are mainly, if not exclusively carried by K^+ and Cl^- ions. This assumption is supported by the results of Hind et al. [41] who showed that at K concentrations above 20 mM the active proton influx is balanced mainly by passive K and Cl fluxes. Ample evidence has been presented [39,67,68] that the active proton influx results in a pH-gradient of about 3 units, indicating that the internal proton concentration increases in the light to about 10^{-4} M. Assuming that 99% of the transported hydrogen ions have been neutralized by internal buffering groups, this would mean that the K and Cl concentration has been changed by an amount of about 5 mM. Concentration changes of this magnitude are likely to occur in association with the approximated passive fluxes, given by Eqn. 7a. According to this equation the initial efflux and chloride influx, driven by an initial electrogenic potential of 50 mV are $\phi_K^p(0) = 2.5 \cdot 10^{-11}$ and $\phi_{Cl}^p(0) = 1.1 \cdot 10^{-11}$ moles.cm^{-2}.s^{-1}, respectively. This would mean that the initial active proton influx $\phi_H^a(0) = 3.6 \cdot 10^{-11}$ moles.cm^{-2}.s^{-1}. The active proton influx under conditions of steady-state electron transport, ϕ_H^a(ss) is lower (e.g. p.64, refs. 88 and 91). According to data of Rumberg and Siggel [68] the steady-state is likely to be reached in a first order reaction with a rate constant $k = 10$ s^{-1} [91]. With $\phi_H^a(ss) = 1 \cdot 10^{-11}$, which would correspond with $\phi_K^p(ss) = 7 \cdot 10^{-12}$ and $\phi_{Cl}^p(ss) = 3 \cdot 10^{-12}$ moles.cm^{-2}.s^{-1}, and a surface to volume ratio, A/V, of a thylakoid of approx. $1.2 \cdot 10^6$ cm^{-1}, the amount of protons which has entered the thylakoid is equal to $(\phi_H^a(0) - \phi_H^a(ss)) \cdot k^{-1} \cdot A/V = 3.1 \cdot 10^{-6}$

moles.cm^{-3}. Assuming that 99% of the transported H^+ are neutralized by internal buffering groups, the entry of these protons has caused a change in the internal proton concentration of about 31 μM, which corresponds with an inner pH of about 4.5. The amount of K^+ and Cl^- transported across the membrane is equal to $2.1 \cdot 10^{-6}$ and $1.0 \cdot 10^{-6}$ moles.cm^{-3} which has caused concentration changes $\Delta c_K = -2.1$ mM and $\Delta c_{Cl} = +1.0$ mM. As the thylakoid to stroma volume ratio is about 0.125 [39] the fluxes will not have caused an appreciable change in the external concentration of the transported ions. Furthermore, as the changes in the internal concentration of K and Cl are small as compared to the assumed actual concentrations (100 to 300 mM), the internal concentrations of these ions can be considered to be approximately constant and volume changes are expected to be negligible. Thus the membrane diffusion potential immediately after the cessation of the electrogenic mechanism, i.e. upon darkening, is given by

$$({}^{D}E_m)_i = \frac{RT}{F} \ln \left[\frac{P_H c_H^o + a}{P_H c_H^i + a}\right] \simeq \frac{RT}{F} \ln \left[\frac{a}{P_H c_H^i + a}\right] \quad (8b)$$

in which $a = (P_K \cdot c_K + P_{Cl} \cdot c_{Cl}) \simeq 1.8 \cdot 10^{-11}$ moles.cm^{-2}.s^{-1}, $c_H^o \simeq 10^{-10}$ and $c_H^i \simeq 10^{-7}$ moles.cm^{-3}.

$({}^{D}E_m)_i \simeq 0$ when $P_H << 10^{-5}$ and will be negative when P_H is of the order of 10^{-5} cm.s^{-1} and higher. As will be shown in a next section, transient negative dark potentials are routinely measured in chloroplasts in situ after illumination, which suggests that the proton permeability coefficient of the thylakoid membrane is considerably higher than of the other anions and cations, and might be of the order of 10^{-5} cm.s^{-1}. However, as the proton concentration is relatively low as compared to the concentration of other permeating ions, the proton conductivity of the thylakoid membrane is comparatively low.

It should be noted that the H^+/K^+ exchange mechanism, illustrated so far for chloroplasts in situ where the actual K^+ (and Cl^-) concentration is assumed to be high, is expected to become competitive with a H^+/Mg^{2+} exchange, under conditions at which the internal K^+-concentration is much smaller (e.g. ref. 41). Preliminary experiments with intact chloroplasts, isolated and suspended in media of variable K^+-concentrations (Bulychev and Vredenberg, 1975, unpublished) are not inconsistent with the results, discussed by Barber in Chapter 3 of this book, that in isolated chloroplasts energization is accompanied by an H^+/Mg^{2+} exchange across the thylakoid membrane.

Under conditions at which an ion is at electrochemical equilibrium and moves independently of other ions, the ionic conductance $g_j \left(= F \frac{\delta \phi_j}{\delta E} \right)$ is given by the following equation

$$g_j = \frac{F^3}{R^2T^2} E \frac{1}{[1 - \exp(-FE/RT)]} P_j c_j \qquad (9)$$

For chloroplasts in situ under non-energized steady-state conditions, i.e. in the dark when $c_j^o = c_j^i$ and $E_m = 0$, we obtain $g_K = {}^Dg_K \simeq (F^2/RT) P_K c_K$ and ${}^Dg_H \simeq (F^2/RT) P_H c_H$. As energization does not cause an appreciable change in the internal and external concentration of potassium ions, we may approximate for the conductance after energization ${}^Lg_K \simeq {}^Dg_K = g_K$. Similarly ${}^Lg_{Cl} \simeq {}^Dg_{Cl} \simeq (F^2/RT) P_{Cl} c_{Cl}$. The passive proton conductance ${}^Lg_H$ after energization can be approximated by differentiating the proton flux equation (Eqn. 7). Under these conditions, i.e. when $c_H^i >> c_H^o$:

$${}^Lg_H = F \frac{\partial \phi_H}{\partial E} = \frac{F^3 E}{2R^2T^2} \frac{1}{[1 - \exp(-FE/RT)]} P_H c_H^i \simeq \frac{F^2}{2RT} P_H c_H^i$$

when $|E| \leqslant 25$ mV. Thus approximately ${}^Lg_H \simeq 5 \cdot 10^2 \cdot {}^Dg_H$. With a = $(P_K \cdot c_K + P_{Cl} \cdot c_{Cl}) = 1.8 \cdot 10^{-11}$, $c_H^i = 10^{-7}$ and $P_H = 2 \cdot 10^{-5}$, we get that ${}^Lg_H \simeq 0.1\,(g_k + g_{Cl})$.

2.2.4. Potential changes associated with energization

The approximations on the magnitude of the membrane parameters discussed so far suggest the following kinetic pattern of the potential changes across the thylakoid membrane during and after energization. Let us consider again the (extended) equivalent electric model circuit of the membrane (Fig. 2.6). The electrogenic pump with e.m.f. E_e and variable conductance g_p operates in parallel with passive diffusion e.m.f.'s. $E_j^N = z_j\,(RT/F)$ ln

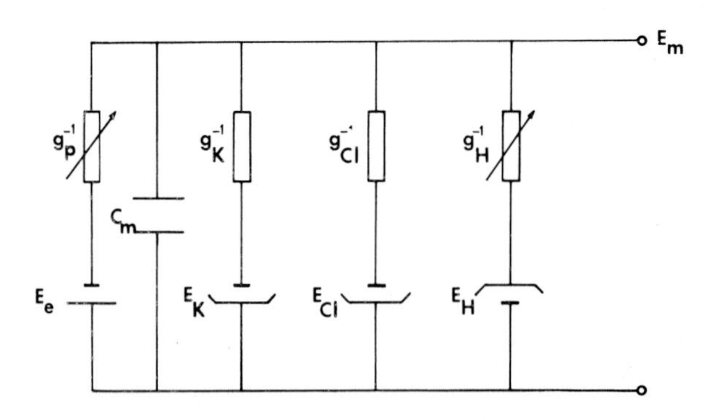

Fig. 2.6. Extended equivalent electric circuit of the thylakoid membrane. Details are given in the text. (See also the legend to Fig. 2.4.)

(c_j^o/c_j^i) and associated conductances g_j of ions j = K, H, and Cl. Under non-energized conditions, i.e. in the dark, all e.m.f.'s are zero and $E_m = 0$.

(*a*) The initial charge separation upon energization (illumination), which occurs within approx. 20 ns, is associated with E_e = 10 to 50 mV. As a consequence the membrane capacitance C_m is charged at a potential $(^LE_m)_i = E_e$ with a relaxation time $\tau_1 = C_m/g_p$.

(*b*) The (change in) potential in continuing illumination is determined by the magnitude of the active influx ϕ_H^a, sustained by the steady-state electron flux through the electron transport chain, by the active and passive membrane conductance g_p and g (= $g_K + g_{Cl} + g_H$), respectively, and by the magnitude of the created passive e.m.f.'s, E_j^N. As we have discussed already g_p and g_H are expected to change appreciably in conjunction with the changes in electron transport rate and in E_H^N. Under steady-state light conditions at which $g_p = {}^Lg_p$, $g_H = {}^Lg_H$ and the passively moved ions have been distributed in equilibrium with the membrane potential $(^LE_m)_{ss}$, this potential is given by

$$(^LE_m)_{ss} = (^Lg_p E_e + g_K E_K^N + g_{Cl} E_{Cl}^N + {}^Lg_H E_H^N)/(^Lg_p + g) \tag{1a}$$

(*c*) Upon shutting off the light, the membrane potential initially reached in the dark $(^DE_m)_i$ is given by:

$$(^DE_m)_i = (g_K E_K^N + g_{Cl} E_{Cl}^N + {}^Lg_H E_H^N)/g \tag{2a}$$

The transition from $(^LE_m)_{ss}$ to $(^DE_m)_i$ occurs with a relaxation time $\tau_2 = C_m/(g_K + g_{Cl} + {}^Lg_H)$. Because E_K^N and E_{Cl}^N were shown to be approximately zero at high K and Cl concentration (100—300 mM for chloroplasts in situ), $(^DE_m)_i$ is mainly determined by ${}^Lg_H$ and E_H^N.

(*d*) The potential change from $(^DE_m)_i$ to the steady-state dark potential $(^DE_m)_{ss} = 0$ will occur with a relaxation time which is determined by the rate of passive diffusion through g_H and $g_K + g_{Cl}$, and by the rate of change in the conductance g_H.

2.3. CHARACTERISTICS OF CHLOROPLAST MEMBRANE POTENTIALS

Bulychev and coworkers [18,19] have been the first to show that energization of chloroplasts is associated with potential changes of chloroplast membranes, as probed by glass micro-capillaries inserted into a chloroplast in a mesophyll cell of a *Peperomia metallica* leaf, or in an isolated chloroplast from the same plant species. Their results were confirmed and extended by Vredenberg and coworkers [88—91], using the same plant material. Light-induced potential changes also have been measured in the chloroplast of a single cell of the hornwort *Phaeoceros leavis* [23]. This section will review the experimental results obtained so far, which suggest that the microelectrode technique is promising in contributing to the knowledge about primary and associated energetic processes occurring at the chloroplast lamellar membranes.

2.3.1. Microelectrode measurements — experimental

Mesophyll cells of *Peperomia metallica* usually contain three to five chloroplasts with a diameter of 15 to 25 μm. Measurements on chloroplasts in situ were done using leaf sections cut at one end with a sharp razor blade. Such section is clamped in position in a sample holder on the stage of a microscope mounted on a rigid and shock-free experimental bench. The specimen is flushed with medium of appropriate osmolarity and ionic composition [19,89].

Potentials are measured by conventional microelectrode techniques [33]. Glass micro-capillaries with a tip diameter of less than 0.5 μm [18], usually between 0.2 and 0.3 μm [91], filled with 2.5 M KCl and in connection with an Ag/AgCl wire, are used as the internal measuring microelectrode. A similar capillary electrode, either located in the cell cytoplasm (micro-capillary) or in the external medium, serves as the reference electrode. An accurate and remote-controlled micromanipulator is used for the positioning of the measuring electrode with respect to a suitable located chloroplast inside a cell. A hydraulic device with a motor driven stepwise motion is used for the unidirectional insertion of the electrode in to the chloroplast. Fig. 2.7 shows the measuring electrode inserted in the chloroplast and the reference electrode situated in the cytoplasm. The potential difference between measuring and reference electrode is displayed, after impedance matching by a high-impedance unity-gain amplifier, on the oscilloscope, or on a recorder.

Three distinct potential steps usually are noticed during the insertion of an electrode in a cell and subsequently in a chloroplast [89] : (*i*) a drop of the potential to about −20 to −30 mV, which has been ascribed to the potential difference between cell wall and surrounding medium; (*ii*) a level of −80 to −125 mV which is the potential of the cytoplasm, and (*iii*) a potential 10 to 25 mV more positive than that of the cytoplasm. The latter potential level, which is characterized by fast responses occurring upon illumination, originally was ascribed to the chloroplast-enclosing membrane (i.e. the envelope) [90].

Bulychev et al. [18] have reported that the potential difference between chloroplast interior and cytoplasm, measured in different cells varied from +15 to −60 mV. For isolated chloroplasts the potential difference in the dark between chloroplast interior and outer medium was found to be in the range between −20 and + 15 mV [18]. The cytoplasmic potential has been observed to change upon illumination with relaxation times of 10 s or more [19] while those originating from the chloroplast interior occurred with reaction times smaller than 1 s. Thus for measuring the latter changes the location of the reference electrode outside the chloroplast is not critical.

Fig. 2.7. Layer of mesophyll cells in a leaf section of *Peperomia metallica*. The photograph shows the situation in which the measuring microelectrode is inserted in a chloroplast and the reference electrode is located in the cytoplasm. (Reprinted from ref. 19).

Fig. 2.8 shows an electron micrograph of the chloroplast in a thin cross section of the *Peperomia* leaf. A large number of appressed thylakoids (granum stacks) are densely distributed parallel to the chloroplast-enclosing membranes, preferentially at the polar ends of the egg-shaped organelle. Similar micrographs of these chloroplasts have been shown by Neumann [59]. The thylakoid lamellae are about 1.0 to 1.5 μm in diameter, and form a stack of about 1.5 μm in height. The information on the internal structure has led to the conclusion that micro-capillaries with a tip diameter of 0.1 to 0.3 μm (see Fig. 2.9) have a much higher probability of hitting and probing a granum stack than of being in the stroma phase, as was originally assumed [89,90]. Bulychev et al. [18] have suggested that in some cases the potential changes observed upon illumination were those of a situation in which the electrode tip was within the lamellar membrane system. Although light-induced potentials are routinely observed after an electrode impalement into the chloroplast, most of them are probably originating from a situation in which the electrode has caused a partial damage of the lamellar membranes and consequently an artificial ionic leak. The transient signals under these conditions usually were lost within 1 to 2 min. Vredenberg [88,91] has summarized the evidence that as an average about 20% of the electrode

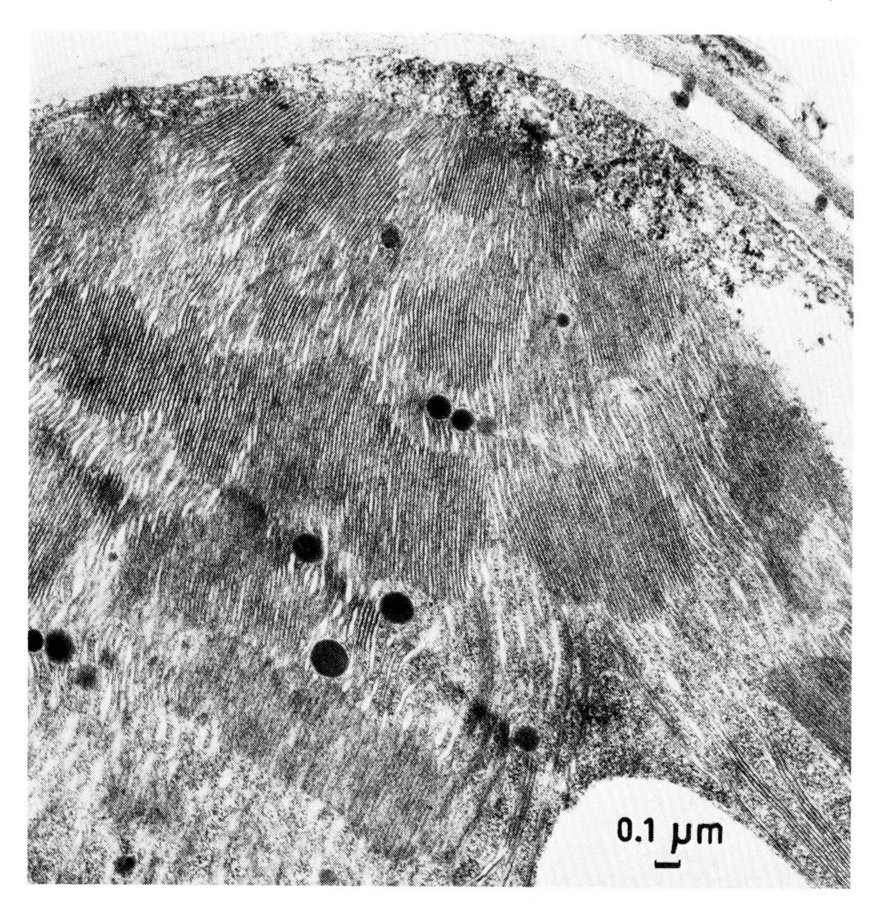

Fig. 2.8. Transmission electron micrograph of a thin cross section of a leaf of *Peperomia metallica* showing the dense distribution of granum stacks in the polar end (i.e. the part that is shown) of the usually egg-shaped chloroplast. The bar corresponds to a length of 0.1 μm. This micrograph was made by the Technical and Physical Engineering Research Service, Electron Microscopy Section (Ing. S. Henstra) at Wageningen, using a Philips EM 300 transmission electron microscope.

impalements, preferentially done at a polar end of the lamellar body, resulted in a situation at which the electrode tip is in a granum stack and isolated from the stroma phase and cytoplasm by an unperturbed thylakoid membrane system and chloroplast envelope. This evidence is based on the fact that in such a situation the light-induced responses, characterized by distinct phases with specific rate constants, are reproducible and stable during an appreciable time of experimentation (10 to 30 min), and are sensitive to agents that are known to modify the membrane conductivity. The signals measured in such an optimal situation will be discussed below.

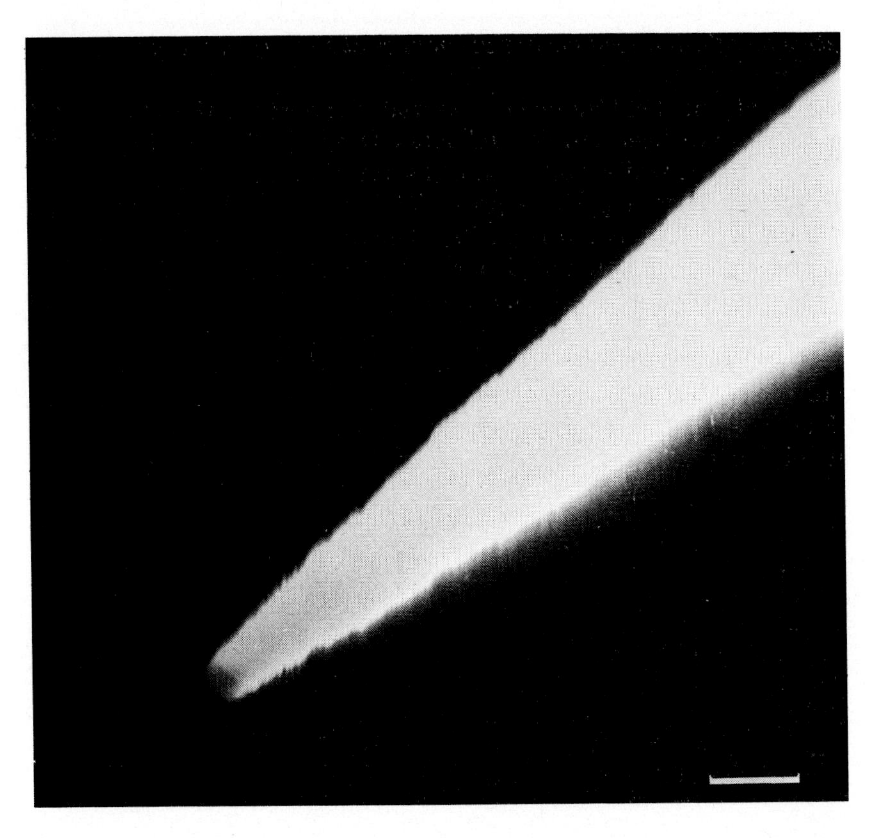

Fig. 2.9. Scanning electron micrograph of the (open) end of the glass microcapillary. The micrograph was made by the Electron Microscopy Section mentioned in the legend of Fig. 2.8, using a JEOL-JSM-U3 scanning electron microscope. The bar corresponds to a length of 1 μm.

It should be noted that in very few cases, which were difficult to reproduce, the multi-phasic potential response in the light was observed to be of a polarity opposite to the normal one, but with the same kinetics and characteristics (see below). Such situation was not stable, and in general changed abruptly in to one with a "normal" response. One might speculate that in such cases the electrode tip temporarily is in contact with the outer membrane surface of a granum stack, or even in the partition phase between granum lamellae. An earlier suggestion [90] that in such a situation the electrode tip is in the interspace between the inner and outer envelope seems to be less probable because potential changes in this space are only to be expected when lamellar membranes are in continuity with the inner enclosing membrane, as was assumed originally. However, electron micrographs of these chloroplasts (cf. Fig. 2.7) have not given evidence for this assumption.

2.3.2. Kinetics of light-induced potential changes at the thylakoid membranes

The kinetics of the light-induced potential response of the chloroplast inner membranes have been studied in detail [88,91]. A typical potential response to a saturating white light flash of 1 s duration is shown in Fig. 2.10. At the start of the flash a positive potential is generated which has been denoted as the phase 1 potential rise [89]. The phase 1 rise is followed by a potential decrease in the light, called phase 2. The steady state potential in the light has been called phase 3, and is smaller than the phase 1 potential for chloroplasts in situ. Upon turning off the light a rapid change occurs to a potential level below the original dark state (transient potential undershoot), which is followed by a slower increase to the steady-state zero dark potential. Magnitude and rate of the potential change in the light were found to be positively correlated with light intensity [19,89]. At sufficiently low

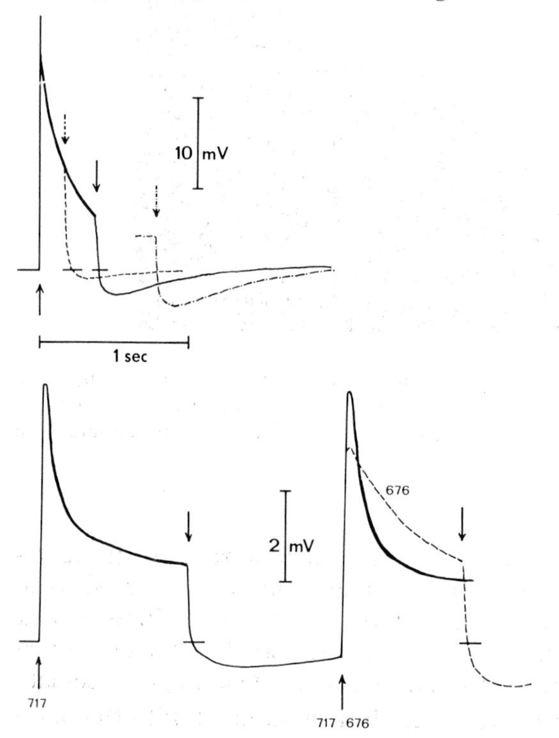

Fig. 2.10. Potential responses of an electrode inserted into a chloroplast of a mesophyll cell of *Peperomia metallica*, in white light flashes of 0.4 (solid trace), 0.2 and 0.8 s (broken traces) duration. The potential change is attributed to the change in potential across the membranes of thylakoids, constituting a granum stack (see text). An upward deflection means an increase in potential (inside more positive). Upward and downward pointing arrows mark the beginning and end, respectively, of illumination. The horizontal dashes in the light-off traces mark the level of the dark potential, and are given to mark the potential undershoot during the light-off reaction. (Reprinted from ref. 88.)

intensities of white, as well as of 676 and 717 nm, light the steady-state phase 3 was reached by a small or even without a transition through the decay phase 2. No appreciable effect of temperature on the initial phase 1 potential was found in the range between 3 and 20°C [18,89]; the temperature coefficient, Q_{10}, within this temperature interval has been reported to be 1.0 to 1.2 [19]. The action spectrum of the phase 1 rise was shown to correspond to the absorption spectrum of the chloroplasts with maxima at 420, 500 and 680 nm [19].

2.3.2.1. Phase 1 potential rise

The characteristics of the phase 1 potential rise have been studied by analysing the potential response in short saturating light flashes under different conditions [90]. The rise in potential was shown to follow a first-order reaction $V(t) = \alpha V_{max} (1 - \exp(-k_1 t))$ with $k_1 = (2.7 \pm 0.3) \cdot 10^3\ s^{-1}$, or $\tau_1 = k_1^{-1} = 0.37 \pm 0.03$ ms, and V_{max} varying for different chloroplasts between 5 and 50 mV. The factor α accounts for the decrease in V_{max} which was observed when the flash-induced response was measured in the absence ($\alpha = 1$) and presence ($\alpha < 1$) of weak background light. The rate constant k_1 was found not to be changed in the presence of background light. The extent V_{max} of the potential rise in consecutive flashes, fired at a dark interval of 250 ms, was shown to be approximately constant in the absence of DCMU, but in the presence of this inhibitor the second and following flashes caused only half of the response ($\alpha = 0.5$) of the first flash. These results have been interpreted to indicate that the phase 1 potential rise is associated with the primary charge separation in the reaction centers which, according to the model proposed by Grunhagen and Witt [38], is followed by an ion binding reaction at the inner and outer membrane surface with concomitant oxidation and reduction of the electron donor and acceptor of both reaction centers. It has been suggested that the exponential rise in potential during phase 1 is determined by the rate constant of this ion binding process [90]. However, this interpretation needs a modification in view of the results of Fowler and Kok [32] and of Witt and Zickler [96]. who showed that the creation of ionic charge imbalance across the membrane occurs within 1 μs. It might be suggested that, as the relatively fast ion binding occurs at fixed (localized) sites on the membrane occupied by the donor and acceptor molecules of the reaction centres, the homogeneous charge distribution occurs by charge diffusion across the membrane surface which is controlled by a passive conduction mechanism. In the electric analogue of the membrane (e.g. Fig. 2.6) this conduction has been represented by g_p, which determines the rate of (homogeneous) charging of the membrane capacitance. This suggestion needs to be confirmed by experiments in which this passive conduction is varied. Junge [47] has discussed the existence of a transversal diffusion barrier for protons which shields the redox reaction sites at the outer surface of the thylakoid membrane from the outer aqueous phase.

The effect of background illumination and of DCMU on the potential rise has been interpreted as evidence that the magnitude of this potential change is linearly dependent on the number of reaction sites at which charge separation and subsequent ion binding can occur [90]. Potentials as high as 75 mV [19] in magnitude have been measured, and are in the range predicted for the potential generation associated with the primary charge separation across the thylakoid membrane, when all available sites can be charged (see section 2.2.1). It has been reported [18,19] that DCMU was inhibitory on the phase 1 potential rise. However, as these experiments were not performed under the necessary conditions of complete darkness (Bulychev, personal communication), the effect of background illumination may have influenced the results under these conditions. The magnitude of the phase 1 potential rise in continuous light has been found not to differ much, if at all, from the one initiated by a short saturating flash. However, the flash response observed shortly after a pre-illumination period was observed to be considerably smaller [89]. It has been suggested that this is due to a change in the membrane conductance which would occur in prolonged illumination. I presume, and will discuss later on, that the change in ionic distributions, notably of protons, across the thylakoid membrane in the light causes a decrease in the conductance of the proton pump, and consequently a decrease in the phase 1 potential when the protons have not been redistributed in the dark after a light period. Rate and magnitude of the phase 1 potential rise were found not to be affected by uncouplers and other membrane modifying agents.

Under optimal conditions at which damage of the thylakoid membrane was likely to be negligible, the exponential decay of the potential in the dark after a short flash has been found to occur with rate constants k_d of about 30 s^{-1} and $3.5 \cdot 10^2$ s^{-1} [88] in the absence and presence, respectively, of valinomycin. According to the electric model (Fig. 2.6) these rate constants would suggest a passive membrane conductance g (= $C_m \cdot k_d$) of 30 and 350 μhmo $\cdot$ cm^{-2} in the absence and presence, respectively of valinomycin, when it is assumed that $C_m = 1\ \mu F \cdot cm^{-2}$. These conductances are in the same range as approximated in the preceding section on basis of membrane permeability coefficients and concentration of ions. Somewhat lower values of k_d ($\sim$15 s^{-1}) have been reported for isolated chloroplasts in the absence of the ionophore [20].

2.3.2.2. Phase 2 potential decrease and phase 3 steady-state potential

The kinetics of the phase 2 potential decrease have been shown to be dependent on the rate of electron transfer through the electron transport chain, and on the permeability coefficients of passively leaking ions [88,91]. This evidence was obtained from experiments in which the potential responses were measured in 1 s illuminations with monochromatic light, absorbed by both pigment systems (676 nm), or exclusively by system one (717 nm), in the absence and presence of membrane modifying agents, like CCCP, valinomycin and ionophore A23187 [13,66]. The kinetics of the potential decrease in 676 and 717 nm light, and their dependence on pre-illumination are shown in Fig. 2.11. The rate of potential decrease is comparatively faster in 717 than in 676 nm light, at least after a dark-adaption. Moreover the rate of potential change in 676 nm light is affected by pre-illumination, 717 nm light causing a decrease and 676 nm light an increase in the rate of this change. These differences have been discussed to result from differences in the rate of electron transfer through system two, which

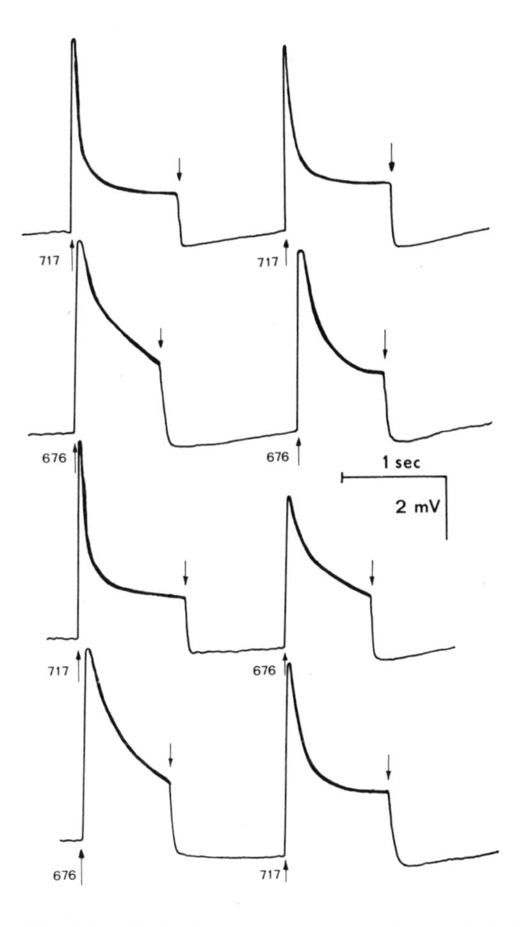

Fig. 2.11. Potential responses of chloroplast (thylakoid) membrane in two sequential monochromatic flashes. The sequence of wavelength of the flashes was: 717—717 (a), 676—676 (b), 717—676 (c) and 676—717 nm (d). The first flash was given after a dark period of approx. 2 min. Intensities were 20 and 6 kerg/(cm^2s^{-1}) for 717 and 676 nm flashes, respectively. Further details are in the text of Fig. 2.10. (Reprinted from ref. 88.)

is known to be dependent on the redox state of an electron carrier pool, probably plastoquinone [5,82], between the two photochemical systems. In agreement with this it was found that in the presence of DCMU, which is known to inhibit the electron transfer between the primary acceptor Q of system two and plastoquinone [1], the reaction kinetics of phase 2 were identical in 676 and 717 nm light, and not affected by preillumination. Moreover, initiation of increased electron flow through system one by adding reduced DCPIP as electron donor to system one in DCMU-inhibited cells, has confirmatory been shown to be associated with a decreased rate of potential decay during phase 2 [88]. Bulychev et al. [20] have reached similar conclusions with respect to the control of the phase 2 decay by electron transfer through the electron transport chain. They have shown in addition that the rate constant of the decay in the light decreases upon lowering the temperature. At 2° C this rate constant was found to be approximately equal to the rate constant of the dark decay after a flash.

It has been found [88] that in the presence of valinomycin, which is known to increase the membrane conductance g_K of potassium ions, the rate of potential decay in the dark after a flash, as well as the phase 2 decrease in the light, is considerably enhanced. This has been interpreted to indicate evidence that the phase 2 decrease is due to a potential driven, charge-dissipating, passive moment of ions, with a substantial contribution of potassium ions [91]. CCCP and ionophore A23187 were hardly found to affect, if at all, the phase 2 decay, suggesting that, at least for chloroplasts in situ, protons and divalent cations do not contribute appreciably to the passive fluxes after the primary phase 1 potential generation.

The apparent fact that a potential decrease occurs during phase 2 indicates that during the initial stage of this phase the magnitude of the charge dissipating passive ion fluxes is larger than that of the charge separating electron flux through the electron transport chain. A stationary situation is reached when both fluxes are equal. With $\phi_H^a \sim 10^{13}$ ions. $cm^{-2}.s^{-1}$ (p.64) and $g \sim \left(\frac{F^2}{RT}\right) \cdot a \sim 80\ \mu hmo.cm^{-2}$ (p.70) this situation, except for the contribution of diffusion potentials, would be associated with a potential of about 20 mV. The steady state (phase 3) finally will be reached when the net flux of passively moving ions is zero, and the active proton influx is balanced by an equal passive efflux of protons through the proton conducting channels, including the ATP-synthesizing site. This steady-state phase is characterized by a membrane potential $({}^LE_m)_{ss}$, of which the magnitude (Eqn. 1a) is determined by the potential maintained by the charge-separating electron flux (electrogenic proton pump) and by a diffusion potential associated with the concentration gradients of the passively distributed ions and actively transported protons.

As we have discussed in a preceding section, it is likely for chloroplasts in situ, in which the potassium concentration is 100 — 300 mM, that the change in internal potassium concentration in the light will be relatively small, $\Delta c_K \sim 5-10$ mM. This means that under steady-state conditions $E_K \left(= \frac{RT}{F} \ln c_k^o/c_k^i\right)$ is close to zero (i.e. with $c_k^i = 290$ and $c_k^o = 300$ mM, $E_K^N \sim +1$ mV). Thus it is predicted that under steady-state conditions $(^LE_m)_{ss} = E_K^N \sim 0$. It has been reported [90], and shown in Fig. 2.10 that at the end of a 1-s illumination period the membrane potential in the light is less than 20% of the initial phase 1 potential, $(^LE_m)_i = E_e$, i.e. less than 10 mV. In the presence of valinomycin $(^LE_m)_{ss}$ has been found to be close to zero at the end of a 1-s light period [88]. Although as yet not studied in the very details the experiments done so far have indicated that in prolonged illumination (2 s or more) the potential slowly decreases in the light to a steady state level close to zero. The decrease in ^LE_m during this slow phase is presumed to be due to a decrease in the pump conductance g_p. This decrease is suggested to occur in association with an increase in the pH gradient across the membrane, i.e. with an increase in E_H^N. As g_p is controlling the active proton influx through the active channel, its decrease reflects the decrease of the proton influx, or consequently the proportional decrease in the rate of electron transfer through the electron transport chain and the charge separating sites. Such feed-back control between internal pH and rate of electron transport has extensively been documented [6,7,67,69]. From Eqn. 1a follows that, when $(^LE_m)_{ss} = E_K^N \sim 0$, $E_N^H = (g_p/g_H) \cdot E_e$. With $E_H^N \sim -175$ mV (Δ pH ~ 3), $E_e \sim 50$ mV and $g_H = {}^Lg_H \sim 0.1\ (g_k + g_{Cl})$ (see p.70), one obtains that under steady-state light conditions $g_p = {}^Lg_p \sim 3.5\ {}^Lg_H \sim 0.35\ (g_K + g_{Cl})$. The decrease in g_p in the light, associated with energization (Δ pH formation), would explain the observation [89] that the phase 1 potential $(^LE_m)_i$ after a pre-illumination is much smaller than after a preceding dark period. According to the equivalent electric circuit (Fig. 2.6) $(^LE_m)_i = E_e \cdot g_p/(g_p + g)$. After darkness, when $g_p = {}^Dg_p >> g$, $(^LE_m)_i \simeq E_e$; after pre-illumination, when $g_p = {}^Lg_p \sim 0.3$ g, $(^LE_m)_i \simeq 0.23\ E_e$.

It will be clear from the foregoing discussion that the phase 2 transition from the initial phase 1 to the steady-state phase 3 potential is determined by more than one process. The initial stage of the decay phase is determined by the relative magnitude of the discharging potential-dependent passive ion fluxes and of the charging electron flux through the electron transport chain. Inductory effects in the latter, due to filling of redox pools, will cause induction effects in the initial phase 2 decay. Such effects have been observed indeed but need further experimental analyses. It has been shown [20] that after completion of these inductory effects, i.e. after about 50 to 100 ms of illumination, this stage of the phase 2 decay, which is completed after about 0.5 to 0.8 s [88], proceeds with first-order kinetics, of which the rate constant is temperature-dependent. It has been found [20, 88] that under conditions of suppressed electron transport the time constant of the phase 2 decay is exclusively determined by the passive membrane conductance, e.g. $\tau \sim C_m/g$, and is equal to the rate constant τ_d of the potential decay after a short flash. In a subsequent stage of the phase 2 decay, at which the magnitude of the passive fluxes is equal to that of the electron flux, the decay is assumed to be controlled by a change in the conductance of the electrogenic pump. The completion of this stage occurs in illumination periods of a few seconds. In the presence of valinomycin at which the phase 2 decay and consequently the rate of formation of the pH-gradient are considerably enhanced (e.g. ref. 88), the steady-state phase 3 potential $({}^{L}E_m)_{ss} \sim 0$ has been found to be reached within 1 s, as would be predicted.

2.3.2.3. Diffusion potentials after illumination

As discussed already (p.69) it is likely that for chloroplasts in situ a pH gradient of about 3 units is built up during the completion of phase 2 due to the light-driven active influx of protons into the thylakoid interior. Such gradient corresponds to a Nernst potential E_H^N of about —175 mV. The transient negative potential observed after illumination has been suggested to be due to the contribution of the proton gradient to the initial diffusion potential $({}^{D}E_m)_i$ in the dark [88,91]. According to Eqn. 2a, and with $E_K \sim E_{Cl} \sim 0$, $({}^{D}E_m)_i \simeq E_H^N \cdot {}^{L}g_H/g \simeq E_e \cdot {}^{L}g_p/g$.

Initial dark potentials between —5 and —15 mV routinely have been observed after light periods of 1 s or more. These values suggest that ${}^{L}g_H \sim (0.03\text{—}0.1) \cdot g$. In the previous section we have approximated that with $g \simeq (g_K + g_{Cl}) \simeq 80\ \mu\text{hmo} \cdot \text{cm}^{-2}$ this proton conductance in the light is associated with a proton permeability coefficient P_H of the order of 10^{-5} cm.s^{-1}.

The potential undershoot was found to be much smaller in the presence of valinomycin at a concentration of 10 μM. CCCP at the same concentration was found to be without an appreciable effect. Valinomycin added in the presence of the uncoupler CCCP also caused a decrease in magnitude of $({}^{D}E_m)_i$ but to a lesser extent than in the absence of the uncoupler. The rise in potential towards the steady-state dark potential $({}^{D}E_m)_{ss} = 0$ was somewhat enhanced after adding valinomycin in the presence of CCCP (Fig. 2.12). In the absence of the latter this effect could not be measured accurately because of the small size of the undershoot under these conditions. As valinomycin and CCCP were found not to influence E_e, and are likely not to affect g_p, but will increase g_K and g_H, respectively, the initial potential undershoot $({}^{D}E_m)_i \simeq E_e \cdot {}^{L}g_p/g$, with $g = (g_K + g_{Cl} + g_H)$, will decrease in the presence of valinomycin, but is likely not to change in the presence of CCCP as long as $g_H < g_K$. A 10-fold increase of g_K in the presence of valinomycin might be expected for chloroplasts in situ, according to an estimated 10-fold increase in the potassium permeability coefficient P_K in the presence of this ionophore [10]. However, an increase in P_H induced by CCCP will not be associated with a proportional increase in g_H. For instance, a concentration of the uncoupler which has caused a 50% inhibition of the pH gradient from Δ pH = 3 to Δ pH = 1.5 at a stroma pH of about 7, which means a 2-fold decrease in $E_H^N \left(\sim \frac{^Lg_p}{^Lg_H} \cdot E_e\right)$, apparently has brought about an approximate 2-fold increase in ^Lg_H. From $E_H^N = RT/F \ln (c_H^o/c_H^i)$ and $^Lg_H = \frac{F^2}{2RT} \cdot P_H \cdot c_H^i$, it follows that these changes in the potential and conductance are due to a 30-fold decrease in c_H^i and a 60-fold increase in P_H, caused by the uncoupler. The slow potential rise from the initial negative level $(^DE_m)_i$ to the zero dark steady-state potential would reflect the dark decay of the pH gradient across the thylakoid membrane which, according to fluorescence measurements with umbelliferone [74], may take a few seconds. The kinetics of the potential change are determined by the capacity of the proton gradient and by the relative magnitudes of the ion conductances in the membrane. As the proton conductance is small as compared to the other conductances, the decay will be determined mainly by g_H. However, g_H will decrease simultaneously from ^Lg_H to ^Dg_H by a factor of about 5.10^2 (p.70), and consequently this change also will affect the kinetic pattern of the dark equilibration process. A quantitative analysis thereof has not been made as yet.

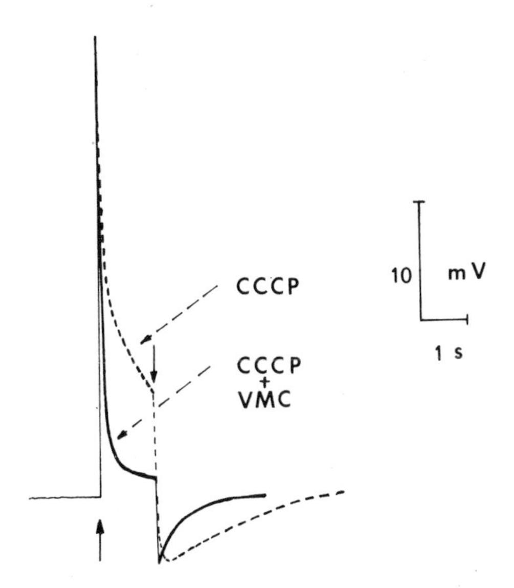

Fig. 2.12. Kinetics of light-induced potential change of chloroplast (thylakoid) membrane in the presence of CCCP (5 μM) and of CCCP (5 μM) + valinomycin (3 μM). Further explanations are in the text, and in the legend to Fig. 2.10.

2.4. CONCLUSIONS

The kinetics of the potential changes measured with a micro-capillary glass electrode during and after energization of the chloroplast appear to be in reasonably good agreement with the changes predicted for the thylakoid membrane. The initial phase 1 potential rise is due to the homogenous charging of the membrane after the light-driven charge separation in the localized reaction centers and the subsequent ion binding processes at these sites. These processes are likely to be associated with potentials of 10 and 50 mV, or even somewhat higher. The potential change in continuing illumination (phase 2) is determined by the rate of electron transfers through the charge-separating sites, and by the ion conductances of the membrane. These conductances are high for chloroplasts in situ, due to the relatively high concentrations of potassium and chloride ions, with the consequence that under steady-state light conditions the potential maintained by the electrogenic hydrogen ion pump is balanced by an (opposite) diffusion potential, mainly determined by the proton gradient across the thylakoid membrane. As a result the steady-state condition in the light is characterized by an electrical potential close to zero and by a proton gradient of about 3 pH units or even more [39,67,68]. Potential measurements on isolated chloroplasts have shown (Bulychev and Vredenberg, unpublished) that the membrane potential under steady-state light conditions can be appreciably under conditions at which the ion conductances in the membrane are low, i.e. for chloroplasts suspended in low salt media. Under these conditions it was conclusively found that the relaxation times of the potential decay after a saturating flash in the dark, or in continuous light during phase 2, were increased.

The kinetics of the 515 nm absorption change in illumination periods of 1 s or more, which have been suggested to indicate a proportional change in the membrane potential [46,94] appear to be quite different from those measured by micro-capillary glass electrodes. However, it must be considered that there is as yet no experimental proof that the 515 nm changes occur in response to a potential difference independently from accompanying conformational and configurational changes in the membrane associated with the ion movements across the membrane [21,93]. One might for instance speculate, after comparing the kinetics of the multiphasic changes in 515 nm absorption in continuous light [95] with the changes in potential and pH gradient predicted by quantitative reasoning, and confirmatory measured with micro-electrodes, that the magnitude of the potential indicating 515 nm change is dependant on the pH, i.e. is increased concomitantly with a decrease in pH of the thylakoid inner space. Unfortunately it has been proved to be difficult to make an experimental mV calibration of the extent of the 515 nm absorption change by measuring its magnitude in response to salt or pH gradients. Such calibration has been made for the carotenoid shifts in bacterial chromatophores [22,42], but similar experiments done with chloroplasts did not allow quantitative conclusions [83]. The slow 3- to 4-fold increase in 515 nm response in 1 s light flashes [95], as well as the slow decay phase (phase d) [45] after a short saturating light flash, measured in *Chlorella* cells, occur with relaxation times which are close to those of the creation and decay, respectively of the pH gradient [74]. Moreover, a steady-state potential in the light of approx. 100 mV, as concluded from the 515 nm responses in *Chlorella* [36], would suggest that the potassium ions, present in the dark at a concentration of approx. 100 mM [9,75], are distributed in the light across the thylakoid membrane at an inside—outside concentration ratio of approx 1 : 100. Based on the approximations made on the magnitude of the steady-state electron flux and on the membrane conductance of potassium ions, it is difficult to see how the potassium concentration in the inner thylakoid space can decrease that much in the light to reach equilibrium at such high potential.

Although further critical experimental tests are needed, it seems justified to conclude that in the intact plant cell the electrical gradients between chloroplasts and cytoplasm are small under steady state conditions, but are effectively controlling the energy production and energy supply in the cellular organization.

REFERENCES

1 Amesz, J. (1964) Biochim. Biophys. Acta, [79], 257—265.
2 Amesz, J. (1973) Biochim. Biophys. Acta, [301], 35—51.
3 Amesz, J. and Visser, J.W.M. (1971) Biochim. Biophys. Acta, [234], 62—69.
4 Amesz, J. and Vredenberg, W.J. (1966) in Currents in Photosynthesis Research (Thomas, J.B. and Goedheer, J.C., eds), pp. 75—81, Donker, Rotterdam.
5 Amesz, J. and Vredenberg, W.J. (1967) in Biochemistry of Chloroplasts (Goodwin, T.W., ed), Vol. 2, pp. 593—600. Academic Press, London.
6 Avron, M. (1972) Proc. IInd Int. Cong. on Photosynthesis Research (Forti, G. Avron, M. and Melandri, A., eds.), vol. 2, pp. 861—871, Junk, The Hague.
7 Avron, M.(1974) in Membrane Transport in Plants (Zimmermann, U. and Dainty, J., eds), pp.249—255, Springer, Berlin.
8 Avron, M. (ed), (1974) Proc. 3rd Int. Cong. on Photosynthesis, Rehovoth (Israel), Elsevier, Amsterdam.
9 Barber, J. (1968) Biochim. Biophys. Acta, [150], 618—625.
10 Barber, J. (1972) Biochim. Biophys. Acta, [275], 105—116.
11 Barber, J. (1972) FEBS Letters, [20], 251-254.
12 Barber, J. and Kraan, G.P.B., (1970) Biochim. Biophys. Acta, [197], 49—59.
13 Barber, J., Telfer, A. and Nicolson, J. (1974) Biochim. Biophys. Acta, [357], 161—165.
14 Barr, C.E. (1965) J. Gen. Physiol., [49], 181—197.
15 Bishop, D.G. (1974) Photochem. Photobiol., [20], 281—291.
16 Briggs, G.E., Hope, A.B. and Robertson, R.N. (1961) In Electrolytes and Plant Cells, James, W.O. (ed), vol. I, Davis, Philadelphia.
17 Bobrov, V.A., Kurella, G.A. and Yaglova, L.G. (1974) Sov. Plant Physiol., [21], 433—440.

18 Bulychev, A.A., Andrianov, V.K., Kurella, G.A. and Litvin, F.F. (1971) Sov. Plant. Physiol., (Eng Transl.), [18], 204—210.
19 Bulychev, A.A., Andrianov, V.K., Kurella, G.A. and Litvin, F.F. (1972) Nature, [236], 175—176.
20 Bulychev, A.A., Andrianov, V.K., Kurella, G.A. and Litvin, F.F. (1973), in Biophysics of Membranes (Proc. Symp. on Molecular Mechanisms of Permeability of Membrane Structures, in Russian) Kaunas, U.S.S.R.
21 Chance, B., Kihara, T., de Vault, D., Hildreth, W., Nishimura, M. and Hiyama, T. (1969) in Progress in Photosynthesis Research (Metzner, H., ed), Vol. III, pp. 1321—1346, Laupp, Tubingen.
22 Crofts, A.R., Jackson, J.B., Evans, E.H. and Cogdell, R.J. (1972) in Proc. 2nd Int. Cong. Photosynthesis Research (Forti, G., Avron, M. and Melandri, A., eds), Vol. 2, pp. 873—902, Junk, The Hague.
23 Davis, R.F. (1974) in Membrane Transport in Plants (Zimmermann, U. and Dainty, J., eds), pp. 197—201, Springer, Berlin.
24 Dilley, R.A. (1971) in Current Topics in Bioenergetics (Sanadi, D.R., ed), vol. 4, pp. 237—271, Academic Press, London.
25 Dilley, R.A., and Rothstein, A. (1967) Biochim. Biophys. Acta, [135], 427—443.
26 Dilley, R.A. and Vernon, L.P. (1965) Arch. Biochim. Biophys, [111], 365—375.
27 Duysens, L.N.M. (1954) Science, [120], 353—354.
28 Eisenman, G., Sandblom, J.P. and Walker, J.L. (1967) Science, [155], 965—974.
29 Emrich, H.M., Junge, W. and Witt, H.T. (1969) Z. Naturforsch., [24b], 1144—1146.
30 Forti, G., Avron, M. and Melandri, A. (eds), (1971) Proc. 2nd Int. Cong. Photosynthesis Research, Stresa, pp. 1—2745, Junk, The Hague.
31 Fowler, C.F. and Kok, B. (1972) in 6th Int. Cong. Photobiology, Bochum, Abstr. No. 417.
32 Fowler, C.F. and Kok, B. (1974) Biochim. Biophys. Acta, [357], 308—318.
33 Frank, K and Becker, M.C. (1964) in Physical Techniques in Biological Reaearch (Nastuk, W.L., ed), vol. V, part A, pp. 23—87, Academic Press, New York.
34 Gimmler, H., Schafer, G. and Heber, U. (1974) in Proc. 3rd Int. Cong. on Photosynthesis, Rehovoth (Israel) (Avron, M., ed), vol. II, pp. 1381—1392, Elsevier, Amsterdam.
35 Goldman, D.E. (1943) J. Gen. Physiol., [27], 37—60.
36 Graber, P. and Witt, H.T. (1974), Biochim. Biophys. Acta, [333], 389—392.
37 Greville, G.D. (1969) in Current Topics in Bioenergetics (Sanadi, D.R., ed), vol. 3, pp. 1—78, Academic Press, London.
38 Grunhagen, H.H. and Witt, H.T. (1970) Z. Naturforsch., [25b], 373—386.
39 Heldt, H.W., Werdan, K., Milovancev, M. and Geller, G. (1973) Biochim. Biophys. Acta, [314], 224—241.
40 Hind, G. and Jagendorf, A.T. (1965), J. Biol. Chem., [240], 3195—3201.
41 Hind, G., Nakatani, H.Y. and Izawa. S. (1974) Proc. Natl. Acad. Sci. USA, [71], 1484—1488.
42 Jackson, J.B. and Crofts, A.R. (1969) FEBS Letters, [4], 185—189.
43 Jagendorf, A.T. and Uribe, E.G. (1966) Proc. Natl. Acad. Sci. USA, [55], 170—177.
44 Jagendorf, A.T. and Uribe, E.G. (1967) Brookhaven Symp. Biol., [19], 215—241.
45 Joliot, P. and Delosme, R. (1974) Biochim. Biophys. Acta, [357], 267—284.
46 Junge, W. and Witt, H.T. (1968) Z. Naturforsch., [23b], 244—254.
47 Junge, W., Auslander, W. and Eckhof, A. (1974) in Membrane Transport in Plants (Zimmermann, U. and Dainty, J., eds), pp. 264—273, Springer, Berlin.
48 Ke, B. (1973) Biochim. Biophys. Acta, [301], 1—33.
49 Kitasato, H. (1968) J. Gen. Physiol., [52], 60—87.
50 Kraan, G.P.B., Amesz, J., Velthuys, B.R. and Steemers, R.G. (1970) Biochim. Biophys. Acta, [223], 129—145.

51 Kraayenhof, R. and Katan, M.B. (1972) in Proc. 2nd Int. Cong. Photosynthesis Research (Forti, G., Avron, M. and Melandri, A., eds), vol. 2, pp. 937—949, Junk, The Hague.
52 Kraayenhof, R. and Slater, E.C. (1974) in Proc. 3rd Int. Cong. on Photosynthesis, Rehovoth (Israel) (Avron, M., ed), pp. 985—996, Elsevier, Amsterdam.
53 Kreutz, W. (1970) Adv. Bot. Res., [3], 53—169.
54 Larkum, A.W.D. (1968) Nature, [218], 447—449.
55 Liberman, E.A. and Skulachev, V.P. (1970) Biochim. Biophys. Acta, [216], 30—42.
56 Ling, G.N. (1973). Biophys. J., [13], 807—816.
57 Menke, W. (1972) Jahrbuch der Max-Planck-Gesellschaft zur Förderung der Wissenschaften E.V. Max-Planck Institut für Züchtungsforschung, Köln-Vogelsang, pp. 132—155.
58 Mitchell, P. (1968), Chemi-osmotic Coupling and Energy Transduction, Glynn Research, Bodmin, Cornwall.
59 Neumann, D. (1973) Protoplasma, [77], 467—471.
60 Neumann, J. and Jagendorf, A.T. (1964) Arch. Biochem. Biophys., [107], 109—119.
61 Nobel, P.S. (1969) Biochim. Biophys. Acta, [172], 134—143.
62 Nobel, P.S. (1974) Introduction to Biophysical Plant Physiology, Freeman, San Francisco.
63 Packer, L., Murakami, S. and Mehard, C.W. (1970) Ann. Rev. Plant Physiol., [21], 271—304.
64 Park, R.B. and Sane, P.V. (1971) Ann. Rev. Plant Physiol., [22], 395—430.
65 Racker, E. (1970) in Membranes of Mitochondria and Chloroplasts (Racker, E., ed), pp. 127—171, Van Nostrand Reinhold, New York.
66 Reed, P.N. and Lardy, H.A. (1972) J. Biol. Chem., [247], 6970—6977.
67 Rottenberg, H., Grunwald, T. and Avron, M. (1972) Eur. J. Biochem., [25], 54—63.
68 Rumberg, B. and Siggel, U. (1969) Naturwissenschaften, [56], 130—132.
69 Rumberg, B., Reinwald, E., Schröder, H. and Siggel, U. (1969) in Progress in Photosynthesis (Metzner, H., ed), vol. 3, pp. 1374—1382, Laupp, Tübingen.
70 Schmidt, S., Reich, R. and Witt, H.T. (1972) in Proc. 2nd Int. Cong. on Photosynthesis Research (Forti, G., Avron, M. and Melandri, A., eds), vol. 2, pp. 1087—1095, Junk, The Hague.
71 Schuldiner, S. and Avron, M. (1971) Eur. J. Biochem., [19], 227—231.
72 Schuldiner, S., Rottenberg, H. and Avron, M. (1973) Eur. J. Biochem., [39], 445—462.
73 Schwartz, M. (1971) Ann. Rev. Plant Physiol., [22], 469—484.
74 Schliephake, W., Junge, W. and Witt, H.T. (1968) Z. Naturforsch., [23b], 1571—1578.
75 Shieh, Y.J. and Barber, J. (1971) Biochim. Biophys. Acta, [233], 594—603.
76 Singer, S.J. (1974) Ann. Rev. Biochem., [43], 805—833.
77 Singer, S.J. Nicholson, G.K. (1972), Science, [175], 720—731.
78 Skulachev, V.P. (1970) FEBS Letters, [11], 301-308.
79 Skulachev, V.P. (1971) in Current Topics in Bioenergetics (Sanadi, D.R., ed), vol. 4, pp. 127—190, Academic Press, London.
80 Slater, E.C. (1966) in Comprehensive Biochemistry (Florkin, M. and Stotz, E.H., ed), vol. 14, pp. 327—396, Elsevier, Amsterdam.
81 Slayman, C.L. (1974) in Membrane Transport in Plants (Zimmermann, U. and Dainty, J., eds), pp. 107—119, Springer, Berlin.
82 Stiehl, H.H. and Witt, H.T. (1969) Z. Naturforsch., [24b], 1588—1598.
83 Strichartz, G.R. and Chance, B. (1971) Biochim. Biophys. Acta, [256], 71—84.
84 Trebst, A. (1974) Ann. Rev. Plant Physiol., [25], 423—458.
85 Trebst, A. and Hauska, G. (1974) Naturwissenschaften, [61], 308—316.
86 Uribe, E.G. (1973) FEBS Letters, [36], 143—147.

87 Vredenberg, W.J. (1973) in Ion Transport in Plants (Anderson, W.P., ed), pp. 153–169, Academic Press, London.
88 Vredenberg, W.J. (1974) in Proc. 3rd Int. Cong. on Photosynthesis, Rehovoth (Israel) (Avron, M., ed), vol. II, pp. 929–939, Elsevier, Amsterdam.
89 Vredenberg, W.J., Homann, P.H. and Tonk, W.J.M. (1973) Biochim. Biophys. Acta, [314], 261–265.
90 Vredenberg, W.J. and Tonk, W.J.M. (1974) FEBS Letters, [42], 236–240.
91 Vredenberg, W.J. and Tonk, W.J.M. (1975) Biochim. Biophys. Acta, [387], 580–587.
92 Walker, N.A. and Hope, A.B. (1969) Aust. J. Biol. Sci., [22], 1179–1195.
93 Walker, D.A. and Crofts, A.R. (1970) Ann. Rev. Biochem., [39], 389–428.
94 Witt, H.T. (1971) Q. Rev. Biophys., 4, 365–477.
95 Witt, H.T. and Moraw, R. (1959) Z. Physikal. Chemie, Neue Folge, [20], 283–298.
96 Witt, H.T. and Zickler, A. (1973), FEBS Letters, [37], 307–310.
97 Wolff, Ch., Buchwald, H.E., Ruppel, H., Witt, K. and Witt, H.T. (1969) Z. Naturforsch., [24b], 1038–1041.
98 Wraight, C.A. and Crofts, A.R. (1971) Eur. J. Biochem., [19], 386–397.

The Intact Chloroplast – edited by J. Barber
© Elsevier/North-Holland Biomedical Press. 1976 — Printed in The Netherlands

Chapter 3

Ionic Regulation in Intact Chloroplasts and its Effect on Primary Photosynthetic Processes

J. BARBER

Department of Botany, Imperial College, London SW7 2BB (Great Britain)

CONTENTS

Abbreviations: CCCP, carbonyl cyanide *m*-chlorophenylhydrazone; DAD, diaminodurene (i.e. 2,3,5,6-tetramethyl-*p*-phenylene diamine); DCMU, 3-(3,4-dichlorophenyl)-1,1-dimethylurea; DCPIP, 2,6-dichlorophenol indophenol; FCCP, fluorocarbonyl cyanide phenylhydrazone; HEPES, *N*-2-hydroxyethyl piperazine-*N*-2-ethanesulphonic acid; HES, high-energy state; OAA, oxaloacetic acid; P680, reaction centre trap for PS2, P700, reaction centre trap for PS1; PGA, phosphoglycerate; PMS, phenazine methosulphate; PS1, photosystem one; PS2, photosystem two; TMPD, *N,N,N'N'*-tetramethyl-*p*-phenylenediamine; Tris, Tris-(hydroxymethyl)-methylamine.

3.1. INTRODUCTION

In this chapter I want to review our present state of knowledge of ionic regulation in intact chloroplasts and discuss how this regulation may exert its influence on primary photochemical processes of photosynthesis. In other chapters in this book, mention is also made of how ionic changes within the chloroplast can affect the biochemical reactions involved in ATP synthesis (Chapter 4) and the fixation of carbon dioxide (Chapter 7). For this reason the more biochemical consequences of ionic regulation in photosynthesis will not be dealt with here in any detail and the reader should refer to the appropriate chapters.

In recent years there have been many observations made with isolated chloroplasts not retaining their outer membranes which indicate that varying the ionic environment of the thylakoids induces structural and functional changes of this membrane system. These changes give rise to effects associated with the properties of the pigment systems and the associated electron carriers both of which are specifically arranged within the thylakoids. But, of course, the physiological medium for the thylakoid is the stroma. Thus there is every reason to suspect that one of the main functions of the outer chloroplast membranes or envelope, is to maintain and carefully control the ionic environment of the stroma so that the photosynthetic machinery can operate efficiently. Recent studies with isolated intact chloroplasts have allowed a start to be made in the interpretation of a number of ionic sensitive photosynthetic reactions in terms of their physiological significance. Although the overall picture is still not clear, it is these recent observations made with the intact isolated organelle which has given the framework to the contents of this chapter.

3.2. IONIC LEVELS AND FLUXES IN CHLOROPLASTS

As stated many times throughout this book, it is important to remember that the intact chloroplast is essentially a two-compartment system. The two outer membranes or envelope enclose the stromal compartment while the thylakoid membrane gives rise to an intrathylakoid or granal space. In the case of the envelope, it seems that the inner membrane constitutes the main permeability barrier between the stroma and cytoplasm (see Chapter 6). Although not always clearly stated it is, I think, true to say that most studies into the ionic relations of chloroplasts have been carried out on preparations in which the outer membranes have not been retained. In Chapter 4 of this volume, Hall has listed the properties expected of a preparation of isolated chloroplasts with intact envelopes (see Ch. 4, Table V). This type of isolated chloroplast has been designated Type A by Hall [56] and is able to use CO_2, but not ferricyanide, as an electron acceptor. In fact it is the inability of Type A chloroplasts to reduce ferricyanide which is used to estimate the "intactness" of a particular preparation (rates of ferricyanide reduction are measured before and after osmotic shock to calculate the percentage intactness of the preparation; see ref. 63). Bearing this in mind, it is pertinent to point out that there are reports in the literature regarding ionic levels and fluxes which have stated that the chloroplasts used were intact. However, it seems unlikely that the chloroplasts used in most of these studies retained a functional outer membrane system (i.e. Type A). It is this inability of most workers to clearly identify the nature of their preparations and also to use varied suspending media, which has brought about a great deal of confusing and conflicting data which is difficult to interpret in relation to the true physiological nature of ionic regulation in chloroplasts.

Much of the work with broken chloroplasts (Type C in Ch. 4, Table V) has been concerned with the relationship between electron transport and H^+ pumping into the intrathylakoid space. It is not my intention in this chapter to discuss the details of light induced electron flow, proton pumping and phosphorylation in terms of Mitchell's chemiosmotic scheme. This has to some extent been covered in other chapters in this volume, particularly Chapters 2 and 4, as well as being extensively reviewed elsewhere [66,74, 166,178]. Nevertheless, in discussing the distribution of ions within the intact chloroplast it is necessary to note that there is a light driven uptake of protons into the intrathylakoid space. Thus H^+ uptake is clearly seen with broken preparations and corresponds to the total movement of about 600 nequiv. H^+/mg chlorophyll [68]. Estimates of the pH of the intrathylakoid space [65,146,147,152,173] suggest that most of these protons become bound [33,68]. The uptake of H^+ by broken chloroplasts can readily be detected as an alkalisation of the suspending media [68,75,130], an observation not seen on illuminating a suspension of isolated Type A chloroplasts [47,64,65]. Heber and Krause [64] were the first to report that very little H^+ movement occurs across the outer chloroplast membranes and in fact the small flux that does occur on illumination is in an outward direction [47, 64,65]. Under these conditions however, Heldt and his colleagues have shown that H^+ pumping does occur at the thylakoids resulting in alkalisation of the stroma and acidification of the intrathylakoid space [65,173]. From studies of the uptake and distribution of radioactive dimethyloxazolidinedione (weak acid) and methylamine (weak base) it has been estimated that within the illuminated intact chloroplast a pH gradient of at least 2.5 units is generated across the in vivo thylakoids. Although this ΔpH value was insensitive to the pH of the suspending medium the actual stromal and intrathylakoid pH were not. With an external pH of 7.6 and steady-state illumination the stromal pH was 7.9 and the pH of the intrathylakoid space 5.4. Moreover, it was estimated that the intrathylakoid compartment acts as a buffer of pK 5.5 while the stroma is a weaker buffer of pK 6.8.

To maintain bulk electrical neutrality the net H^+ uptake into the intrathylakoid space must be balanced either by anion influx or cation efflux. The former would also tend to give rise to an influx of water with concomitant swelling while proton/cation exchanges may not necessarily be accompanied by osmotic volume changes. There have been many attempts with broken chloroplasts to identify the nature of the coions associated with the proton pump. These studies have been complicated by the use of various bathing media which have given rise to conflicting results. For example, Dilley and Vernon [36] found that under their experimental conditions the combined K^+ and Mg^{2+} efflux was about equal to the H^+ uptake into the intrathylakoid space and that Cl^- movement was insignificant. On the other hand, Deamer and Packer [30] and others [27] have presented evidence that Cl^- is the major coion for the proton pump. The physiological significance of these findings are doubtful since they almost certainly reflect the presence of weak bases or acids in the suspending media. Perhaps the most thorough study of this question has been made by Hind et al. [68]. They simultaneously monitored light-induced changes of Cl^-, H^+, Na^+, Mg^{2+} and Ca^{2+} activities in a suspension of broken Type C chloroplasts using ion-specific electrodes. Strong evidence was presented that under certain ionic conditions the proton pump was balanced partly by Mg^{2+} efflux and partly by Cl^- uptake. However, the Mg^{2+} efflux could be replaced to some extent by K^+ efflux when the K^+/Mg^{2+} activity ratio in the suspension was high or by Ca^{2+} if the thylakoids were initially equilibrated with a Ca^{2+}-containing suspending medium. Their work emphasises how the composition of the suspending medium can govern the nature of the fluxes across the thylakoids. Not until we have a clear understanding of the ionic activities in the stroma can studies with broken Type C chloroplasts give precise information about the nature of the light-induced fluxes which occur across the thylakoids within the intact organelle.

Unfortunately our knowledge of the ionic composition of intact chloroplasts is still unsatisfactory. Menke [104] was the first to report that spinach chloroplasts contain K^+, Mg^{2+} and Ca^{2+}. Later the work of Stocking and Ongun [158], Saltman et al. [149], Larkum [90] and Nobel [135] as well as others [59,60,84,91,127] gave rise to the general conclusion that chloroplasts contain high levels of K^+, low levels of Na^+ and intermediate levels of Mg^{2+} and Ca^{2+} with Cl^- as the major diffusible anion. However, as far as I can judge none of the above analyses have been carried out on isolated chloroplasts retaining their outer membranes in a functional state and showing the ability to fix CO_2 at high rates (that is, falling into the category of Class A chloroplast as defined in Ch. 4, Table V). Even analyses of non-aqueous isolated chloroplasts [90,91] are suspect because of cytoplasmic and vacuolar contamination [136].

Perhaps the most satisfactory analysis of the ionic composition of chloroplasts has only recently been achieved by Gimmler et al. [47]. They have used isolated intact spinach chloroplast suspensions showing high rates of CO_2, PGA and OAA reduction and estimated to be 70 to 95% Class A by the ferricyanide test [63]. The cation composition of these isolated intact chloroplasts was essentially independent of the ionic levels of the external medium indicating a very low cation permeability of the envelope membranes. The ratio of K^+:Na^+:Mg^{2+} was found to be about 2:1:2 with the K^+ level in the region of 0.5 to 0.8 μmoles/mg chlorophyll. When recalculated in terms of osmotic space (obtained by the difference of tritiated water and [^{14}C]sorbitol free space of the chloroplasts rapidly spun through an inert silicone oil layer; see Chapter 6) these values correspond to an internal concentration of 20 to 30 mM. In my opinion the use of these biochemically autonomous preparations (with regard to carbon fixation) is far more satisfactory than the earlier analyses on broken or partially broken chloroplasts avoiding the inevitable problems of contamination, leakage and influence of the suspending media.

Gimmler et al. [47] were also able to demonstrate the impermeability of the envelope to cations by osmotic response studies monitored as absorption changes at 535 nm and by studies with radiotracers. With regard to anions, osmotic response measurements with KCl and K acetate in the presence of valinomycin indicated that the outer membranes are permeable to Cl^- and acetate [47].

As mentioned above, when isolated intact chloroplasts are illuminated a light induced H^+ efflux occurs across the envelope from the stroma. Gimmler et al. have reported [47] that this efflux is partly associated with a specific K^+ exchanger dependent in some way on the electron flow processes of the thylakoids. The properties of this exchanger are not as yet fully understood, but there was no evidence for similar Mg^{2+}/H^+ or Na^+/H^+ exchanges. Apparently the remaining light induced H^+ efflux is balanced by concomitant Cl^- movement. The inability of Gimmler et al. [47] to demonstrate net Mg^{2+} movement across the envelope supports the results of Pfluger [144] but contrasts with the earlier work of Lin and Nobel [95]. Moreover, Nobel [136] has reported a light induced net KCl efflux from isolated chloroplasts thought to be intact and has correlated this with water loss and chloroplast shrinkage. Clearly, these results conflict with those of Gimmler et al. [47] and emphasise the necessity for further studies into the ionic permeability and regulatory functions of the envelope membrane. However, it is worth pointing out here that although the extensive studies from Nobel's laboratory [136] have been carried out on rapidly isolated pea chloroplasts showing high rates of endogenous photophosphorylation [129,132] they almost certainly did not have, as assumed by Nobel [95,131,135,136], their outer functioning membranes. These chloroplasts were able to reduce added ferricyanide [94,135] showed low rates of CO_2 fixation and probably fall into the category of Type B chloroplasts as defined in Table V of Chapter 4.

From the observations outlined above, particularly the more recent work of Hind et al. [68] and Gimmler et al. [47] we can attempt to guess at the nature of the ionic distribution and regulation within the intact chloroplast. Gimmler et al. [47] have shown that the K^+ and Mg^{2+} concentrations of isolated intact chloroplasts are about the same being approximately 1 μmole/ mg chlorophyll (estimated to be 30 mM). According to Hind et al. [68] under conditions when the K^+/Mg^{2+} concentration ratio is about one, the light-induced proton uptake into the intrathylakoid space is partly balanced by Mg^{2+} efflux and there is no K^+ movement. With broken chloroplasts, the remaining part of the H^+ flux was accompanied by Cl^- uptake. Therefore the question arises as to whether Cl^- plays an important role as a coion to the proton pump within the intact chloroplast. With broken chloroplasts, when Cl^- uptake is clearly seen there is also concomitant swelling [30]. However, in vivo, the intrathylakoid compartment shrinks on illumination [62,87,105,136,140]. Such a shrinkage is difficult to reconcile with a net Cl^- uptake. Moreover, if the net H^+ uptake into the intrathylakoid space (estimated to be 3.3 μl/mg chlorophyll [65]) is in the region of 600 nmoles/ mg chlorophyll and the internal pH does not drop below 4 [65,146,147,152, 173] then over 99% of the H^+ become bound (buffered). Since no swelling occurs then this binding would presumably be to fixed negative charges such as carboxyl groups on the inner side of the thylakoid membrane [12,33] and would indicate significant cation/proton exchange. In addition to this it seems that the whole organelle tends to behave as a negatively charged Donnan phase [34,47]. This is probably because about 50% of the dry weight of the intact organelle is made up of protein which at physiological pH has surplus negative charges [47]. With about 0.2 mM KCl in the external medium, Gimmler et al. [47] have estimated the internal K^+ level of valinomycin-treated chloroplasts to be 11 mM. This is then the passive equilibrium distribution of K^+ between the chloroplast and the medium and corresponds to a Donnan potential of 58 log 0.2/11 mV or -100 mV. Under these conditions the Cl^- level in the chloroplast would be very small indeed, about 0.004 mM. Naturally as the external KCl level is raised the Donnan potential would decrease and the Cl^- level in the chloroplast rise. Accepting that the envelope membranes are permeable to Cl^-, as found by Gimmler et al. [47] then the level of this anion in the in vivo chloroplast will be determined by the negative Donnan potential and the Cl^- level in the cytoplasm. In fact the free cytoplasmic chloride level of plant cells may not be as high as originally thought [128,162].

Accepting that there is no active Cl^- pumping mechanism operating on the outer chloroplast membranes, for which there is no evidence and bearing in mind the above points, it seems very unlikely that Cl^- fluxes play any significant role within the intact chloroplast. The evidence as it stands at present, suggests that Mg^{2+} acts as the main counter ion to the proton pump (see also section 3.7.3. and contrast this conclusion with the assumptions used in Chapter 2, section 2.2.3.). The energy-dependent shrinkage of the intrathylakoid compartment seen on illumination can be explained if the H^+ taken up becomes more firmly bound (less osmotically active) than the effluxing Mg^{2+} ions [33].

One may ask the question; why do chloroplasts, unlike mitochondria and bacterial systems [51], convert a significant portion of their high energy state or proton motive force [109] into a pH gradient [65,146,147,152, 173] (see also Chapters 2 and 4). Bearing in mind the concepts of Mitchell's chemiosmotic scheme (see Chapter 2), the establishment and maintenance of a significant electrical gradient ($\Delta\psi$) could serve equally well as an energy precursor for photophosphorylation. However, the conversion of $\Delta\psi$ to ΔpH (see Chapter 4) has an advantage in that it results in secondary ion movement. Above I have argued that Mg^{2+} is a strong candidate for the main counter ion to the light-driven proton uptake into the intrathylakoid space. As a consequence of this H^+/Mg^{2+} exchange the levels of these two cations alter in the intrathylakoid and stromal compartments. In particular the stroma becomes more alkaline, rising to a pH of about 8.0 [67,173] while a net efflux of Mg^{2+} corresponding to 300 nmoles/mg chlorophyll (assuming only Mg^{2+}/H^+ exchange) into a stromal space of 23 μl/mg chlorophyll [65] would result in stromal Mg^{2+} increase of 13 mM. In this way pH and Mg^{2+}-sensitive reactions in these compartments could be regulated. In fact, as mentioned previously (section 3.1) there is increasing evidence that this type of cation regulation plays an important part in controlling several stages in the overall reactions of photosynthesis. As yet we do not know the subtleties of the light-induced cation regulation processes within the intact chloroplasts nor do we have a full understanding of the part played by the outer membranes which constitute the envelope. In vivo volume studies indicate that not only do the intrathylakoid spaces shrink on illumination but also that the whole organelle reduces its volume by a flattening effect [133,136]. Heldt et al. [65] did not observe this volume change with illuminated intact isolated chloroplasts and although Nobel has detected in vitro shrinkage [131,132,134,136] it seems likely, for the reasons stated above, that this represented net water flux from the intrathylakoid space. Thus it seems that in vivo the complex traffic of organic and inorganic molecules between the cyostol and the stroma can also bring about net solute loss from the stroma. As pointed out by Nobel [136] the important consequence of this reduction of volume, which corresponds to a loss of 40% of the total chloroplastic water, is to concentrate impermeant metabolites and ions in the stromal phase. Thus the stromal Mg^{2+} level may rise by a total of 22 mM in the light although of course the activity of this cation may be much lower. Further the displacement of Mg^{2+} from the intrathylakoid space would be expected to bring about conformational changes in that compartment and within the thylakoid membrane itself. In this way the primary photochemical and electron flow processes of photosynthesis may be modified and regulated. In the remaining part of this chapter I will review the properties of those primary photosynthetic processes which are cation sensitive and attempt to relate their sensitivity to the light-induced ionic changes which seem to occur in the stromal and intrathylakoid compartments of the intact organelle.

3.3. CHLOROPHYLL FLUORESCENCE

Fluorescence from in vivo chlorophyll is a sensitive intrinsic probe of the primary photochemical reactions of photosynthesis. Changes in these primary reactions including those brought about by conformational changes of the thylakoid membranes are often reflected as changes in fluorescence yield. Before considering the inter-relationships between fluorescence and ionic regulation within the chloroplast I have thought it appropriate to briefly outline the basic properties of in vivo chlorophyll fluorescence. I have not attempted to duplicate the excellent recent reviews on this subject [49,142] but simply furnish the reader who is unfamiliar with this aspect of photosynthesis with sufficient background knowledge to appreciate the discussions in later sections of this chapter.

3.3.1. General properties

When oxygen-evolving organisms are illuminated, fluorescence emission occurs in the red region of the spectrum. The peak of the emission is about 685 nm and is almost entirely due to those chlorophyll *a* molecules (chl a_2) which act as light harvesting pigment for photosystem two (PS2). The actual chlorophyll species involved in this emission is probably that which has its maximum red absorption at 678 nm (chl a_{678}). At room temperature the chlorophyll molecules (chl a_1) which serve as light-gathering antennae for photosystem one (PS1) show very weak emission in the region of 730 nm. However, if the sample is frozen to liquid nitrogen temperature (77°K) the emission spectrum for in vivo chlorophyll fluorescence becomes more complex with chl a_2 peaks at about 685 nm and 695 nm and a chl a_1 peak in the region of 720 to 740 nm [22,49,124,142] (and see section 3.5). Typical emission spectra for isolated spinach chloroplasts measured at room temperature and 77°K are shown in Fig. 3.1.

Fluorescence results as a consequence of de-excitation of the excited singlet state and the observed yield for the emission is governed by all the other competing pathways available for de-excitation. Thus the fluorescence yield (ϕ) is given by

$$\phi = \frac{I_f}{I} = \frac{k_f}{k_f + \Sigma_i k_i} \tag{1}$$

where I_f is the intensity of the fluorescence emission, I is the intensity of the absorbed excitation light, k_f is the rate constant for fluorescence emission

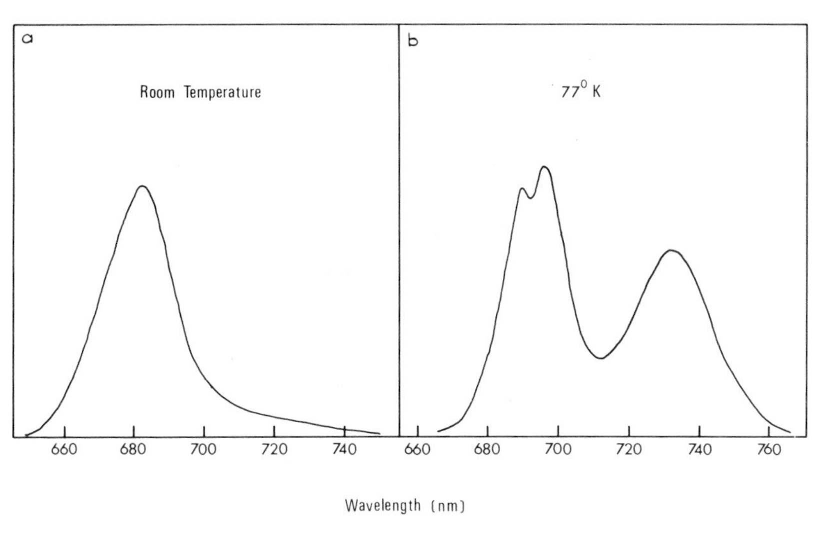

Fig. 3.1. Fluorescence emission spectra from isolated intact chloroplasts measured at (a) room temperature, (b) liquid N_2 temperature. Excitation monochromator was set at 440 nm with a slit width of 40 nm and the emission detected with a monochromator set at a slit width of 4 nm. (The relative intensities at the two spectra are not comparable.)

and k_i is the rate constant for other modes of de-excitation.

For in vivo chl a_2 the fluorescence yield may be given by

$$\phi = \frac{k_f}{k_f + k_h + k_t + k_p\ [P]} \tag{2}$$

where k_h, k_t and k_p are rate constants for: radiationless de-excitation to heat, transfer from chl a_2 to chl a_1 and photosynthetic trapping respectively. [P] is the concentration of open PS2 traps and in this equation the traps are assumed to be uniformly distributed throughout the light-harvesting chl a_2 (i.e. the lake or statistical model [93,167]).

Clearly, if there are changes in any of the above rate constants or in [P] then there will be changes in fluorescence yield.

It is not surprising that when dark-adapted algae or a leaf is exposed to light the yield of chl a_2 fluorescence changes with time. In fact characteristic changes can be observed which can be divided into two main components, fast changes and slow changes. The characteristics of these changes have been discussed in detail in recent reviews by Govindjee and Papageorgiou [49] and Papageorgiou [142] and here I will only outline their basic properties.

Essentially the fast fluorescence changes reflect changes in [P] and occur within a second or two after initiation of illumination while the slower components of the fluorescence induction are mainly associated with changes in rate constants of Eqn. 2 and occur in the seconds to minutes time scale.

Consider the following sequence of events brought about by illuminating a dark-pretreated photosynthetic system. The trapping molecule at the reaction centre of photosystem two, P, which is thought to be a special chl a_2 molecule, often called P680, is able to accept a photon. Z and Q are the primary electron donor and acceptor molecules respectively in very close proximity to P680 and whose exact chemical identities are not yet known. On excitation with a photon the following sequence of events occurs (the exact rates for these reactions are not yet established and may well depend on several factors but the numbers shown give some idea of the times involved).

$$\underset{\text{open trap}}{\text{Z P Q}} \xrightarrow{h\nu} \text{Z P*Q} \xrightarrow[\text{[102,177]}]{< 20\text{ nsec}} \underset{\text{closed trap}}{\text{Z }P^{+}Q^{-}} \quad \text{(i)}$$

Normally Z rapidly reduces P^+

$$\text{Z }P^{+}Q^{-} \xrightarrow[\text{[48,102]}]{1\text{–}3\ \mu\text{sec}} Z^{+}\text{P }Q^{-} \quad \text{(ii)}$$

Secondary electron flow then opens the trap

(a) electron flow from H_2O

$$Z^{+}\text{P }Q^{-} \xrightarrow[\text{[169,177]}]{\sim 600\ \mu\text{sec}} \text{Z P }Q^{-} \quad \text{(iii)}$$

(b) electron flow to secondary acceptor

$$\text{Z P }Q^{-} \xrightarrow[\text{[44,157,177]}]{\geqslant 600\ \mu\text{sec}} \underset{\text{open trap}}{\text{Z P Q}} \quad \text{(iv)}$$

Thus when the trap is closed and [P] = 0 the fluorescence yield should be at a maximum. There is evidence that under certain conditions fluorescence quenching still occurs even when the traps are closed [42,57,58,102,139]. Usually, however, when the trap is in the form Z^+ P Q^- or Z P Q^-, then fluorescence is at a maximum. The kinetics and extent of the rise in the fluorescence due to trap closure is affected both by the rate of secondary electron flow and by the degree of excitation transfer from photosynthetic units with closed traps to those with open traps; a point which will be considered later.

The opening of the closed trap is dependent on the supply and removal of electrons by secondary electron flow. Since Z^+ is rapidly reduced it is the removal of electrons from Q^- which normally controls the open trap concentration [40,44,97] although under some experimental conditions the rate of trap opening can be controlled by Z^+ reduction [20,58]. The idea that the redox state of Q is the main factor controlling fluorescence yield has been, and still is, the rationale behind fluorescence studies designed to understand the photochemical reactions of photosynthesis [49,93,142]. A similar series of electron transfer reactions are thought to occur in PS1 but in this case there are no associated changes in the yield of chl a_I fluorescence [171].

The slower fluorescence changes observed with intact photosynthetic systems seem to some extent to be independent of k_p and [P]. Kautsky [83] reported many years ago just how complex the time course of in vivo chlorophyll fluorescence can be during the first few minutes of illuminating algae or leaves. The time-dependent changes can be described using a notation introduced by Lavorel [92] and extended by Munday and Govindjee [113]. On illuminating a dark-treated spinach leaf the fluorescence rises very rapidly (limited by the shutter speed but in principle equal to the fluorescence lifetime of the chlorophylls involved) to an initial level o which is also often called the f_o level. This is then followed by a relatively rapid rise, the rate of which is dependent on light intensity, to a level p showing in some cases a slight step or dip id (see Fig. 3.2). Then a decline often occurs from p to s. With a light intensity that saturates photosynthesis, the ops phases are over

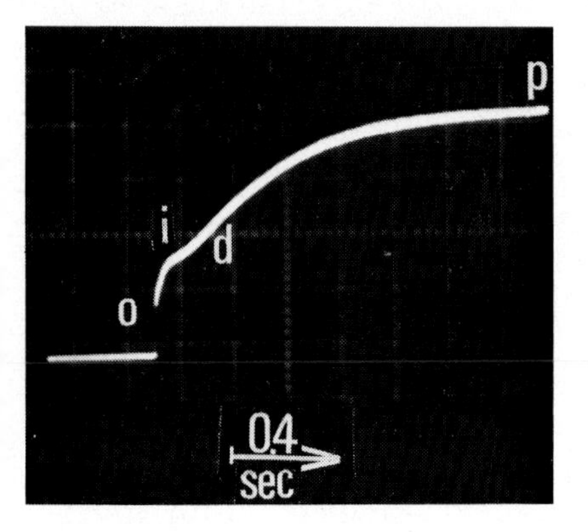

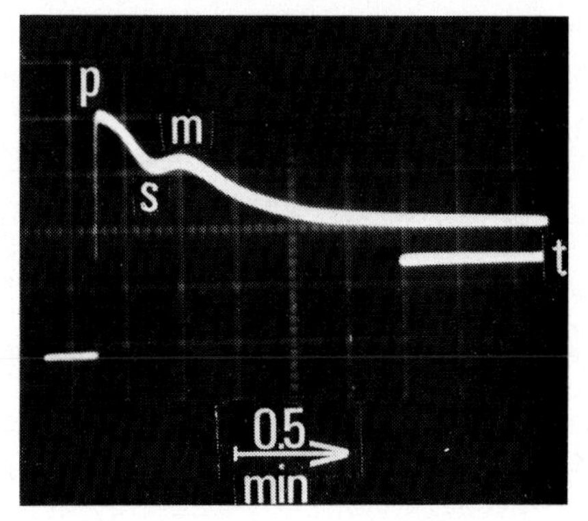

Fig. 3.2. Chlorophyll fluorescence transients measured on illuminating a dark-pretreated spinach leaf showing fast and slow components. The final t level was recorded after 5 min of illumination. The excitation was by broad blue-green light (4 mm Schott BG18) at 20 kerg/cm² s and the emission measured at 686 nm. (Synowiec and Barber, unpublished result.)

within a few seconds and mainly seem to represent the fast trap-dependent phase of the fluorescence induction as discussed above [49]. The o or f_o level remains more or less constant with time and its significance is not clear. It could originate from non-photosynthetically active chl *a*, not associated with a trap (cf. ref. 25) or could be due to the reversibility or inefficiency of the PS2 trapping processes [18,22,85,99]. Alternatively this constant fluorescence could be derived partly from chl a_I [93]. Recent psec fluorescence lifetime measurements under open trap conditions indicates that the initial o level is very low (yield of less than 1%) and originates from the light-harvesting chlorophylls of PS2 and PS1 [8a].

With intact algae and leaves the fast o to s phases are followed by slower fluorescence changes which vary with different organisms. However, the general features of the slow phase are a rise in the fluorescence yield from s to m followed by a significant decline to a low terminal level t. It seems that these slow changes are not controlled by the redox state of the PS2 trap but are under the control of energy-dependent conformation changes of the thylakoids (for extensive discussion, see reviews refs. 49, 142). The significance of these trap-independent changes is considered in the following section.

3.3.2. State one—state two transitions

Until 1969 attempts to explain the slow fluorescence yield changes had been made in terms of the generation of quenchers other than Q [40,41]. It was, however, the elegant studies of Bonaventura and Myers [14] and Murata [119] which produced a model for the slow changes. These experiments indicated that the fluorescence quenching giving rise to the mt phase was associated with a redistribution of the incoming quanta in favour of PS1. The existence of a mechanism which could control quantal distribution to PS2 and PS1 would be advantageous to the organism under conditions when there is an imbalance between the light incident on the two photosystems. Thus the mechanism could function, under a particular light condition, to optimise electron transport and would explain the fact that the quantum yields for photosynthesis are surprisingly constant and maximal over a wide spectral range where PS2 absorption predominates [96,126]. This process may play an important role in maintaining the efficient use of radiant energy through a leaf canopy where the lower leaves are subjected to light of a reduced intensity and different spectral content compared with the less shaded upper leaves.

Bonaventura and Myers [14] used modulated PS2 light* to simultaneously record the effect of superimposing non-modulated PS1 light* on the modulated fluorescence and O_2 evolution signals obtained with *Chlorella*. In

*The terms PS2 and PS1 light are used in the sense that they are wavelengths of light preferentially absorbed by PS2 and PS1 respectively.

this way they were able to measure directly the effect of PS1 light on the quantal efficiency of fluorescence and oxygen production. The effect of PS1 light is to draw electrons from PS2 resulting in an increased rate of O_2 evolution and a decrease of chlorophyll fluorescence intensity ([P] becomes larger in Eqn. 2). The increase in quantal efficiency for O_2 emission brought about by PS1 light is usually known as the Emerson enhancement effect [126] (section 3.4.2.5.) and can be measured as an enhancement ratio (E) where

$$E = \frac{\text{Rate of } O_2 \text{ evolution with both PS1 and modulated PS2 light}}{\text{Rate of } O_2 \text{ evolution with modulated PS2 light only}}$$

In fact it was found that the value of E decreased as fluorescence decreased from m to t. If, however, the non-modulated PS1 light was left on for 4 to 5 min there was an increase in fluorescence and a corresponding increase in enhancement. The low fluorescence-low enhancement state induced by PS2 light was called *State 2* while the maximum enhancing state induced by PS1 light and by dark treatment, was called *State 1*. Therefore the State 1 to State 2 shift represents an adaption process whereby the initial imbalance of photon delivery to the two photosystems, due to excess PS2 light, is minimised giving rise to a change from a maximum to a minimum enhancement situation as defined above. On the other hand, a shift from State 2 to State 1 occurs when PS1 is initially overexcited relative to PS2. These slow changes then seem to control the flow of quanta to the two photosystems and optimise the rate of electron flow for a particular illumination condition. Bonaventura and Myers [14] argued that the transition from State 1 to State 2 was brought about by diverting 10% of the light normally absorbed by PS2 in the State 1 condition to PS1.

A similar model was independently suggested by Murata [119]. Using the red alga *Porphyridium cruentum* he showed that the relative heights of the chl a_2 and chl a_1 peaks in the 77°K emission spectrum varied depending on whether the algal suspension had been preilluminated or not prior to freezing. He was able to correlate the slow chl a_2 fluorescence decrease with an increase in the chl a_1 emission at 77°K suggesting that the decrease in chl a_2 yield was accompanied by an increase in chl a_1 yield. Like Bonaventura and Myers [14], he was also able to demonstrate the reverse of this if prolonged illumination with PS1 light was given.

Although these observations indicated the existence of a control process for optimising energy distribution to the two photosystems they did not give any information about the mechanism involved. Myers [126] and Duysens [39] considered the possibility that the effect was due to a shift in the initial quantal distribution between the two photosystems while Murata [120] was more specific and considered spillover of energy from PS2 to PS1 to be due to an increase in the rate constant k_t for the bulk light-harvesting chl a_2 molecules (see Eqn. 2). Either could be brought about by light-induced membrane conformational changes altering both distance and orientation between pigments. More detailed indications of the mechanism involved have come from studies with isolated chloroplasts.

3.4. EFFECTS OF CATIONS ON ISOLATED BROKEN CHLOROPLASTS

3.4.1. Chlorophyll fluorescence

3.4.1.1. Fluorescence yield

A hint of the mechanism which may control the in vivo State 1—State 2 shifts stems from the initial observations of Homann [70] that the addition of Mg^{2+} or other cations to DCMU-poisoned broken chloroplasts isolated from *Phytolocca americana* and suspended in a low salt medium, induces a rise in the yield of chl a_2 fluorescence. The action of DCMU is to block electron flow from Q^- to the secondary electron acceptors [40] and as a consequence keep the PS2 traps closed. Under these conditions changes in fluorescence yield must be caused by changes independent of the redox condition of the PS2 trap. Murata in collaboration with colleagues [120,121, 125] carried out a series of experiments designed to characterise the properties of this Mg^{2+}-dependent fluorescence change. Using isolated spinach chloroplasts it was shown that a number of divalent and monovalent cations could bring about the fluorescence yield increase but in general divalent cations were effective at much lower concentrations (about 1 mM) than monovalent cations (about 100 mM). Among the cations which have been tested, the ability to increase the fluorescence yield is remarkably unspecific. All the alkaline earth cations together with Co^{2+} and Mn^{2+} will at about equally low concentrations increase the fluorescence yield although Zn^{2+}, Cd^{2+}, Pb^{2+}, Cr^{3+} and La^{3+} have very little effect [94,125,163]. At a hundred times higher concentration the alkali metal cations are about equally as effective and also can be substituted by the methylamine [101] and the choline cations [108]. The extent of the yield increase varies from one chloroplast preparation to another and seems to be a function of the condition of the thylakoid membranes. Small cation-induced changes in yield are associated with poor chloroplast preparations. In fact fixation of the thylakoids with glutaraldehyde completely abolishes the cation-induced fluorescence increase [77,110] as does the preparation of PS1 and PS2-rich particles [110, 121].

Murata extended the room temperature observations by investigating the effect of cations on the 77° K emission spectra and found that the cation-induced increase in PS2 fluorescence occurred at the apparent expense of PS1 fluorescence [120,125]. From this important observation, which has been confirmed by others [5,22,53], Murata was the first to suggest that the presence of cations decrease spillover between PS2 and PS1 and thus gave the basis to the formulation of possible mechanisms for controlling the in vivo State 1—State 2 transition.

3.4.1.2. *Fluorescence lifetime*

The State 1—State 2 concept requires that a fraction of light delivered to PS2 in State 1 is diverted to PS1 when the system is driven into State 2 by PS2 light. Such a change could occur either from a pigment shift associated with a small number of chlorophylls which diverts a fraction of the light absorbed by PS2 to PS1 or from an increase in intersystem exciton transfer from all the bulk chlorophylls of PS2 to PS1 (increase in k_t of Eqn. 2).

Simultaneous measurement of chloroplast fluorescence intensity and lifetime can distinguish between these two possibilities.

Consider the following

$$\phi = \frac{\tau}{\tau_0} \tag{3}$$

where ϕ is the fluorescence yield, τ is the observed fluorescence lifetime of chl a_2 and τ_0 is the intrinsic lifetime and equals $\frac{1}{k_f}$

$$\phi = \frac{k_f}{k_f + k_h + k_t + k_p\ (P)} \tag{2}$$

and

$$\tau = \frac{1}{k_f}\phi \tag{4}$$

since

$\phi = \frac{I_f}{\alpha I}$ where α is the initial fraction of light absorbed by chl a_2, then

$$I_f = \alpha\ I\ k_f\ \tau \tag{5}$$

If αI is constant then I_f will vary linearly with τ due to changes in k_t (or k_h). If, however, αI varies then I_f could change without any change in τ.

As far as I know this experiment has not been carried out with intact organisms but Briantais et al. [19] have tested Eqn. 5 using isolated lettuce chloroplasts treated with and without Mg^{2+}. As shown in Fig. 3.3 an approximately linear relationship was found between fluorescence yield and lifetime of chl a_2. This observation supports the idea that controlled spillover between chl a_2 and chl a_1 occurs via changes in k_t and that this is the means of regulating the State 1—State 2 shifts seen in vivo (for more detailed discus-

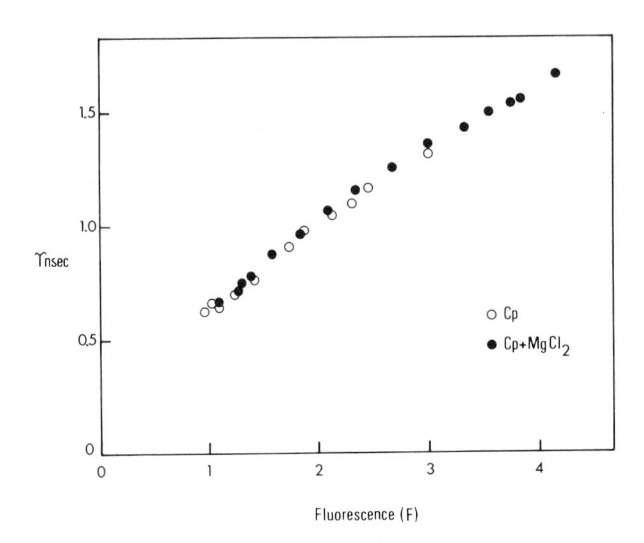

Fig. 3.3. The relationship between the fluorescence lifetime (τ) and intensity (F) during PS2 induction measured on isolated chloroplasts using a phase fluorimeter. The open circles indicate chloroplasts not treated with Mg^{2+} while measurements made in the presence of 7.5 mM Mg^{2+} are indicated by closed circles (reproduced with permission from ref. 19).

sion of the implication of this type of experiment, see ref. 112).

3.4.1.3. Fast fluorescence induction

Several workers have studied the effect of Mg^{2+} on the fast PS2 trap-dependent fluorescence rise [10,19,22,69,70,77,110,120]. Using isolated broken chloroplasts treated with DCMU it has been found that Mg^{2+} addition only slightly increases the initial o or f_o fluorescence level. At room temperature the main effect of Mg^{2+} addition is on the yield of the variable component dependent on the redox state of the PS2 traps. According to the theory of Malkin and Kok [97] and Murata et al. [123,124] the area over the fast induction curve gives a measure of the electron acceptor pool on the reducing side of PS2. In the presence of DCMU, the effect of Mg^{2+} is to increase this area but when normalised by dividing by the relative variable fluorescence yields the two areas are almost identical [69,120]. It was this calculation which lead Murata [120] to favour the spillover model (i.e. change in k_t) in preference to the pigment shift model, since the latter predicts considerable changes in the normalised areas [120]. Bennoun [10], Briantais et al. [19], Malkin and Siderer [98] and Li [94] came to the same conclusion that the yield changes are due to changes in the rate constants controlling radiation-less de-excitation of chl a_2 but also considered that changes in k_h as well as k_t may be involved.

Very recently an elegant analysis of the fast induction curves of fluorescence measured at 77°K has been carried out by Butler and Kitajima [22]. Although at this low temperature chl a_1 fluoresces at 730 nm it shows no yield changes due to changes in the redox state of the PS1 trap [171]. Butler and Kitajima have taken advantage of this to compare fluorescence changes at 730 nm with chl a_2 yield changes (at 690 nm) in the presence and absence of Mg^{2+}. Their analysis was based on an assumed model [22,85] and support the general concept that Mg^{2+} can influence the distribution of excitation energy between the two photosystems. However, they argued that this cation acts in two ways, to control the initial distribution of quanta to the two photosystems (i.e. change α in Eqn. 3.5) and also alter k_t in Eqn. 3.2. They estimate from their studies with broken spinach chloroplasts that the addition of Mg^{2+} increases α (defined as β by Butler and Kitajima) by 9% and decreases the yield of energy transfer from the bulk light-harvesting chl a_2 to PS1 from 0.12 to 0.065 for open PS2 traps and 0.28 to 0.23 for closed PS2 traps.

In addition to the increase in fluorescence yield, the presence of low concentrations of Mg^{2+} or high concentrations of monovalent cations causes a change in the kinetics of the fast fluorescence rise. Several investigators have reported that the rise is sigmoidal when cations are present but in their absence tends to be exponential [10,19,69]. These cation-induced changes in kinetics can be interpreted in terms of the model of Joliot and Joliot [80] which relates the non-exponential rise of fluorescence with the degree of exciton transfer between PS2 photosynthetic units. According to this model the relationship between variable fluorescence yield ϕ, the fraction of closed PS2 traps (1—Q) and the probability of transfer of an exciton from a PS2 unit with a closed trap to another PS2 unit (ρ) is given by:

$$\phi = \frac{(1-\rho)\ (1-Q)}{1-\rho\ (1-Q)} \qquad (6)$$

Hipkins and Barber [69] have used this expression to analyse the fast induction kinetics measured with broken pea chloroplasts with and without Mg^{2+} in the suspending medium. This analysis indicated that in the presence of Mg^{2+} the probability of transfer between PS2 units is 0.55 while in the absence of the cation, ρ reduces to a low value suggesting that Mg^{2+} deficiency creates a condition in which the PS2 units act more like separate entities.

Overall these results indicate that in media containing low levels of cations there is minimum transfer of energy between PS2 units but maximum spillover and delivery of quanta to PS1. The addition of low concentrations of divalent cations or high concentrations of monovalent cations creates the opposite situation, maximum transfer between PS2 units and minimum spillover of energy from PS2 to PS1. As discussed in more detail in later sections these changes in energy distribution and transfer between like and unlike units are almost certainly due to conformational changes within the thylakoid membrane induced by the addition of cations.

3.4.2. Electron transport

3.4.2.1. General effects

Over the years there have been many reports of various salt effects on electron transport measured with broken chloroplast preparations (e.g. refs. 35,55,71,76,155). This is not surprising since certain ions like Mn^{2+} and Cl^- are required for normal rates of electron transport [17,23,24,61, 67,73] while the establishment of the high energy state [35,111] and photophosphorylation [55,76,155] depend on ions in the suspending media. Many of these salt effects arise because of pretreatments given to the isolated preparations or because of the non-physiological nature of the suspending media. Nevertheless, they do serve to emphasise the close relationships between the ionic environment of the thylakoids and the electron transport reactions which take place on them. Rather than cover all the various reports of ionic and osmotic effects on electron transport I will confine myself to those which may be involved in the control of energy distribution to the two photosystems and related to the chlorophyll fluorescence observation reported above in section 3.4.1.

3.4.2.2. Partial PS1 and PS2 reactions

If changes in cation levels regulate the State 1–State 2 transition then it should be possible to observe cation induced changes in the quantal efficiencies of PS1 and PS2 mediated electron transfer reactions. The fluorescence studies outlined in section 3.4.1. suggest that cations inhibit intersystem exciton transfer between PS2 and PS1 and/or alter the distribution of quanta in favour of PS2. Thus under limiting light conditions the addition of cations should increase the quantal efficiency of PS2 at the expense of decreasing PS1 efficiency. In many cases this anticipated result has been observed but there are some exceptions. In Table I, I have listed the observations of a number of investigators. It should be said that in most cases, except where stated, it was found that monovalent alkali metal cations were as effective as Mg^{2+} or Ca^{2+} but, like cation-sensitive fluorescence, required much higher concentrations (above 100 mM for full effect).

As Table I shows in all cases except one, it has been found that Mg^{2+} increases the quantal efficiency of electron donation from H_2O to a PS2 artificial electron acceptor. On the other hand Mg^{2+} addition has been found to decrease the PS1-mediated flow of electrons from reduced DCPIP to methyl viologen although there are some reports of no effects of cations on this reaction. At first sight the situation with PS1 and PS1 + PS2 mediated $NADP^+$ reduction is more complex. Some workers have reported a stimulation by Mg^{2+} [16,101,148,161] and some an inhibition [9,101,120]. As

TABLE I

Effect of Mg^{2+} on electron transport

Reaction	Illumination wavelength (nm)	(Mg) Added (mM)	+Mg/ —Mg	Reference
A. PS2 + PS1				
H_2O — $NADP^+$	650—700[a]	3.0	1.4 to 2.0	[101]
	530—640	5.0	2.00	[16] (Table X)
	696	5.0	2.30	[16] (Table XI)
	530—645	5.0	1.50	[148]
H_2O — MeV	650	3.0	1.00	[101]
	690	3.0	1.37	
B. PS2				
H_2O — FeCN	640	2.0	1.72	[9]
	650	3.0	1.00	
	690	3.0	1.20	[101]
	710	3.0	1.20	
H_2O — DCPIP	480	3.0	1.12	[120]
	695	3.0	1.38	
	647	7.5	1.42	[19]
	695	7.5	1.35	
H_2O — phenylene-diamine	638	2.5	1.30	[78]
C. PS1 (+DCMU)				
$DCPIPH_2$ — MeV	640	2.0	0.82	[9]
	647	7.5	0.67	[19]
	695	7.5	0.85	
	650	3.0	0.80	[101]
	710	3.0	1.00	
	638	2.5	1.00	[78]
$DCPIPH_2$ — $NADP^+$	480	3.0	0.71	[120]
	640	3.0	0.74	[9]
	650—700[a]	3.0	1.4—2.0	[101]
	650[b]	3.0	0.80	
	530—640	5.0	1.05	[16] (Table X)
	696	5.0	1.77	[16] (Table XI)

a Stimulation also seen with 30 mM monovalent cations.
b Measured in presence of 30 mM monovalent cations.

far as one can judge at present the controversy can be partly explained by the observations of Marsho and Kok [101] and Harnischfeger and Shavit [58a]. They have shown that the ferredoxin-linked reduction of NADP has an additional cation requirement which is probably due to an increase in ferredoxin activity. With a 3 mM Mg^{2+} addition the reduction of NADP involving PS1 or both photosystems was stimulated in agreement with the observations from other laboratories [16,148]. However, a similar stimulation could be induced by relatively low concentrations of monovalent cations (30 mM) and contrasts with the high levels of monovalent cations required for the State 1→State 2 transition. Marsho and Kok found that if 30 mM K^+ was initially introduced into the medium then a further addition of 3 mM Mg^{2+} decreased the quantum yield for the PS1-dependent reduction of NADP (see Table I). In addition to this Bose [16] has found that the effect of Mg^{2+} on the PS1-mediated reduction of NADP to be pH sensitive with a stimulation below pH 7.6. while above this value there was an inhibition.

Even bearing these effects in mind, additional care must be taken in interpreting the results obtained for cation induced changes in the rate of electron transport through both photosystems. In the case of the partial reactions, no matter what wavelengths are used for excitation and as long as the intensity is rate-limiting, the effect of cations can be directly interpreted as changes in quantal efficiency of the particular reaction. However, with the open system the situation is more complex. If, for example, the wavelength used for irradiation is preferentially absorbed by PS2 and cation addition inhibits spillover from PS2 to PS1, then the overall reaction is inhibited since PS1 becomes even more rate-limiting. On the other hand, with PS1 rich light, cation addition will, by inhibiting spillover, increase PS2 efficiency at the expense of PS1 and thus stimulate the overall reaction. It is conceivable that in this case the gain in PS2 efficiency could be exactly balanced by the decrease in PS1 efficiency in which case no overall change will occur in the rate of electron transport through both photosystems (see also ref. 101). Because of the problems mentioned above of using $NADP^+$ as an electron acceptor, the effect of cations of electron flow through both photosystems awaits critical investigation. Marsho and Kok [101] have, however, used the H_2O to methyl viologen reaction and found no cation effect at 650 nm but a stimulation of the rate at 690 nm (see Table I). These results can be interpreted by the above arguments if 690 nm and to a lesser extent 650 nm light are preferentially absorbed by PS1 in the absence of cations. The fact that spillover seems to occur even in far red light led Marsho and Kok to favour a mechanism of control which involves changes in α rather than k_t since under these conditions the PS2 traps are open and k_p would be expected to compete very effectively with k_t.

3.4.2.3. Redox pool between PS2 and PS1

Studies of the rate of photooxidation of P700 by PS1 light, after a pre-illumination with non-saturating PS2 light, can be used to estimate the extent of reduction of the intermediate redox pool between PS2 and PS1 [100]. Using this procedure, Marsho and Kok [101], found that addition of low levels of Mg^{2+}, Ca^{2+} or higher levels of monovalent cations increased the extent of reduction of the intermediate pool (see Fig. 3.4) giving support to the concept that cations increase PS2 efficiency.

3.4.2.4. Effect of cations on transfer between PS2 units

In section 3.4.1.3. it was reported that the addition of cations seems to increase exciton transfer between PS2 units as indicated from the kinetic analyses of the fast fluorescence induction. A similar analysis is possible by comparing the rate of oxygen evolution (V_{O_2}) as a function of the fraction of open traps denoted by Q.

In this case the expression is

$$V_{O_2} = \frac{Q}{1-\rho\,(1-Q)} \qquad (7)$$

where, as in Eqn. 6, ρ is the probability of exciton transfer between PS2 units. In the original experiments of Joliot et al. [82] with broken chloroplasts suspended in a cation-containing medium (100 mM K^+) and also with dark-adapted *Chlorella* cells, the values of ρ were found to be 0.5 to 0.6. Recently, Marsho and Kok [101], used Eqn. 7 to investigate the effect of cations on ρ. Using broken spinach chloroplasts and the flash-induced oxygen yield technique for measuring open trap concentration under a particular continuous light condition, they found as shown in Fig. 3.5 that ρ = 0.55 in the presence of 3 mM Mg^{2+} but decreased to zero in the absence of

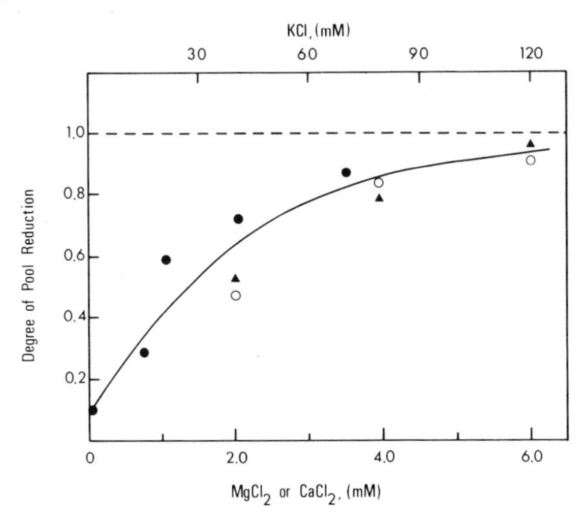

Fig. 3.4. Influence of cations on the reduction of the intermediate electron carrier pool between PS2 and PS1 when isolated broken spinach chloroplasts are illuminated by weak 650 nm light. The chloroplasts were suspended in varying concentrations of $MgCl_2$ (solid circles), $CaCl_2$ (triangles) or KCl (open circles) (reproduced with permission from ref. 101).

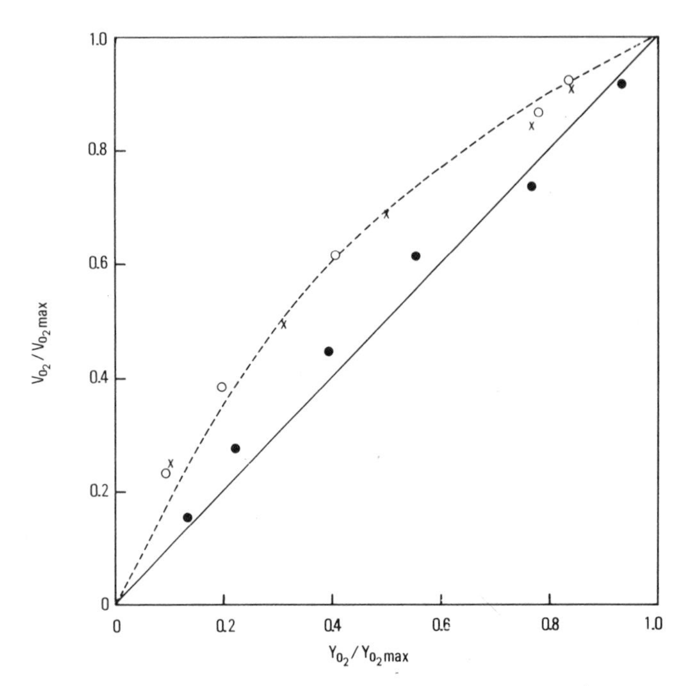

Fig. 3.5. The effect of $MgCl_2$ on the relative rates (V_{O_2}/V_{O_2max}) vs. relative flash yield (Y_{O_2}/Y_{O_2max}) of O_2 evolution from broken spinach chloroplasts. The reaction medium contained 50 mM tris (pH 7.5), 10 mM KCl, no electron acceptor and 3 mM $MgCl_2$ (crosses), 3 mM $MgCl_2$ + 100 mM KCl (open circles) and no additional cations (closed circles). Solid line corresponds to ρ = 0 and dashed line ρ = 0.55; see Eqn. 7 (reproduced with permission from ref. 101).

this cation. This change in the value of ρ could also be induced by concentrations of K^+ in excess of 100 mM. This experiment has been repeated by Bennoun [10] who measured ρ = 0.54 in the presence of 8 mM Mg^{2+} but found that in the absence of this cation ρ = 0.22.

Thus both fluorescence and oxygen yield studies indicate that the probability of transfer between PS2 units is controlled by the presence of cations in the medium.

3.4.2.5. Enhancement

Essentially enhancement is the increase in photosynthetic quantal efficiency of PS2 or PS1 when PS1 or PS2 light is superimposed respectively (for a complete description of enhancement see ref. 126). It is a consequence of the difference in the absorption spectra for PS1 and PS2 and the push-pull nature of the Z scheme (see Chapter 2). Enhancement can be detected by separately measuring the rates of photosynthesis in light preferentially absorbed by PS1 and PS2 and comparing their sum with the enhanced rate measured with the combined PS1 and PS2 lights. Alternatively, as explained in section 3.3.2. enhancement studies can be conveniently carried out by using a modulated PS2 or PS1 light beam and superimposing a non-modulated PS1 or PS2 beam, respectively, and following the changes in the modulated signal using a lock-in amplifier. This method gives a direct measure of any increase in the quantal efficiency of the reactions giving rise to the modulated signal [81].

Until 1972 there was some confusion in the literature regarding the ability of broken chloroplasts to show enhancement. Some workers reported enhancement [4,13,50,150] and others did not [46,103,129,151] and it was not until the work of Sun and Sauer [161] and Sinclair [156] that it became clear that it was necessary to have a certain level of cations in the medium before enhancement could be detected. By following $NADP^+$ reduction spectroscopically Sun and Sauer found that the addition of Mg^{2+} in excess of 3 mM to broken spinach chloroplast preparations changed them from a non-enhancing to an enhancing condition. Using the modulated oxygen electrode technique of Joliot [81], Sinclair also showed the same Mg^{2+} dependence for enhancement and also demonstrated that the effect could be observed with high concentrations of K^+ (see Fig. 3.6). As ex-

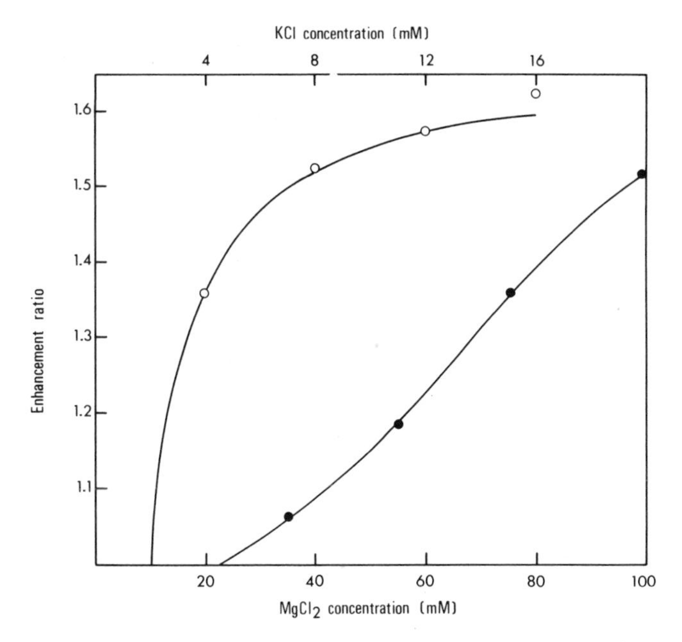

Fig. 3.6. Dependence of enhancement on the concentration of $MgCl_2$ (open circles) and KCl (closed circles). Measurements made with broken spinach chloroplasts suspended in 33 mM Tris-NaOH, pH 8.2, 3.3 mM $NADP^+$ and 3.3 mM KH_2PO_4. The basic illumination was a modulated 650 nm light beam and enhancement was detected as an increase in the modulated signal produced by superimposing constant intensity 710 nm light (Redrawn and reproduced from ref. 156).

plained in the legend of Fig. 3.6, Sinclair measured the modulated oxygen evolution signal brought about by electron donation from H_2O to $NADP^+$ using modulated PS2 light (650 nm) and non-modulated PS1 light (710 nm). Marsho and Kok [101] have used the same technique as Sinclair except they used methyl viologen as the electron acceptor and modulated PS1 light (720 nm). As expected, under these conditions, non-modulated PS2 light (650 nm) caused enhancement of the modulated oxygen signal but the effect was greatest in the presence of 6 mM $MgCl_2$ in accordance with Mg^{2+}-induced increase in the quantal efficiency of the PS2 photoreaction.

These enhancement studies are consistent with the effect of cations on chlorophyll fluorescence and electron transport discussed above. They show that the addition of Mg^{2+} increases the quantal efficiency of PS2 relative to PS1. Unfortunately, at present there are no detailed quantitative comparisons of all these effects but, in my opinion, the bulk of the qualitative evidence supports a model in which the presence of low concentrations of divalent cations or high concentrations of monovalent cations increases the efficiency of PS2 at the expense of PS1. Jennings and Forti [77] have cast doubts on a simple relationship between the various effects based on different concentration requirements. However their conclusion may not be correct since as discussed later (section 3.7.) the presence of a high-energy state may dramatically change the concentration requirement for a particular cation effect. It is usually assumed, as in the discussion above, that the action of cations is to decrease the spillover from PS2 to PS1. However, there is an alternative argument that cations increase PS1 to PS2 transfer. Sun and Sauer [160,161] prefer this latter model and although energetically less likely it is difficult to disprove experimentally. Whether cations decrease PS2 to PS1 or increase PS1 to PS2 transfer, the overall effect is to increase the efficiency of PS2 at the expense of PS1 and it seems likely that it is this cation control of quantal distribution which is in some way responsible for the in vivo State 1—State 2 shift.

3.4.4. Thylakoid stacking and conformational changes

If the addition of cations to broken chloroplasts can affect the distribution of quanta to the two photosystems and bring about a change from State 2 to State 1 then the question is raised as to the mechanism involved. Clearly conformational changes of the thylakoid membranes could alter interactions between PS1 and PS2 pigments. In fact there are many reports of osmotic and non-osmotic effects on thylakoid organisation and structure [34,52,118,141]. Izawa and Good [72] and also Ohki et al. [139a] observed that when broken chloroplasts were isolated into low salt media the thylakoids become unstacked and loosely held together. Addition of salts induced stacking like that observed in the intact organelle. Murakami and

Packer [117] correlated these stacking changes with changes in light scattering and chlorophyll fluorescence. They and others [11,66] have considered that this reorganisation of stacking could be the controlling mechanism for the State 1—State 2 transitions although more recent work on in vivo systems does not support this [163,170] (but also see ref. 11).

Variations in the cation composition of the suspending medium not only affect thylakoid stacking but also the volume of the intrathylakoid space and the thickness of the thylakoid membrane itself. When Mg^{2+} is added to broken chloroplasts suspended in a medium of low ionic strength there is a partial reduction in the intrathylakoid volume at neutral pH but not at low pH. This is because on lowering the pH of the suspending medium to about 5.0 the chloroplasts undergo even further shrinkage [34,115] which also results in a reduction of the thylakoid membrane thickness by 20 to 30% [116]. As discussed in section 3.2. and in detail by Murakami et al. [118] this pH-induced shrinkage, which is also brought about by light-induced H^+ pumping [114], can be explained by protonation of fixed negative charges within the intrathylakoid compartment coupled to the displacement of osmotically active counterions, suggested to be Mg^{2+} in section 3.2. In particular, as argued by Murakami et al. [118], protonation of negative sites in the thylakoid membrane could increase the membrane hydrophobicity and bring about a decrease in its thickness. Although as yet this decrease in intrathylakoid volume and membrane thickness has not been directly correlated either with State 1—State 2 shifts or with the various cation-induced effects outlined in sections 3.4.1. and 3.4.2., it has been shown that the Mg^{2+} induced chlorophyll fluorescence yield increase is totally inhibited at about pH 5.0 which is the isoelectric point for the above pH-induced shrinkage [34,106]. This indicates that the Mg^{2+}-sensitive sites responsible for the State 1—State 2 transition become protonated at low pH. It has also been shown that illumination of chloroplasts in vivo not only results in shrinkage of the intrathylakoid space as mentioned earlier (section 3.2.) but also that the thylakoid membranes move closer together and become thinner [141,115]. Under certain conditions it has been found that in *Porphyra* the membrane thickness decreased from 6.4 ± 0.8 nm to 4.9 ± 0.4 nm while in another marine alga, *Ulva*, the thickness changed from 7.0 ± 0.9 nm to 5.3 ± 0.7 nm [115]. Bearing in mind that the thylakoids are the sites of the primary photochemical reactions then it is not difficult to imagine that a light or pH induced reduction in membrane thickness of about 20% would cause changes in the photochemical properties of the light-harvesting pigments. It is not clear at present whether the thickness change is due to water loss from the membrane or simply results from cation-induced conformation changes of macromolecules within or on the membrane and there is a report of no membrane thickness changes occurring at all in vivo [105]. Nevertheless it seems very probable that it is the cation and pH-induced microstructural changes in the thylakoid membrane and not gross morphological changes associated with chloroplast volume and thylakoid reorganisation which are primarily responsible for the State 1—State 2 transitions [110,168,170].

3.4.5. *Antagonism of monovalent and divalent cations on spillover*

An interesting observation which may be pertinent to the in vivo mechanism controlling the State 1—State 2 interconversions was initially made by Gross and Hess [53] and subsequently studied by others [54,168,180]. In a sense the main observation was an extension of Murata's earlier work. Gross and Hess found that when DCMU-treated chloroplasts were suspended in a sugar medium of high osmotic strength containing no cations the fluorescence was already high and far less sensitive to divalent cations. Addition of 2 to 10 mM monovalent cations such as K^+, Na^+ or tetraethylammonium, decreased the fluorescence level which could then be restored to the high level by addition of low concentrations of Ca^{2+} or Mg^{2+} or high concentrations of K^+ or Na^+. This antagonism of low concentrations of monovalent and divalent cations on fluorescence yield is shown in Fig. 3.7.

When Gross and Hess measured the emission spectra at 77°K for the various ionic conditions they observed relative intensity shifts in the PS1 chl a_1 735 nm peak and the PS2 chl a_2 685 and 695 nm peaks which indicated that the change from the high to the low fluorescence state was accompanied by an increase in spillover between PS2 and PS1. It seems that low concen-

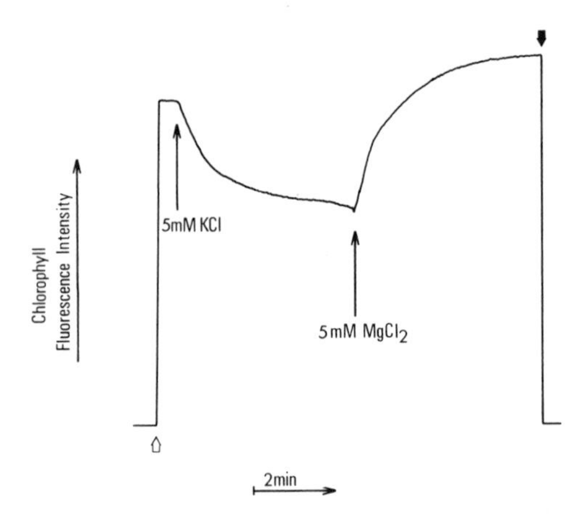

Fig. 3.7. Antagonism of monovalent and divalent cations on chlorophyll fluorescence yield measured with broken spinach chloroplasts suspended in 0.33 M sorbitol brought to pH 8.0 with Tris base. 5 mM KCl and 5 mM $MgCl_2$ were injected into the suspension as shown. Fluorescence was excited with a broad band blue-green actinic beam (80 kerg/cm^2 s) and the emission measured at right angles at 686 nm (Mills, Telfer and Barber, unpublished).

trations of monovalent cations increase spillover while divalent cations reverse this process. In a further study Gross and Prasher [54] reported that both structural and stacking phenomena also showed monovalent/divalent cation antagonism. In particular, they showed that under conditions of high chlorophyll fluorescence, i.e. with no cations in the medium or in the presence of divalent cations, the grana were stacked. However, lowering the fluorescence yield by additon of low concentrations of monovalent cations caused unstacking. These results contrast with the earlier work reported above [72,117] where unstacking occurred in "low ionic strength" medium. Closer inspection of the conditions used in these cases indicate that both osmotic and monovalent cation effects can account for the unstacking observed. Although these changes in the organisation of the thylakoid are of interest they are, as pointed out above, unlikely to be primarily involved in the State 1—State 2 shifts. Such drastic changes in stacking apparently do not occur in vivo [170] and attempts to correlate the kinetics of light scattering changes with the antagonistic cation-induced fluorescence changes do not implicate gross morphological changes of the thylakoids with the mechanism involved [168].

3.4.6. Cation binding properties

If the control of spillover is related to subtle cation-induced microconformational changes in the thylakoid membrane then it is necessary to investigate these changes at the molecular level. Glutaraldehyde fixation inhibits these structural changes and also the cation induced fluorescence effects [77,110] and at one time it was thought that conformational changes in the ATPase system controlled the state of spillover [110] although this has been shown not to be the case [77]. Gross and her colleagues [28, 145] have recently made a serious start at trying to elucidate the cation binding properties of the thylakoids. Using isolated broken spinach chloroplasts they have shown that there are two divalent cation binding sites. Site 1 binds cations to the extent of 0.65 μmoles/mg chl with a dissociation constant of 8 μM while Site 2 binds divalent cations to the extent of 0.5 μmoles/mg chl and has a dissociation constant of 51 μM. The latter site seems to control both structural and fluorescence changes and can also be occupied by monovalent cations [145]. Prochaska and Gross [145] have shown that these binding sites are inhibited after treating chloroplasts with 1-ethyl-3-(3-dimethylaminopropyl) carbodiimide hydrochloride plus a suitable nucleophile, such as glycine ethyl ester. This suggests that the cation-sensitive binding sites are either carboxyl groups or the negative groups of the membrane phospholipids. Gross and her colleagues favour carboxyl groups since studies with $^{45}Ca^{2+}$ indicate that 80% of the label is extractable as chloroplast protein [29]. Moreover, Davis and Gross [28] have found that the protein—pigment complex which represents 50% of the membrane protein and is thought to be the main component of the light-harvesting chlorophyll complex (called light-harvesting chlorophyll *a/b*-protein complex by Thornber [165] and denoted as Chl LH in Fig. 3.9) is able to effectively bind Ca^{2+} and in so doing undergoes association (binds 9.5 μmoles Ca^{2+}/mg chl with Kd = 32 μM).

If the addition of divalent cations to chloroplasts results in binding to the carboxyl groups of membrane proteins then there should also be a loss of cations to counter balance the charge exchange. Bose [16] has shown that Mg^{2+} binding to broken spinach chloroplasts is associated with H^+ displacement. As can be seen in Fig. 3.8 maximum proton displacement occurred on adding 5 mM $MgCl_2$ to chloroplasts suspended in an unbuffered medium at pH 6.5 and amounted to 600 nmoles H^+/mg chl.

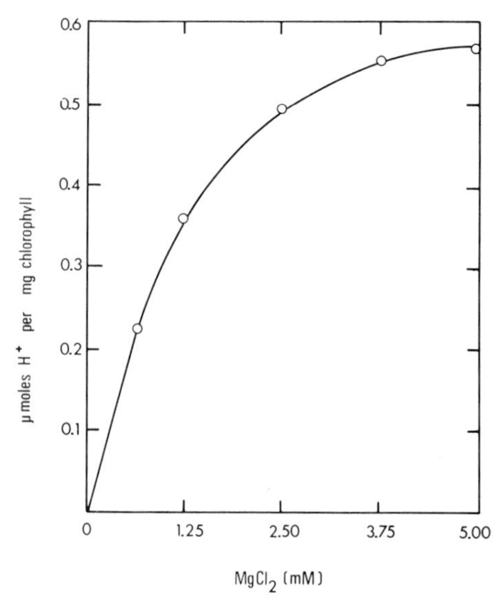

Fig. 3.8. Proton extrusion in the dark from broken spinach chloroplasts on the addition of various amounts of $MgCl_2$. The initial pH was 6.5. and the experiment was conducted at 12°C (Fig. 6 from ref. 16, reproduced with permission).

3.4.7. Conclusions

In this section, I have outlined many different types of studies with isolated broken chloroplasts which indicate that cation induced structural changes of the thylakoid membranes, possibly involving the main light harvesting protein-pigment complex, bring about the State 2→State 1 transition. Although it has been considered that this transition is due to changes in the equilibrium constant for the electron transfer reactions between PS2 and PS1 [32] most evidence points to a direct control on quantal delivery to the two photosystems [172]. The direction and mechanism of transfer is still to some extent under discussion, but it seems likely that cations exert control both on the initial distribution of quanta (through changes in α in Eqn. 5) and on the yield of energy transfer from PS2 to PS1 (by causing changes in k_t in Eqn. 2). Both effects may be the consequence of the same conformational change of the membrane and the degree of redistribution will partly be determined by the state of the traps [22]. Cations also seem to regulate the extent of transfer between PS2 units. This could also result from changes in spatial separation of adjacent PS2 units due to conformational changes or simply be explained by changes in the extent of exciton transfer through the bulk light-harvesting chl a_2 as a consequence of changes in spillover to PS1 (i.e. an increase in k_t shortens the lifetime of the excited bulk chl a_2 molecules and therefore decreases the exciton diffusion path length [69]). It now remains, in the following sections, to consider how cation-induced conformational changes can alter energy transfer between PS2 and PS1 pigment molecules and reconcile this with our knowledge of ionic control within the intact chloroplast and with the in vivo State 1–State 2 shifts.

3.5. PHOTOCHEMICAL MODEL

There are several possible ways in which energy may be transferred between adjacent pigment molecules within the photosynthetic units [15]. The most widely accepted mode of transfer, however, is inductive resonance or resonance transfer [38]. The probability of transfer (P_t) by this process is a function of distance and orientation of the adjacent molecules as well as the degree of overlap of their absorption and fluorescence spectra. The appropriate equation for this type of transfer is given below. It was derived by Forster [45] and initially applied to photosynthetic systems by Duysens [37] and more recently by Knox [86].

$$P_t = \frac{Af(\phi\theta)}{\tau_0 R^6} \int_\nu \frac{\epsilon(\nu) \cdot f(\nu) d\nu}{\nu^4} \quad (8)$$

where $\epsilon(\nu)$ is the extinction coefficient of the acceptor, $f(\nu)$ is the spectral distribution of emission of the donor, R is the distance between the donor and acceptor molecules, A is characteristic of the environment and dependent on the refractive index, $f(\phi\theta)$ is a function of the mutual orientation of the molecules and τ_0 is the intrinsic fluorescence lifetime of the donor. For transfer between chlorophyll *a* this process becomes efficient when R is less than 7 nm. In fact for the light-harvesting in vivo chlorophylls the distance between adjacent molecules is probably about 2 nm indicating that resonance transfer could act as a very efficient mechanism of delivering quanta to the reaction centres [38].

Chemical treatments and separation procedures of thylakoid membranes have yielded 3 major pigmented fractions [2,165,174]; Chl LH, which is a light-harvesting chlorophyll *a*/*b*-protein complex serving both photosystems, and PS1 and PS2 complexes. The latter two complexes contain antenna chlorophyll chl a_1 and chl a_2 and their respective reaction centre traps P700 and P680. The three low temperature emission bands at 685, 695 and 730 nm have been attributed to chl a_2, Chl LH and chl a_1, respectively [22]. A simple model incorporating these three components is shown in Fig. 3.9 [22]. Bearing in mind the above equation (Eqn. 8) it is not difficult to imagine that changes in conformation could alter distances and/or orientations of chlorophyll molecules at the interfaces of the three complexes and thus change the degree of transfer between them.

A more detailed model of the possible inter-relationships between the light harvesting chlorophylls has been proposed by Seely [153,154]. He has taken into account the fact that there are at least six different types of chl a species in vivo C663, C670, C678, C685, C693 and C700 and produced a planar model array of these chlorophylls which allows for efficient energy transfer to the reaction centres and for changes in spillover of energy from PS2 to PS1. His model is shown in Fig. 3.10 in which the long wavelength chlorophyll-absorbing forms are associated mainly with PS1 while the shorter-wavelength forms serve both photosystems. He has assumed chlorophyll *b* transfers its energy preferentially to PS2 and that energy is efficiently transferred to the longer-wavelength forms of chlorophyll and finally to the reaction centres P680 and P700. Direction of transfer to one or other of the reaction centres is governed by the molecular orientation as shown by arrows in Fig. 3.10. The total array consists of 344 chlorophyll molecules arranged in various sized sets to give the minimum number of transfers for a high probability of trapping. The model also incorporates a channel (sets 13,14

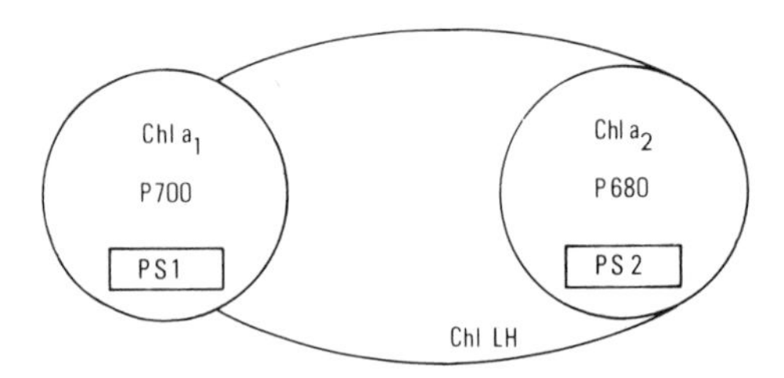

Fig. 3.9. A model given in ref. 22 of the possible interactions of the three major pigment—protein complexes isolated from chloroplasts [165]. The PS1 and PS2 complexes each contain appreciable amounts of antenna chlorophyll, chl a_1 and chl a_2 and their respective reaction centre traps, P700 and P680. The two photosystem complexes are separated by a light-harvesting chlorophyll complex, Chl LH, which can transfer excitation energy to either of the two photosystem complexes. Chl LH is the light-harvesting chlorophyll *a*/*b*-protein complex isolated by Thornber [165].

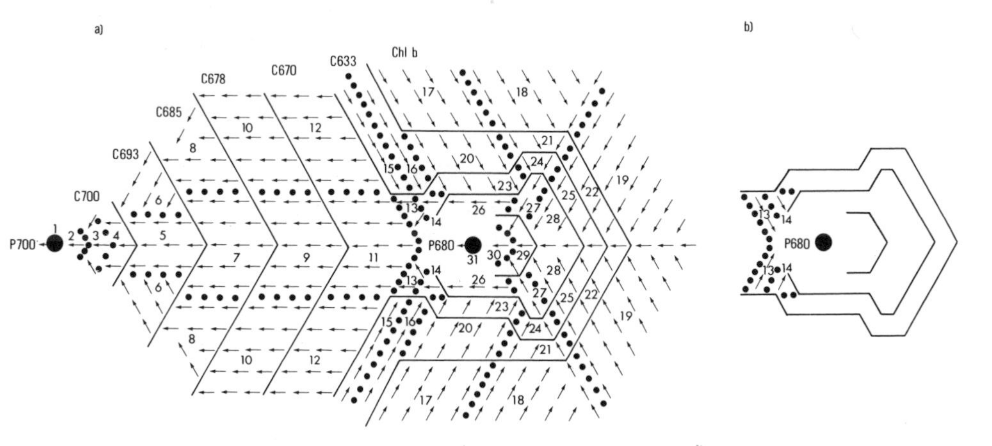

Fig. 3.10. (a) Model array of a photosynthetic unit, with the traps for PS1 (P700) and PS2 (P680). Arrows indicate directions of molecule transition moment vectors, numbers designate sets. Regions of different spectral chlorophyll *a* types are separated by solid lines; different sets of the same spectral type are separated by dotted lines. (b) Shows the reorientation of the transition moment vectors of the six chlorophylls in sets 13 and 14 which would correspond to an increase in energy transfer of 16% to P680 at the expense of a decrease in transfer to P700 (reproduced with permission from ref. 154).

and 26) for passage of energy between PS2 and PS1. By changing the orientation of the six chlorophyll molecules in sets 13 and 14 in the way shown in Fig. 3.10b the transfer of energy to the PS2 trap (P680) is increased by 16% as a result of a decrease of transfer to PS1. Recognising that Seely's model will need modification as more details of the molecular structure of the in vivo pigment complexes become known it does serve to illustrate how a change in orientation of only a small number of chlorophyll molecules may affect the distribution of energy to the two photosystems.

3.6. EFFECT OF THE HIGH-ENERGY STATE ON CHLOROPHYLL FLUORESCENCE

Since Strehler [159] first proposed a relationship between photophosphorylation and in vivo fluorescence several workers have demonstrated that the slow smt phase of the induction (see section 3.3.) is sensitive to uncouplers of phosphorylation [49,142]. Studies into this relationship using isolated chloroplasts were initiated by Murata and Sugahara [122] who found that PMS-stimulated coupled cyclic electron flow brought about a lowering of the chlorophyll fluorescence yield of about 20 to 30%. This decline, which was independent of the PS2 trap conditions since the chloroplasts had been treated with DCMU, was reversed by darkness or by the addition of uncouplers such as CCCP and methylamine. Wraight and Crofts

[179] observed similar HES fluorescence quenching using DAD as the cofactor for cyclic electron flow. They also reported HES quenching with non-cyclic electron flow and concluded that the fluorescence lowering was due to the net uptake of H^+ into the intrathylakoid space and reflected the establishment of ΔpH across the thylakoid membrane. Although this view has been given support by others [26,110] it has been difficult to relate these observations with the in vivo State 1—State 2 shifts or with the fluorescence changes induced on addition of metal cations to isolated chloroplasts. Murata and Sugahara [122] had concluded from low temperature fluorescence emission studies that the HES quenching observed with PMS was not related to the cation-induced changes in spillover between PS2 and PS1. The HES quenching has also been reported to be independent of the presence of cations in the medium [26,110]. However, the dismissal of an interrelationship between the HES and cation-induced fluorescence changes may be incorrect. It is questionable if the low temperature fluorescence measurements are meaningful for the dynamic HES quenching process since the gradients involved could have been partially dissipated before measurement. Moreover, the lack of Mg^{2+} sensitivity on HES quenching reported by Mohanty et al. [110] and Cohen and Sherman [26] may simply reflect the poor response of their particular preparations to this cation. Perhaps of more importance is the problem that compounds like PMS may induce chlorophyll fluorescence quenching by more direct interaction with the pigment bed and that this process could be enhanced by the establishment of a pH gradient across the thylakoid membrane (for a complete description of the various ways PMS can quench in vivo chlorophyll fluorescence see ref. 143). Such an effect could explain the low temperature fluorescence observations of Murata and Sugahara [122] which suggested to them that PMS-induced quenching was independent of spillover.

Mills and Barber [106] have reinvestigated the cation requirement for HES quenching. Using DAD and DCMU treated chloroplasts they have shown that with low concentrations of the cofactor the uncoupler sensitive quenching of fluorescence only occurs in the presence of Mg^{2+} or high concentrations of K^+ when initial fluorescence is high (see Fig. 3.11). A similar requirement for cations was found for DCMU treated chloroplasts when the ΔpH was created by electron donation from reduced DAD or DCPIP to methyl viologen [107]. On the other hand when reduced TMPD was used as PS1 electron donor to methyl viologen no uncoupler-sensitive fluorescence quenching was observed which is consistent with the inability of this electron flow system to create ΔpH [107]. Uncoupler-sensitive fluorescence quenching was observed even in the absence of divalent cations when the concentration of the electron flow mediators was increased [107]. However, the quenching was more dependent on the concentration of the cofactor than on the pH gradient induced, again suggesting that these cofactors interact directly with the chlorophyll molecules of the pigment bed.

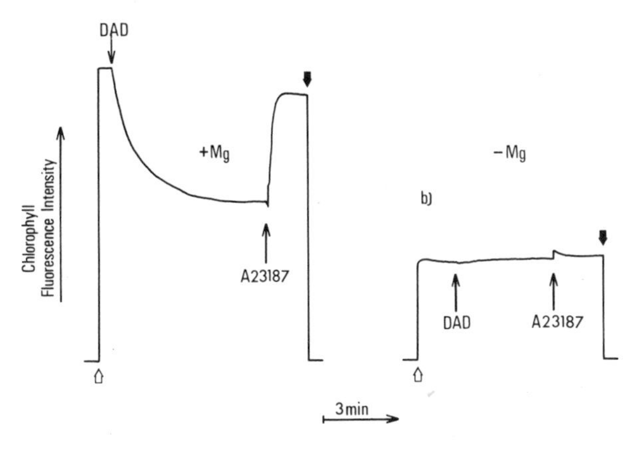

Fig. 3.11. DAD induced high energy state quenching of chlorophyll fluorescence in presence of 10 μM DCMU and showing the requirement for Mg^{2+}. Isolated broken pea chloroplasts suspended in 0.33 M sorbitol, 10 mM HEPES brought to pH 7.6 with Tris base, Additions made were 17 μM DAD and 10 μM A23187 in presence and absence of 10 mM Mg^{2+}. A23187 is, in the presence of Mg^{2+}, an uncoupler of electron flow [7]. Lighting conditions as for Fig. 3.7 (Mills and Barber, unpublished).

In view of the results and arguments presented in the above sections and also in the following section (section 3.7.) it would be logical to accept that the HES quenching is due to the protonation of internal negative groups on the thylakoid membrane coupled to the displacement of Mg^{2+} or some other cation from those sites. The cation independent quenching observed with high concentrations of cyclic electron flow mediators may be due to additional quenching processes which are enhanced by the presence of a ΔpH across the thylakoid membranes as suggested by the work of Papageorgiou [143].

3.7. INTACT CHLOROPLASTS AND RECONSTITUTED SYSTEMS

3.7.1. Chlorophyll fluorescence properties of isolated intact chloroplasts

Although as outlined in sections 3.4. and 3.6. isolated broken chloroplasts show cation and HES-sensitive chlorophyll fluorescence changes which are independent of the redox state of the PS2 trap, they had not, until recently been found to show the slow fluorescence induction comparable to that seen in vivo and attributed to the State 1→State 2 transitions. A direct correlation of the slow in vivo fluorescence changes and the cation/HES effects has come from studies with isolated Class A spinach chloroplasts. These intact chloroplasts show slow fluorescence changes similar to those observed with the leaves from which they were isolated [5,87]. The main characteristics of the slow fluorescence decay observed on illuminating a dark pretreated suspension of intact chloroplasts are shown in Fig. 3.12. The rate of quenching was found to be dependent both on the rate of electron flow and on the lag period between turning on the illumination and observing O_2 evolution [5, 8] and, although not shown in Fig. 3.12 sometimes showed a shoulder or slight "bump" similar to the sm phase in Fig. 3.2 during the initial stages of the decay. By using OAA, which unlike PGA and CO_2 does not require ATP for its reduction, it has been shown that the slow quenching can be independent of the redox state of the PS2 trap [5]. Addition of DCMU relieves the quenched state at rates comparable with the dark reversal ($t\frac{1}{2}$~1.5 min) except for an initial fast phase which probably reflects rapid DCMU closing of the few PS2 traps that are open [5,8]. That this quenching is dependent on the HES was shown by the ability of uncouplers such as nigericin and CCCP to induce a relatively rapid ($t\frac{1}{2}$~2 s) return from the low to the high fluorescence state [5,8,88]. It has been further shown that the light-induced fluorescence quenching is due to the build up of the ΔpH component of the HES [8] and that this process results in a decrease in enhancement in accordance with a shift from State 1 to State 2 [176]. After treating intact chloroplast with DCMU a similar light-induced uncoupler-sensitive quenching can be catalysed by low concentrations of DAD as shown in Fig. 3.13.

There is also a similarity between the fast induction curves measured with cells and with intact isolated chloroplasts [10,69,79]. Using Eqn. 6 Hipkins and Barber [69] found for DCMU treated intact chloroplasts that the probability of transfer between PS2 units (ρ) is 0.63±0.05 in the high

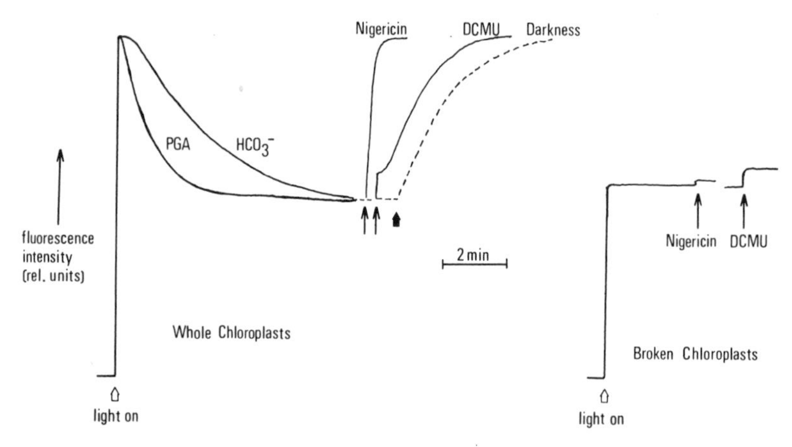

Fig. 3.12. An illustration of some of the properties of the slow fluorescence changes observed on illuminating a suspension of dark-pretreated isolated intact spinach chloroplasts. The experiments were carried out in 3 ml of medium containing 0.33 M sorbitol and 50 mM HEPES adjusted to pH 7.6 with KOH with a chlorophyll concentration of about 30 μg/ml. The medium also contained either 5 mM HCO_3^- or 1 mM PGA and the injections were 0.1 μM nigericin of 3.3. μM DCMU. The broken chloroplasts were obtained by osmotic shock in 1.5 ml H_2O followed by the addition of 1.5 ml of the above medium at double strength. The illumination and measurement as for Fig. 3.7.

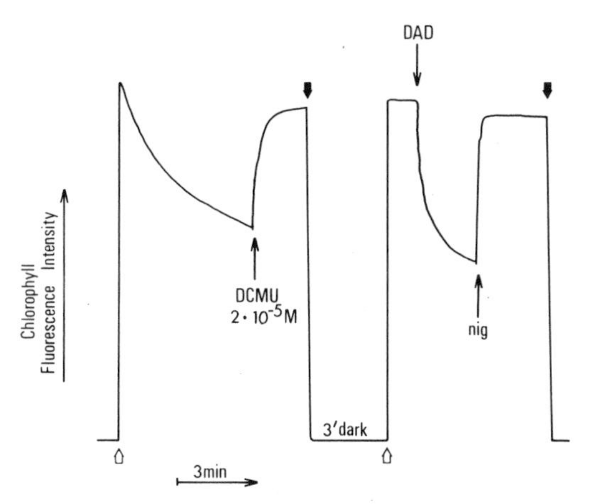

Fig. 3.13. Comparison of chlorophyll fluorescence quenching in isolated intact pea chloroplasts induced by electron flow to carbon dioxide (before addition of 20 μM DCMU) or by DAD-mediated cyclic electron flow (after DCMU addition). The additions were 1.3 · 10^{-7} M nigericin (nig) and 40 μM DAD and the chloroplasts were suspended in the medium given in Fig. 3.12 but also containing 10 mM $NaHCO_3$ and 1 mM ribose-5-phosphate. Other conditions as for Fig. 3.12. [164a].

fluorescence state but after light-induced quenching decreased to 0.38±0.07. Using Chlorella cells Bennoun [10] carried out the same experiment and found that ρ decreases from 0.63 in State 1 to 0.45 in State 2.

3.7.2. Reconstituted systems

As shown in Figs. 3.12 and 3.14, when intact chloroplasts are subjected to osmotic shock to remove their outer membranes and resuspended in a low cation-containing medium, the chlorophyll fluorescence yield decreases to the low level and the slow induction effect is lost. These broken chloroplasts were unable to reduce CO_2 or PGA but did show a slow well coupled endogenous electron flow of the Mehler type [88] and were able to establish a pH gradient across the thylakoid [88]. Apparently removal of the outer chloroplast membranes allows release of cations since addition of low concentrations of Mg^{2+} or high concentrations of K^+ to these envelope-free chloroplasts restored the high yield and subsequent light-induced fluorescence quenching. As can be seen in Fig. 3.14 a dark pretreatment was necessary before the cation effect could be observed and the fluorescence from this "reconstituted" system showed similar properties to that seen with intact chloroplasts. These results link together the HES and cation-induced fluorescence changes reported in earlier sections of this chapter. They suggest that in vivo the high fluorescence state (State 1) exists when there are sufficient metal cations in the intrathylakoid space. On initiation of the

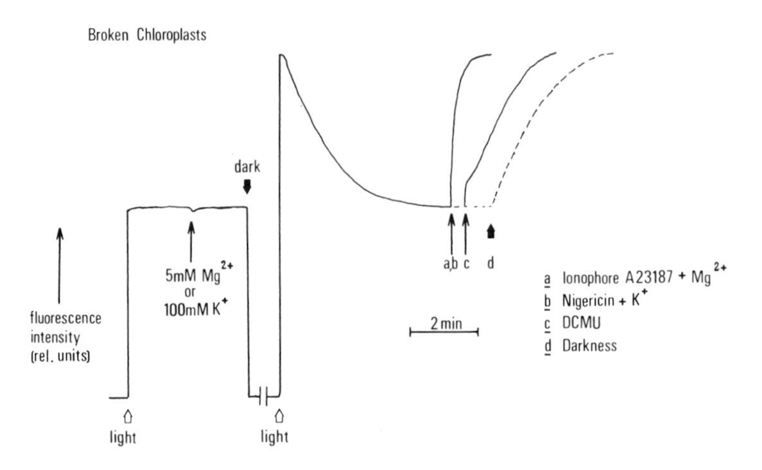

Fig. 3.14. An illustration of some of the properties of cation-dependent fluorescence changes. The method of preparing the broken chloroplasts, concentrations of nigericin and DCMU and conditions of measurement are the same as Fig. 3.12. Ionophore A23187 was added to give a final concentration of 1 μg/ml (i.e. 2 μM).

proton pump by illumination, metal cations are displaced from inside the thylakoid compartment by H^+, which are then effluxed into the stroma and the fluorescence is quenched to the State 2 level. This would account for the ability of cations to induce the high fluorescence state in the dark but not in the light when a ΔpH exists, and incidentally, emphasises the problem of comparing the concentration requirements of the various cation-induced effects mentioned in previous sections of this chapter where some have, and some have not, been measured in the presence of HES (see arguments in ref. 77). Another important observation to emerge from studies with reconstituted "chloroplasts" was made by Krause [89]. He had found that with intact chloroplasts there seemed to be a correlation between the slow fluorescence changes and low angle light scattering changes (measured as absorbance at 535 nm). However, as shown in Fig. 3.15a this correlation was fortuitous since the cation-induced scattering change occurred in the light when no fluorescence change is observed. When the experiment was carried out at low light intensity or in the presence of uncoupler (Fig. 3.15b) when no ΔpH existed, then the scattering and fluorescence changes showed essentially the same kinetics and cation requirement. Thus it seems that metal cations binding to the outer surface of the thylakoid membrane induce the scattering change while cation binding to the internal surface controls the fluorescence yield changes. The situation is probably more complicated than this and we will return again to the problem of "sidedness" in section 3.7.4.

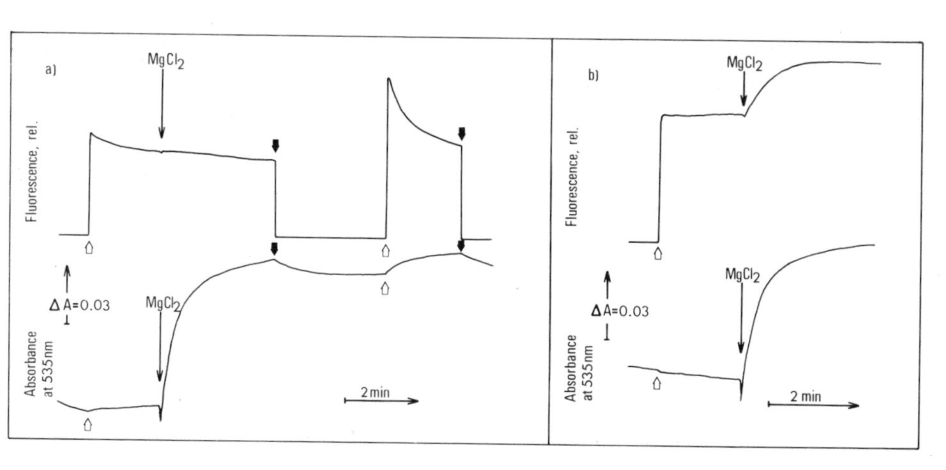

Fig. 3.15. (a) Effect of Mg^{2+} on chlorophyll fluorescence and apparent absorbance of broken spinach chloroplasts. Light on, upward arrows; light off, downward arrows. 5 mM $MgCl_2$ was injected in the light as indicated and was accompanied by an absorbance but not a fluorescence change. (b) As (a) except that 1 μM FCCP was added before the start of illumination (reproduced with permission from ref. 89).

3.7.3. The nature of the in vivo cation

The cation requirement for the reconstituted fluorescence quenching has been found to be identical [163] with that originally reported by Murata and others (see section 3.4.11.) for the cation-induced fluorescence increase with DCMU-poisoned broken chloroplasts. In an attempt to determine the nature of the cation involved in the in vivo State 1—State 2 shifts a number of different cation-selective ionophores have been employed [6—8,89,106, 164]. The experiments were conducted with intact isolated spinach chloroplasts suspended in a medium containing no metal cations. Under these conditions they were still able to reduce CO_2, PGA and OAA and showed the slow light-induced fluorescence quenching. It was found that ionophores which require monovalent cations for their operation, like nigericin and monensin, were ineffective both at reversing the light-induced fluorescence quenching and at uncoupling electron flow until small amounts of K^+ or Na^+ were introduced into the medium [6,8]. On the other hand, the divalent requiring ionophore A23187 did reverse the fluorescence quenching [6—8] and uncouple electron flow without the addition of divalent cations to the medium [164]. This ionophore does not select between Ca^{2+} and Mg^{2+} but further experiments with the Ca^{2+} requiring ionophore beauvaricin indicated that Mg^{2+} but no Ca^{2+} was available for the uncoupling action of A23187 within the intact organelle [6]. Another series of experiments based on the same idea of using cation-selective ionophores also supported the above conclusions but this time the experiments were conducted with

DCMU-treated intact chloroplasts showing DAD catalysed light-induced chlorophyll fluorescence quenching [106]. From a slightly different approach, Krause came to the same conclusion that divalent cations were most probably involved in the in vivo fluorescence yield change [89].

These results are consistent with the arguments presented in section 3.2. that within the intact chloroplast the light-induced uptake of H^+ across the thylakoid membrane is balanced by Mg^{2+} efflux and further suggests that it is this cation exchange which drives the in vivo State 1--State 2 transitions.

3.7.4. Sidedness of the cation effects

The above observations suggest that the light-induced changes in chlorophyll fluorescence yield are controlled by Mg^{2+}/H^+ binding sites on the inner side of the thylakoid membrane. It is not difficult to visualize that the displacement of Mg^{2+} by H^+ at these binding sites causes a pigment-protein conformational change which maximises PS2 to PS1 spillover and generates the low fluorescence state. However, the situation may be more complex than this [108] since the rate of rise in the fluorescence yield induced by adding metal cations to DCMU-treated broken chloroplasts is never as fast as the rise induced when the HES is dissipated by uncouplers (i.e. the rate of reversal of the light-induced quenching by injection of uncoupler into a reconstituted chloroplast suspension). This difference is not due to differences in rate of transport of cations across the membrane to the internal binding sites since the rate of rise in the absence of a HES cannot be stimulated by addition of appropriate ionophores [108]. Moreover, with broken chloroplasts the high fluorescence state can be induced by the addition of the non-permeant monovalent cation, choline [108]. These results suggest that external sites on the thylakoid membranes can also control the conformational changes which alter the fluorescence yield [108]. It now seems that cation binding is required on both sides of the membrane to create the high fluorescence state (State 1) and that their removal from the outer surface by osmotic shock and resuspension in low salt media, or from the inner surface by H^+ pumping, induces the low fluorescence state (State 2). Thus, the kinetics of the cation-induced fluorescence rise in the absence of a HES (e.g. when pretreated with DCMU) is not controlled by the membrane permeability but by the external binding sites while the fast kinetics observed when the HES is dissipated are controlled by the exchange of H^+ for metal cation at sites on the inner surface of the thylakoids.

3.7.5. Conclusions

Clearly studies with isolated Class A intact chloroplasts are helping to bridge the gap between observations made with intact cells and with isolated broken chloroplasts. I hope, without too much bias, that the conclusions drawn from these studies can be sensibly reconciled with many of the various observations made with broken chloroplast preparations in drawing up an overall picture of the possible mechanism of cation control of the State 1—State 2 transitions.

It seems that State 1 exists when there is maximum Mg^{2+} binding to the thylakoid membrane. Evidence is emerging that this Mg^{2+} binding is to carboxyl groups on the light harvesting chlorophyll *a/b* pigment—protein complex which represents the bulk of the membrane protein and light-harvesting chlorophylls. It could be that this pigment—protein complex spans the entire membrane exposing cation binding sites on both sides. Such a model has recently been suggested by Anderson [1,2]. Removal of Mg^{2+} from either side causes a conformational change which decreases the fluorescence yield of chl a_2 and enhances the delivery of excitons to PS1 at the expense of PS2. In the intact organelle, the shift from State 1 to State 2 is probably due only to the displacement of Mg^{2+} from the inner sites by the light induced proton pump.

Mg^{2+} binding to the external sites almost certainly plays an important role in thylakoid stacking and explains the close correlation between Mg^{2+}-induced stacking and fluorescence changes observed with broken chloroplasts suspended in low salt media. Moreover this correlation again implicates the light harvesting chlorophyll *a/b*-protein since the presence of this complex seems to be required for stacking to occur (cf. refs. 2,165). In the intact organelle, however, the changes in fluorescence do not correlate with stacking changes and further support the concept that it is the inner thylakoid cation binding sites which normally govern the State 1—State 2 transitions.

The independence of chlorophyll fluorescence changes and thylakoid stacking has recently been verified from electron microscopy studies of broken chloroplasts. It was found that unlike fluorescence, the ability of Mg^{2+} to induce thylakoid stacking was independent of the existence of the HES. Moreover no significant change in stacking was found to occur on illuminating isolated intact chloroplasts although the fluorescence decreased in the usual way [164b].

The above model is consistent with the observations and arguments presented in the earlier sections of this chapter. Since light of wavelengths below 680 nm is preferentially absorbed by PS2 [43,175] it is usually advantageous to have a shift from State 1 to State 2. This optimisation in the distribution of incoming quanta would maximise the rate of electron transport and associated proton pumping and thus create a large enough

ΔpH to maintain the State 2 condition. However, in far red light, where PS1 is preferentially excited, the situation may favour a shift towards State 1. According to the scheme presented in Chapters 2 and 7 (Figs. 2.2 and 7.28) PS1 cannot pump H^+ without the cooperation of PS2 unless cyclic electron flow occurs via the plastoquinone pool. If no PS1 proton pumping occurs then the ΔpH would drop and spillover from PS2 to PS1 would be minimised in an attempt to optimise non-cyclic electron flow. Unfortunately the picture cannot be this simple since experimentally it is possible to reverse the State 1 to State 2 shift by simply superimposing PS1 light on light preferentially absorbed by PS2 [14]. With the above model it is hard to explain how PS2 and PS1 can show this antagonism. The antagonistic effects of monovalent and divalent cations as shown by Gross and colleagues [53,54] may be important in this respect, but it seems unlikely that the ionic status of the stroma can be compared with the non-ionic media used to demonstrate the monovalent-divalent cation antagonism. It could be that the key to the switch between State 1 and State 2 is directly through the extent of the HES and/or the ATP level in the stroma. It is, however, difficult to know just how these two parameters vary under different conditions. Superimposing PS1 light on PS2 light could result in an increase in the stromal ATP level due to PS1 mediated cyclic phosphorylation and that the State 2 to State 1 transition seen under these conditions is brought about by pumping Mg^{2+} into the intrathylakoid space against the gradient induced by the proton pump [8]. There is as yet no evidence for an ATP dependent Mg^{2+} pump of this sort on the thylakoid membranes and in any case the picture is made more complex by the fact that ATP will tend to chelate the stromal Mg^{2+}.

Obviously we still do not know all the subtleties of the regulatory mechanism which drive the State 1—State 2 transitions. There is a need to gain more understanding of the possible role of ion transport processes across the outer membranes and to know how light affects the activity coefficients of the stromal cations. As yet it is not clear whether the State 1—State 2 changes occur only in the granal regions and that the stromal lamellae are not involved (see ref. 3 giving evidence for and against the existence of both PS1 and PS2 on stromal lamellae). Moreover, it has been strongly argued that Mg^{2+} plays the major role as a counterion for the H^+ pump and is involved in the cation induced control of spillover. It is difficult to reconcile this with the high K^+ content of chloroplasts and I would at this stage of our knowledge be unhappy to dismiss the importance of this cation in the controlling photochemical processes within the intact chloroplast.

3.8. FINAL COMMENTS

To some extent I have throughout this chapter tended to favour Mg^{2+}/H^+ exchange as an important process in controlling quantal distribution to the two photosystems. In my opinion there is a great deal of evidence to favour such a relationship but I am also very conscious that many of the comparisons I have made between different effects have been qualitative. There is a serious need to undertake a far more rigorous and quantitative approach to the concept of cation control of spillover of the type discussed above.

In the end such measurements should be done with "physiological" systems and isolated intact chloroplasts are ideal for this purpose. As with studies into the process of CO_2 fixation as reported in Chapter 7, isolated intact chloroplasts have the advantage in that the simple removal of the outer membranes can be compensated by "reconstitution" of the suspending media to a condition which mimics the stroma. The role of the outer chloroplast membranes and stromal ion levels in controlling and regulating primary reactions of photosynthesis should not stop at studies related to the State 1—State 2 phenomenon but also be extended to both structural and electron transport aspects of photosynthesis with the overall aim of increasing our knowledge of how these processes may function in the intact cell.

3.9. ACKNOWLEDGEMENTS

I would like to thank my colleagues, Alison Telfer, Jennifer Nicolson, Geof Searle, Mike Hipkins and John Mills for their invaluable discussions and help during the preparation of this chapter.

REFERENCES

1 Anderson, J.M. (1975) Nature, [253], 536—537.
2 Anderson, J.M. (1975) Biochim. Biophys. Acta, [416], 191—235.
3 Arntzen, C.J. and Briantais, J.M. (1975) in Bioenergetics of Photosynthesis (Govindjee, ed) pp. 51—111. Academic Press, New York.
4 Avron, M. and Ben-Hayyim, G. (1969) in Progress in Photosynthesis Research (Metzner, H. ed), Vol. 3. pp. 1185—1196. IUBS, Tübingen.
5 Barber, J. and Telfer, A. (1974) in Membrane Transport in Plants (Dainty, J. and Zimmerman, U., eds), pp. 281—288. Springer, Berlin.
6 Barber, J., Mills, J. and Nicolson, J. (1974) FEBS Letters, [49], 106—110.
7 Barber, J., Telfer, A. and Nicolson, J. (1974) Biochim. Biophys. Acta, [357], 161—165.
8 Barber, J., Telfer, A., Mills, J. and Nicolson, J. (1974) in Proc. 3rd Int. Cong. Photosynthesis (Avron, M., ed), Vol. 1, pp. 53—63. Elsevier, Amsterdam.
8a Beddard, G.S., Porter, G., Tredwell, C.J. and Barber, J. (1975) Nature, [258], 166—168.

9 Ben-Hayyim, G. and Avron, M. (1971) Photochem. Photobiol., [14], 389—396.
10 Bennoun, P. (1974) Biochim. Biophys. Acta, [368], 141—147.
11 Bennoun, P. and Jupin, H. (1974) in Proc. 3rd Int. Cong. Photosynthesis (Avron, M., ed) Vol. 1, pp. 163—169, Elsevier, Amsterdam.
12 Berg. S., Dodge, S., Krogmann, D.W. and Dilley, R.A. (1974) Plant Physiol., [53], 619—627.
13 Bishop, P.M. and Whittingham, C.P. (1963) in Studies on Microalgae and Photosynthetic Bacteria. (Ashida, J., ed). pp. 291—296. Univ. of Tokyo Press, Tokyo.
14 Bonaventura, C. and Myers, J. (1969) Biochim. Biophys. Acta, [189], 366—383.
15 Borisov, A. Yu and Godik, V.I. (1973) Biochim. Biophys. Acta, [301], 227—248.
16 Bose, S.K. (1975) Ph. D. Thesis, University of Rochester, U.S.A.
17 Bove, J.M., Bove, C., Whatley, F.R. and Arnon, D.I. (1963) Z. Naturforsch., [18b], 683—688.
18 Briantais, J.M., Merkelo, H. and Govindjee (1972) Photosynthetica, [6], 133—141.
19 Briantais, J.M., Vernotte, C. and Moya, I. (1973) Biochim. Biophys. Acta, [325], 530—538.
20 Butler, W.L. (1972) Proc. Natl. Acad. Sci. USA, [69], 3420—3422.
21 Butler, W.L. and Kitajima, M. (1975) Biochim. Biophys. Acta, [376], 116—125.
22 Butler, W.L. and Kitajima, M. (1975) Biochim. Biophys. Acta, [396], 72—85.
23 Cheniae, G.M. (1970) Ann. Rev. Plant Physiol., [21], 467—498.
24 Cheniae, G.M. and Martin, I.F. (1969) Biochim. Biophys. Acta, [153], 819—837.
25 Clayton, R.K. (1969) Biophys. J., [9], 60—76.
26 Cohen, W.S. and Sherman, L.A. (1971) FEBS Letters, [16], 319—323.
27 Crofts, A.R., Deamer, D.W. and Packer, L. (1967) Biochim. Biophys. Acta, [131], 97—118.
28 Davis, D.J. and Gross, E.L. (1975) Biochim. Biophys. Acta, [387], 557—567.
29 Davis, D.J. and Gross, E.L. quoted in Prochaska, L.J. and Gross, E.L. (1975) Biochim. Biophys. Acta, [376], 126—135.
30 Deamer, D.W. and Packer, L. (1969) Biochim. Biophys. Acta, [172], 539—545.
32 Delrieu, M.J. (1972) Biochim. Biophys. Acta, [256], 293—299.
33 Dilley, R.A. (1971) Current Topics in Bioenergetics. (Sanadi, D.R. ed.), Vol. 4. pp. 237—271. Academic Press, New York.
34 Dilley, R.A. and Rothstein, A. (1967) Biochim. Biophys. Acta, [135], 427—443.
35 Dilley, R.A. and Shavit, N. (1968) Biochim. Biophys. Acta, [162], 86—96.
36 Dilley, R.A. and Vernon, L.P. (1965) Arch. Biochem. Biophys., [111], 365—375.
37 Duysens, L.N.M. (1952) Ph.D. Thesis, Utrecht.
38 Duysens, L.N.M. (1964) Prog. Biophys. Mol. Biol., [14], 1—104.
39 Duysens, L.N.M. (1972) Biophys. J., [12], 858—863.
40 Duysens, L.N.M. and Sweers, H.E. (1963) in Studies on Microalgae and Photosynthetic Bacteria. (Ashida, J., ed), pp. 353—372. University of Tokyo Press, Tokyo.
41 Duysens, L.N.M. and Talens, A. (1969) in Progress in Photosynthesis Research. (Metzner, H.H., ed), Vol. 2. pp. 1073—1081, University of Tübingen, Tübingen.
42 Duysens, L.N.M., den Haan, G.A. and van Best, J.A. (1974) in Proc. 3rd Int. Cong. Photosynthesis. (Avron, M., ed), Vol. 1. pp. 1—12. Elsevier, Amsterdam.
43 Emerson, R. (1958) Ann. Rev. Plant Physiol., [9], 1—24.
44 Forbush, B. and Kok, B. (1968) Biochim. Biophys. Acta, [162], 243—253.
45 Forster, Th. W. (1948) Ann. Phys., [2], 55—75.
46 Gibbs, M., Fewson, C.A. and Schulman, M.D. (1963) Carnegie Inst. Wash. Year Book, [62], 352—375.
47 Gimmler, H., Schäfer, G. and Heber, U. (1975) in Proc. 3rd Int. Cong. Photosynthesis (Avron, M., ed), Vol. II, pp. 1381—1392. Elsevier, Amsterdam.

48 Gläser, M., Wolff, C. and Renger, G. (1975) 5th Int. Biophys. Congr., Copenhagen, p.57 (Abstr.).
49 Govindjee and Papageorgiou, G. (1971) in Photophysiology (Giese, A.C., ed), Vol. VI, pp. 1—46, Academic Press, New York.
50 Govindjee, R., Govindjee and Hoch, G. (1964) Plant Physiol., [39], 10—14.
51 Greville, G.D. (1969) in Current Topics in Bioenergetics (Sanadi, D.R., ed.), Vol. 3, pp. 1—78, Academic Press, New York.
52 Gross, E.L. and Packer, L. (1967) Arch. Biochem. Biophys., [121], 779—789.
53 Gross, E.L. and Hess, S. (1973) Arch. Biochem. Biophys., [159], 832—836.
54 Gross, E.L. and Prasher, S.H. (1974) Arch. Biochem. Biophys. [164], 460—468.
55 Gross, E.L., Dilley, R.A. and San Pietro, A. (1969) Arch. Biochem. Biophys., [134], 450—462.
56 Hall, D.O. (1972) Nature, [235], 125—126.
57 Haan, G.A. den, Warden, J.T. and Duysens, L.N.M. (1973) Biochim. Biophys. Acta, [325], 120—125.
58 Haan, G.A. den, Duysens, L.N.M. and Egberts, D.J.N. (1974) Biochim. Biophys. Acta, [368], 409—421.
58a Harnischfeger, G. and Shavit, N. (1974) FEBS Letters, [45], 286—290.
59 Harvey, M.J. and Brown, A.P. (1967) Biochem. J., [105], 30—31.
60 Harvey, M.J. and Brown, A.P. (1969) Biochim. Biophys. Acta, [172], 116—125.
61 Heath, R.L. and Hind, G. (1969) Biochim. Biophys. Acta, [189], 222—233.
62 Heber, U. (1969) Biochim. Biophys. Acta, [180], 302—319.
63 Heber, U. and Santarius, K.A. (1970) Z. Naturforsch., [25b], 718—728.
64 Heber, U. and Krause, G.H. (1971) in Photosynthesis and Photorespiration (Hatch, M.D., Osmond, C.B. and Slatyer, R.O. eds), pp. 218—223, Wiley-Interscience, New York.
65 Heldt, H.W., Werden, K., Milovancev, M. and Geller, G. (1973) Biochim. Biophys. Acta, [314], 224—241.
66 Hind, G. and McCarty, R.E. (1973) in Photophysiology (Giese, A.C., ed), Vol. VIII, pp. 113—156, Academic Press, New York.
67 Hind, G., Nakatani, H.Y. and Izawa, S. (1969) Biochim. Biophys. Acta, [172], 277—289.
68 Hind, G., Nakatani, H.Y. and Izawa, S. (1974) Proc. Natl. Acad. Sci. USA, [71], 1484—1488.
69 Hipkins, M.F. and Barber, J. unpublished.
70 Homann, P. (1969) Plant Physiol., [44], 932—936.
71 Izawa, S. and Good, N.E. (1966) Plant Physiol., [41], 533—543.
72 Izawa, S. and Good, N.E. (1966) Plant Physiol., [41], 544—552.
73 Izawa, S., Heath, R.L. and Hind, G. (1969) Biochim. Biophys. Acta, [180], 388—398.
74 Jagendorf, A.T. (1975) in Bioenergetics of Photosynthesis (Govindjee, ed.), pp. 413—492, Academic Press, New York.
75 Jagendorf, A.T. and Hind, G. (1963) in Photosynthetic Mechanisms of Green Plants, Natl. Acad. Sci-Nat. Res. Counc. Publ., [1145], 599—610.
76 Jagendorf, A.T. and Smith, M. (1962) Plant Physiol., [37], 135—141.
77 Jennings, R.C. and Forti, G. (1974) Biochim. Biophys. Acta, [347], 299—310.
78 Jennings, R.C. and Forti, G. (1974) Plant Sci. Letters, [3], 25—33.
79 Jennings, R.C. and Forti, G. (1975) Biochim. Biophys. Acta, [396], 63—71.
80 Joliot, A. and Joliot, P. (1964) C. R. Acad. Sci. Paris, [258], 4622—4625.
81 Joliot, P. and Joliot, A. (1968) Biochim. Biophys. Acta, [153], 625—634.
82 Joliot, P., Joliot, A. and Kok, B. (1968) Biochim. Biophys. Acta, [153], 635—652. 652.

83 Kautsky, H., Appel, W. and Amann, H. (1960) Biochem. Z., [332], 277—292.
84 Kishimoto, U. and Tazawa, M. (1965) Plant Cell Physiol., [6], 507—528.
85 Kitajima, M. and Butler, W. (1975) Biochim. Biophys. Acta, [376], 105—115.
86 Knox, R.S. (1975) in Bioenergetics of Photosynthesis (Govindjee, ed.), pp.183—221, Academic Press, New York.
87 Krause, G.H. (1973) Biochim. Biophys. Acta, [292], 715—728.
88 Krause, G.H. (1974) Biochim. Biophys. Acta, [333], 301—313.
89 Krause, G.H. (1974) in Proc. 3rd Int. Cong. Photosynthesis (Avron, M. ed.), pp.1021—1030, Elsevier, Amsterdam.
90 Larkum, A.W.D. (1968) Nature, [218], 447—449.
91 Larkum, A.W.D. and Hill, A.E., (1970) Biochim. Biophys. Acta, [203], 133—138.
92 Lavorel, J. (1959) Plant Physiol., [34], 204—209.
93 Lavorel, J. and Joliot, P. (1972) Biophys. J., [12], 815—831.
94 Li, Y.S. (1975) Biochim. Biophys. Acta, [376], 180—188.
95 Lin, D.C. and Nobel, P.S. (1971) Arch. Biochem. Biophys., [145], 622—632.
96 Malkin, S. (1967) Biophys. J., [7], 629—649.
97 Malkin, S. and Kok, B. (1966) Biochim. Biophys. Acta, [126], 413—432.
98 Malkin, S. and Siderer, Y. (1974) Biochim. Biophys. Acta, [368], 422—431.
99 Mar, T., Govindjee, Singhal, G.S. and Merkelo, H. (1972) Biophys. J., [12], 797—808.
100 Marsho, T.V. and Kok, B. (1970) Biochim. Biophys. Acta, [223], 240—250.
101 Marsho, T.V. and Kok, B. (1974) Biochim. Biophys. Acta, [333], 353—365.
102 Mauzerall, D. (1972) Proc. Natl. Acad. Sci. USA, [69], 1358—1362.
103 Mayne, B.C. and Brown, A.H. (1963) in Studies on Microalgae and Photosynthetic Bacteria (Ashida, J., ed), pp. 347—352, University of Tokyo Press, Tokyo.
104 Menke, W. (1940) Z. Physiol. Chem., [263], 104—106.
105 Miller, M.M. and Nobel, P.S. (1972) Plant Physiol., [49], 535—541.
106 Mills, J. and Barber, J. (1975) Arch. Biochem. Biophys., [170], 306—314.
107 Mills, J. and Barber, J. (1975) unpublished.
108 Mills, J., Telfer, A. and Barber, J. (1975) unpublished.
109 Mitchell, P. (1966) Biol. Rev., [41], 445—502.
110 Mohanty, P., Braun, B.Z. and Govindjee (1973) Biochim. Biophys. Acta, [292], 459—476.
111 Molotkovsky, Y.G. and Dzyubenko, V.S. (1968) Nature, [219], 496—498.
112 Moya, I. (1974) Biochim. Biophys. Acta, [368], 214—227.
113 Munday, J.C. and Govindjee (1969) Biophys. J., [9], 1—35.
114 Murakami, S. and Packer, L. (1969) Biochim. Biophys. Acta, [180], 420—423.
115 Murakami, S. and Packer, L. (1970) Plant Physiol., [45], 289—299.
116 Murakami, S. and Packer, L. (1970) J. Cell Biol., [47], 332—351.
117 Murakami, S. and Packer, L. (1971) Arch. Biochem. Biophys., [146], 337—347.
118 Murakami, S., Torres-Pereira, J. and Packer, L. (1975) in Bioenergetics of Photosynthesis (Govindjee, ed), pp.555—618, Academic Press, New York.
119 Murata, N. (1969) Biochim. Biophys. Acta, [172], 242—251.
120 Murata, N. (1969) Biochim. Biophys. Acta, [189], 171—181.
121 Murata, N. (1971) Biochim. Biophys. Acta, [226], 422—432.
122 Murata, N. and Sugahara, K. (1969) Biochim. Biophys. Acta, [189], 182—192.
123 Murata, N., Nishimura, M. and Takamiya, A. (1966) Biochim. Biophys. Acta, [120], 23—33.
124 Murata, N., Nishimura, M. and Takamiya, A. (1966) Biochim. Biophys. Acta, [126], 234—243.
125 Murata, N., Tashiro, H. and Takamiya, A. (1970) Biochim. Biophys. Acta, [197], 250—256.

126 Myers, J. (1971) Ann. Rev. Plant Physiol., [22], 289—312.
127 MacRobbie, E.A.C. (1964) J. Gen. Physiol., [47], 859—877.
128 MacRobbie, E.A.C. (1970) Quart. Rev. in Biophys., [3], 251—294.
129 MacSwain, B.D. and Arnon, D.I. (1968) Proc. Natl. Acad. Sci. USA, [61], 989—996.
130 Neumann, J. and Jagendorf, A.T. (1964) Arch. Biochem. Biophys., [107], 109—119.
131 Nobel, P.S. (1967) Plant Physiol., [42], 1389—1394.
132 Nobel, P.S. (1968) Plant and Cell Physiol., [9], 499—509.
133 Nobel, P.S. (1968) Plant Physiol., [43], 781—787.
134 Nobel, P.S. (1968) Biochim. Biophys. Acta, [153], 170—182.
135 Nobel, P.S. (1969) Biochim. Biophys. Acta, [172], 134—143.
136 Nobel, P.S. (1975) in Ion Transport in Plant Cells and Tissues (Baker, D.A. and Hall, J.L. eds), pp.101—124, North Holland, Amsterdam.
137 Nobel, P.S., Chang, D.T., Wang, C., Smith, S.S. and Barcus, D.E. (1969) Plant Physiol., [44], 655—661.
139 Okayama, S. and Butler, W.L. (1971) Biochim. Biophys. Acta, [267], 523—529.
139a Ohki, R., Kuneida, R. and Takamiya, A. (1971) Biochim. Biophys. Acta, [226], 144—153.
140 Packer, L., Barnard, A.C. and Deamer, D.W. (1967) Plant Physiol., [42], 283—293.
141 Packer, L., Murakami, S. and Mehard, C.W. (1970) Ann. Rev. Plant Physiol., [21], 271—304.
142 Papageorgiou, G. (1975) in Bioenergetics of Photosynthesis (Govindjee, ed), pp.319—371, Academic Press, New York.
143 Papageorgiou, G. (1975) Arch. Biochem. Biophys., [166], 390—399.
144 Pflüger, R. (1973) Z. Naturforsch., [28c], 779—780.
145 Prochaska, L.J. and Gross, E.L. (1975) Biochim. Biophys. Acta, [376], 126—135.
146 Rottenberg, H. and Grunwald, T. (1972) Eur. J. Biochem., [25], 71—74.
147 Rumberg, B. and Siggel, U. (1969) Naturwissenschaften, [56], 130—132.
148 Rurainski, H.J. and Hoch, G.E. (1972) in Proc. 2nd Int. Congr. Photosynthesis Research (Forti, G., Avron, M. and Melandri, A. eds), Vol. 1, pp.133—141, Junk, The Hague.
149 Saltman, P., Forte, J.G. and Forte, G.M. (1963) Exp. Cell Res., [29], 504—514.
150 Sane, P.V. and Park, R.B. (1971) Biochem. Biophys. Res. Commun., [44], 491—496.
151 Sauer, K. and Park, R.B. (1965) Biochemistry, [4], 2791—2798.
152 Schuldiner, S., Rottenberg, H. and Avron, M. (1972) Eur. J. Biochem., [25], 64—70.
153 Seely, G.R. (1973) J. Theor. Biol., [40], 173—188.
154 Seely, G.R. (1973) J. Theor. Biol., [40], 189—199.
155 Shavit, N. and Avron, M. (1967) Biochim. Biophys. Acta, [131], 516—526.
156 Sinclair, J. (1972) Plant Physiol., [40], 778—783.
157 Stiehl, H. and Witt, H. (1969) Z. Naturforsch., [b.24.], 1588—1598.
158 Stocking, C.R. and Ongun, A. (1962) Amer. J. Bot., [49], 284—289.
159 Strehler, B.L. (1953) Arch. Biochem. Biophys., [43], 67—79.
160 Sun, A.S.K. and Sauer, K. (1971) Biochim. Biophys. Acta, [234], 399—414.
161 Sun, A.S.K. and Sauer, K. (1972) Biochim. Biophys. Acta, [256], 409—427.
162 Tazawa, M., Kishimoto, U. and Kikuyama, M. (1974) Plant and Cell Physiol., [15], 103—110.
163 Telfer, A. and Barber, J. (1976) unpublished.
164 Telfer, A., Barber, J. and Nicolson, J. (1975) Biochim. Biophys. Acta, [396], 301—309.
164a Telfer, A., Barber, J. and Nicolson, J. (1975) Plant Sci. Letters, [5], 171—176.
164b Telfer, A., Nicolson, J. and Barber, J. (1976) FEBS Letters, [65], 77—85.

165 Thornber, J.P. (1975) Ann. Rev. Plant Physiol., [26], 127—158.
166 Trebst, A. (1974) Ann. Rev. Plant Physiol., [25], 423—458.
167 Tumerman, L.A. and Sorokin, E.M. (1967) Mol. Biol. (USSR), [1], 628—638.
168 Vandermeulen, D.L. and Govindjee (1974) Biochim. Biophys. Acta, [368], 61—70.
169 Vater, J., Renger, G., Stiehl, H. and Witt, H. (1968) Naturwissenschaften, [55], 220—221.
170 Vernotte, C., Briantais, J.M., Armond, P. and Arntzen, C.J. (1975) Plant Sci. Letters, [4], 115—123.
171 Vredenberg, W.L. and Slooten, L. (1967) Biochim. Biophys. Acta, [143], 583—594.
172 Wang, R.T. and Myers, J. (1974) Biochim. Biophys. Acta, [347], 134—140.
173 Werden, K., Heldt, H.W. and Milovancev, M. (1975) Biochim. Biophys. Acta, [396], 276—292.
174 Wessels, J.S.C. and Borchert, M.T. (1975) Proc. 3rd Int. Cong. Photosynthesis (Avron, M., ed), Elsevier, Amsterdam.
175 Williams, W.P., Murty, N.R. and Rabinowitch, E. (1969) Photochem. Photobiol., [9], 455—469.
176 Williams, W.P., Solomon, Z., Maullem, A., Barber, J. and Mills, J. (1976) Biochim. Biophys. Acta, [430], 300—311.
177 Witt, H.T. (1971) Quart. Rev. of Biophys., [4], 365—477.
178 Witt, H.T. (1975) in Bioenergetics of Photosynthesis (Govindjee, ed), pp. 493—554, Academic Press, New York.
179 Wraight, C.A. and Crofts, A.R. (1970) Eur. J. Biochem., [17], 319—327.
180 Wydrzynski, T., Gross, E.L. and Govindjee (1975) Biochim. Biophys. Acta, [376], 151—161.

The Intact Chloroplast — edited by J. Barber
© Elsevier/North-Holland Biomedical Press. 1976 — Printed in The Netherlands

Chapter 4

The Coupling of Photophosphorylation to Electron Transport in Isolated Chloroplasts

D.O. HALL

University of London King's College, 68 Half Moon Lane, London SE24 9JF (Great Britain)

CONTENTS

Abbreviations: AP, aminophenol; AQS, anthraquinone sulphonate; BBMD, benzyl-α-bromomalodinitrile; BQ, benzoquinone; BZ, benzidine; C_{550}, absorbance change at 550 nm associated with PSII; Chl, chlorophyll; Cyt, cytochrome; DAB, diaminobenzidine; DAD, diaminodurene (2,3,5,6-tetramethyl-*p*-phenylenediamine); DAT, diaminotoluene; DBMIB, 2,5-dibromo-3-methyl-6-isopropyl-*p*-benzoquinone; DCMU, 3-(3,4-dichlorophenyl)-1,1-dimethylurea; DCPIP, 2,6-dichlorophenol indophenol; DHDP, dihydroxydiphenyl; DMPD, dimethyl-*o*-phenylenediamine; DMQ, dimethyl-*p*-benzoquinone; DPC, diphenyl carbazide; DPHZ, diphenylhydrazine; DSPD, disalicylidene propane diamine; EDAC, l-ethyl-3-(3-dimethyl aminopropyl)-carbodiimide; FD, ferredoxin; FeCy, ferricyanide (Fe^{3+}); Fe-S, iron-sulphur protein; FMN, flavin mononucleotide; FP, flavoprotein (FD-NADP reductase); HQ, hydroquinone; K_3, Vitamin K_3 (menadione); ΔE, redox potential difference; ΔG, phosphate potential; ΔG_0, standard free energy of reaction; $\Delta\psi$, membrane potential (component of PMF); MMPD, 2-methyl-5-methoxy-*p*-phenylenediamine; MV, methyl viologen; NQ, naphthoquinone; O_2^-, superoxide radical; PADR, phosphoadenosine diphosphate ribose; Pc, plastocyanin (in electron transport chain); PC, photosynthetic control (State 3/State 4); PD, phenylene diamine; PMF, proton motive force (electrochemical gradient); PMS, *N*-methylphenazonium methosulphate (phenazine methosulphate); PQ, plastoquinone; PSI(II), photosystem I (II); Q, primary electron acceptor PSII (quencher of fluorescence); ST, silicotungstate (Silicomolybdate also functions: Fig. 1); TCHQ, tetrachlorohydroquinone; TMB, tetramethylbenzidine; TMI, tetramethylindamine; TMPD, *N*-tetramethyl-*p*-phenylenediamine; Tris, Tris-(hydroxymethyl)aminomethane.

4.1. INTRODUCTION

How phosphorylation is coupled to electron transport has been a problem in bioenergetics for many decades. In chloroplast membranes we now have a clearer indication of this coupling mechanism, even though it is often erroneous to say that we have "solved" a scientific problem. One of the reasons for this is probably that the transport of electrons and protons, and therefore the formation of ATP, is very closely coupled to light reactions: thus in chloroplasts one can initiate electron transport just by the manual flick of a light switch, or, at much shorter times, timing mechanisms which give very short impulses of substrate, i.e. a flash of light. Thus photosynthetic systems have a distinct advantage over other bioenergetic systems and this has been applied to great advantage in both chloroplasts (plants and algae) and with chromatophores from photosynthetic bacteria.

In this chapter I shall discuss electron transport and coupled phosphorylation mainly in in vitro chloroplasts under conditions which try to emulate the in vivo chloroplast regime as closely as possible. This is obviously very difficult to accomplish but a start has to be made in this direction in order to relate in vitro biochemical type investigations to in vivo physiological conditions.

It will be seen that the structural state of the isolated chloroplast is important in determining its coupling efficiency. This may also apply in vivo but we know very little about this aspect of photophosphorylation. In addition I shall discuss the problems in trying to determine the phosphate potential in chloroplasts and how this relates to endogenous concentrations of P_i, ADP, ATP, Mg, other salts and to pH.

The proposed schemes for electron transport which show the orientation of components within the membrane and how this may be coupled to electron transport, proton translocation and ATP information, will be presented. This presupposes a scheme for photosynthetic phosphorylation which is very much akin to Mitchell's chemiosmotic hypothesis. Even this hypothesis may not be strictly correct, since conformational charges in proteins or localized proton pools within the membrane might have to be accommodated within Mitchell's general chemiosmotic hypothesis; however, for our purposes this does not matter too much, since the overall idea of a chemiosmotic hypothesis seems to form the closest approximation to the "truth" that we are able to ascertain at this stage.

There are two other chapters in this book which are directly relevant to my discussions. These are Chapter 5 by Krause and Heber on the energetics of intact chloroplasts and Chapter 3 by Barber on ionic relations of intact chloroplasts and regulations of electron flow. I will try to steer clear of overlap with these chapters, but inevitably this shall happen.

Excellent reviews on photosynthetic phosphorylation have recently been published by Trebst [147] and also chapters by Jagendorf [76], Avron [8] and Witt [169] in a recent book edited by Govindjee [46]. In addition the Proceedings of the 3rd International Congress on Photosynthesis Research (1974) has been published [7], which makes a very useful adjunct to the Proceedings of the 2nd [31] and 1st [98] International Congresses on Photosynthesis Research. These reviews and Proceedings should be consulted for much more detailed references than I am able to give.

4.2. ELECTRON TRANSPORT SCHEMES IN THE MEMBRANE

We owe most of our modern concepts on electron transport in photosynthesis in plants to the so-called Z-scheme of Hill and Bendall [66] (postulated in 1960) which proposed two-light reactions in series activating the transport of electrons from H_2O to NADP via changes in redox potentials of components of the eletron transport chain. Two-light reaction proposals of Emerson and Rabinowitch [27], Kok and Hoch [86], and Duysens et al. [24,25] have all contributed to the scheme we now use so freely (see reviews mentioned above for discussions of evidence [46,155]). The beauty of this scheme was that it provided a clear description of how electron transport might occur and was able to be tested by very many different types of experimental techniques. It has stood up remarkably well to the test of time over the last fifteen or so years and provides an explanation for most of the photosynthetic reactions which have been studied, whether in whole cells or in isolated chloroplast or sub-chloroplast particles. A linear equivalent of this scheme is shown in Fig. 4.1.

Since the adoption of Mitchell's chemiosmotic hypothesis [104] to explain the coupling of electron transport, proton translocation and ATP formation, we have been able to incorporate the Z-scheme of Hill and Bendall with Mitchell's hypothesis and this is shown in Fig. 4.2. which is based on the work for the groups of Trebst [148], Junge [80], Witt [169], and Good [75]. There is still the question of the complete acceptance of the Mitchell theory for the translocation of protons across chloroplast membranes; this is a consequence of the work of Good [75], Junge [80,81] and Kok [33a]. Their modification of the Mitchell scheme is to include a rapidly reacting proton pool within the membrane rather than considering only the proton pools on either side of the membrane (see also ref. 164). However, for our purposes there are probably no great differences between these schemes, since we are primarily going to be concerned with the stoichiometry of ATP, protons and electrons which may be unaffected by the precise localisation of the components within the membrane itself.

Without being too biased, I think it is safe to say that there are no other schemes of electron transport in chloroplast membranes which are considered

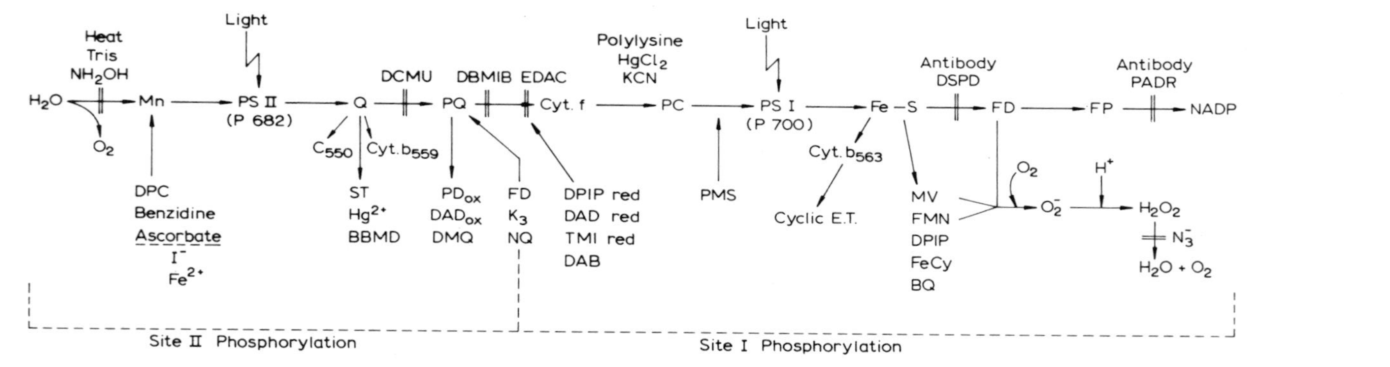

Fig. 4.1. Scheme for non-cyclic electron transport and phosphorylation. Figs. 4.1 and 4.2 derived from numerous references [8,9, 13,36,37,52,75,76,80,104,147,169].

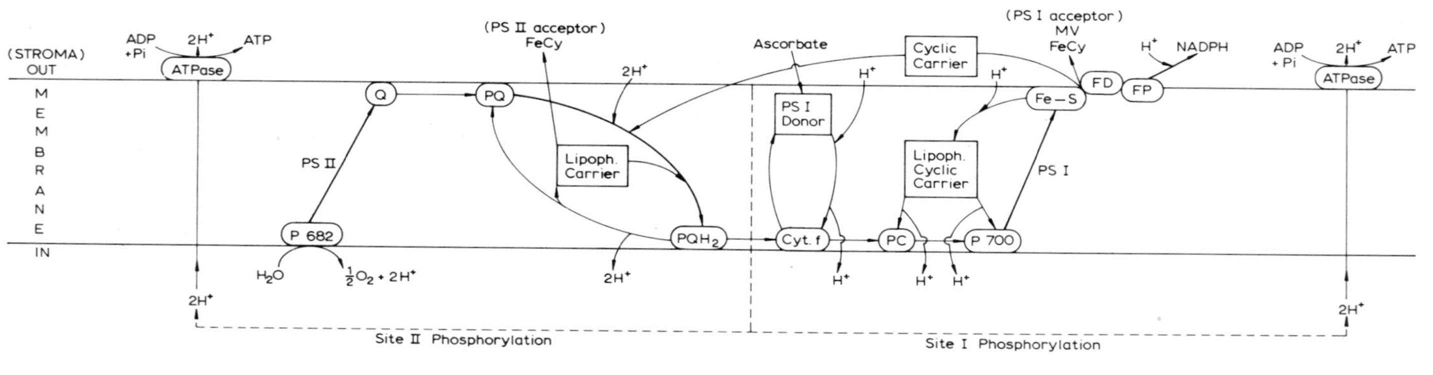

Fig. 4.2. Non-cyclic electron transport and phosphorylation in and across a chloroplast membrane.

seriously at this time. However, we may be wrong and be over-confident, but besides the conformational hypothesis which will not be discussed here (see reviews and discussion [104,164]) the author is not aware of any valid alternative scheme to explain ATP formation which is coupled to electron transport in photophosphorylation.

Since electron transport is thought to occur across membranes and proton movements to occur from one side of the membrane to the other (ultimately giving rise to ATP formation) it is obvious that the structure of the membrane is going to be important in determining the efficiency of coupling of these various mechanisms. Thus the control of the structure of the membrane is probably important and is a topic which will be touched upon briefly. It is a subject which is not very well known at present and deserves much greater recognition than it is now receiving. What we do know about are gross morphological changes which occur in chloroplasts, but we do not know very much about the microstructure changes that occur in the chloroplast membranes themselves. We also know much more about the movements of ions across membranes, whether they be protons, or counter-ions like magnesium, but again it is only recently that the influence of these ion movements on the structure of the membrane has been recognised (see review [105]).

The efficiency of ATP formation and electron transport and its coupling is an important factor in the overall process of carbon dioxide fixation. I mean here the stoichiometry of the phosphorylation coupled to the transport of electrons through the electron transport chain to reduce NADP i.e. non-cyclic photophosphorylation [3] (ratios expressed as ATP/2e ≡ P/O ≡ ADP/O i.e. μmoles ATP formed or P_i taken up, or ADP phosphorylated, per pair of electrons transported to form one μmole NADPH or to evolve one μatom oxygen). Since NADPH and ATP are required for CO_2 fixation and the stoichiometry of these requirements is well known, i.e. one NADPH + 1½ ATP are required for each carbon dioxide fixed, the importance of this coupling is quite evident. If photosynthesis is to operate efficiently, the degree of coupling between NADPH formation and ATP formation should be optimal. If this value is such that only one ATP is formed for every NADPH, then additional ATP would have to be synthesised in order to fix CO_2. However, if two ATP's are made for every NADPH then there would be sufficient ATP; in fact even more than required in order to just fix a molecule of CO_2 to the level of sugars.

4.3. THE STOICHIOMETRY OF ELECTRON TRANSPORT, PROTON EXCHANGE AND ATP FORMATION

Until a few years ago, it was not considered that more than one ATP could be made for every pair of electrons transported from water to NADPH in non-cyclic electron transport. Evidence had been provided as early as 1965 by Winget et al. [165] that this so-called ATP/2e was in fact greater than one and was likely to be two [72]. Further evidence to support this was slow in accumulating and there was much discussion between the groups that thought that there was less than or more than one ATP formed for every pair of electrons transported from water to NADP. However, in the last few years a considerable body of evidence has accumulated that this stoichiometry is greater than 1.0 and indeed may be greater than 1.5 (and in fact may be 2.0).

This implies that if the ATP/2e ratio is greater than one then there are two "sites" of ATP formation in the non-cyclic electron transport scheme; and evidence has now been provided that there are indeed two separate sites of ATP formation. Firstly there is the work of Good's group with different classes of electron acceptors [134], i.e. one type (Class III) accepting at Photosystem II and the other type (Class I) accepting at Photosystem I, each of the photosystems having its own ATP formation site. Secondly the evidence of Trebst's group [153,154] who introduced the electron transport inhibitor, DBMIB, which functions between Photosystem II and Photosystem I; they were able to show that Photosystem I catalyses an ATP formation which is insensitive to DBMIB (as also does photosystem II, again insensitive to DBMIB). Under a wide range of experimental conditions Good's group has found that subtraction of the basal, non-phosphorylating rate of electron transport from the overall rate of electron transport gave an ATP/2e ratio of two for the phosphorylating part of the electron transport. Thirdly, the evidence from Tris-treated chloroplasts (no water-splitting) where ascorbate plus PD, TMPD or DCPIP restored the electron transport to NADP. In the presence and absence of DCMU two phosphorylating sites were distinguished — one associated with each of the photosystems [171]. Fourthly, the identification of four protolytic reaction sites attributed to the oxidation of water and plastoquinone on the inner side of the thylakoid membrane and the reduction of plastoquinone and the terminal electron acceptor of PSI at the outer side [5,33,80,81] (see Fig. 4.2).

Problems related to the exact amounts of ATP formed with every pair of electrons transported are still considerable. Table I shows the importance of knowing the H^+/e and H^+/ATP ratios as they affect the overall ATP/2e. Directly related to this is the fifth line of evidence which comes from our group which has concentrated over the last 7 years or more on trying to isolate chloroplasts that function in vitro such that the maximum possible amounts of ATP could be formed for every pair of electrons transferred [51,69,124—127]. We believe that we have shown that two ATPs can be formed in non-cyclic electron transfer from water to NADP (or ferricyanide or methyl viologen) and that one ATP can be formed in

TABLE I

Calculated stoichiometries in non-cyclic photophosphorylation (PSI + PSII)

H^+/e	$H^+/2e$	H^+/ATP	ATP/H^+	e/ATP	ATP/2e
2	4	2	1/2	1.0	2.0
2	4	3	1/3	1.5	1.33
2	4	4	1/4	2.0	1.0

PSI and another in PSII. We have studied many variables in the isolation and assay of chloroplasts in order to establish a maximum ATP/2e ratio. A summary is presented in Table II which shows the data from different groups of workers who have studied the problem of the stoichiometry of phosphorylation and the assays that they have used.

Work with living systems such as algae has also substantiated the fact that there might be two sites of ATP formation in non-cyclic electron transfer. This is the work on ATP levels by Gimmler with *Dunaliella* [38] and Kylin and coworkers with *Scenedesmus* [88,89], and of Senger studying the quantum efficiency of O_2 evolution by synchronised *Scenedesmus* [136] (he obtained a value of 8.0; Emerson earlier obtained a value of 8.8 with *Chlorella* [26]). (See also review by Kok [86a] for extensive discussion of quantum requirements between 8 and 10 in photosynthesis.)

4.3.1. ATP/2e and chloroplast integrity

After the intial discovery of Winget et al. [165] in 1965 that the stoichiometry of photosynthetic phosphorylation could be greater than one (in this case between 1.1 and 1.3) we decided to investigate in some detail the conditions of isolation and assay of chloroplasts which would give a maximum possible photosynthetic ATP/2e ratio. Our initial data [69] in 1968 showed that we could obtain ATP/2e ratios of 1.5 with ferricyanide and 1.6 with NADP as the electron acceptors. This was accomplished by using chloroplasts isolated in a slightly different way from Winget et al. [165], and indicated that one could obtain ATP/2e ratios greater than 1.5. Izawa and Good [72] in 1968 in a detailed analysis of the ATP/2e ratio, showed that the so-called basal electron transport, that is the rate of electron transport which occurs in the presence of ADP and absence of exogenous phosphate, could be subtracted from the overall ATP/2e ratio to give a net ATP/2e ratio of 2.0. The problem here is to establish what the basal electron transport is and whether it remained unaltered, or increased or decreased, during the process of phosphorylation.

In Table III we show that the ATP/2e ratio can vary from a value of about 1.0 to 1.6 depending on how the chloroplasts were isolated from the

TABLE II

Reported ATP/2e stoichiometries ⩾ 1.0 for PSII + PSI in non-cyclic photophosphorylation

(1)	$H_2O \rightarrow$ FeCy	0.9—1.1	Yin et al. [173]
		1.1—1.3	Winget et al. [165]
		1.5 (av.)	Horton and Hall [69]
		1.0—1.3	Shavit and Avron [137]
		1.1	Izawa and Good [72]
		1.1—1.2	Gromet-Elhanan [48]
		1.0—1.4	West and Wiskich [160]
		1.0—1.3	Smith and West [143]
		1.1—1.3	Frackowiack and Kaniuga [34]
		1.5—1.6	Hall et al. [51]
		1.0—1.5	Reeves et al. [124]
		1.1—1.3	Saha et al. [134]
		1.8	Lin and Nobel [92]
		1.1	Hauska et al. [54]
		1.5—1.7	Reeves and Hall [126]
		1.2—1.7	West and Wiskich [161]
		1.6—1.8	Heathcote and Hall [58]
(2)	$H_2O \rightarrow$ NADP	1.4	Yin et al. [173]
		1.6 (av.)	Horton and Hall [69]
		1.1—1.3	Gromet-Elhanan [48]
		1.3—3.3	Miginiac-Maslow and Moyse [99]
		1.1—1.4	Mathieu et al. [97]
		1.3—3.2	Miginiac-Maslow [100]
		1.4—2.2	Forti [29]
		1.6—2.1	Hall et al. [51]
		1.8—2.2	Nobel et al. [111] (calculated from Figs. 4 and 5)
		2.1—2.4	Heber [61] (calculated from Figs. 1, 4 and 6)
		1.7—2.2	Forti and Rosa [30]
		1.7	Reeves and Hall [126]
(3)	Ascorbate/HQ (BZ) → NADP (DCMU sensitive)	0.9—1.0	Yamashita and Butler [172]
(4)	$H_2O \rightarrow$ MV	1.3—2.1	Hall et al. [51]
		0.9—1.1	Whitehouse et al. [162]
		1.7	Reeves and Hall [126]
		1.7	Allen and Hall [1]
		1.6	Heathcote and Hall [58]
		1.2	Giaquinta et al. [36]
		1.0—1.2	Ort [116]
(5)	$H_2O \rightarrow$ Quinones	1.0	Trebst and Eck [149]
(6)	$H_2O \rightarrow$ FMN	1.0—1.1	Whitehouse et al. [162]
(7)	Catechol, Hydroquinone, Benzidine, Aminophenol, Dihydroxybiphenyl → MV	1.0—1.1	Izawa et al. [75]
(8)	Glycerate and *P*-glycerate reduction	1.2—1.4	Heber and Kirk [62]

TABLE III

ATP/2e ratios of ferricyanide reduction by sucrose- and salt-prepared spinach chloroplasts in the presence and absence of sucrose and salt in the reaction mixture [124]

Grinding and washing medium	Resuspending medium	Reaction mixture	Ferricyanide reduction (μmoles/2 · mg chl^{-1} · h^{-1})	ATP formation (nmoles · mg chl^{-1} · h^{-1})	ATP/2e
(1) 0.4 M Sucrose	0.4 M Sucrose	+ Sucrose	78	120	1.5
(2) 0.4 M Sucrose	0.35 M NaCl	+ Sucrose	109	156	1.4
(3) 0.4 M Sucrose	0.035 M NaCl	+ Sucrose	125	165	1.3
(4) 0.4 M Sucrose	0.4 M Sucrose	— Sucrose	106	138	1.3
(5) 0.4 M Sucrose	0.35 M NaCl	— Sucrose	156	173	1.1
(6) 0.4 M Sucrose	0.035 M NaCl	— Sucrose	150	184	1.2
(7) 0.35 M NaCl	0.4 M Sucrose	+ Sucrose	142	177	1.3
(8) 0.35 M NaCl	0.35 M NaCl	+ Sucrose	137	177	1.3
(9) 0.35 M NaCl	0.035 M NaCl	+ Sucrose	154	189	1.2
(10) 0.35 M NaCl	0.4 M Sucrose	— Sucrose	209	211	1.0
(11) 0.35 M NaCl	0.35 M NaCl	— Sucrose	206	230	1.1
(12) 0.35 M NaCl	0.035 M NaCl	— Sucrose	221	216	1.0

TABLE IV

A comparison of non-cyclic electron transport and phosphorylation with three electron acceptors [126]

	Electron transport (μmoles O_2/mg chlorophyll per h)		
	Ferricyanide	Methyl viologen	$NADP^+$
State 2	22	24	11
State 3	55	59	38
State 4	10	12	7
State 4 + NH_4Cl	60	61	44
Photosynthetic control	5.5	4.9	5.4
ADP/O ratio	1.5	1.7	1.7

leaf, how they were resuspended, and how they were assayed in the test tube. At this time, West and Wiskich [160] introduced the concept of photosynthetic control, which is analogous to respiratory control in mitochondria; in both cases a State 3—State 4 transition in the rate of electron transport is observed after the limiting quantities of ADP have been completely phosphorylated to ATP.

This enabled us then to use this technique for a much closer investigation of the ATP/2e ratio. In this system, the photosynthetic control is measured as the ADP/O ratio since limiting amounts of ADP are added in an oxygen electrode; that is, one measures the difference between the oxygen formed in

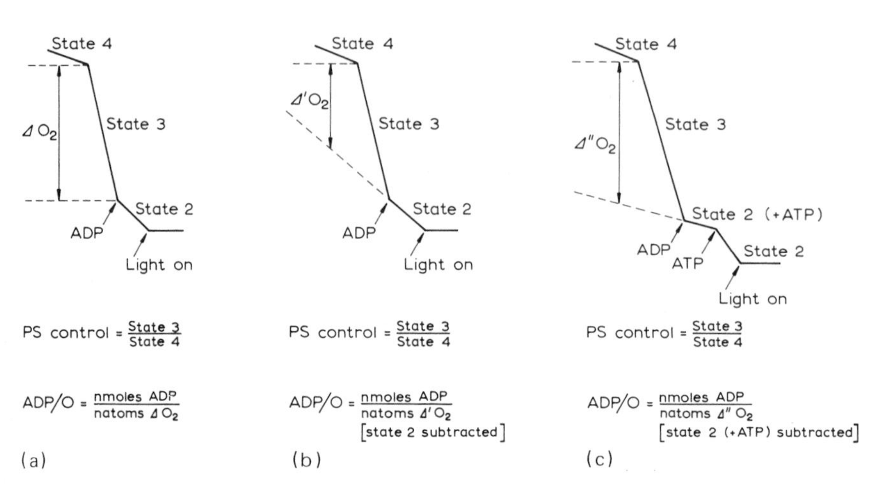

Fig. 4.3. Diagrammatic representation of photosynthetic control of electron transport (a) without subtraction of basal electron transport from State 3, (b) with subtraction, (c) with subtraction of ATP-inhibited basal electron transport from State 3. Oxygen evolution traces proceed from bottom to top, with time from right to left (see Fig. 4.4).

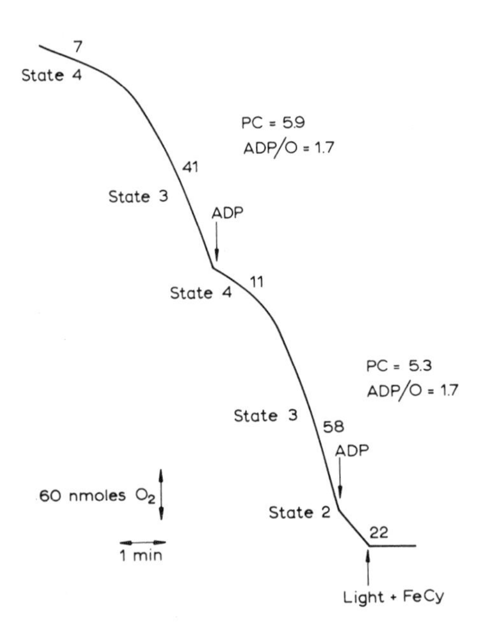

Fig. 4.4. Examples of photosynthetic control and ADP/O ratios with ferricyanide as the electron acceptor [125].

the absence (State 2) and the presence (State 3) of ADP; this is shown in Fig. 4.3. In Fig. 4.4 we show that ADP/O ratios between 1.5 and 2.0 can be obtained with ferricyanide as the electron acceptor. Similar values are shown in Table IV for NADP and MV (in the case of MV, oxygen uptake is measured rather than oxygen evolution as in the case with ferricyanide or NADP). Measurement of such ADP/O ratios with three different electron acceptors is a useful technique since it shows that electron acceptors with redox potentials between +440 and −430 mV and studying either oxygen evolution or oxygen uptake can give stoichiometries of ADP/O greater than 1.5. We were also able to confirm these ADP/O values by measuring the amount of ATP^{32} formed, and the amount of $NADPH_2$ or ferrocyanide formed spectrophotometrically; this gave confirmatory evidence to the oxygen electrode experiments.

In our investigations of the optimum conditions for the isolation and assay of chloroplasts [124—127], we studied the following type of experimental variables (following the basic techniques of Walker [156]): (*a*) sorbitol concentrations varied in the grinding, resuspending, and reaction mixture media; (*b*) varying the pH of these three different media (*c*) the addition of bovine serum albumin, (*d*) various calcium concentrations, (*e*) varying the magnesium concentration in the three different types of media (*f*) pH optima and (*g*) temperature optima. In using these different techniques, we have derived a reproducible experimental scheme which in our hands we think is optimum for obtaining high ATP/2e ratios (with rapid isolation techniques and the assay of the chloroplasts using an electrode). We routinely attained values of between 1.5 and 1.8 with ferricyanide, NADP or methyl viologen as electron acceptors (i.e. non-cyclic electron transport systems through both PSII and PSI). In addition we have studied the effect of varying concentrations of ADP and ferricyanide as the electron acceptor in order to see if this affects the stoichiometry of photophosphorylation. It has also been shown [8a] that well coupled chloroplasts exhibit a reasonable photosynthetic control of the 518 nm shift which is considered indicative of the membrane potential [168]. The structural and metabolic state of the chloroplasts used in vitro is an important factor in obtaining optimal phosphorylation efficiencies. An attempt (albeit imperfect) to characterize isolated chloroplasts is shown in Table V.

It is seen that the structural state of the chloroplast is probably important in ascertaining its coupling efficiency. There is quite a lot of evidence now [14,18,48a,142] that the extent of granal stacking influences the ability of chloroplasts to capture light (especially at low intensities) even though there is probably little or no difference between the biochemical activities of stacked (granal) and unstacked (stromal) lamellae (see also review [4]). It is likely that the grana are continuous membrane systems contiguous with the stroma lamellae [65,102,107,163]. The

TABLE V

Types of chloroplast preparations after the classification of Hall [50] together with an outline of their main properties

Chloroplast type	Description	Preparation method	Appearance under phase contrast microscopy	Envelope	Rate of CO_2 fixation μmoles CO_2 / mg chlorophyll/h	Exogenous substrate penetration and requirements	Electron transport and photophosphorylation capacity
A	Complete chloroplasts	Rapid, in isotonic or hypertonic sugar, one centrifugation	Outer mobile jacket present. Bright and high reflecting. Grana not obvious	Intact	50—250	NADP and ferricyanide do not penetrate. Slow uptake of ATP, ADP and P_i	Presumed to be unimpaired. (ATP/2e approaching 2.0 when assayed in hypotonic medium; good photosynthetic control)
B	Unbroken chloroplasts	In isotonic or hypertonic sugar or salt, with 2 or 3 centrifugations	Bright and highly reflecting. Smooth outline. Grana not obvious ("Class I")	Morphologically but not functionally intact	<5	NADP, ferricyanide and ADP penetrate. Ferredoxin may not be necessary to add for NADP reduction	Good ATP/2e (greater than 1.0 and often approaching 2.0 when assayed in hypotonic medium). Good photosynthetic control
C	Broken chloroplasts	Vigorous in isotonic sugar or salt	Not bright in appearance. 2 to 3 times larger than types A and B. Grana conspicuous (granular). ("Class II")	Broken and usually lost in preparation (stroma also lost)	Little or none	NADP, ferricyanide and ADP penetrate. Ferredoxin needs to be added for NADP reduction	ATP/2e >1.0 or <1.0 depending on isolation and assay conditions. Some photosynthetic control

D	Free-lamellar chloroplasts	Osmotic shock of type A chloroplasts immediately followed by return to isotonic medium	—	Lost from chloroplasts but retained in medium	High rates if carbon-pathway intermediates and chloroplast extract added	—	Good photosynthetic control
E	Chloroplast fragments	Resuspend chloroplasts in hypotonic medium	—	Lost	None	Ferredoxin needs to be added for NADP reduction	Higher rates of electron transport and lower rates of photophosphorylation than Types B and C. ATP/2e <1.0 No photosynthetic control
F	Sub-chloroplast particles	By sonication or detergent treatment or French press. Fractionation by centrifugation	—	Lost	None	Ferredoxin and plastocyanin (and sometimes reductase) need to be added for NADP reduction	Photophosphorylation (cyclic only) low or absent. Limited electron transport when electron donors added

efficiency of transfer of electrons between PSII and PSI is also greatly affected by the structural state of the membrane and is probably also correlated with the ionic environment prevailing at any given time (see Chapter 3 by Barber). It seems that we now should pay much more attention to the microstructure of chloroplasts at the membrane level in order to determine its influence on photophosphorylation.

4.3.2. Contributions from cyclic and pseudocyclic phosphorylation

One of the problems of obtaining high ATP/2e ratios is over the possibility that cyclic phosphorylation could contribute the extra ATP, over and above a stoichiometry of 1.0. However, two lines of argument have been used against this idea of a contribution from cyclic phosphorylation. Firstly, using isolated chloroplasts in the presence of ferricyanide it has been calculated [72,126] that cyclic electron transport would not contribute more than 5% to 8% of the total amount of ATP formed in the presence of ferricyanide (which is a very oxidising reagent). In addition, since we obtained stoichiometries higher than 1.5 with methyl viologen, NADP or ferricyanide (with their greatly different redox potentials and thus their greatly different effects on cyclic electron transport) we feel that cyclic electron transport cannot make a contribution in isolated chloroplasts. Secondly, in whole algae or in intact chloroplasts where CO_2 fixation reactions are being studied, it has also been deduced by a number of workers [12,17,85,101,122] that cyclic phosphorylation cannot contribute to steady state CO_2 fixation. It is possible, however, that cyclic electron flow may contribute extra ATP in the early stages of CO_2 fixation where there is a lag of up to 1 min or more before steady-state photosynthesis is obtained. However, beyond this contribution it is considered that cyclic phosphorylation itself does not contribute in whole cell photosynthesis or whole chloroplast photosynthesis (see discussion later for these latter points and contrast with arguments presented in Chapter 12 of this volume).

This conclusion on the lack of a contribution from cyclic phosphorylation does not eliminate a possible contribution from a pseudo-cyclic phosphorylation to provide the extra ATP for CO_2 fixation. In the case of pseudo-cyclic phosphorylation there is no net oxygen evolution or uptake since the reaction operates by a Mehler-type reaction [1, 149, 162]; thus ATP is formed by a non-cyclic system where oxygen is taken up and evolved (in a ratio of 1:1). This is very difficult to distinguish from the open chain type of non-cyclic phosphorylation, i.e. $H_2O \longrightarrow NADP$. However, in the case of the use of methyl viologen as the electron acceptor, this problem of pseudo-cyclic phosphorylation would not be evident since the oxygen interacts with reduced methyl viologen to produce hydrogen peroxide which in turn is trapped (since any catalase is inhibited by the addition of sodium azide to the reaction mixture); in this case the stoichiometry of ATP/2e is again close to 2.0. How a contribution of pseudo-cyclic phosphorylation in CO_2 fixing systems can be eliminated is still a problem. However, if we believe that two ATPs can be formed for every pair of electrons transferred to make NADPH, pseudo-cyclic phosphorylation would not be a necessary requirement for CO_2 fixing systems. The evidence of Jennings and Forti [78] on the role of pseudo-cyclic phosphorylation in the lag phase of CO_2 fixation is quite interesting, since this may be the case in which a pseudo-cyclic system actually operates.

4.3.3. CO_2-fixing chloroplasts and algae

Another problem alluded to earlier is the turnover of ATP and the levels of ATP in CO_2-fixing algae and chloroplasts. The turnover of ATP in chloroplasts during photosynthesis must be quite great, since there are only approx. 15 to 50 nmoles of ATP per mg of chlorophyll in chloroplasts and an optimum rate of CO_2 fixation might be between 100 and 200 μmoles of CO_2 fixed per mg chlorophyll per hour. An interesting recent observation [32] is that small quantities of ammonia can actually increase CO_2 fixation rates when simultaneously the ATP concentration is actually lowered in the first 5 min (or less) of CO_2 fixation; this experiment shows that one can have lower concentrations of ATP and stimulate CO_2 fixation. Thus it seems as if the ATP itself may not be a limiting factor in CO_2 fixation catalysed by endogenous phosphorylation reactions.

Other experiments on CO_2-fixing isolated chloroplasts have also indicated that extra ATP (other than non-cyclic photophosphorylation) may only be required during the initial stages of illumination when the pool of phosphorylated intermediates is low after a dark period (see review on induction phenomenon [157] and Chapter 7 of this volume). The experiments of Miginiac-Maslow and Champigny [101] show that during steady-state CO_2 fixation (after an induction period of 2 min) "higher rates of carboxylation are always connected with lower levels of ATP and vice versa". Nobel et al.'s work [111] on initial ATP and NADPH levels in CO_2 fixing chloroplasts upon illumination shows that the ATP level reaches a maximum 10 to 16 s (depending on light intensity) before the maximum NADPH level is reached (the time difference is a factor of about two); only when the NADPH level reaches its maximum does the CO_2 fixation commence. Work with CO_2 fixing *Anacystis* also shows that CO_2 fixation is much more sensitive to inhibitors (e.g. CCCP) than the levels of ATP in the cells, indicating the availability of excess ATP during CO_2 fixation, from which Bornefield and Simonis [17] concluded that "ATP itself cannot be the limiting factor for CO_2 fixation". Biggins [12] was able to show that with saturating light intensities the alga *Porphyridium* does not require any contribution (not more than 2%) from cyclic electron transport in order to assimilate CO_2.

The work of Heber and Kirk [62] suggests that ATP is a rate-limiting factor in CO_2 fixation by intact chloroplasts implying an ATP/2e ratio of less than 1.5. They have in fact calculated ATP/2e ratios of 1.2 to 1.4 ("and occasionally higher") from quantum requirements in phosphoglycerate and glycerate reduction experiments. They provide arguments (also outlined in Chapter 5) in favour of a flexible coupling between ATP formation and electron transport and conclude that the extent of coupling decreases with increasing phosphorylation potential $\frac{(ATP)}{(ADP)\,(P_i)}$. However, our experiments [126] on the influence of ATP and ADP on the ATP/2e ratio show (Figs. 4.5 and 4.6) that varying the ATP concentration 1000-fold (from 10^{-6} to 10^{-3} M) and the ADP concentration 20 fold (from $2.5 \cdot 10^{-5}$ to $5 \cdot 10^{-4}$ M) has no effect on the ADP/O ratio as measured in the reaction mixtures. ATP itself at 1 mM decreases the rate of State 3 (phosphorylating) electron transport by only about 20% (but again having no influence on the ATP/2e ratio). This conclusion is also substantiated by the work of Kraayenhof [87] who varied the ATP concentration 347-fold and that of Pick et al. [119] who varied the ATP concentration 20-fold and showed that the phosphate potential was not altered. Work with mitochondria by Wilson et al. [117,166] also showed that the ADP/O ratio was independent of ATP (varied 11-fold) and ADP (varied 42-fold) concentrations (see later for further discussion).

The idea of flexible coupling has appeal if it is implied in a different sense i.e. at the reducing side of PSI, chloroplasts can divert electrons away from NADP reduction to a pseudo-cyclic electron flow (no net O_2 uptake or evolution) when they require extra ATP e.g. during the initial stages of CO_2 fixation or for strictly ATP-requiring synthesis reactions. However, the author does not think that during the transport of pairs of electrons through the phosphorylating non-cyclic electron transport chains (whether to NADP, O_2, MV or any other electron acceptor) that less than two ATP molecules will be formed — unless there is an experimentally induced flow of electrons (or protons?) through a non-phosphorylating "basal" electron transport chain (see discussion by Good [72] and others [13,52,76] on this point). In order to accept the idea of flexible coupling in the sense of Heber and colleagues we should have evidence that functional chloroplasts (in vitro and in vivo) can divert electrons in a controlled manner through both phosphorylating and non-phosphorylating electron transport chains. It would be very interesting to obtain such evidence of precise control by chloroplast membranes but until it is forthcoming the author withholds judgement and opts for an ATP/2e of 2 for non-cyclic phosphorylating electron transport.

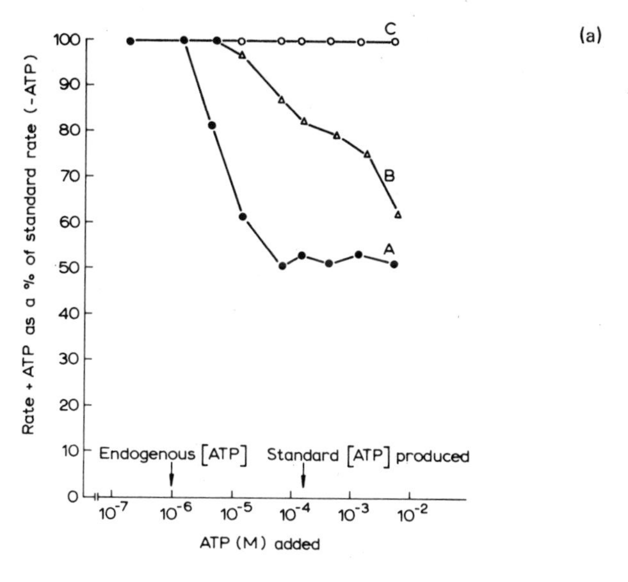

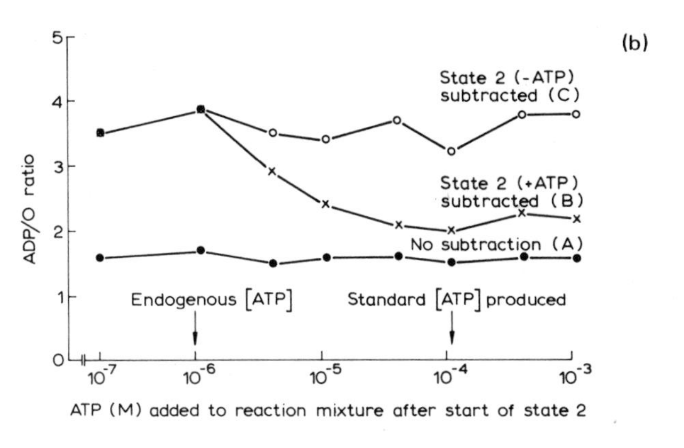

Fig. 4.5. (a) Effect of exogenous ATP on electron transport; curve (A) State 2; curve (B) State 3; curve (C) State 4. (b) Effect of exogenous ATP on ADP/O ratios. (Both graphs from ref. 126.)

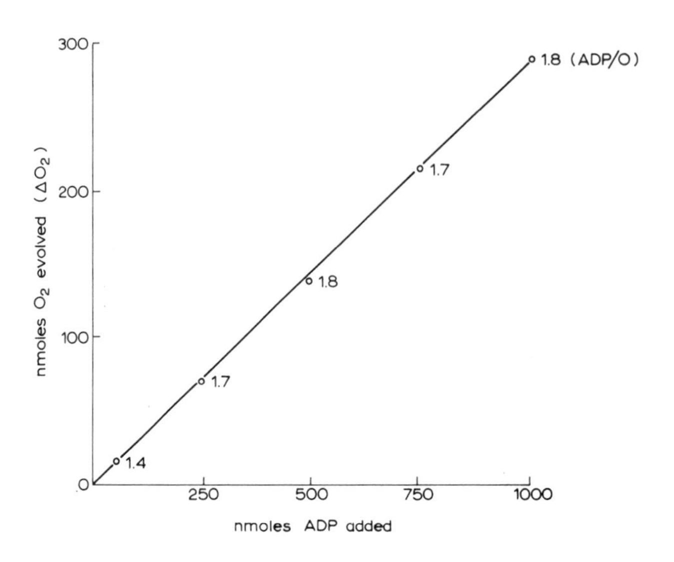

Fig. 4.6. Effect of exogenous ADP on the ADP/O ratio. Figures on line are the ADP/O ratios at given ADP concentrations [126].

4.3.4. ATP/2e in photosystems I and II

If the ATP/2e ratio of PSII and PSI in the overall non-cyclic electron transport is two and we assume two "sites" of phosphorylation, then the ATP/2e ratio for each photosystem should be one. Reported stoichiometries with various electron donors and acceptors are given in Table VI for PSI reactions and in Table VII for PSII reactions. It should be noted that a given electron carrier may donate electrons and not protons to a certain PS and thus not catalyze a phosphorylation e.g. TMPD in the case of PSI, and TMB or I^- in PSII. Generally speaking it is easier to observe ATP/2e ratios approaching one in PSI than it is in PSII. This is probably due to the greater inherent stability of PSI in isolated chloroplasts compared to PSII and this may also apply in vivo. Moreover ATP formation by PSII has only recently been recognised and we may not yet understand the optimum conditions for its assay. In calculating the overall stoichiometry of phosphorylation there seems general agreement that the H^+/e ratio is 2 for PSII + PSI and 1 for PSII or PSI alone (see Table VIII).

TABLE VI

Reported ATP/2e stoichiometries for PSI

	System	ATP/2e	Reference
(1)	Ascorbate/DCPIP → MV	1.0	Strottman and von Gosseln [145]
		0.5—0.6	Gould and Izawa [41]
		0.6	Ort and Izawa [114]
		1.0	(approx.) Allen and Hall [1]
		0.9	Heathcote and Hall [58]
(2)	Ascorbate/ {DAD, DAT, PD, DPHZ, *o*-Tolidine} → MV	0.5—0.6	Ort and Izawa [115]
(3)	Ascorbate/DAD → MV	0.4	Ort and Izawa [114]
(4)	Ascorbate → MV	0.6	Ort and Izawa [114]
(5)	Ascorbate → AQS	0.5	Bohme and Trebst [15]
(6)	Ascorbate/ {DCPIP, DAT, DAD} → MV	0.6—0.7	Gould [45]
(7)	Ascorbate/ {DCPIP, DAD, DAT, TCHQ, DMPD} → MV	0.5—0.7	Ort [116]
(8)	DAB → MV	0.7—0.8	Goffer and Neumann [40]
		0.4—0.5	Ort [116]
(9)	{Fe^{2+} Cy, I^-, Fe (dipyridyl), Mn (oxine), Mn $(dipyridyl)_2$} → MV	0.4—0.6	Izawa and Ort [74]
			Izawa et al. [75]
(10)	Ascorbate/DCPIP → NADP	1.1	Losada et al. [93]
		0.7—1.4	Trebst and Pistorius [150]
		1.2	Trebst et al. [151]
		0.5	Trebst and Pistorius [152]
		0.5	Hauska et al. [56]
(11)	Ascorbate/ {PD, DCPIP} → NADP (+ DCMU; Tris-washed)	0.2—0.4	Yamashita and Butler [171]
(12)	Ascorbate/DAD → NADP	0.5—1.0	Trebst and Pistorius [150]
		1.0	Trebst et al. [151]
		0.9—1.0	Hauska et al. [56]
(13)	Ascorbate/Tetramethylbenzidine → NADP	0.5—0.6	Harth et al. [53]
(14)	Ascorbate/ {DAD, Indamine, TMI, (+ DCMU + DBMIB)} → NADP	0.8—1.1	Trebst [148]; Oettmeier et al. [113]
(15)	Ascorbate/ {*N,N,N'*-TMPD, *N*-Phenyl PD, (+ DCMU + DBMIB)} → NADP	0.6—0.9	Hauska et al. [56]

TABLE VII

Reported ATP/2e stoichiometries for PSII

(1)	H_2O	→	PD, DAD, DMQ /FeCy	0.6	Saha et al. [134]
(2)	H_2O	→	DMQ, BQ /FeCy (+DBMIB)	0.3	Trebst and Reimer [154]
(3)	H_2O	→	PD/FeCy (+DBMIB)	0.5	Trebst and Reimer [154]
(4)	H_2O	→	DBMIB/FeCy	0.3—0.4	Gould and Izawa [42] Izawa et al. [73]
(5)	H_2O	→	DMQ, PD /FeCy (+DBMIB)	0.4	Gould and Ort [43]
(6)	H_2O	→	PD, DAD, DMQ, DAT /FeCy (+DBMIB)	0.4—0.5	Izawa et al. [73]
(7)	H_2O	→	DAD, DMQ, PD /FeCy (+KCN)	0.3—0.4	Ouitrakul and Izawa [118]
(8)	H_2O	→	PD, DAT, MMPD, TMPD /FeCy (+DBMIB)	0.4—0.5	Trebst [148]
(9)	H_2O	→	PD, DAD, DMQ /FeCy (±DBMIB)	0.6—0.9	Heathcote and Hall [58]

TABLE VIII

Reported H^+/e ratios for non-cyclic photosystems

PSI + PSII	2—3	Crofts [22]
	2	Schwartz [141]
	2	Izawa and Hind [70]
	2	Schliephake et al.[138]
	2	Rumberg et al. [130]
	2	Rumberg et al. [131]
	2	Schröder et al. [139]
	2	West and Wiskich [160]
	1.7	Gould and Izawa [44]
	2	Junge and Auslander [81]
	2	Fowler and Kok [33a]
PSII	1	Schliephake et al. [138]
	0.5	Gould and Izawa [44]
	1	Auslander et al. [6,81]
PSI	1	Schliephake et al. [138]
	1	Izawa et al. [75]
	1	Strottman and van Gosseln [145]
Mitochondria	2	Mitchell and Moyle [103]
Chromatophores	2	Walker and Crofts (review) [158]

4.3.5. H^+/ATP ratio

A major disagreement comes in calculating the H^+/ATP ratio; this value is important in estimating the ATP/2e ratio since the H^+/e ratio is thought to be 2 (see Table I). Values of 2, 3 and 4 have been reported for the H^+/ATP ratio — even from the same laboratory (see Table IX for reported values from various laboratories). The "correct" H^+/ATP value is also important in discussions on the extent of the redox potential spans in the electron transport chain which would be required in order to synthesize ATP.

TABLE IX

Reported H^+/ATP ratios for photosynthetic phosphorylation

Chloroplasts	2	Izawa and Good [72]
	2	Schwartz [141]
	2	Gould and Izawa [44]
	2.4	Izawa [71]
	3	Junge et al. [79]
	2	Rumberg et al. [131]
	3	Rumberg [132]
	3	Schröder et al. [139]
	4	Rumberg and Schröder [133]
	3—4	Graber and Witt [47]
	3	Portis and McCarty [120]
	4	Schröder et al. [140]
	1.7—1.8	Fiolet and van de Vlught [28]
	2.4	Graber and Witt [47a]
Chromatophores	1.7	Jackson et al. [75a]
Chloroplast ATPase	2(approx.)	Carmeli [19]
Algal plasma ATPase	2	Walker and Smith [159]
Chloroplast ATP formation in dark (acid base transition with succinate)	1.4—1.6	Nishizaki [108]
Mitochondrial ATPase	2	Mitchell and Moyle [104]
Submitochondrial ATPase	2	Moyle and Mitchell [105]
Submitochondrial ATPase	2	Thayer and Hinkle [146]

4.3.6. Phosphate potentials

The energy required to synthesize an ATP molecule can be calculated as the phosphate potential (ΔG) from the values of the standard free energy (ΔG_o) of the reaction ADP + $P_i \longrightarrow$ ATP + H_2O and the concentration of the reactants and products [65,75]:

$$\Delta G = \Delta G_o + 1.36 \log \frac{(ATP)}{(ADP)\ (P_i)}$$

(in this equation the concentration of water is taken to be so large as to be constant).

The value of ΔG_o varies according to the pH, temperature, free Mg, and ionic strength of the medium and thus, as would be expected, the reported values have varied considerably e.g. from about 7.0 to about 10.1 kcal/mole at pH 7 and Mg = 0, or 6.7 to 8.7 kcal/mole at Mg = 10^{-3} M (see various articles [23,49,63,87,128,167]). (Note: 1 kcal = 4.187 kJoules.)

The experimental determination of the phosphate potential (ΔG) has been made for chloroplasts, chromatophores and mitochondria [23,76,126,167, 170]. The rational is to determine the concentrations of ATP and ADP at the point at which ATP synthesis ceases when studying photosynthetic or respiratory control i.e. at the State 3—State 4 transition (see Fig. 4.3). At this juncture the driving force for ATP synthesis is thought to be at its upper limit. Kraayenhof [87] has measured initial and final concentrations of ATP, ADP and P_i in a phosphorylating chloroplast reaction mixture. The free initial ATP/ADP varied between 0.1 and 34.4 while the final ATP/ADP was between 25 and 67; the P_i concentrations used were between 775 μM and 414 μM. He obtained phosphate potentials (ΔG) between 15.15 and 15.61 kcal/mole using ΔG_o values of 9.25 and 9.40 kcal/mole.

Since the concentrations of ATP, ADP and P_i and Mg in the chloroplast will have a great influence on the calculations of any phosphate potential, I have attempted to tabulate the values which have been reported. Table X shows the chloroplast volumes and chlorophyll concentrations of chloroplasts; from this I think a value of 26 μl volume osmotic space per mg chlorophyll (for Type A, complete chloroplasts) is a reasonably well documented figure and will be used in subsequent calculations. The concentrations of nucleotides, P_i and Mg are shown in Table XI. A value for ATP in illuminated intact chloroplasts of about 40 nmoles/mg chlorophyll (equivalent to 1.5 mM) seems reasonable. The ADP value is about 12 nmoles/mg chlorophyll (equivalent to 0.5 mM). Measured ATP/ADP ratios in light in vivo and in vitro range from 1.6 to 6.5 for various plants — a value of 3 is thus also not unreasonable.

Estimations of the P_i levels of intact chloroplasts are more difficult since P_i is easily lost from chloroplasts during their isolation in aqueous media while with non-aqueous techniques there is often the problem of cytoplasmic contamination. Reeves [125] has estimated the amount of P_i available for phosphorylation in "Type A, complete" chloroplasts as 1.4 μmoles/mg chlorophyll (= 54 mM) using the techniques shown in Fig. 4.7. Reeves and Hall [126] reported a value of 3.6 μmoles P_i mg chlorophyll (= 138 mM) using total perchlorate extraction. My thought is that the higher figure rather than the lower is closer to the correct value, but the important consideration is that the P_i concentration of complete, intact chloroplasts is high and this concentration will have a significant influence on the calculation of the phosphate potential. In algal chloroplasts the

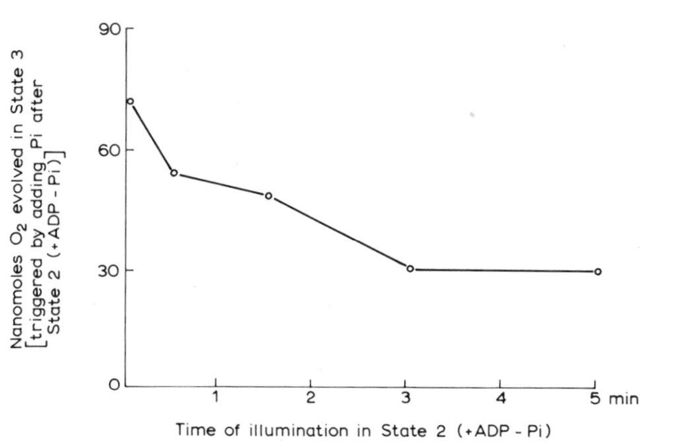

Fig. 4.7. Estimation of endogenous P_i content of Type A chloroplasts. The chloroplasts were illuminated for various times in State 2 in the presence of ADP, but with no added P_i. State 3 was then triggered by adding P_i. The amount of O_2 evolved in State 3 was dependent on the time of illumination in State 2 i.e. the amount of ADP remaining after phosphorylation minus the total amount of ADP added in State 2 is equivalent to the amount of endogenous P_i [125].

TABLE X

Chloroplast and chlorophyll quantities

Chloroplast volumes		Ref. No.
	17 μl/mg chlorophyll	[123]
	41 μm^3 (= $4.1 \cdot 10^{-11}$ ml)	[123]
	25 μm^3	[21]
	24 μl/mg chlorophyll (solute available space)	[90]
	31 μm^3 (in light)	[109]
	26 μl/mg chlorophyll (sucrose impermeable space)	[64]
	26 μl/mg chlorophyll (osmotic space)	[39]
Chlorophyll		
	$2.4 \cdot 10^{-9}$ mg/chloroplast	[123]
	$1.3 \cdot 10^{9}$ molecules/chloroplast	[123]
	10^{-3} moles/l of leaf	[112]
	25mM in chloroplast	[112]
	65 μg/mg chloroplast dry wt.	[110]
	4—8% of dry wt. of chloroplast	[123,82,83]
	1.2% of dry wt. of leaf	[123,83]
	0.25% of wet wt. of leaf	[123]
	2—4 mg/g fresh wt. of leaf	[57,83]
	1 mg = 1.13 μmole *or* 0.89 mg = 1 μmole	

TABLE XI

Concentrations of nucleotides and ions in chloroplasts

ATP levels (nmoles/mg chlorophyll)	
14—20:	in non-aqueous isolated chloroplasts; illuminated leaves [82]
46:	difference after illumination [111]
3—5:	in aqueous isolated chloroplasts [60]
30—55:	in non-aqueous isolated chloroplasts [60]
27:	in leaf disc chloroplasts (calculated after illumination) [96]
14—56:	in chloroplasts after illumination [61]
16:	in "Type A" chloroplasts [126]
36:	in non-aqueous chloroplasts after illumination [16]
25:	in illuminated chloroplasts [78]
14:	in illuminated chloroplasts [32]
15—20:	in illuminated chloroplasts [101]
ADP levels (nmoles/mg chlorophyll)	
7—11:	in non-aqueous isolated chloroplasts; illuminated leaves [82]
15:	in aqueous isolated chloroplasts [60]
22—29:	in non-aqueous isolated chloroplasts [60]
9:	in "Type A" chloroplasts [126]
22:	in leaf disc chloroplasts (calculated after illumination) [96]
13:	in non-aqueous chloroplasts after illumination [16]
3—5:	in illuminated chloroplasts [101]
ATP/ADP ratio in light	
$1.6 \rightarrow 3.1$:	Wheat [16]
2.8:	Chard [135]
3.9:	Spinach [135]
$1.7 \rightarrow 2.2$:	Tobacco [82]
5.0:	Elodea [60]
P_i levels	
ca. 75 mM:	from 1.5% P_i in dry matter and 9% of ash in chloroplast [123]
ca. 46 mM:	from $3 \cdot 10^{-11}$ mg P/chloroplast [77]
4—25 mM:	trichloroacetate extraction [135]
18 mM:	CO_2 fixation requirements; 0.5 μm/mg chlorophyll [20]
34 mM:	trichloroacetate extraction; 0.88 μm/mg chlorophyll [34]
62 mM:	unwashed chloroplasts (1.6 μm/mg chlorophyll) [144]
25 mM:	2 x washed or broken chloroplasts (0.7 μm/mg chlorophyll) [144]
54 mM:	"Type A" chloroplasts, then broken and calculated from photosynthetic control; 1.4 μm/mg chlorophyll [125]
138 mM:	"Type A" chloroplasts, extracted with perchlorate; 3.6 μm/mg chlorophyll [126]
Mg levels	
14 mM:	*Tolypella* [90]
10 mM:	rise in level on illumination [92]
15 mM:	rise in level in stroma on illumination [68]
27 mM:	non-aqueous isolated chloroplasts [39]
Ionic strength	
0.20 M:	Kishimoto and Tazawa [84]
0.16 M:	Nobel [110]
0.13 M:	Gimmler et al. [39]

estimate of endogenous P_i concentrations is much lower, ranging from 4 to 15 mM (see reviews [11,121]); the discrepancy, if real, is difficult to reconcile with higher plant estimates.

The level of Mg in the chloroplast is also higher than thought a few years ago. A value between 15 mM and 30 mM Mg in the stroma is probably quite close to the correct concentration (see Chapter 3). Again this amount of Mg, whether "free" or "bound", will have an influence on the calculated phosphate potential.

The ionic strength within the chloroplast also influences the phosphate potential — and this and other factors such as Mg concentration, pH and temperature have all been tabulated [42,106]. Calculated ionic strengths of isolated chloroplasts vary between 0.13 M and 0.20 M. I have used a value of 0.2 M in subsequent calculations as this gives slightly higher rather than lower values for the phosphate potential.

In Table XII are listed the phosphate potentials calculated at various ATP, ADP and P_i concentrations and at four different pHs (one without Mg). In section *i* the ATP/ADP ratio is 3; at 58 mM P_i the phosphate potential varies between 8.6 and 11.2 kcal/mole; it is also seen that the higher the P_i concentration the lower the phosphate potential. Changing the ATP/ADP ratio to 5 (section *ii*) only increases ΔG by about 0.3 kcal/mole. In section *iii* the final concentrations of ATP, ADP and P_i determined by Kraayenhof [87] in his reaction mixture containing broken chloroplasts are given. At low P_i concentrations (0.667 and 4.011 mM) the ΔG varies between 12.1 and 15.1 kcal/mole. However, if one uses the high P_i concentration of 58 mM or 116 mM, values of ΔG of 10.1 to 13.1 kcal/mole are obtained — a difference of at least 2 kcal/mole (at an ATP/ADP ratio of 67). In sections *iv* and *v* of Table XII are listed values of ΔG calculated from ATP/ADP ratios of 200 and 10^4. The variation in pH values given are those which may be expected in the intrathylakoid and stromal spaces of chloroplasts. Values lower than pH 6.0 can however be measured and thus would give a ΔG_o value lower than 6.3 kcal/mole.

The phosphate potential values listed in Table XII emphasize the problems of estimating the actual value of ΔG in intact chloroplasts. My inclination would be to use the values in section *i* or *ii* of Table XII (ATP/ADP = 3 or 5; and P_i concentration = 58 mM or 116 mM) thus giving phosphate potential values between about 8.2 and 11.5 kcal/mole, depending on the pH thought to be operating at the site of phosphorylation. Again, if I were to choose, my prejudice as to what was happening in the in vivo chloroplast would be for the lower value of 8.2 kcal/mole i.e. operating close to a pH of about 6, but this is purely a personal feeling with no "facts" to support it.

In order to calculate the redox potential difference necessary to synthesize an ATP with the above phosphate potential values we use the formula:

$$\Delta G = nF\Delta E$$

TABLE XII

Calculation of phosphate potential (ΔG) in kcal/mole (Note: 1 kcal = 4.187 kJoules)

$$\Delta G = \Delta G_o + 1.36 \log \frac{(ATP)}{(ADP)(P_i)}$$

$$= \Delta G_o + \Delta G_e$$

ΔG = phosphate potential
ΔG_o = free energy of ATP hydrolysis (formation)
$\Delta G_e = 1.36 \log \frac{(ATP)}{(ADP)(P_i)}$

ATP concentration (after illumination)	=	40 (30—50) nmoles/mg chlorophyll [1.5 (1.2—1.9) mM]
ADP concentration (after illumination)	=	12 (10—15) nmoles/mg chlorophyll [0.5 (0.4—0.6) mM]
P_i concentration	=	1.5 (0.5—3.6) μmoles/mg [58 (19—138) mM]
Mg concentration	=	20 (14—27) mM
Chloroplast solute volume	=	26 μl/mg chlorophyll
ΔG_o at 25° C, Mg = 20 mM, ionic strength	=	0.2, pH 6 = 6.3 pH 7 = 7.0 pH 7 = 7.9 (at Mg = 0) pH 8.5 = 8.9

	(ATP) μM	(ADP) μM	ATP/ADP ratio	(P_i) mM	(ADP)(P_i) M^{-1}	ΔG_e	ΔG (ΔG_o = 6.3)	ΔG (ΔG_o = 7.0)	ΔG (ΔG_o = 7.9)	ΔG (ΔG_o = 8.9)
(i)	1500	500	3	15	$2.301 \cdot 10^2$	3.13	9.4	10.1	11.0	12.0
	1500	500	3	29	$1.034 \cdot 10^2$	2.74	9.0	9.7	10.6	11.6
	1500	500	3	58	$5.172 \cdot 10^1$	2.33	8.6	9.3	10.2	11.2
	1500	500	3	116	$2.586 \cdot 10^1$	1.92	8.2	8.9	9.8	10.8
(ii)	2500	500	5	15	$3.333 \cdot 10^2$	3.43	9.7	10.4	11.3	12.3
	2500	500	5	29	$1.724 \cdot 10^2$	3.04	9.3	10.0	10.9	11.9
	2500	500	5	58	$8.621 \cdot 10^1$	2.63	8.9	9.6	10.5	11.5
	2500	500	5	116	$4.310 \cdot 10^1$	2.22	8.5	9.2	10.1	11.1
(iii)	1150	46	25	0.667	$3.748 \cdot 10^4$	6.22	12.5	13.2	14.1	15.1
	1819	27	67	4.011	$1.679 \cdot 10^4$	5.75	12.1	12.8	13.7	14.7
	1819	27	67	15	$4.491 \cdot 10^3$	4.97	11.3	12.0	12.9	13.9
	1819	27	67	29	$2.323 \cdot 10^3$	4.58	10.9	11.6	12.5	13.5
	1819	27	67	58	$1.162 \cdot 10^3$	4.17	10.5	11.2	12.1	13.1
	1819	27	67	116	$5.808 \cdot 10^2$	3.76	10.1	10.8	11.7	12.7
(iv)	—	—	200	4	$5.000 \cdot 10^4$	6.39	12.7	13.4	14.3	15.3
	—	—	200	15	$1.333 \cdot 10^4$	5.61	11.9	12.6	13.5	14.5
	—	—	200	58	$3.448 \cdot 10^3$	4.81	11.1	11.8	12.7	13.7
	—	—	200	116	$1.724 \cdot 10^3$	4.40	10.7	11.4	12.3	13.3
(v)	—	—	10^4	4	$2.500 \cdot 10^6$	8.70	15.0	15.7	16.6	17.6
	—	—	10^4	15	$6.667 \cdot 10^5$	7.92	14.2	14.9	15.8	16.8
	—	—	10^4	58	$1.724 \cdot 10^5$	7.12	13.4	14.1	15.0	16.0
	—	—	10^4	116	$8.621 \cdot 10^4$	6.71	13.0	13.7	14.6	15.6

where ΔG = phosphate potential, n = number of electrons, F = Faraday constant (23 kcal/eV equivalent), ΔE = redox potential difference. Thus

$$\Delta E = \frac{\Delta G}{nF} = \frac{8.2}{2(23)} \text{ to } \frac{11.5}{2(23)} = 178\text{mV to } 250\text{mV (for } n = 2 \text{ and } \Delta G = 8.2 \text{ or } 11.5 \text{ kcal/mole).}$$

In order to see how the ΔE is related to the proton motive force (PMF) postulated by Mitchell (see Chapter 2) we start by using the equation [76]

$$\text{PMF} = \Delta\psi + Z(\Delta\text{pH}) \equiv \Delta G$$

$\Delta\psi$ = membrane potential, Z = 59 mV, ΔpH = pH difference.

pH gradients as high as 3.5 units and membrane potentials of 100 mV have been reported or calculated for illuminated chloroplasts. If the PMF is made up entirely of a ΔpH and $\Delta\psi = 0$ then:

$$\text{PMF} = 59\ (3.5) = 206 \text{ mV}$$

(in practise the PMF is probably made up of both a ΔpH and a $\Delta\psi$ component [168] but see [91]).

In the above equation ΔG is the quantitative expression of the PMF term since

$$\Delta G = 1.36\ (\Delta\text{pH}) + 1.36 \frac{(\Delta\psi)}{(59)}$$

The energy required to synthesize one ATP equals the PMF times H^+/ATP ratio. Thus if the PMF is made up solely of a $\Delta\psi$

PMF required to synthesize one ATP molecule $= \Delta\psi = \dfrac{\Delta G \times 59}{1.36 \times (H^+/\text{ATP ratio})}$

= 178 mV for $\Delta G = 8.2$ and $H^+/\text{ATP} = 2$
= 250 mV for $\Delta G = 11.5$ and $H^+/\text{ATP} = 2$
= 119 mV for $\Delta G = 8.2$ and $H^+/\text{ATP} = 3$
= 167 mV for $\Delta G = 11.5$ and $H^+/\text{ATP} = 3$

Comparing these PMF values with the previously calculated ΔE it will be seen that numerically the PMF required for ATP synthesis has the same values as the ΔE requirement.

If the $\Delta\psi$ is zero the PMF required to synthesize an ATP will give the following values:

$$\Delta\text{pH} = \frac{\Delta G}{1.36 \times (H^+/\text{ATP ratio})}$$

= 3.0 pH units if ΔG = 8.2 and H^+/ATP = 2
= 4.3 pH units if ΔG = 11.5 and H^+/ATP = 2
= 2.0 pH units if ΔG = 8.2 and H^+/ATP = 3
= 2.8 pH units if ΔG = 11.5 and H^+/ATP = 3

A pH gradient of 3.5 units (Heldt et al. [64] have measured light-dark ΔpH values of 2.5) could provide a potential difference of 206 mV which is sufficient for the synthesis of one ATP molecule if H^+/ATP = 2 and a phosphate potential (ΔG) of 8.2 kcal/mole is used (since the derived ΔE = 178 mV) without the requirement for a membrane potential contribution. If ΔG = 11.5 kcal/mole (with a derived ΔE = 250 mV) then a potential difference of 206 mV made up solely of a pH gradient would be insufficient; in this latter case a membrane potential contribution of at least 44 mV would be required to synthesize an ATP i.e. the total PMF is made up of both a ΔE and a ΔpH component.

It is interesting that Kraayenhof [87] has shown that the phosphate potential is independent of the initial concentrations of ATP (12—4158 μM) and ADP (104—212 μM), and that it is the same with both non-cyclic and cyclic electron transport. Recent calculations of the ΔpH by Pick et al. [119] using the fluorescence of 9 amino-acridine gave a threshold or ΔpH value of 2.7 below which no phosphorylation occurred. Interestingly they showed that this threshold value was independent of changes in the initial ATP/ADP ratios between 0.05 and 20. They also showed that at a constant ΔpH, phosphorylation was optimal at lower pHs (around pH 7) and that the rate of ATP formation was dependent both on the extent of the ΔpH and the external pH of the reaction medium.

Of equal relevance are the experiments of Wilson et al. [166] on dog heart mitochondria which show very high respiratory control values (greater than 25 in their work). They were thus able to follow the control of electron transport, ADP/O ratios and phosphate potentials very closely. They showed that the respiratory rate is dependent on the phosphate potential (ΔG) and not on the concentrations of the individual reactants or on the ATP/ADP ratio; thus they showed that it is the (ATP)/(ADP) (P_i) ratio which is important. The P_i concentration will thus be of prime significance in calculating phosphate potentials and on the effects of ATP or ADP on electron transport e.g. they showed that increasing the P_i concentration from 3 mM to 20 mM reversed the inhibitory effect of 3 mM ATP on electron transport (see previous discussion).

There has been much discussion on the magnitude of the redox potential difference required to synthesize ATP (see reviews [68,76,129,158]). The above calculations show that the phosphate potential and H^+/ATP values are not outrageous and need to be carefully considered in calculating phosphorylation potentials. If the additional 100 mV of a $\Delta\psi$ is also taken into account there would seem little difficulty in synthesizing an ATP molecule with a ΔG = 11.5 kcal/mole giving a required ΔpH of 2.5 units (PMF = 250 mV = $\Delta\psi$ + Z (ΔpH); therefore 250 = 100—59 (ΔpH); or even less difficult with a ΔG = 8.2 kcal/mole giving a required ΔpH of 1.3 units (PMF = 178 mV when $\Delta\psi$ = 100 mV).

In conclusion it would seem that there is sufficient reason to believe that a ΔE of 250 mV (and possibly less) is sufficient to synthesize one ATP even if relatively high values of the ΔG, low values of the $\Delta\psi$, low value of the H^+/ATP, and a reasonable ΔpH are used. Whether these values are used to interpret ATP synthesis via a conventional Mitchell chemiosmotic hypothesis or a micro-chemiosmotic hypothesis (reactions and pools within the membrane) it is still a problem to decide what these values are at the actual site of ATP synthesis. However, there is no room for pessimism!

4.4. ADDITIONAL CHLOROPLAST REACTIONS INVOLVING OXYGEN

When studying light-induced oxygen uptake reactions (usually associated with PSI, but possibly also other partial reactions) careful attention needs to be paid to the stoichiometry of oxygen molecules consumed per pair of electrons transported. From Fig. 4.8 this value is now thought to be 2.0, and not 1.0 as was previously supposed, which sometimes lead to incorrect calculations of ATP/2e ratios for PSI. The presence of variable amounts of endogenous superoxide dismutase in the chloroplast preparation and contaminating amounts of (cytoplasmic) catalase can confuse studies of oxygen uptake.

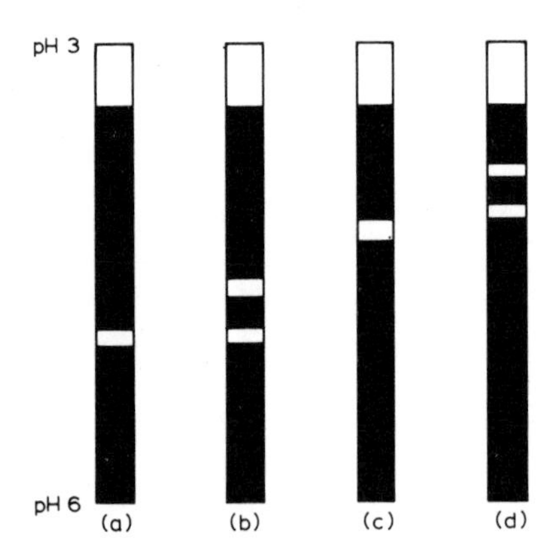

Fig. 4.8. Soluble superoxide dismutases — polyacrylamide gel electrofocussing of (a) spinach chloroplasts stroma (Cu/Zn type); (b) whole spinach leaf (Bu/Zn types); (c) blue-green alga *Spirulina* — major isozyme (Fe type); (d) *Spirulina* — minor isozymes (probably Fe type). A chloroplast membrane-bound, cyanide-insensitive superoxide dismutase is also detectable [94,95].

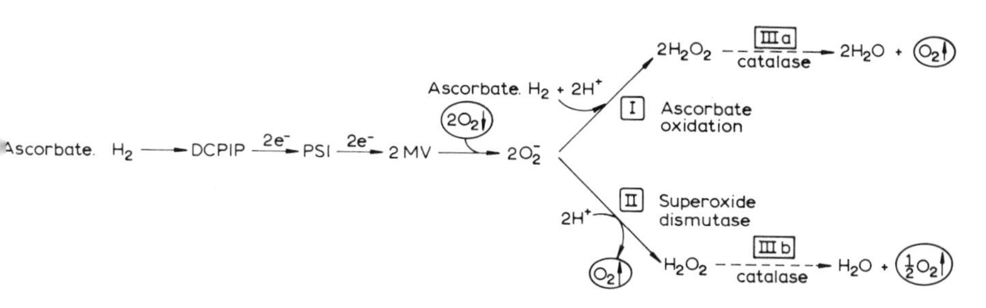

Additions	Reactions involved	O_2 molecules consumed per $2e^-$ transferred	Product (oxygen balance)
None	I	2	H_2O_2 ($2O_2\downarrow$)
SOD	II	1	H_2O_2 ($2O_2\downarrow + O_2\uparrow$)
Catalase	I + IIIa	1	H_2O ($2O_2\downarrow + O_2\uparrow$)
SOD + catalase	II + IIIb	½	H_2O ($2O_2\downarrow + 1½O_2\uparrow$)

Fig. 4.9. Scheme for O_2 uptake by Photosystem I with reactions involving ascorbate [I] superoxide dismutase [II], and catalase [IIIa], [IIIb] from ref. 2).

Since O_2 is the initial product of photosynthesis it seems natural to assume that the chloroplast has developed means of protecting itself from the known harmful effects of O_2. There seem to be at least two ways in which the chloroplast can protect itself — using its high endogenous concentrations of ascorbate (about 50 mM) and its endogenous superoxide dismutase enzymes (Fig. 4.8). Oxygen interacts with high energy free radicals e.g. produced at PSI (and PSII?) to form superoxide (O_2^-) which is as very toxic substance. This O_2^- is reduced by ascorbate or dismutated by superoxide dismutase (see Fig. 4.9). It will thus be seen that although O_2 uptake and exchange techniques are useful in various types of photosynthesis research, care needs to be exercised in interpreting results, because of the variable presence of enzymes like superoxide dismutase and catalase and reducing agents like ascorbate.

4.5. ACKNOWLEDGEMENTS

The author thanks the following for reading and criticizing this article, but he accepts full responsibility for the heresies perpetrated: N.K. Boardman, A.R. Crofts, N.E. Good, R. Hill, A.J. Jagendorf, W. Junge, J. Raven, A. Trebst, D.A. Walker and J. Wiskich.

REFERENCES

1 Allen, J.F. and Hall, D.O. (1973) Biochem. Biophys. Res. Commun., [52], 856—862.
2 Allen, J.F. and Hall, D.O. (1974) Biochem. Biophys. Res. Commun., [58], 579—585.
3 Arnon, D.I., Whatley, F.R. and Allen, M.B. (1958) Science, [127], 1026—1029.
4 Arntzen, C.J. and Briantais, J.M. (1975) in Bioenergetics of Photosynthesis, (Govindjee, ed), pp. 52—113, Academic Press, New York.
5 Auslander, W. and Junge, W. (1974) Biochim. Biophys. Acta, [357], 285—298.
6 Auslander, W., Heathcote, P. and Junge, W. (1974) FEBS Lett., [47], 229—235.
7 Avron, M. ed. (1975) Proc. 3rd Int. Cong. Photosynthesis, Vols. I, II and III, Elsevier, Amsterdam.
8 Avron, M. (1975) in Bioenergetics of Photosynthesis (Govindjee, ed), pp. 374—386, Academic Press, New York.
8a Baltscheffsky, M. and Hall, D.O. (1974) FEBS Lett., [39], 345—348.
9 Barr, R., Crane, F.L. and Giaquinta, R.T. (1975) Plant Physiol., [55], 460—462.
10 Bauer R. and Wijnands, M.J. (1974) Z. Naturforsch., [29c], 725—732.
11 Biels, I.R. (1973) Ann. Rev. Plant Physiol., [24], 225—252.
12 Biggins, J. (1973) Biochemistry, [12], 1165—1170.
13 Bishop, D.G. (1974) Photochem. Photobiol., [20], 281—299.
14 Boardman, N.K., Bjorkman, O., Anderson, J.M., Goodchild, D.J. and Thorne, S.W. (1975) in Proc. 3rd Int. Cong. Photosynthesis (Avron, M., ed), pp. 1809—1827, Elsevier, Amsterdam.
15 Bohme, H. and Trebst, A. (1969) Biochim. Biophys. Acta, [180], 137—148.
16 Bomsel, J.L. and Sellami, A. (1975) in Proc. 3rd Int. Cong. Photosynthesis (Avron, M., ed), pp. 1363—1367, Elsevier, Amsterdam.
17 Bornefeld, T. and Simonis, W. (1975) in Proc. 3rd Int. Cong. Photosynthesis (Avron, M., ed), pp. 1557—1565, Elsevier, Amsterdam.
18 Brown, J.S. (1973) in Photophysiology (Giese, A.C., ed), Vol. VIII, pp. 97—112, Academic Press, New York.
19 Carmeli, C. (1970) FEBS Lett., [7], 297—300; also Carmeli, C. et al. (1975) Biochim. Biophys. Acta, [376], 249—258.
20 Cockburn, W., Baldry, C.W. and Walker, D.A. (1967) Biochim. Biophys. Acta, [143], 614—624.
21 Dilley, R.A. (1970) Arch. Biochem. Biophys., [137], 270—283.
22 Crofts, A.R. (1967) J. Biol. Chem., [242], 3352—3359.
23 Dutton, P.L. and Wilson, D.F. (1974) Biochim. Biophys. Acta, [346], 165—212
24 Duysens, L.M.N., Amesz, J. and Kemp, B.M. (1961) Nature, [190], 510—511
25 Duysens, L.M.N. (1964) Progr. Biophys. Mol. Biol., [14], 1—104.
26 Emerson, R. (1956) quoted by Rabinowitch, E.I. in Photosynthesis and Related Processes, p.1946, Interscience, New York.
27 Emerson, R. and Rabinowitch, E. (1960) Plant Physiol., [35], 477—485.
28 Fiolet, J.W.T. and van de Vlught, F.C. (1975) FEBS Lett., [53], 287—291.
29 Forti, G. (1968) Biochem. Biophys. Res. Commun., [32], 1020—1024.
30 Forti, G. and Rosa, L. (1972) in Proc. 2nd Int. Cong. Photosynthesis (Forti, G. Avron, M. and Melandri, A., eds), pp. 1261—1270, Junk, The Hague.
31 Forti, G., Avron, M. and Melandri, A., eds. (1972) in Proc. 2nd Int. Cong. Photo synthesis, Vols. I, II and III, Junk, The Hague.
32 Forti, G., Rosa, L., Fuggi, A. and Garlaschi, F.M. (1975) in Proc. 3rd Int. Cong Photosynthesis (Avron, M., ed), pp. 1499—1505, Elsevier, Amsterdam.
33 Fowler, C.F. and Kok, B. (1974) Biochim. Biophys. Acta, [357], 299—307
33a Fowler, C.F. and Kok, B. (1976) Biochim. Biophys. Acta, [423], 510—523.
34 Frackowiak, B. and Kaniuga, Z. (1971) Biochim. Biophys. Acta, [226], 360—365.
35 Gaensslen, L.E. and McCarty, R.E. (1971) Arch. Biochem. Biophys., [147], 55—61.
36 Giaquinta, R.T., Dilley, R.A., Crane, F.L. and Barr, R. (1974) Biochem. Biophys Res. Commun., [59], 985—991.

37 Girault, G. and Galmiche, J.M. (1974) Biochim. Biophys. Acta, [333], 314—319.
38 Gimmler, H. (1973) Z. Pflanzenphysiol., [68], 289—307.
39 Gimmler, H., Schäfer, G. and Heber, U. (1975) in Proc. 3rd Int. Cong. Photosynthesis (Avron, M., ed), pp. 1381—1392, Elsevier, Amsterdam.
40 Goffer, J. and Neumann, J. (1973) FEBS Lett., [36], 61—64.
41 Gould, J.M. and Izawa, S. (1973a) Biochim. Biophys. Acta, [314], 211—223.
42 Gould, J.M. and Izawa, S. (1973b) Eur. J. Biochem., [37], 185—192.
43 Gould, J.M. and Ort, D.R. (1973) Biochim. Biophys. Acta, [325], 157—166.
44 Gould, J.M. and Izawa, S. (1974) Biochim. Biophys. Acta, [333], 509—524.
45 Gould, J.M. (1975) Biochim. Biophys. Acta, [387], 135—148.
46 Govindjee, ed. (1975) Bioenergetics of Photosynthesis, Academic Press, New York.
47 Graber, P. and Witt, H.T. (1975) in Proc. 3rd Int. Cong. Photosynthesis (Avron, M., ed), pp. 427—436, Elsevier, Amsterdam.
47a Graber, P. and Witt, H.T. (1976) Biochim. Biophys. Acta, [423], 141—163.
48 Gromet-Elhanan, Z. (1968) Arch. Biochem. Biophys., [123], 447—456.
48a Gross, E.L. and Prasher, S.H. (1974) Arch. Biochem. Biophys., [164], 460—468.
49 Guynn, R.W. and Veech, R.L. (1973) J. Biol. Chem., [248], 6966—6972.
50 Hall, D.O. (1972) Nature [235], 125—126.
51 Hall, D.O., Reeves, S.G. and Baltscheffsky, H. (1971) Biochem. Biophys. Res. Commun., [43], 359—366.
52 Hall, D.O. and Evans, M.C.W. (1972) Sub-cell. Biochem., [1], 197—206.
53 Harth, E., Ottmeier, W. and Trebst, A. (1974) FEBS Lett., [43], 231—234.
54 Hauska, G., Trebst, A. and Draber, W. (1973) Biochim. Biophys. Acta, [305], 632—637.
55 Hauska, G., Reimer, S. and Trebst, A. (1974) Biochim. Biophys. Acta, [357], 1—13.
56 Hauska, G., Oettmeier, W., Reimer, S. and Trebst, A. (1975) Z. Naturforsch., [30c], 37—45.
57 Heath, O.V.S. (1969) Physiological Aspects of Photosynthesis, pp. 198—203, Heinemann, London.
58 Heathcote, P. and Hall, D.O. (1974) Biochem. Biophys. Res. Commun., [56], 767—774.
59 Heathcote, P. and Hall, D.O. (1975) in Proc. 3rd Int. Cong. Photosynthesis (Avron, M., ed) pp. 463—471, Elsevier, Amsterdam.
60 Heber, U. and Santarius, K.A. (1970) Z. Naturforsch., [25b], 718—728.
61 Heber, U. (1973) Biochim. Biophys. Acta, [305], 140—152.
62 Heber, U. and Kirk, M.R. (1975) Biochim. Biophys. Acta, [376], 136—150.
63 Heldt, H.W., Klingenberg, M. and Milovancev, M. (1972) Eur. J. Biochem., [30], 434—440.
64 Heldt, H.W., Werdan, K., Milovancev, M. and Geller, G. (1973) Biochim. Biophys. Acta, [314], 224—241.
65 Henriques, F. and Park, R.B. (1974) Plant Physiol., [54], 386—391.
66 Hill, R. and Bendall, F. (1960) Nature, [186], 136—137.
67 Hind, G. and McCarty, R.E. (1973) in Photophysiology (Giese, A.C., ed), Vol. VIII, pp. 113—156, Academic Press, New York.
68 Hind, G., Nakatani, H.Y. and Izawa, S. (1974) Proc. Natl. Acad. Sci. USA, [71], 1484—1488.
69 Horton, A.A. and Hall, D.O. (1968) Nature, [218], 386—388.
70 Izawa, S. and Hind, G. (1967) Biochim. Biophys. Acta, [143], 377 —390.
71 Izawa, S. (1970) Biochim. Biophys. Acta, [197], 328—331.
72 Izawa, S. and Good, N.E. (1968) Biochim. Biophys. Acta, [162], 380—391.
73 Izawa, S., Gould, J.M., Ort, D.R., Falker, P. and Good, N.E. (1973) Biochim. Biophys. Acta, [305], 119—128.
74 Izawa, S. and Ort, D.R. (1974) Biochim. Biophys. Acta, [357], 127—143.
75 Izawa, S., Ort, D.R., Gould, J.M. and Good, N.E. (1975) in Proc. 3rd Int. Cong. Photosynthesis (Avron, M., ed), pp. 449—461, Elsevier, Amsterdam.
75a Jackson, J.B., Saphon, S. and Witt, H.T. (1975) Biochim. Biophys. Acta, [408], 83—92.

76 Jagendorf, A.T. (1975) in Bioenergetics of Photosynthesis (Govindjee, ed), pp. 413—492, Academic Press, New York.
77 Jagendorf, A.T. and Wildman, S.G. (1954) Plant Physiol., [29], 270—279.
78 Jennings, R.C. and Forti, G. (1975) in Proc. 3rd Int. Cong. Photosynthesis (Avron, M., ed) pp. 735—743, Elsevier, Amsterdam.
79 Junge, W., Rumberg, B. and Schröder, H. (1970) Eur. J. Biochem., [14], 575—581.
80 Junge, W. (1975) in Proc. 3rd Int. Cong. Photosynthesis (Avron, M., ed) pp. 273—286, Elsevier, Amsterdam.
81 Junge, W. and Auslander, W. (1973) Biochim. Biophys. Acta, [333], 59—70.
82 Keys, A.J. (1968) Biochem. J., [108], 1—8.
83 Kirk, J.T.O. and Tilney-Basset, R.A.E. (1967) The Plastids, pp.18, 477—478, Freeman, London.
84 Kishimoto, U. and Tazawa, M. (1965) Plant Cell Physiol., [6], 507—518.
85 Klob, W., Kandler, O. and Tanner, W. (1973) Plant Physiol., [51], 825—827.
86 Kok, B. and Hoch, G. (1961) in Light and Life (McElroy, W.D. and Glass, B., eds), pp. 397—416, Johns Hopkins, Baltimore.
86a Kok, B. (1960) in Encylcopedia of Plant Physiology (Ruhland. W., ed), Vol. 5, pp. 566—633, Springer, Berlin.
87 Kraayenhof, R. (1969) Biochim. Biophys. Acta, [180], 213—215.
88 Kylin, A., Sundberg, I. and Tillberg, J.-E. (1972) Physiol. Plant., [27], 376—383.
89 Kylin, A. and Okkeh, A. (1974) Physiol. Plant., [30], 58—63.
90 Larkum, A.W.D. (1968) Nature, [218], 447—449.
91 Larkum, A.W.D. and Boardman, N.K. (1974) FEBS Lett., [40], 229—232.
92 Lin, D.C. and Nobel, P.S. (1971) Arch. Biochem. Biophys., [145], 622—632.
93 Losada, M., Whatley, F.R. and Arnon, D.I. (1961) Nature, [190], 606—610.
94 Lumsden, J. and Hall, D.O. (1974) Biochem. Biophys. Res. Commun., [58], 35—41.
95 Lumsden, J. and Hall, D.O. (1975) Biochem. Biophys. Res. Commun., [64], 595—602.
96 Lüttge, U., Ball, E. and von Willert, K. (1971) Z. Pflanzenphysiol., [65], 326—335.
97 Mathieu, Y., Miginiac-Maslow, M. and Remy, R. (1970) Biochim. Biophys. Acta, [205], 95—101.
98 Metzner, H. ed. (1969) Proc. 1st Int. Cong. Photosynthesis, Vols. I, II and III, IUBS, Tübingen.
99 Miginiac-Maslow, M. and Moyse, A. (1969) in Proc. 1st Cong. Photosynthesis (Metzner, H., ed), pp. 1203—1212, IUBS, Tübingen.
100 Miginiac-Maslow, M. (1971) Biochim. Biophys. Acta, [234], 353—359.
101 Miginiac-Maslow, M. and Champigny, M-L. (1974) Plant Physiol., [53], 856—862.
102 Miller, K.R. and Staehelin, L.A. (1973) Protoplasma, [77], 55—78.
103 Mitchell, P. and Moyle, J. (1968) Eur. J. Biochem., [4], 530—539.
104 Mitchell, P. and Moyle, J. (1974) Biochem. Soc. Spec. Publ., [4], 91—110.
105 Moyle, J. and Mitchell, P. (1973) FEBS Lett., [30], 317—320.
106 Murakami, S., Torres-Pereira, J. and Packer, L. (1975) in Bioenergetics of Photosynthesis (Govindjee, ed), pp. 555—618, Academic Press, New York.
107 Nir, I. and Pease, D.C. (1973) J. Ultrastruct. Res., [42], 534—550.
108 Nishizaki, Y. (1973) Biochim. Biophys. Acta, [314], 312—319.
109 Nobel, P.S. (1968) Plant Physiol., [43], 781—787.
110 Nobel, P.S. (1969) Biochim. Biophys. Acta, [172], 134—143.
111 Nobel, P.S., Chang, D.T., Wang, C., Smith, S.S. and Barcus, D.E. (1969) Plant Physiol., [44], 655—661.
112 Nobel, P.S. (1974) Biophysical Plant Physiology, pp. 97 and 201, Freeman, San Francisco.
113 Oettmeier, W., Reimer, S. and Trebst, A. (1974) Plant Sci. Lett., [2], 267—271.
114 Ort, D.R. and Izawa, S. (1973) Plant Physiol., [52], 595—600.

115 Ort, D.R. and Izawa, S. (1974) Plant Physiol., [53], 370—376.
116 Ort, D.R. (1975) Arch. Biochem. Biophys., [166], 629—638.
117 Owen, C.S. and Wilson, D.F. (1974) Arch. Biochem. Biophys., [161], 581—591.
118 Ouitrakul, R. and Izawa, S. (1973) Biochim. Biophys. Acta, [305], 105—118.
119 Pick, U., Rottenberg, H. and Avron, M. (1975) in Proc. 3rd Int. Cong. Photosynthesis (Avron. M., ed), pp. 967—974, Elsevier, Amsterdam.
120 Portis, A.R. and McCarty, R.E. (1974) J. Biol. Chem., [249], 6250—6254.
121 Raven, J.A. (1976) Encyclopedia of Plant Physiology, New Series, Vol. 2A, Springer, Berlin (in press).
122 Raven, J.A. (1974) in Algal Physiology and Biochemistry (Stewart, W.D.P., ed), pp. 391—423 and 434—455, Blackwell, Oxford.
123 Rabinowitch, E.I., ed. (1945) Photosynthesis and Related Processes, Vol. I, pp. 270—391; 408—412, Interscience, New York.
124 Reeves, S.G., Hall, D.O. and West, J. (1972) in Proc. 2nd Int. Cong. Photosynthesis (Forti, G., Avron, M. and Melandri, A., eds), pp. 1357—1369, Junk, The Hague.
125 Reeves, S.G. (1972) Ph.D. Thesis, University of London King's College.
126 Reeves, S.G. and Hall, D.O. (1973) Biochim. Biophys. Acta, [314], 66—78.
127 Reeves, S.G., Heathcote, P. and Hall, D.O. (1972) Abstr. 6th Int. Cong. Photobiology, Bochum, Germany, p.284.
128 Rosing, J. and Slater, E.C. (1972) Biochim. Biophys. Acta, [267], 275—290.
129 Rottenberg, H. (1975) J. Bioenergetics, [7], 61—74.
130 Rumberg, B., Reinwald, E., Schröder, H. and Siggel, U. (1968) Naturwissenschaften, [1968], 77—79.
131 Rumberg, B., Reinwald, E., Schröder, H. and Siggel, U. (1969) in Proc. 1st Int. Cong. Photosynthesis (Metzner, H., ed), pp. 1374—1382, IUBS, Tübingen.
132 Rumberg, B. (1972) Abstr. 6th Int. Cong. Photobiology, Bochum, Germany, p. 036.
133 Rumberg, B. and Schröder, H. (1974) in Progress in Photobiology; Proc. 6th Int. Cong. Photobiology (Schenck, G.O., ed), pp. 036—037, Deut. Ges. f. Lichtforschung, Frankfurt.
134 Saha, S., Ouitrakul, R., Izawa, S. and Good, N.E. (1971) J. Biol. Chem., [246], 3204—3209.
135 Santarius, K.A. and Heber, U. (1965) Biochim. Biophys. Acta, [102], 39—54.
136 Senger, H. (1971) in Proc. 2nd Int. Cong. Photosynthesis (Forti, G., Avron, M. and Melandri, A., eds), pp. 723—730, Junk, The Hague.
137 Shavit, N. and Avron, M. (1967) Biochim. Biophys. Acta, [131], 516—525.
138 Schliephake, W., Junge, W. and Witt, H.T. (1968) Z. Naturforsch., [23b], 1571—1578.
139 Schröder, H., Mühle, H. and Rumberg, B. (1972) in Proc. 2nd Int. Cong. Photosynthesis (Forti, G., Avron, M. and Melandri, A., eds), pp. 919—930, Junk, The Hague.
140 Schröder, H., Siggel, U. and Rumberg, B. (1975) in Proc. 3rd Int. Cong. Photosynthesis (Avron, M., ed), pp. 1031—1039, Elsevier, Amsterdam.
141 Schwartz, M. (1971) Ann. Rev. Plant. Physiol., [22], 469—484.
142 Smillie, R.M., Nielsen, N.C., Henningsen, K.W. and von Wettstein, D. (1975) in Proc. 3rd Int. Cong. Photosynthesis (Avron, M., ed), pp. 1841—1860, Elsevier, Amsterdam.
143 Smith, F.A. and West, K.R. (1969) Aus. J. Biol. Sci., [22], 351—363.
144 Strotmann, H. (1972) in Proc. 2nd Int. Cong. Photosynthesis (Forti, G., Avron, M., and Melandri, A., eds) pp. 1319—1328, Junk, The Hague.
145 Strotmann, H. and von Gosseln, C. (1972) Z. Naturforsch., [27b], 445—455.
146 Thayer, W.S. and Hinkle, P.C. (1973) J. Biol. Chem., [248], 5395—5402.
147 Trebst, A. (1974) Ann. Rev. Plant Physiol., [25], 423—458.
148 Trebst, A. (1975) in Proc. 3rd Int. Cong. Photosynthesis (Avron, M., ed), pp. 439—448, Elsevier, Amsterdam.
149 Trebst, A. and Eck, H. (1961) Z. Naturforsch., [16b], 455—461.

150 Trebst, A. and Pistorius, E. (1965) Z. Naturforsch., [20b], 143—147.
151 Trebst, A., Pistorius, E. and Elstner, E. (1966) in Currents in Photosynthesis (Thomas, J.B. and Goedheer, J.C., eds), pp. 409—416, Donker, Rotterdam.
152 Trebst, A. and Pistorius, E. (1967) Biochim. Biophys. Acta, [131], 580—582.
153 Trebst, A. and Reimer, S. (1973) Biochim. Biophys. Acta, [305], 129—139.
154 Trebst, A. and Reimer, S. (1973) Biochim. Biophys. Acta, [325], 546—557.
155 Vernon, L.P. and Avron, M. (1965) Ann. Rev. Biochem., [34], 269—296.
156 Walker, D.A. (1965) Biochem. J., [92], 22c—23c; [101], 636—641.
157 Walker, D.A. (1973) New Phytol., [72], 209—235.
158 Walker, D.A. and Crofts, A.R. (1970) Ann. Rev. Biochem., [39], 389—428.
159 Walker, D.A. and Smith, F.A. (1975), Plant Sci. Lett., [4], 125—132.
160 West, K.R. and Wiskich, J.T. (1968) Biochem. J., [109], 527—532.
161 West, K.R. and Wiskich, J.T. (1973) Biochim. Biophys. Acta, [292], 197—205.
162 Whitehouse, D.G., Ludwig, L.J. and Walker, D.A. (1971) J. Exp. Bot., [22], 772—791.
163 Wildman, S.G. (1974), Personal communication.
164 Williams, R.J.P. (1975) FEBS Lett., [53], 123—125.
165 Winget, G.D., Izawa, S. and Good, N.E. (1965) Biochem. Biophys. Res. Commun., [21], 438—443.
166 Wilson, D.F., Owen, C., Mela, L. and Weiner, L. (1973) Biochem. Biophys. Res. Commun., [53], 326—333.
167 Wilson, D.F., Erecinska, M. and Dutton, P.L. (1974) Ann. Rev. Biophys. Bioengin., [3], 203—230.
168 Witt, H.T. (1971) Quart. Rev. Biophysics, [4], 365—477.
169 Witt, H.T. (1975) in Bioenergetics of Photosynthesis (Govindjee, ed), pp. 493—554, Academic Press, New York.
170 Wraight, C., Crofts, A.R. and Fleischmann, S. (1971) FEBS Lett., [15], 89—99.
171 Yamashita, T. and Butler, W.L. (1968) Plant Physiol., [43], 1978—1986.
172 Yamashita, T. and Butler, W.L. (1969) Plant Physiol., [44], 435—438.
173 Yin, H.C., Shen, Y.K., Shen, G.M., Yong, S.Y. and Chiu, K.S. (1961) Scientia Sinica, [10], 976—984; 1098—1106.

The Intact Chloroplast — edited by J. Barber

Chapter 5

Energetics of Intact Chloroplasts

GOTTHARD H. KRAUSE and ULRICH HEBER

Botanisches Institut der Universität Düsseldorf, 4 Düsseldorf (Germany)

CONTENTS

Abbreviations: DHAP, dihydroxyacetone phosphate; 1,3-DiPGA, 1,3-diphosphoglycerate; FCCP, carbonyl cyanide 4-trifluoromethoxyphenylhydrazone; FDP, fructose-1,6-diphosphate; GAP, glyceraldehydephosphate; HMP, hexose and heptose monophosphates; KG, α-ketoglutarate; OAA, oxaloacetate; PEP, phosphoenolpyruvate; P_i, inorganic phosphate; PGA, 3-phosphoglycerate; PN^+, oxidized pyridine nucleotides; PNH, reduced pyridine nucleotides; RuDP, ribulose-1,5-diphosphate; SDP, sugar diphosphates.

5.1. INTRODUCTION

Three aspects of chloroplast bioenergetics will be dealt with in this chapter; *(i)* energy export by chloroplasts in the light and energy import in the dark, *(ii)* energy coupling and *(iii)* energy loss during photorespiration. Being confronted with the alternative of confusing the reader with an impartial presentation of the literature or rather presenting our own biased views, we admit to have yielded in part to the temptation posed by the latter.

5.1.1. Energy transfer

Intact chloroplasts possess an envelope consisting of two membranes. Rupture releases soluble constituents including enzymes and chlorophyll-containing membrane systems called thylakoids (from ϑυλακοειδής, sack-like in view of the vesicular nature of the membranes [120]). Isolated thylakoids supplied with the soluble protein ferredoxin are capable of evolving oxygen from water in a rapid light-driven reaction that leads to NADP reduction, proton uptake into the thylakoid interior and phosphorylation of ADP. In the intact cell, the energy conserved in the form of NADPH and ATP is used for CO_2 reduction and thus constitutes the energetic basis of autotrophic plant life, and, through this, also of heterotrophic life on earth.

When chloroplasts, which have preserved their integrity, are supplied with NADP, ADP and phosphate and are then illuminated, little oxygen evolution is observed in the absence of CO_2. These chloroplasts neither exhibit the fast light-driven phosphorylation of added ADP shown by the internal membrane systems nor reduction of exogenous NADP or proton uptake from the medium. However, in the presence of substrates such as bicarbonate or 3-phosphoglycerate, which for their reduction need reduced pyridine nucleotide and ATP, illumination results in fast oxygen evolution demonstrating complex photosynthetic activity and, indeed, functional integrity of the isolated organelles. The conclusion is that the chloroplast envelope prevents free exchange of protons, adenylates and pyridine nucleotides, but not of CO_2 and phosphoglycerate between reactive sites in the chloroplast interior and the environment, which in the cell is the cytosol. Fast reduction of CO_2 and phosphoglycerate inside the chloroplasts is indicative of rapid turnover of small internal adenine and pyridine nucleotide pools which are kept separated from external nucleotides.

The simple view of a barrier for energy exchange with the chloroplast exterior is immediately challenged by observations showing that photosynthetically active light stimulates energy-requiring cell reactions which are located outside the chloroplasts [117]. Also, the cytoplasmic phosphorylation potential (ATP)/(ADP) (P_i) rises on illumination of leaves along with the chloroplast phosphorylation potential. If adenylates cannot readily pass across the chloroplast envelope, this must indicate the operation of indirect transfer systems for photosynthetically generated ATP. Such transfers would permit utilization of photosynthetic energy for cytoplasmic reactions in the light and supply chloroplasts with energy in the dark (see Chapter 12 in this volume).

5.1.2. Coupling

In order to reduce one mole of carbon dioxide to the sugar level via the reactions of the Calvin cycle three moles of ATP and two moles of NADPH are needed. In plant species fixing CO_2 via the C_4 carboxylic acid cycle an even higher ATP/NADPH ratio is required for CO_2 reduction. Additional ATP may be utilized by other energy-requiring reactions in the photosynthesizing cell. Considerable effort has been spent in the past to answer the question of stoichiometry of coupling between photosynthetic electron transport and photophosphorylation of ADP. If a sufficiently high ATP/NADPH ratio is not attained in the normal non-cyclic electron transport, reduction of oxygen, instead of $NADP^+$, or cyclic electron transport via photosystem I are possible means of supplying the additional ATP without further $NADP^+$ reduction [5,7]. From studies on O_2 evolution with substrates requiring different ratios of ATP/NADPH for their reduction, it is now apparent that in intact chloroplasts the ratio of ATP molecules formed per transported pair of electrons is not a fixed number, and that coupling possesses a certain flexibility allowing adjustment to the changing demand for ATP. Even so, phosphorylation during electron transport to $NADP^+$ alone seems insufficient to sustain ATP consumption in chloroplasts actively fixing CO_2.

5.1.3. Photorespiratory energy conversion

An important factor in the bioenergetics of chloroplasts in intact cells is the photorespiratory oxidation of carbohydrates to glycolate. Photosynthetic glycolate formation using intermediates of the carbon reduction cycle assumes significant proportions in intact isolated chloroplasts when the supply of CO_2 is limited or exhausted. While chloroplasts cannot metabolize glycolate, it does not accumulate in intact cells of higher plants as an end product. Glycolate conversion and reentry of conversion products into the carbon reduction cycle require energy. There is the paradoxical situation that oxidation of carbohydrates does not conserve but dissipates energy. The possible significance of this process is that it can use up excess energy and thereby prevent damage to the photosynthetic apparatus when sufficient CO_2 is not available during illumination, e.g. when stomata of leaves are closed under water stress in the light. While this is a beneficial effect, photorespiration also reduces plant productivity under conditions frequently found in the natural environment which permit both assimilation of atmospheric CO_2 and the operation of the glycolate pathway.

5.2. ADENINE AND PYRIDINE NUCLEOTIDES IN INTACT CHLOROPLASTS

5.2.1. Pool sizes

For direct photophosphorylation or reduction, small pools of adenylates and pyridine nucleotides are available in the stroma of intact chloroplasts. Their size has been determined after isolation of chloroplasts in aqueous or non-aqueous media [146,158]. Both methods have advantages and drawbacks. In the latter, leaves are rapidly frozen and freeze-dried, and chloroplasts are isolated from the dried leaf material in mixtures of the non-polar solvents petroleum ether and carbon tetrachloride. This avoids secondary translocation and leakage of hydrophilic solutes such as adenylates and pyridine nucleotides, but results in the loss of functional integrity of the chloroplasts. Cytoplasmic contamination is also a problem [22,63]. Aqueously isolated chloroplasts, on the other hand, remain functionally intact during isolation, but are likely to suffer loss of water-soluble components. Preparations of such chloroplasts are usually not homogeneous and also contain envelope-free chloroplasts, which have lost all water-soluble constituents. Proper corrections of analytical data must therefore be made. Surprisingly, in aqueously isolated chloroplasts, the pyridine nucleotide content was very similar to that in non-aqueous chloroplasts [74,168], indicating that pyridine nucleotides are effectively retained. On the other hand, some loss of adenine nucleotides takes place during aqueous isolation, probably due to leaching across the chloroplast envelope into the medium [76]. This may be seen from Table I which shows some published data on nucleotide levels in isolated chloroplasts. The in vivo concentration of adenylates (AMP+ADP+ATP) in spinach chloroplasts is between 1 and 3 mM. Levels of pyridine nucleotides, NADP(H) and NAD(H), are lower and range from 0.5 to 1.5 mM. Since on illumination endogenous ADP becomes phosphorylated and $NADP^+$ reduced, large transient changes in the proportions of the single nucleotides occur. Increased levels of NADPH and ATP then drive reversible biochemical reactions in the chloroplast stroma in the direction of sugar phosphate synthesis.

The non-aqueous cell fractionation procedure permits an approximate comparison of nucleotide levels in chloroplasts and cytoplasm of leaf cells. Based on dry weight, the total adenylate content of the "cytoplasmic" cell fraction was similar to that of the chloroplasts although ATP/ADP ratios differed [99]. The levels of NADPH were higher in the cytoplasm than in the chloroplasts both in the dark and in the light. Scarcely any $NADP^+$

TABLE I

Nucleotide levels in isolated chloroplasts

Values are from several sources [62,74,75,99,136,168] and were in part recalculated. Concentrations were calculated on the basis that under isotonic conditions the osmotic space of chloroplasts containing 1 mg chlorophyll ($4 \cdot 10^8$ chloroplasts [131]) is 30 μl.

Nucleotides	Plant species	Isolation technique	Nucleotide level (nmoles/ mg chl/h)	Approximate concn. (mM)
ATP+ADP+AMP	*Spinacia oleracea*	non-aqueous	70—90	2.5
	Beta vulgaris	non-aqueous	90	3
	Nicotiana tabacum	non-aqueous	60	2.0
ATP+ADP+AMP	*Spinacia oleracea*	aqueous	37	1.2
ATP (dark)	*Spinacia oleracea*	non-aqueous	44	1.5
ADP (dark)	*Spinacia oleracea*	non-aqueous	25	0.8
AMP (dark)	*Spinacia oleracea*	non-aqueous	4	0.1
ATP (dark)	*Spinacia oleracea*	aqueous	4	0.1
ADP (dark)	*Spinacia oleracea*	aqueous	15	0.5
AMP (dark)	*Spinacia oleracea*	aqueous	18	0.6
$NADPH+NADP^+$	*Nicotiana tabacum*	non-aqueous	10	0.3
$NADH + NAD^+$	*Nicotiana tabacum*	non-aqueous	4	0.1
$NADPH+NADP^+$	*Spinacia oleracea*	non-aqueous	9—19	0.3—0.6
$NADH+ NAD^+$	*Spinacia oleracea*	non-aqueous	15—28	0.5—1
$NADPH+NADP^+$	*Spinacia oleracea*	aqueous	8	0.3
$NADH+NAD^+$	*Spinacia oleracea*	aqueous	9—15	0.3—0.5
NADPH (light)	*Spinacia oleracea*	aqueous	8—20	0.3—0.6

could be detected in the cytoplasm. In contrast to the cytoplasmic NADP system, which was mainly reduced, the NAD system was largely oxidized in the cytoplasm. Only a small percentage of the NAD was in the reduced state. The cytoplasmic NAD level was higher than the chloroplast level by a factor of more than two [75]. The latter appeared to fluctuate, very probably because of a light-dependent phosphorylation of NAD to NADP [125,126].

5.2.2. Changes in the levels of NADPH and NADH in the dark—light transition

In aqueously isolated intact chloroplasts kept in the dark, a surprisingly large percentage of pyridine nucleotides is in the reduced state even in the presence of air levels of oxygen. Between 5 and 25% of the endogenous NADP and about 10% of the NAD were reduced [62,74,168]. A similar proportion of NADP (20%) or NAD (5%) was found reduced, when leaves were frozen in the dark and the chloroplasts were subsequently isolated non-aqueously [75]. On illumination of isolated chloroplasts in aqueous media, large increases in the percentage of reduced NADP, but much less dramatic changes in the redox state of NAD were observed [74]. During illumination in the absence of electron acceptors such as phosphoglycerate, CO_2 or oxaloacetate, usually about 60%, but occasionally up to 90%, of the NADP became reduced. At the same time the level of reduced NAD rose to only 30% of the total NAD in the system. Depending on the light intensity, the percentages of reduced pyridine nucleotides were lower in the presence of electron acceptors such as phosphoglycerate or CO_2 [168]. They were calculated after acidic or alkaline extraction of NAD^+, NADH, $NADP^+$, and NADPH from the chloroplasts. Since it is known that binding of pyridine nucleotides to proteins occurs [87], the percentages of free pyridine nucleotides, which are in the reduced or oxidized state and are thermodynamically available to drive reductive or oxidative reactions, may be very different from those determined after extraction. If, as equilibrium calculations indicate [75], a large part of NADPH and NADH is bound in the dark, the changes in the redox state of free pyridine nucleotides occurring on illumination are much larger than calculated on the basis of total pyridine nucleotide content.

From work by Shin and Arnon [142,143] it is known that electrons are transferred from the photosynthetic electron transport chain much more readily to NADP+ than to NAD+. However, intact chloroplasts contain malate dehydrogenases [60,73,94] and a glyceraldehyde phosphate dehydrogenase [116] which can react both with NAD(H) and NADP(H). Together with endogenous malate and oxaloacetate or glyceraldehyde phosphate and 1,3-diphosphoglycerate, respectively, these enzymes should be expected to equalize the redox state of the NAD and the NADP system and thus exhibit transhydrogenase function. Since, contrary to this expectation, differences in the redox states of the $NADPH/NADP^+$ and $NADH/NAD^+$ couples are maintained, it appears that these systems are kinetically restricted. One of the enzymes in question, NADP-specific malate dehydrogenase, is inactive in the dark and becomes activated in the light [94]. Thus a controlled "transhydrogenase" may function as follows:

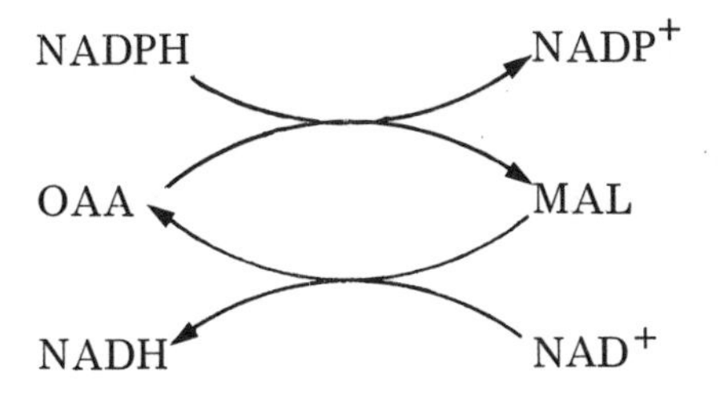

NADP-specific enzyme (controlled)

NAD-specific enzyme (non-limiting activity)

Fast freezing of leaves and subsequent non-aqueous cell fractionation allows analysis of transient changes in the levels of single nucleotides not only in the chloroplasts but also in the cytoplasmic compartment of leaf cells [75]. In contrast to the behaviour of chloroplasts, no significant changes in the redox state of cytoplasmic pyridine nucleotides could be shown to occur in the dark—light and in the light—dark transition. However, binding of NADH could mask transient changes in the proportion of free cytoplasmic NADH.

5.2.3. Changes in the levels of ATP, ADP and AMP in the dark—light transition

Illumination of intact leaves or intact isolated chloroplasts in aqueous media increases the ATP level with similar kinetics as the level of NADPH [62,116,136]. The rise in ATP is accompanied by a drop in ADP and, in many cases, in AMP. The latter indicates the presence of an active adenylate kinase. The total adenylate contents remain constant. For ATP-utilizing processes in the chloroplasts, e.g. the phosphoglycerate reduction, the ratio (ATP)/(ADP) (P_i) is of importance. This term has been designated as phosphate potential or phosphorylation potential [31]. It determines the magnitiude of the free energy change of ATP hydrolysis, or "group transfer potential" of ATP, according to the equation

$$\Delta G' = \Delta G'_0 - RT\ln \frac{(ATP)}{(ADP)\,(P_i)}$$

Chloroplasts are capable of synthesizing ATP with a phosphorylation potential of about 30 000 and a group transfer potential of about 15 kcal/ mole [102]. The equation for ATP hydrolysis

$$ATP^{4-} + H_2O \rightarrow ADP^{3-} + HPO_4{}^{2-} + H^+$$

shows that the proton concentration plays a role in the free energy change of ATP hydrolysis. As reported by Heldt et al. [84], the pH of the chloroplast stroma rises on illumination by almost one pH unit, while that of the intrathylakoid space drops by approx. 2 pH units. In view of the pH dependence of the free energy change of ATP hydrolysis, the alkalization of the chloroplast stroma in the light thermodynamically favours the ATP-consuming reactions of the photosynthetic carbon cycle.

Another term describing the phosphorylation power is the "energy charge" [10] given by the ratio

$$\frac{1}{2} \quad \frac{(ADP) + 2(ATP)}{(AMP) + (ADP) + (ATP)}$$

It describes the availability of high energy phosphate in the adenylate system and can assume all values between 1 (all adenylates in form of ATP) and 0.0 (all adenylates in form of AMP). It provides a measure of the energy status of intact chloroplasts, if there is enough adenylate kinase activity present to allow fast equilibration between ATP, ADP and AMP. This does not seem always to be the case [136]. As experimental data for AMP are often lacking and, on the other hand, the P_i concentration in chloroplasts is usually rather high (in the order of 10^{-2} M) and therefore does not change much during fluctuations of the phosphorylation state of the adenylate system (see however ref. 76), the ATP/ADP ratio is probably the most conveniently available measure of the phosphorylation power of the adenylate system in intact chloroplasts and green cells.

As shown after non-aqueous isolation from intact leaves, the ATP/ADP ratio in chloroplasts of *Spinacia*, *Beta* and *Nicotiana* increases on illumination, even in the presence of CO_2, from values below 1 up to 2 or 3, i.e. by a factor of 3 to 4 [136,100]. Similarly, in leaves of *Elodea* the chloroplast ATP/ADP ratio was about 1.5 in the dark and rose to about 5 in the light

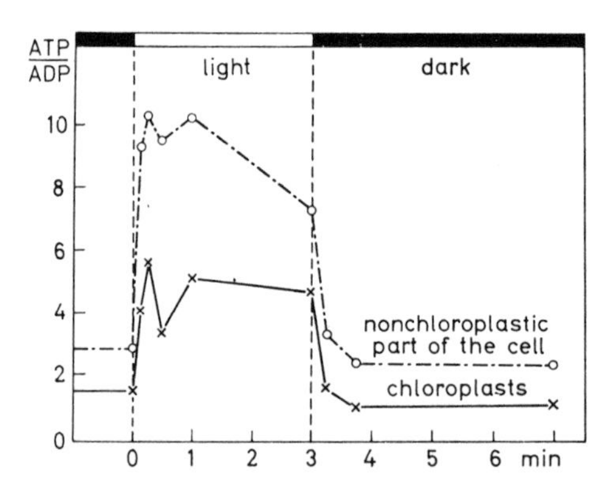

Fig. 5.1. Light-induced changes in the ratio of ATP/ADP in chloroplasts and cytoplasm of leaf cells of *Elodea densa*. (From ref. 76, with permission.)

[76]. Interestingly, the dark—light increase of the ATP/ADP ratio in the chloroplasts is mimicked in the cytoplasm (Fig. 5.1). This clearly indicates a fast transfer of photosynthetic phosphorylation power from chloroplasts to cytoplasm. From Fig. 5.1 the apparent transfer rate can be calculated to be almost 10 μmoles/mg chlorophyll/h. This is a minimal estimate as it neglects ATP utilization in the cytoplasm. The real transfer of phosphate energy may be much faster.

Moreover, Fig. 5.1 shows that the cytoplasmic ATP/ADP ratio is higher by a factor of about 2, both in the light and in the dark. Even larger differences between chloroplastic and cytoplasmic ATP/ADP ratios have been observed by Keys and Whittingham [100] in tobacco leaves. Assuming that the P_i level is nearly the same in chloroplasts and cytoplasm, it must be concluded that the phosphorylation potential of the cytoplasm is higher than that of the chloroplasts even in the light. This would exclude unrestricted permeation of adenylates across the chloroplast envelope. From work of Klingenberg et al. [101] it is known that respiring mitochondria also maintain an internal phosphorylation potential which is lower than that of the surrounding medium. They export ATP against a concentration gradient. While the redox state of pyridine nucleotides in isolated chloroplasts appears to be similar to that in chloroplasts of an intact leaf both in the dark and in the light, it is remarkable that the energy state of the adenylate system of darkened isolated chloroplasts is very different from that of chloroplasts in situ [76]. Table I shows that in vivo the energy charge of chloroplasts is kept at a high value (about 0.8) even in the dark, whereas in aqueously isolated chloroplasts it falls to about 0.3. In these chloroplasts AMP represents about 50% of the total adenylates as compared to about 6% in chloroplasts isolated non-aqueously from darkened leaves. Light, therefore, increases the ATP level in aqueously isolated chloroplasts by a much larger factor than in vivo [62]. Without doubt, the adenylate pools found in non-aqueous chloroplasts represent the in vivo situation much more closely than the adenylates in aqueously isolated chloroplasts. In the latter, not only is some adenylate lost, but the phosphorylation potential is lowered in the dark by ATP requiring reactions or ATPase activity to a value never attained in the intact leaf cell. In the dark, isolated chloroplasts are unable to synthesize ATP from endogenous energy sources. The high energy charge of the adenylate system of chloroplasts in situ then suggests import of energy from the cytoplasm in the dark.

Changes in the phosphorylation potential of intact chloroplasts can conveniently be followed both in vitro and in vivo by recording the chlorophyll fluorescence and photo-induced increase in light scattering by chloroplasts [61,104]. Both phenomena reflect in part energy-dependent conformational changes of the thylakoid membranes. The former also reflects the redox state of the electron transport chain [35], the latter shrinkage of the thylakoids. Usually the increase in scattering of a 535 nm measuring beam seen after onset of actinic illumination is accompanied by chlorophyll fluorescence quenching. The kinetics of both processes are very similar. Recently it has been shown that both phenomena are indicative of cation fluxes across the thylakoid membranes [14–16,105–107]. Light-induced H^+ uptake by the thylakoids is thought to be largely compensated for by counterflow mainly of divalent cations such as Mg^{2+} (see Chapter 3). In intact chloroplasts high ion gradients are maintained by light between thylakoid and stroma space, when the phosphorylation potential is high, i.e. when insufficient ADP is available to dissipate the H^+ gradient by fast phosphorylation. This state corresponds with a low fluorescence yield and comparatively high scattering. When ATP is utilized in fast biochemical reactions, the phosphorylation potential is lowered and, as more ADP is now available for phosphorylation, the ion gradients become smaller [85]. This is reflected by a fluorescence increase and scattering decrease. Direct comparisons have, indeed, shown a close correlation between ATP level and light scattering by chloroplasts of the green alga *Dunaliella parva* [54]. Fluorescence and light scattering thus permit monitoring of light-dependent changes in the phosphorylation potential of intact chloroplasts, both in preparations of the isolated organelles and in situ, without disturbance of the systems studied.

5.2.4. ATP and NADPH supply in chloroplasts during darkness

Precise rates of ATP consumption by chloroplast metabolism in darkened leaves are unknown. As indicated by the rate of respiratory oxygen uptake, about 30 μmoles/mg chlorophyll/h of ATP are supplied by glycolysis and respiration in dark-treated spinach leaves at 20°C. Assuming that ATP utilization by endergonic reactions corresponds to protein distribution in green cells, the chloroplasts would consume close to 20 μmoles ATP/mg chlorophyll/h, as they contain about 60% of the total cell protein [60,145]. This is probably an overestimate, but it should be safe to conclude that ATP consumption by chloroplast metabolism in the dark is not negligible. As direct ATP transport across the chloroplast envelope is very slow (see section 5.3.2), the necessary import of energy from the cytoplasm appears to proceed mainly via an import of triosephosphate, which can be oxidized inside the chloroplasts to yield ATP, reduced pyridine nucleotide and phosphoglycerate, which is returned to the cytoplasm. This energy import system is discussed in Chapter 6. A full glycolytic reaction sequence is not possible in chloroplasts as they lack glycolytic enzymes such as pyruvate kinase [68].

It has been mentioned that the chloroplast NADP system is kept reduced to some extent even during darkness. This suggests that chloroplasts possess an enzyme system capable of producing NADPH in the dark. Indeed, not only the enzymes of the reductive pentose phosphate cycle of photosynthesis, which operates under NADPH oxidation, but also those of the oxidative pentose phosphate pathway are present in chloroplasts. The oxidative reactions catalyzed by glucose-6-phosphate dehydrogenase and by 6-phosphogluconate dehydrogenase, which were demonstrated to occur in chloroplasts [68,cf.3], produce NADPH and proceed with a large change in free energy. Actually, under the substrate conditions found in chloroplasts in situ, NADP should be almost completely reduced at the equilibrium of these reactions [21]. Since only some 20% of the NADP is reduced in the dark, action of the chloroplast enzymes appears to be under kinetic restriction. Control of these enzymes is indeed evident, as in *Chlorella* cells measurable levels of 6-phosphogluconate, which together with NADPH is the product of glucose-6-phosphate oxidation, appear only after darkening. This indicates light inactivation of glucose-6-phosphate dehydrogenase [18]. In intact isolated chloroplasts addition of Vitamin K_5 inhibits photosynthesis apparently by preventing NADP reduction. High levels of 6-phosphogluconate can be observed immediately after the inhibitor is added [109]. This suggests that the site of glucose-6-phosphate dehydrogenase activation is the chloroplast and that the enzyme is controlled by the $NADP^+$/NADPH ratio. More recent studies by Lendzian and Ziegler [112] confirmed that chloroplast glucose-6-phosphate dehydrogenase, but not 6-phosphogluconate dehydrogenase, is inactivated by light. NADPH appeared to be the natural inhibitor of glucose-6-phosphate oxidation. Thus it seems that during periods of darkness chloroplasts depend on external phosphorylation energy but are able to produce NADPH for synthetic reactions by oxidation of endogenous carbohydrates, This oxidation is turned off in the light, possibly by allosteric action of the increased level of NADPH. Alternatively, NADPH could act indirectly by reducing disulphides. Recently, Anderson et al. [3] provided evidence that glucose-6-phosphate dehydrogenase is inhibited by dithiothreitol. As this agent could also substitute for light in activating certain enzymes of the carbon reduction cycle, they suggested that a reductive process taking place in the light is responsible for activation of these enzymes and for inactivation of glucose-6-phosphate dehydrogenase.

5.3. THE CHLOROPLAST ENVELOPE AS A TRANSPORT BARRIER

The chloroplast envelope, as a lipid-protein biomembrane, has a low permeability towards ionic substances. However, a limited number of ionic compounds are allowed to permeate the envelope by means of specific translocators situated in the inner membrane of the envelope. The outer membrane appears to be open to unspecific diffusion. The details of these transport systems operating across the chloroplast envelope are discussed by Heldt in Chapter 6.

5.3.1. *Pyridine nucleotides*

Several lines of evidence indicate that the chloroplast envelope does not permit penetration of pyridine nucleotides. Intact chloroplasts do not lose pyridine nucleotides during isolation, nor can added pyridine nucleotides enter the chloroplast stroma. Thus, exogenous $NADP^+$ does not support light-dependent oxygen evolution [32,68,75,114,135]. In broken, envelope-free chloroplasts $NADP^+$ quenches the chlorophyll *a* fluorescence peak seen during the fluorescence induction period, because $NADP^+$ stimulates non-cyclic electron transport. No such effect is observed if $NADP^+$ is added to intact chloroplasts [108]. Furthermore, NAD- or NADP-dependent enzymes of intact chloroplasts are "cryptic"; they do not react with added cofactors and show activity only after destruction of the chloroplast envelope [68]. This clearly shows that there is neither significant unspecific permeation nor specific exchange of oxidized and reduced pyridine nucleotides across the chloroplast envelope.

5.3.2. *Adenine nucleotides*

The permeability of the chloroplast envelope towards adenylates, though larger than the permeability to pyridine nucleotides, is small in proportion to normal rates of ATP formation and breakdown in green plant cells. Several experiments suggest some penetration of adenylates, although the rates measured were usually very low.

There are always exceptions to the rule. Kraayenhof [102] and West and Wiskich [166] reported fast phosphorylation of external ADP and photosynthetic control in what they termed intact chloroplasts. At first sight this suggests fast adenylate penetration. As electron acceptor, they used ferricyanide, for which the envelope of intact chloroplasts is impermeable [76]. It therefore appears that the response they observed was caused by broken chloroplasts contaminating their preparations. In a recent report, Avron and Gibbs [11] concluded that penetration of ATP was not rate-limiting for phosphorylation of ribulose-5-phosphate by intact chloroplasts which proceeded at high rates. Again it is quite clear from the data presented that their preparations were heavily contaminated by broken chloroplasts and freed stroma enzymes. It is therefore probable that the reactions observed did not occur inside the intact chloroplasts but in the medium. The results reported by Santarius et al. [137] on a fast labelling of chloroplast ATP and ADP in the dark by $H^{32}PO_4^{2-}$ in intact leaves now have to be explained by ATP production from triosephosphate and adenylate kinase action inside the chloroplasts (see section 5.4) and can no longer be regarded as indicating fast direct adenylate transfer. Bassham et al. [20] reported excretion of ^{32}P-labelled ATP from illuminated chloroplasts and its reentry on darkening. ATP, added in the dark, increased the level of ribulose diphosphate, which indicated its utilization in chloroplast phosphorylation reactions [93]. No penetration rates were calculated.

Externally added ATP was found to support a slow hexokinase reaction in intact chloroplasts [68]. Photophosphorylation of external ADP by intact chloroplasts was also very slow if observed rates were corrected for the activity of broken chloroplasts which always contaminate preparations of intact chloroplasts. The percentage of broken chloroplasts can be determined by measuring the rate of the Hill reaction with ferricyanide, $Fe(CN)_6^{3-}$, as this reagent does not penetrate the chloroplast envelope [76,113]. Exogenous ADP was phosphorylated by intact chloroplasts at a rate of about 2 μmoles/mg chlorophyll/h. Similarly, in the dark intact chloroplasts converted added ATP+AMP to ADP with rates between zero and 3 μmoles/mg chlorophyll/h. The reaction is catalysed by chloroplast adenylate kinase which, after rupture of the envelope, formed 50 to 120 μmoles ADP/mg chlorophyll/h. Thus the rates seen with intact chloroplasts represent the permeation of the slowest-moving adenylate.

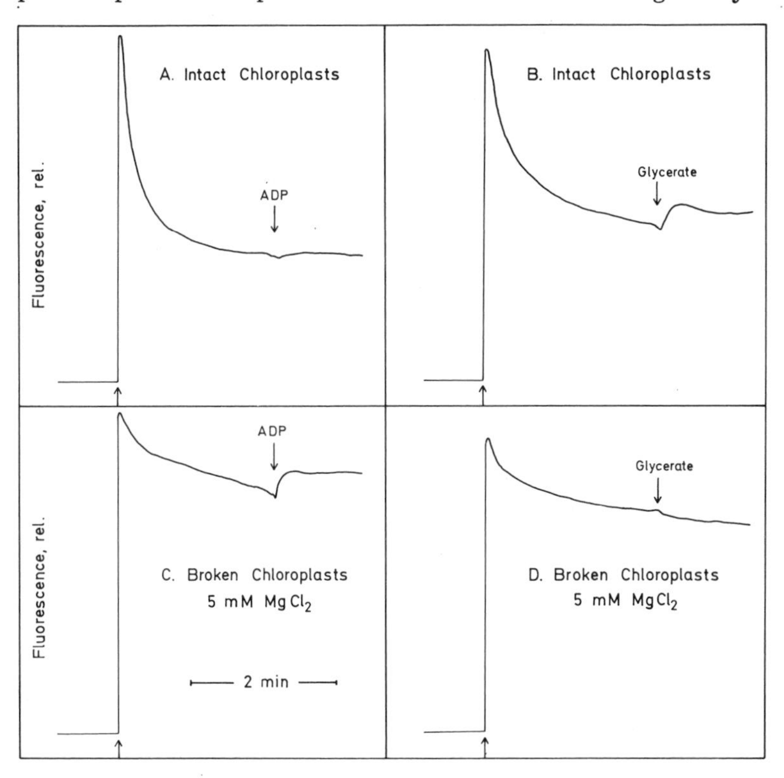

Fig. 5.2. Effects of ADP and glycerate on 740 nm chlorophyll fluorescence of intact and envelope-free chloroplasts. Upward arrows denote start of illumination (10 W/m^2 ; half band width ranging from 620 to 680 nm). (From ref. 64, with permission.)

According to Strotmann and Heldt [150], Strotmann and Berger [149] and Heldt [78], a specific exchange translocation mechanism facilitates counter-exchange of ATP, ADP and AMP between intact chloroplasts and their surrounding medium. The translocator has only a very low capacity. Exchange of external ATP against internal adenylates proceeded with a maximal rate of 5 μmoles/mg chlorophyll/h. ADP and AMP uptake were much slower. The absence of fast ADP—ATP exchange across the chloroplast envelope is also evident from chlorophyll fluorescence measurements (Fig. 5.2). The energy-dependent fluorescence quenching which follows the rise to the peak of fluorescence induction, reflects export of cations from the thylakoids in exchange for protons taken up [105,106] (and see Chapter 3 of this volume). Photophosphorylation of ADP lowers the proton gradient [85] and this should partly reverse the fluorescence quenching. As seen from Fig. 5.2, ADP addition in the light increases only the fluorescence of broken chloroplasts. The fluorescence of intact chloroplasts, on the other hand, is affected by glycerate which penetrates the envelope. Its phosphorylation by ATP increases the internal ADP level and thereby increases fluorescence. This evidence shows that external ADP is unable to increase significantly the internal ADP pool.

Exogenous ATP cannot be utilized at high rates in phosphoglycerate reduction by intact chloroplasts. This has conclusively been shown by Stokes and Walker [148]. After uncoupling of photophosphorylation, photosynthetic phosphoglycerate reduction could be restored by ATP added to re-constituted systems consisting of envelope-free chloroplasts, $NADP^+$, ferredoxin and chloroplast extract. However, in uncoupled intact chloroplasts, added ATP was unable to restore phosphoglycerate reduction.

Thus, in vivo the direct transfer of ATP, ADP and AMP, resulting from diffusion and specific translocation between chloroplasts and cytoplasm, probably is limited to rates lower than 5 μmoles/mg chlorophyll/h. This is insufficient to account for fast dark—light transient changes in cytoplasmic ATP levels observed in vivo and for energy-dependent glucose uptake by cells and green algae [96]. It is doubtful, whether it can account for ATP utilization by chloroplasts in darkness. Since direct transfer of adenylates is very slow, indirect transport systems of higher capacity appear to be involved in the transfer of phosphate energy in vivo (see section 5.4.1).

5.3.3. *Protons*

Illumination of chloroplasts leads to uptake of protons into the thylakoids [92]. The resulting proton gradient represents, according to Mitchell's chemiosmotic hypothesis [122], the driving force of photophosphorylation (see Chapters 2 and 4 of this volume). The proton uptake causes the thylakoid exterior, which in intact chloroplasts is the stromal space, to become alkaline. This alkalization is not transmitted to the cytoplasm [73]. The lower

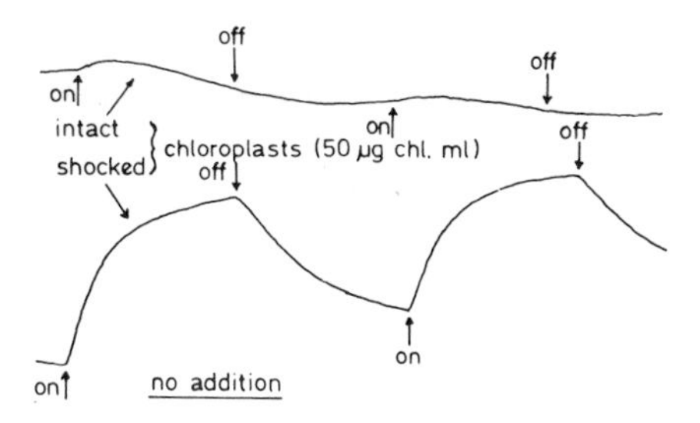

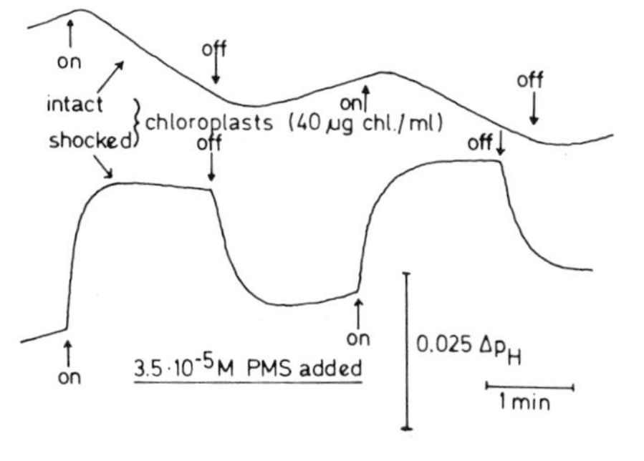

Fig. 5.3. Light-induced pH changes in suspensions of intact and envelope-free (osmotically shocked) chloroplasts. Light on, upward arrows; light off, downward arrows. Upper part, no addition; lower part, with PMS added. (From ref. 73, with permission.)

traces in Fig. 5.3 depict the reversible light-induced alkalization in the medium containing broken chloroplasts. However, before rupture of the envelope (Fig. 5.3, upper traces) no alkalization of the medium is observable, but rather some excretion of protons from the chloroplasts is apparent in the light (the slight upward deflection of the trace at the beginning of illumination is due to proton uptake by a small percentage of broken chloroplasts in the preparation). Heldt et al. [84] calculated the excretion of protons by intact chloroplasts to be less than 10% of the proton uptake by thylakoids. The pH of the stroma increased during illumination by almost one pH unit. It is presently unknown whether connections between thylakoid space and the intermembrane space of the envelope are responsible for light-dependent proton efflux from intact chloroplasts, or whether an independent proton pump operates in the envelope. Preparations of the envelope have been shown to exhibit ATPase activity [34]. Recently Gimmler et al. [55] demonstrated that light-induced excretion of protons from intact chloroplasts is accompanied by a specific K^+ uptake. These measurements confirmed indirect observations by Krause [105,106] and Barber and Telfer [14] and a report by Pflüger [129] that the chloroplast envelope represents a barrier not only to protons, but also to metal cations. Consequently, in vivo light creates a proton gradient between chloroplast stroma and cytoplasm in addition to the gradient between thylakoids and stroma. Due to secondary transport processes these proton gradients are probably to a large extent balanced by metal cation gradients. It should be noted that the stroma pH is not entirely independent of the pH of the surrounding medium, as pH changes in the latter are transmitted to the stroma of isolated intact chloroplasts. This may occur by slow permeation of protons or buffer substances [72].

5.4. SHUTTLE MECHANISMS FOR INDIRECT TRANSFER OF ATP AND REDUCING POWER BETWEEN CHLOROPLASTS AND CYTOPLASM

5.4.1. Indirect ATP transfer by the dihydroxyacetone phosphate/phosphoglycerate shuttle

In the dark, energy demands of leaf cells are met by glycolytic and respiratory ATP production. Respiratory activity is known to be controlled by the phosphorylation potential of mitochondria and of the cytosol [31]. Respiration is increased, when the ratio (ATP)/(ADP) (P_i) falls, and is decreased when it rises. The onset of photophosphorylation in the light leads to a fast increase not only of the chloroplast phosphorylation potential but also of that in the cytoplasm. Apparently, this has a twofold effect on cytoplasmic metabolism. First, it stimulates ATP-consuming reactions. Second, it inhibits glycolysis and mitochondrial respiration at the ATP/ADP-sensitive control points [86,92,96,133]. During illumination, the cytoplasm utilizes ATP at rates possibly severalfold higher than in the dark. The increased ATP/ADP ratio drives ATP-consuming reactions of normal dark metabolism faster in the light. In addition, primary photosynthetic products imported from the chloroplasts are metabolized in the cytoplasm in energy-requiring reactions. For instance sucrose is synthesized in the cytoplasm from photosynthetically generated triosephosphate by consuming phosphate energy [20,63,160]. The observation by some workers [44,51] that sucrose can also be formed by isolated chloroplasts probably may be explained by cytoplasmic contamination of the chloroplast preparations used. Recent enzyme localization studies have led to the conclusion that in leaves sucrose cannot be synthesized in the chloroplasts [23]. The chloroplast envelope also does not permit passage of sucrose [83,123,163]. The largest part of the phosphate energy for cytoplasmic sucrose synthesis would have to come from the chloroplasts, since respiratory activity is controlled and decreased by the rise in the phosphorylation potential. The required energy transfer to the cytoplasm probably exceeds by far the capacity of direct ATP transport. The most efficient indirect transport system for phosphorylation energy appears to be a shuttle system involving the transport of 3-phosphoglycerate and dihydroxyacetone phosphate across the chloroplast envelope [63,47,161]. Transfer of these phosphate esters can be mediated at high rates by a special carrier which is situated in the inner membrane of the envelope [80,83,164] (and see Chapter 6 of this volume). In the light 3-phosphoglycerate is reduced inside the chloroplasts by consuming ATP and reducing equivalents. The product of the reaction, glyceraldehyde-3-phosphate, is isomerized to dihydroxyacetone phosphate, which is exported from the chloroplasts. It can, after reconversion into glyceraldehyde-3-phosphate, be oxidized outside through part of the glycolytic sequence to yield phosphoglycerate. Thereby ADP is phosphorylated and NAD^+ becomes reduced. PGA may return to the chloroplasts and its photosynthetic reduction completes the cycle depicted in Fig. 5.4. The result is export of phosphorylation energy along with reducing equivalents in a ratio of 2H/ATP. In the dark the shuttle would operate in the opposite direction and serve for energy transport from the cytoplasm to the chloroplasts. A simpler version of an ATP transfer shuttle, with 1,3-diphosphoglycerate instead of dihydroxyacetone phosphate serving as the metabolite to be exported from the chloroplasts, probably does not operate because movement of 1,3-diphosphoglycerate across the envelope is too slow [76,64]. Stocking and Larson [147] first reported photosynthetic reduction of extraplastidic NAD^+ in vitro by the dihydroxyacetone phosphate/phosphoglycerate shuttle system. Heber and Santarius [76] found rates between 40 and 50 μmoles/mg chlorophyll/h of light-dependent phosphorylation of external ADP, if isolated chloroplasts were supplied with phosphoglycerate, ADP, P_i and the required cytoplasmic enzymes. In this case the phosphoglycerate is reduced photosynthetically in the chloroplasts.

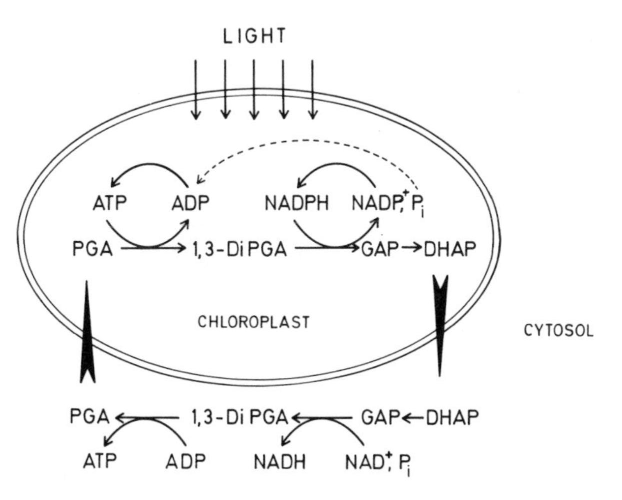

Fig. 5.4. Scheme for the dihydroxyacetone phosphate/phosphoglycerate shuttle.

Dihydroxyacetone phosphate is exported and is oxidized in the external medium. Oxidation is accompanied by substrate level phosphorylation of ADP. In isolated chloroplasts actively fixing CO_2, part of the dihydroxyacetone phosphate formed in the carbon reduction cycle is needed, together with additional ATP, for regeneration of the CO_2 acceptor molecule ribulose-1,5-diphosphate (see Chapter 7 for details of the carbon reduction cycle). In the absence of added phosphoglycerate, these chloroplasts still were able to export more than 20 μmoles ATP/mg chlorophyll/h via the shuttle system [103]. It could be shown that the rate of energy transfer is controlled by mass action through the ATP/ADP and NADH/NAD^+ ratios in the external medium. Thus an increasing phosphorylation potential and NADH/NAD^+ ratio in the cytoplasm would slow down further transfer to the cytoplasm, as the free energy change of the triosephosphate oxidation then approaches zero. So the shuttle would adjust itself to the energy requirement of the cytoplasm. A major obstacle for smooth operation of the dihydroxyacetone phosphate/phosphoglycerate shuttle for energy export to the cytoplasm in the light, however, seems to be the stoichiometric coupling between phosphorylation and NAD^+ reduction resulting in the cytoplasm. As mitochondrial NADH oxidation is slowed down by the increased cytoplasmic ATP/ADP ratio, the only way to remove excess reducing equivalents appears to be their reimportation into the chloroplasts and consumption in the carbon reduction cycle. Given the impermeability of the chloroplast envelope toward pyridine nucleotides, an indirect transfer system for reducing power, independent of the dihydroxyacetone phosphate/phosphoglycerate shuttle, is required.

5.4.2. Indirect transfer of reducing power by the malate/oxaloacetate shuttle

A shuttle transfer of glycolate/glyoxylate has been visualized to serve for indirect transport of reduced pyridine nucleotides [154,155]. However, experiments designed to prove operation of such a shuttle in vitro gave no indication that intact chloroplasts can reduce glyoxylate or oxidize glycolate at significant rates [74]. However, in the presence of oxaloacetate, intact chloroplasts were shown to evolve oxygen on illumination. They are capable of reducing oxaloacetate to malate. Transfer of oxaloacetate and of malate across the chloroplast envelope is catalyzed by a carrier which is specific for the transport of dicarboxylates [81,83] (and see Chapter 8 of this volume). While well-washed intact chloroplasts cannot reduce NAD^+ in the light, photosynthetic reduction of external NAD^+ can be observed, when intact isolated chloroplasts are supplied with NAD^+, malate and NAD-specific malate dehydrogenase [69,73,74]. In the dark, a small part of the added malate will be oxidized by the NAD^+ to oxaloacetate, until equilibrium is achieved. On illumination the oxaloacetate is reduced in the chloroplasts and the malate so formed is reexported to the medium where it again reduces NAD^+ (Fig. 5.5). As in the case of the dihyroxyacetone phosphate/phosphoglycerate shuttle, external $NADH/NAD^+$ ratios rising above a certain limit (about 0.4) will stop further export of reducing equivalents from the chloroplasts. In the presence of phosphoglycerate, which utilizes endogenous reducing power, the shuttle can be reversed, i.e. in the light external NADH is oxidized. This observation, which at first sight seems paradoxical, clearly proves that, depending on the redox potential of pyridine nucleotides inside and outside of the chloroplasts, the shuttle can operate in both directions. The capacity of the malate/oxaloacetate shuttle for transport of reducing equivalents is not high under physiological conditions, even though at 20°C the carrier can transfer about 200 μmoles oxaloacetate or malate per mg chlorophyll/h under substrate saturation. In contrast to this high value, only about 10 μmoles/mg chlorophyll/h reduction or oxidation of external NAD have been observed under conditions resembling those existing in leaves. The rate-limiting step in the shuttle probably is the transfer of oxaloacetate [64]. Its concentrations (0.01 to 0.02 mM in spinach leaves) are well below the K_m of the translocator for oxaloacetate (0.05 mM) and therefore may limit its transfer.

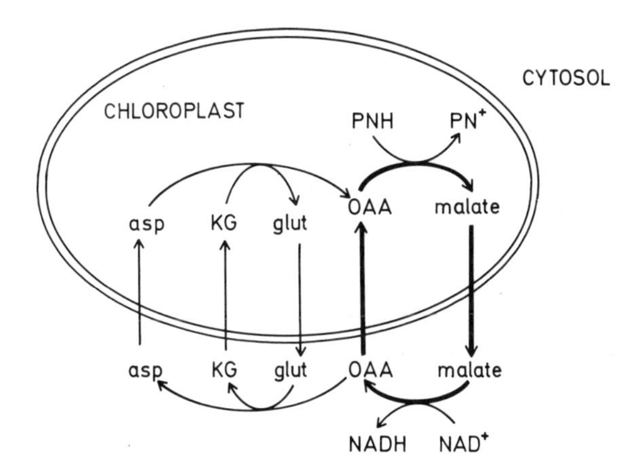

Fig. 5.5. Shuttle systems for transfer of reducing equivalents across the chloroplast envelope: the malate/oxaloacetate and malate/aspartate systems; schematic representation shows transfer from chloroplast stroma to the cytoplasm. The shuttle is reversible and can, depending on conditions, operate in the opposite direction.

5.4.3. *The malate/aspartate shuttle*

Because of the restriction of oxaloacetate transfer across the chloroplast envelope, in vivo the system for transport of reducing equivalents probably includes the reactants of the glutamate oxaloacetate transaminase reaction, aspartate, glutamate and α-ketoglutarate. The transaminase is present both in chloroplasts and cytoplasm [60,138] and all of the carboxylic anions taking part in the reaction can be transferred by the dicarboxylate translocator (see Chapter 6). As under equilibrium conditions the aspartate concentration is much higher than the oxaloacetate concentration, the extended shuttle system depicted in Fig. 5.5 would be more efficient than the simpler malate/oxaloacetate shuttle. Indeed, stimulation of external NAD^+ reduction by illuminated intact chloroplasts has been observed, when the components of the transaminase reaction were added to the medium [64].

5.4.4. *Indirect NADPH transfer to the cytoplasm in the dihydroxyacetone phosphate/phosphoglycerate shuttle*

Recently, Kelly and Gibbs [98] have demonstrated a dihydroxyacetone phosphate shuttle capable of transferring reducing equivalents from the chloroplasts into the cytoplasm. The cytosol of leaf cells contains a non-phosphorylating NADP-specific glyceraldehyde phosphate dehydrogenase which catalyzes the reaction.

$$\text{glyceraldehyde phosphate} + NADP^+ \rightleftharpoons \text{phosphoglycerate} + NADPH + H^+.$$

Since equilibrium is far on the right side, oxidation of triosephosphate exported from the chloroplasts by this enzyme would produce phosphoglycerate, which can be returned to the chloroplasts for reduction, and result in practically complete reduction of cytoplasmic NADP. It has, in fact, long been known that only a small percentage of cytoplasmic NADP occurs in the oxidized state [75]. However, $NADP^+$ reduction in the cytoplasm can also be explained by action of glucose-6-phosphate dehydrogenase, which occurs with high activity in the cytoplasm of leaf cells [68]. Both reactions may compete to keep the cytoplasmic NADP reduced, while the cytoplasmic NAD system is largely oxidized. No connection appears to exist in the cytosol between the two pyridine nucleotide systems.

5.4.5. *Integration of shuttle systems*

Fig. 5.6 shows a scheme for the transfer of phosphorylation energy in the light by cooperation of the dihydroxyacetone phosphate/phosphoglycerate shuttle with the malate/oxaloacetate and the malate/aspartate shuttles.

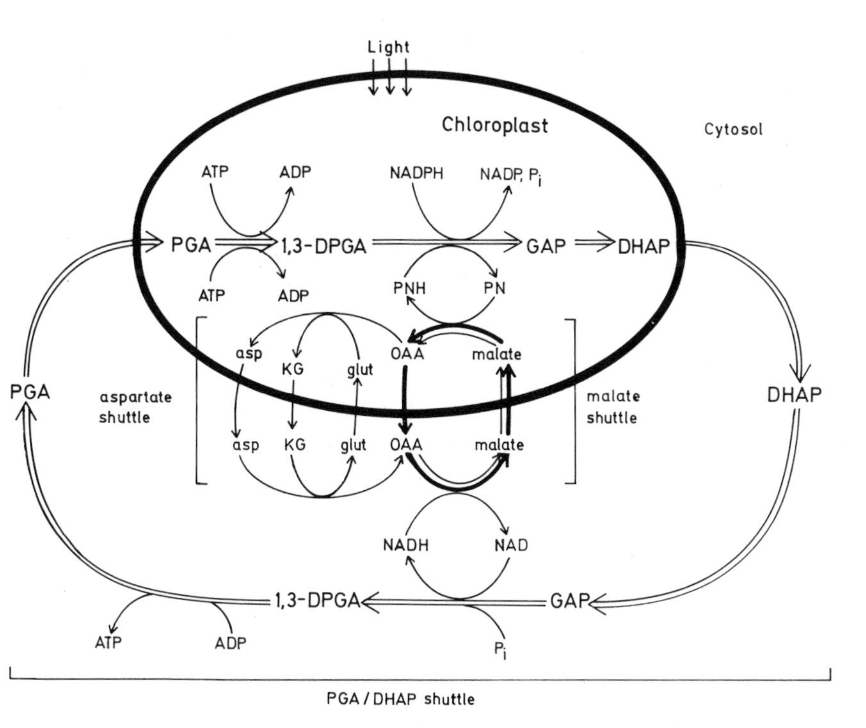

Fig. 5.6. Schematic representation of metabolite transfer between chloroplasts and cytosol designed to export ATP from chloroplasts in the light. Shuttle transfer of dihydroxyacetone phosphate and phosphoglycerate mediates indirect transport of NADH and ATP. Back transfer of NADH is possible by cyclic transfer of malate and oxaloacetate or of malate, glutamate, α-ketoglutarate and aspartate. (From ref. 64, with permission.)

In this system excess reducing equivalents which are transferred to the cytoplasm along with ATP by the dihydroxyacetone phosphate/phosphoglycerate shuttle, are reimported into the chloroplasts for consumption in photosynthetic carbon reduction.

At first sight, the complex shuttle interplay should be expected to equalize differences between chloroplasts and cytoplasm in the phosphorylation potential, as well as in the redox potential of the NADH/NAD$^+$ couple. However, this is not the case. As mentioned above, in vivo a higher phosphorylation potential is maintained in the cytoplasm in the light and in the dark [76,100]. Also, in the light, the NADH/NAD$^+$ ratio is lower in the cytoplasm than in the chloroplasts [75]. Consequently, the indirect transfer of ATP to the cytoplasm and back transfer of reducing equivalents to the chloroplasts, as postulated in Fig. 5.6 must occur "uphill" against existing gradients. Evaluation of published data suggests that, as the "driving force" for such an uphill transport, ion gradients maintained between the two compartments come into play [64]. Protons take part in the reactions of the shuttles. The condition for thermodynamic equilibrium (net flux = zero) of the malate/oxaloacetate shuttle is

$$\frac{(\mathrm{OAA})(\mathrm{PNH})(\mathrm{H^+})}{(\mathrm{malate})(\mathrm{PN^+})}_{\mathrm{chloropl.}} = \frac{(\mathrm{OAA})(\mathrm{NADH})(\mathrm{H^+})}{(\mathrm{malate})(\mathrm{NAD^+})}_{\mathrm{cytopl.}} = \mathrm{const.} \tag{1}$$

If the ratio OAA/malate is similar in chloroplasts and cytoplasm, alkalization of the chloroplast stroma observed under illumination [73,79,84] would, as a consequence of the participation of protons in the malate dehydrogenase reaction, lead to and then maintain different redox ratios in chloroplast and cytoplasmic pyridine nucleotides. Generation of NADH in the cytoplasm above the equilibrium ratio of $NADH/NAD^+$ must result in a flux of reducing equivalents into the chloroplasts even if the ratio of reduced to oxidized pyridine nucleotide is higher there than in the cytoplasm. The driving force for uphill transfer of reducing equivalents is the proton gradient across the chloroplast envelope.

Considerations similar to those valid for the malate shuttle also apply to other shuttle systems which use protons as a reactant.

The equilibrium condition of the reactions of the DHAP/PGA shuttle is

$$\frac{(\mathrm{PGA})(\mathrm{ATP})(\mathrm{PNH})(\mathrm{H^+})}{(\mathrm{DHAP})(\mathrm{ADP})(\mathrm{P_i})(\mathrm{PN^+})}_{\mathrm{chloropl.}} = \frac{(\mathrm{PGA})(\mathrm{ATP})(\mathrm{NADH})(\mathrm{H^+})}{(\mathrm{DHAP})(\mathrm{ADP})(\mathrm{P_i})(\mathrm{NAD^+})}_{\mathrm{cytopl.}} = \mathrm{const.} \tag{2}$$

Since the concentrations of pyridine nucleotides available as reactants for the DHAP/PGA shuttle are the same as those for the malate shuttle, it is possible to divide Eqn. 2 by Eqn. 1 and thereby eliminate the pyridine nucleotides from the following consideration. The resultant equation is

$$\frac{(\mathrm{PGA})(\mathrm{ATP})(\mathrm{malate})}{(\mathrm{DHAP})(\mathrm{ADP})(\mathrm{P_i})(\mathrm{OAA})}_{\mathrm{chloropl.}} = \frac{(\mathrm{PGA})(\mathrm{ATP})(\mathrm{malate})}{(\mathrm{DHAP})(\mathrm{ADP})(\mathrm{P_i})(\mathrm{OAA})}_{\mathrm{cytopl.}} \tag{3}$$

This shows that under simultaneous operation of the two shuttles the phosphorylation potential $(\mathrm{ATP})/(\mathrm{ADP})(\mathrm{P_i})$ in chloroplasts and cytoplasm at equilibrium, is independent of the H^+ gradient as well as of the redox state of the pyridine nucleotides. It is a function of the distribution of phosphoglycerate, dihydroxyacetone phosphate, malate and ocaloacetate between chloroplasts and cytoplasm. The known properties of the dicarboxylate translocator, which is responsible for the transfer of oxaloacetate, malate and the components of the aspartate cycle, suggest that OAA/malate ratios are not very different in chloroplasts and cytoplasm.

ATP/ADP ratios are known to be higher in the cytoplasm than in the chloroplasts. If the shuttle systems really play the role in energy transfer assigned to them, it must be postulated that PGA/DHAP ratios should be higher in the chloroplast. Indeed, in illuminated *Elodea* leaves the DHAP/PGA ratio in the cytoplasm was 3 to 5 times higher than in the chloroplasts (Fig. 5.7). As seen from Eqn. 3 such a high gradient does not only explain the high cytoplasmic ATP/ADP ratio which was usually twice the ratio found in the chloroplasts, but would actually constitute the driving force for the transport of phosphate energy. So far, it is unknown how the unequal distribution of DHAP/PGA is achieved. Certain properties of the translocators involved may come into play. Alternatively, PGA^{3-}, which probably is transported as PGA^{2-} [63], might be trapped according to the Henderson/Hasselbach equation in the stroma, which in the light is more alkaline than the cytosol of leaf cells. The size of the light-dependent pH-difference across the chloroplast envelope is sufficient to explain PGA accumulation in the stroma. If this interpretation is correct, the proton gradient would be responsible for both uphill transfer of reducing equivalents back into the chloroplasts in the light and for uphill export of ATP into the cytoplasm.

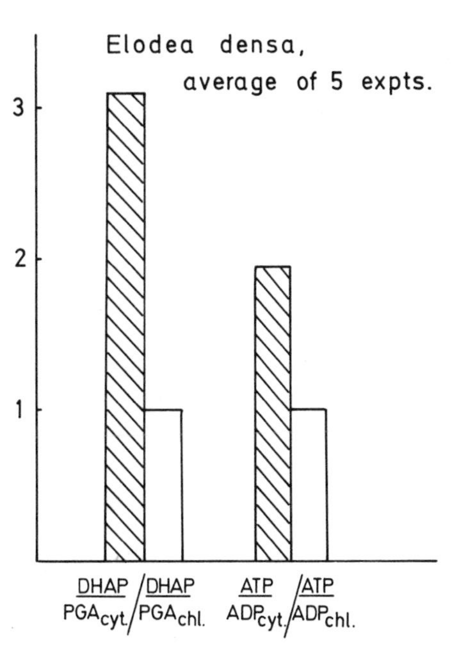

Fig. 5.7. Differences between the ratios of DHAP/PGA and ATP/ADP in cytoplasm and chloroplasts. (From Ref. 64, with permission.)

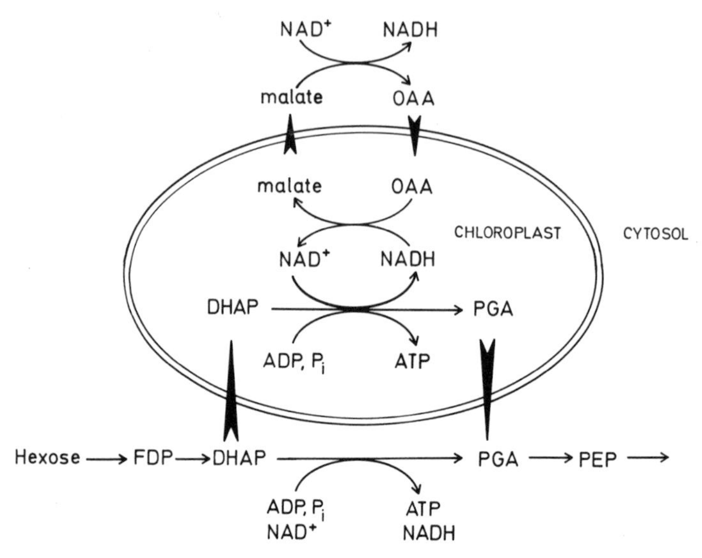

Fig. 5.8. Pathway of ATP formation by chloroplasts in the dark.

The shuttle depicted in Fig. 5.6 must also be seen in relation to export of photosynthates from the chloroplasts in the light. Exported dihydroxyacetone phosphate probably is utilized to synthesize sucrose in the cytoplasm. The required ATP then can be formed by oxidation of part of the dihydroxyacetone phosphate to phosphoglycerate. The latter in not necessarily reimported in total but may in part follow the glycolytic pathway to pyruvate, which includes another phosphorylation step. Pyruvate may serve for synthesis of other cell components such as alanine or, via the citrate cycle, of C_4 compounds as far as this pathway is not inhibited by suppression of respiratory activity in the light. Finally, in the dark, part of the glycolytic sequence, the conversion of dihydroxyacetone phosphate to phosphoglycerate, appears to supply the chloroplasts with ATP and reducing equivalents (Fig. 5.8). This is actually a partial reversal of the shuttle since dihydroxyacetone phosphate comes from the cytoplasm and phosphoglycerate is returned to it. It is also a reversal of photosynthetic phosphoglycerate reduction. This reaction is known to operate close to thermodynamic equilibrium during steady-state photosynthesis [21]. Its reversal due to changed concentrations of its components in the dark is, therefore, easily possible. Accumulation of reducing equivalents in the chloroplasts can be prevented by operation of the malate/oxaloacetate (or malate/aspartate) shuttle as shown in Fig. 5.8. In this version of the shuttle system, the NAD-specific chloroplast enzymes should be acting, because NADP-dependent malate dehydrogenase is inactive in the dark [94]. NADPH for synthetic reactions can be derived, as has been discussed, from the pentose phosphate cycle.

5.5. FLEXIBILITY OF PHOTOPHOSPHORYLATION IN INTACT CHLOROPLASTS

5.5.1. Reduction of substrates requiring different ATP/NADPH ratios

After Arnon discovered photophosphorylation in 1954 [8], it was assumed for several years that one ATP molecule is synthesized during the transfer of two electrons to a suitable acceptor molecule [6,7]. Winget et al. [170] were the first to question this stoichiometry. As discussed in detail in Chapter 4 it is now known that two coupling sites exist in the electron transport chain of chloroplasts [26,54,57,130,157,170]. A number of workers have reported ATP/2e ratios above 1, but below 2 in broken chloroplasts [46,48,58,88,89,121,132,141,170] (and see Chapter 4, Table II). As explained in Chapter 4, by subtracting the electron transport rate observed in the absence of phosphorylation (e.g. in absence of ADP) from the rate under conditions optimal for phosphorylation, ATP/2e ratios of 2 were computed [132,89]. Thermodynamically, a maximal coupling ratio of 2 is just possible [85]. However, it is uncertain whether the corrections of experimental data, which are necessary to obtain a coupling ratio of 2, are justified. They rest on the assumption, that coupling follows a strict stoichiometry and that non-phosphorylating electron transport exists side by side with phosphorylating electron transport. It is also uncertain whether the conditions for phosphorylation in intact chloroplasts favour the attainment of maximal coupling ratios. Therefore, determinations of coupling ratios in intact and functional chloroplasts appear to be necessary to answer the question whether, and to what extent, auxiliary ATP synthesizing reactions participate in photosynthesis. If the ATP/NADPH ratio achieved during non-cyclic electron transport is lower than 1.5, it would be insufficient to support CO_2 reduction by the Calvin cycle [17]. In the Hatch—Slack—Kortschak pathway of C4 photosynthesis, an ATP/NADPH ratio of 2.5 appears to be required [115]. Because of the restricted exchange of ADP and ATP across the chloroplast envelope, photophosphorylation cannot be measured directly in intact chloroplasts. However, substrate-dependent oxygen evolution reveals rates of photophosphorylation in these organelles if the stoichiometry of ATP consumption by the substrates is known. Moreover, the stoichiometry of substrate reduction informs us about the ratios of ATP and NADPH production maintained during steady-state electron transport. In other words, high steady rates of photosynthetic reduction of a certain substrate demonstrate that the photosynthetic apparatus adjusts to the demanded ATP/NADPH ratio. In Table II the required ATP/NADPH ratios of different substrates are given, together with observed rates of reduction by intact chloroplasts. These were fully functional without added cofactors. Their photosynthetic capacity was comparable to that of the parent leaves as shown by comparable rates of CO_2 -dependent oxygen evolution. Reduction of substrates whose ATP requirements differ testifies to the flexibility of chloroplast metabolism. It shows that chloroplasts can, according to demand, produce ATP and NADPH at ratios lower or higher than the ratios necessary for CO_2 reduction. However, it must be emphasized that the observations, though giving information on the ratio of ATP and NADPH production during the photosynthetic process, do not elucidate the true ATP/2e ratio of electron transport, as long as we do not know whether, or to what extent, ATP is contributed by auxiliary reactions. More refined experiments are neccessary to differentiate between such reactions and ATP formation during electron transport to NADP.

TABLE II

ATP/NADPH ratios required for photosynthetic substrate reduction and observed rates of reduction by isolated spinach chloroplasts

Substrate	ATP/NADPH	Observed reduction rate [a] (μmoles/mg chlorophyll/h)
Oxaloacetate	0.0	30—130
Phosphoglycerate	1.0	120—450
CO_2	1.5	70—350[b]
Glycerate	2.0	15—40

[a] Measured in the absence of added cofactors or uncoupling agents.

[b] Rates of net CO_2 fixation by intact spinach leaves are usually 80—150 μmoles/mg chlorophyll/h.

5.5.2. *ATP as a limiting factor in photosynthetic CO_2 reduction*

When CO_2 was added to illuminated intact chloroplasts, the internal ATP level was decreased while the NADPH level remained constant [62]. This observation is evidence against an ATP/2e ratio higher than 1.5 in intact chloroplasts, because such a ratio should maintain the high phosphorylation potential during CO_2 reduction. Similarly, a sudden lowering of the light intensity during CO_2 assimilation lowered the ATP but not the NADPH level in the chloroplasts. Thus, under limiting light intensities ATP and not NADPH seems to limit CO_2 reduction. This is also apparent from the effect of oxaloacetate which was shown to stimulate O_2 evolution when added to chloroplasts during CO_2 fixation [62]. Obviously the oxaloacetate reduction was based on consumption of excess NADPH. In leaf cells assimilating CO_2 at light intensities below saturation, ATP apparently is also the limiting factor. Light-dependent chloroplast shrinkage and chlorophyll fluorescence quenching which can be viewed as indicators of the phosphorylation potential in intact leaves (see section 5.3.2), are reversed by CO_2 [61,104]. The extent of reversal is a function of light intensity.

5.5.3. *Quantum requirements of photosynthetic substrate reduction*

The experiments discussed under sections 5.5.1 and 5.5.2 indicate that, first, the coupling ratio ATP/2e in intact chloroplasts is not fixed but flexible since it can even be zero, as in the case of oxaloacetate reduction. Second, the highest possible ratio is below 2, and probably below 1.5. Otherwise a high phosphorylation potential should be maintained during CO_2 fixation even under low light intensity.

Quantitative estimates of maximal ATP/2e ratios are based on measurements of the quantum requirement of oxygen evolution in different chloroplast reactions at low light intensities. In glycerate-dependent O_2 evolution, for instance, 1 mole of O_2 indicates reduction of 2 moles of glycerate to triosephosphate [63]. Added glycerate enters the chloroplasts where it is first phosphorylated to phosphoglycerate, which then undergoes photosynthetic reduction. This requires a total of 4 ATP and 2 NADPH per O_2 evolved. If all the ATP were formed during electron transport to NADP (ATP/2e = 2) the quantum requirement for liberation of 1 O_2 molecule should be 8, as 4 electrons have to be elevated by the two photoreactions. The observed quantum requirement, however, was between 12 and 13 (Table III). This shows that ATP formation coupled to photosynthetic NADP reduction is insufficient for glycerate reduction. Further light quanta are needed in electron transfer reactions to supply extra ATP. Provided the ATP/2e ratios of the latter are the same as in NADP reduction, a coupling ratio of 1.2 to 1.4 can be calculated for phosphorylation during glycerate reduction [69].

The same calculation shows that the ATP/2e ratio in intact chloroplasts reducing CO_2 is between 1.0 and 1.2 [65]. For phosphoglycerate-dependent O_2 evolution the theoretical quantum requirement of 8 was found, indicating an ATP/2e ratio of 1.0. Interestingly, the coupling ratio during phosphoglycerate reduction can rise again to its maximum value of 1.1 to 1.4, when reducing equivalents are transferred into the chloroplasts by the malate/oxaloacetate shuttle [69,70]. The shuttle provides non-photosynthetic NAD(P)H which, together with additional photosynthetic ATP, stimulates phosphoglycerate reduction [73,74].

TABLE III

$ATP/2e^-$ ratios of electron transport by intact chloroplasts, as calculated from quantum requirements of substrate reduction

Substrate	Observed quantum requirement [a] (quanta per O_2 molecule evolved)	calculated $ATP/2e^-$ ratio
Phosphoglycerate	8	1.0
CO_2	10–12	1.0–1.2
Glycerate	12–13	1.2–1.4

[a] Light intensity was extrapolated to zero.

5.5.4. Flexibility of ATP/2e ratios

The foregoing considerations suggest maximum ATP/2e ratios below 1.5. Moreover, the photosynthetic apparatus obviously can adjust itself to lower values, when low stoichiometric amounts of ATP are needed for photoreductions. This indicates flexible ATP/2e ratios. If coupling of phosphorylation to electron transport were tight, as is known from mitochondria, slow ATP consumption in proportion to NADPH reoxidation would inhibit electron transport, as ADP would become unavailable for phosphorylation. Such "photosynthetic control" is, indeed, apparent to some extent in intact isolated chloroplasts during oxaloacetate reduction which does not consume ATP [69]. The theoretical quantum requirement of 8 for oxaloacetate-dependent oxygen evolution is observed only at very low intensities. With increasing intensity the quantum requirement also increases. This is due to restriction of electron transport by the high phosphorylation potential in the chloroplasts, since uncoupling lowers the quantum requirement to values not far from 8. Very low concentrations of uncouplers stimulate phosphoglycerate reduction at high, but not at low intensities [95,165]. This may also be an expression of control by the phosphorylation potential. However, a similar phenomenon has even been reported for CO_2 reduction under light saturation [47]. Since under these conditions auxiliary ATP producing reactions will be fully activated, interpretation in terms of coupling ratios is difficult.

The high rates of O_2 evolution observed both with phosphoglycerate and oxaloacetate (see Table II) show that photosynthetic control is not strict. Uncoupling during oxaloacetate reduction at high light intensity always stimulated O_2 evolution. But the stimulation was remarkably small in chloroplasts showing the highest CO_2 fixation rates [69]. As such preparations, in comparison with chloroplasts in situ, appear functionally unimpaired, the relatively small stimulation of oxaloacetate reduction by uncoupling cannot be regarded as being due to damage of the phosphorylation system.

If the ATP/2e ratio is flexible, the question as to its regulation is of interest. A simple view of flexible coupling is that as long as ADP is available and the phosphorylation potential is low, the high-energy condition of the thylakoid system is used for phosphorylation. During CO_2 or glycerate reduction under limiting light intensities, the consumption of reducing equivalents produced by the electron transport chain appears to be limited by the availability of ATP. Under these conditions coupling of electron flow to phosphorylation is maximal and NADP remains largely reduced. Additional ATP still needed for reduction would come from auxiliary ATP generation possibly triggered by the redox state of the NADP system [65]. During oxaloacetate reduction, on the other hand, ATP is not consumed. High phosphorylation potentials would to some extent control electron flow to oxaloacetate, but at the same time permit non-phosphorylating dissipation of the high-energy condition. Increased proton efflux from thylakoids due to high proton gradients is a possible means of such dissipation and would permit electron flow in the absence of phosphorylation. However, it is also possible that there is a special mechanism for flexible coupling which permits a more active adjustment of phosphorylation to ATP requirements. It might be noted that in our calculations of ATP/2e ratios in intact chloroplasts no account is taken of "uncoupled" or "basal" electron flow. Corrections for such electron transport have been considered to be necessary by other authors [89,132] in order to obtain "true" ATP/2e ratios, and we refer the reader to Hall's discussions in Chapter 4. However, it appears that electron transport in the absence of phosphorylation only reflects the capability of the membranes to dissipate the high-energy condition if conservation of energy in the form of ATP is not possible.

5.5.5. Phosphorylation without NADP reduction: Cyclic electron transport or "pseudocyclic" oxygen reduction?

If maximum ATP/2e ratios in intact chloroplasts are below 1.5, additional electron transfer reactions for provision of extra ATP must be considered even for CO_2 assimilation in the Calvin cycle. Most modern text-books include cyclic photophosphorylation, which is supported by cyclic electron flow in photosystem I, as a means of producing ATP in addition to that synthesized during linear electron flow to NADP. Still there is no compelling evidence to show that cyclic photophosphorylation plays a significant role in higher plants under natural aerobic conditions [144,152]. Rather, in intact leaves or in chloroplasts oxygen can readily react with an electron carrier beyond photosystem I [45,61,67,91]. In a nitrogen atmosphere intact leaves exhibit strong chloroplast shrinkage under far-red illumination, which excites preferentially photosystem I, indicating phosphorylation by cyclic electron transfer. As very low levels (0.1%) of O_2 reversed this effect it seems that oxygen drains electrons from the cyclic pathway (Fig. 5.9) and thereby inhibits cyclic photophosphorylation [61].

On the other hand, oxygen supports phosphorylation in light which excites both photosystems. Apparently, electrons transferred through the two photosystems can move either to $NADP^+$ or, alternatively, when this is reduced, are transferred to molecular oxygen (Mehler reaction). In this reaction [118,119] H_2O_2 is formed, with the superoxide radical as an intermediate [1,2,40–43]. H_2O_2 can leave the chloroplasts by diffusion and be decomposed by external catalase. Such a system (Fig. 5.9) would be self-regulating with respect to photophosphorylation, if the affinity of the electron transport chain for $NADP^+$ is much higher than for O_2. There are indications that this is, in fact, the case [61,39]. When during induction of CO_2 fixation ATP becomes limiting, NADPH accumulates and less $NADP^+$

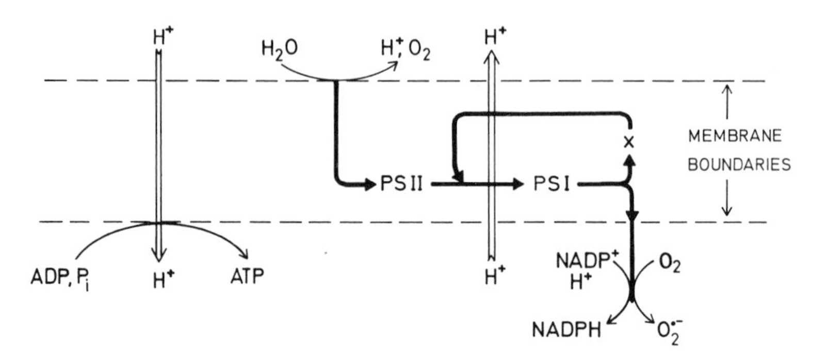

Fig. 5.9. Competition between different acceptors for electrons in photosynthetic electron transport.

is available for reduction. Electrons are then diverted to oxygen which results in additional phosphorylation without NADP reduction. For this process the term "pseudocyclic" electron transport has been used, as oxygen evolution and oxygen uptake balance each other and net gas exchange is zero:

$$H_2O \rightarrow \frac{1}{2}O_2 + 2e^- + 2H^+$$

$$2e^- + O_2 + 2H^+ \rightarrow H_2O_2$$

$$H_2O_2 \rightarrow H_2O + \frac{1}{2}O_2$$

Experiments with intact chloroplasts revealed similar effects of O_2 as discussed above. However, the reactivity of the electron transport chain with oxygen in the chloroplast preparations examined appeared diminished in comparison with intact leaves [110]. In isolated chloroplasts and under aerobic conditions, in addition to pseudocyclic, some true cyclic photophosphorylation may therefore take place. Still, Whitehouse et al. [167] observed high rates of oxygen uptake by intact chloroplasts in the absence of catalase. The oxygen exchange between water and molecular O_2 as indicated above can be measured by mass spectrometry. Recent experiments with $^{18}O_2$ show that during photosynthesis of intact chloroplasts oxygen is not only evolved but also taken up [36]. The rate of oxygen uptake by far exceeded the rate expected from oxygen incorporation into glycolate which is formed during CO_2 reduction. Oxygen uptake was much smaller during reduction of phosphoglycerate than during reduction of CO_2 indicating that it plays a role only in the latter reaction, which has a higher ATP requirement than phosphoglycerate reduction.

Intact chloroplasts do not contain catalase [153,167]. In chloroplast preparations which were carefully washed to remove exogenous catalase the rate of CO_2-dependent oxygen evolution usually was slow. It increased dramatically on addition of catalase (Fig. 5.10). No such effect was apparent when phosphoglycerate was the substrate [36]. This indicates that, indeed, H_2O_2 is formed during CO_2 fixation. It inhibits photosynthesis reversibly and can easily be decomposed by exogenous catalase. Little H_2O_2 seems to be formed during phosphoglycerate reduction. The observations suggest that CO_2 reduction by intact chloroplasts indeed requires auxiliary ATP. At least under low light intensities, this appears to be provided by electron flow from water to oxygen. Only when the oxygen reducing reaction is saturated, may electrons be diverted into the cyclic pathway.

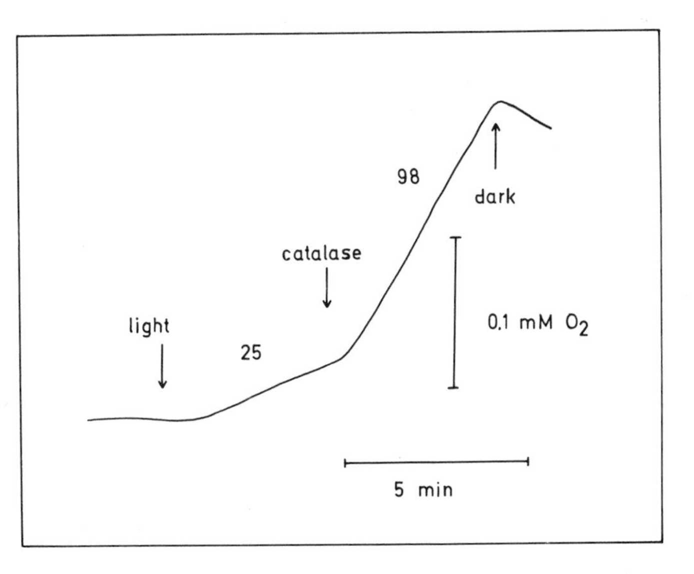

Fig. 5.10. Bicarbonate-dependent oxygen evolution by intact washed chloroplasts before and after addition of catalase. Numbers denote rates of photosynthesis in μmoles O_2/mg chlorophyll/h. Illumination with 200 W/m^2 red light. (From ref. 36, with permission.)

5.6. GLYCOLATE SYNTHESIS AND PHOTORESPIRATION AS ENERGY-CONSUMING PROCESSES

5.6.1. Glycolate formation by intact chloroplasts

During photosynthesis by intact chloroplasts in the presence of rate-limiting concentrations of bicarbonate, glycolate is a major product [20,30, 38,49,156]. It cannot be further metabolized in the chloroplasts and is excreted into the medium [20,30,63,97]. However, in intact leaves it does not accumulate. Obviously it is converted in vivo outside the chloroplasts into other products. Synthesis and metabolism of glycolate are believed to be closely related to photorespiration [56,91,154,171]. This process decreases photosynthetic productivity in a large number of plants whose primary product of CO_2 fixation is 3-phosphoglycerate (C_3-plants), by almost one third as compared with productivity in an atmosphere of nitrogen containing a very low percentage of oxygen and 300 ppm CO_2 [24]. Glycolate synthesis from CO_2 in isolated chloroplasts has been studied by a number of workers [38,50,52,134]. Like photorespiration it is oxygen-dependent. In a competitive fashion it can be suppressed by bicarbonate. After a lag period reflecting autocatalytic build up of photosynthetic intermediates [159,160,162], illumination of intact chloroplasts causes oxygen evolution which indicates carbon dioxide assimilation. When the concentration of CO_2 has fallen to a threshold value due to its uptake, O_2 evolution ceases. The threshold value is determined by the partial pressure of oxygen. It is very low when the oxygen concentration is low, and rises with the oxygen concentration. During the period of steady oxygen evolution, a large percentage of the photosynthetic products, dihydroxyacetone phosphate and 3-phosphoglycerate, leave the chloroplasts. Glycolate appeared in significant amounts in the experiment shown in Fig. 5.11 only after the CO_2 concentration had decreased considerably and oxygen evolution began to decline. It continued to be formed after the synthesis of other photosynthetic products had ceased. In fact, glycolate accumulated at the expense of these products. In the presence of very low CO_2 concentrations, glycolate finally was the only product of CO_2 assimilation.

Thus the chloroplasts were able to oxidize, in the absence of significant CO_2, not only all internal sugar phosphate, but also the triosephosphate and phosphoglycerate, which previously had been exported. This is obviously possible only after these transport metabolites had reentered the chloroplasts.

Glycolate has first been considered to be formed by oxidation of an intermediate of the Calvin cycle, notably of the "active" glycolaldehyde intermediate of the transketolase reaction [19,29,49,140,169]. It was suggested that H_2O_2 was involved in the oxidation [114]. Recently evidence has mounted in favour of glycolate as a product of the oxygenation of

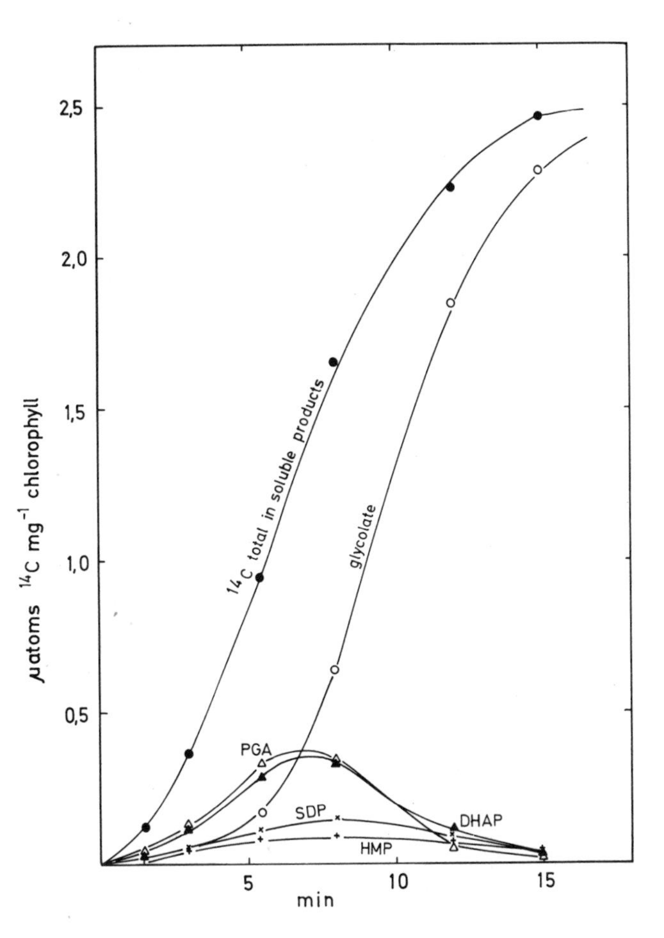

Fig. 5.11. $^{14}CO_2$ fixation and formation of labelled products by intact chloroplasts [71]. The pH was 7.2, initial $H^{14}CO_3^-$ and oxygen concentrations were 0.25 mM and 0.2 mM, respectively.

ribulose diphosphate, the primary acceptor for CO_2 in C_3 photosynthesis [4,13,27,28,66]. Ribulose diphosphate carboxylase has been shown to have two enzymic functions. With CO_2 as the substrate, ribulose diphosphate is carboxylated to yield two molecules of 3-phosphoglycerate. In the oxygenase reaction, phosphoglycolate, which subsequently is hydrolyzed to glycolate, and 3-phosphoglycerate are the products:

$$\text{ribulose-1,5-diphosphate} + O_2 \rightarrow \text{phosphoglycolate} + \text{3-phosphoglycerate}$$

Competition of CO_2 and oxygen for the same substrate actually would explain why glycolate formation in intact chloroplasts is competitively inhibited by CO_2 and proceeds at high rates only in the presence of high concentrations of oxygen or low concentrations of CO_2. Under appropriate conditions, maximal rates of glycolate formation by intact chloroplasts were about 30 μmoles/mg chlorophyll/h. While more than one reaction capable of producing glycolate may exist in chloroplasts [37,171], it should be possible to distinguish between various proposals on the basis of the ATP requirements of ribulose disphosphate formation. The substrate of the oxygenase reaction needs ATP for synthesis and cannot be formed by intact chloroplasts in the presence of an uncoupler of phosphorylation. Glycolate formation from existing sugar phosphate in the transketolase reaction should, on the other hand, be insensitive to uncouplers. It should, in fact, be stimulated by uncoupling, which increases electron flow to oxygen and stimulates H_2O_2 formation in the Mehler reaction. If the oxygenase reaction were to lead to glycolate, it should be inhibited by uncoupling. Fig. 5.12 shows that the uncoupler carbonyl cyanide 4-trifluoromethoxyphenylhydrazone added to actively photosynthesizing chloroplasts immediately inhibits glycolate formation. This strongly suggests that the major part of the glycolate sythesized by chloroplasts during photosynthesis derives from the reaction of oxygen with ribulose diphosphate.

It is important to note that glycolate formation at the expense of sugar phosphate consumes energy. One product in the oxygenase reaction is phosphoglycerate, which is then reduced to the sugar level by utilising ATP and reducing equivalents. Recycling to ribulose diphosphate along the routes of the Calvin cycle consumes further ATP in the phosphoribulokinase reaction. The ATP/NADPH ratio required for complete oxidation of 1 mole of hexose monophosphate to 3 moles of glycolate is 2:

$$\begin{aligned} & \text{F—6—P}^{2-} + 3\ O_2 + 3\ \text{NADPH} + 6\ \text{ATP}^{4-} + 6\ \text{OH}^- \\ & \rightarrow 3\ \text{glycolate}^- + 3\ \text{NADP}^+ + 6\ \text{ADP}^{3-} + 7\ \text{HPO}_4{}^{2-} + 2\ H_2O \end{aligned}$$

In contrast to glycolate formation photosynthetic reduction of CO_2 in the Calvin cycle consumes ATP and NADPH with a ratio of 1.5.

It is a paradoxical situation and most unusual in metabolism that oxidative breakdown of a sugar molecule does not produce but actually consumes energy. Its physiological significance will be discussed in the following section.

5.6.2. Glycolate conversion and reentry of products into the chloroplasts

Glycolate production by isolated chloroplasts does not present an artificial situation but is, in view of the enzymic properties of the photosynthetic apparatus, the inevitable consequence of photosynthesis in an oxidative atmosphere under limiting CO_2 concentrations. Still it does not accumulate in leaves. So its further metabolism is as fast as its rate of formation. Using tracer methodology, Tolbert and associates [154,155] have followed the

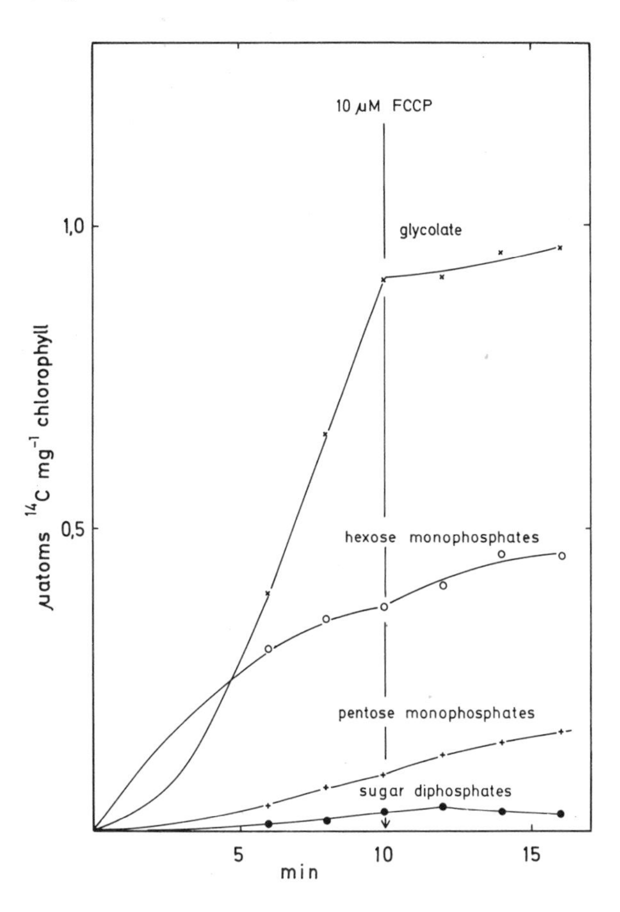

Fig. 5.12. Effects of FCCP on labelling of compounds formed during photosynthesis by intact isolated chloroplasts in the presence of 2 mM $H^{14}CO_3^-$ [71]. Illumination with 100 W/m^2 red light. Rate of CO_2 reduction 210 μmoles/mg chlorophyll/h (in the presence of 3 mM dithiothreitol). Initial O_2 concentration 0.29 mM.

fate of glycolate in leaves. Enzyme localization studies complemented the emerging picture [154,171]. After excretion from the chloroplasts, glycolate appears to be oxidized to glyoxylate in the peroxisomes. The latter compound is converted into glycine. Presumably in the mitochondria, 2 molecules of glycine form, with CO_2 evolution, one molecule of the 3-carbon compound serine. The CO_2 evolved can, by consuming energy, be reduced in the chloroplasts. From serine, hydroxypyruvate is formed which, in turn, is reduced to glycerate in the peroxisomes. Glycerate has been shown to be able to move into the chloroplasts, which contain glycerate kinase [63]. By consuming ATP, glycerate is phosphorylated to phosphoglycerate and then can enter the reactions of the Calvin cycle. After infiltration of radioactive serine into leaves, label indeed appeared rapidly in sucrose in the light, but not in the dark [127]. Even though the proposed reaction sequence may not be finally established in detail and alternative routes are possible, it is entirely clear from a bioenergetic point of view that conversion of glycolate to the sugar level, as it occurs in the light, is not possible without expense of energy in the form of ATP and reducing equivalents.

Light-dependent CO_2 evolution by photorespiration is easily apparent in leaves of C3-plants. It lowers the net CO_2 uptake in photosynthesis and can be observed directly in CO_2-free air. There is a CO_2 concentration without net loss or gain in organic matter (CO_2 compensation point [171]). Photosynthesis and photorespiration then balance each other with an overall consumption of energy, which is dissipated as heat. Fig. 5.13 shows such a situation. Out of three molecules of ribulose diphosphate two are oxygenated, one is carboxylated. The CO_2 for the carboxylation comes from the glycolate pathway. In total, the process utilizes ATP and reduced pyridine nucleotides at the same ratio of 1.5 as the carbon reduction cycle.

Excessive loss of energy under conditions which stimulate photorespiration is indeed apparent in leaves as observed from the response of 535 nm light scattering and chlorophyll fluorescence, both indicators of the phosphorylation potential [61,64,104,111]. During an increase in oxygen concentration from about 5 to 20% or more, which increases rates of photorespiration, energy-dependent fluorescence quenching is relieved and chloroplast shrinkage is lowered indicating a decrease in the energy level of the chloroplast system. As should be expected, these effects can be seen in the absence of CO_2 and become smaller with increasing CO_2 concentrations [108].

There is the question as to whether the light-dependent dissipation of energy which is unaccompanied by a gain and actually results in a loss of

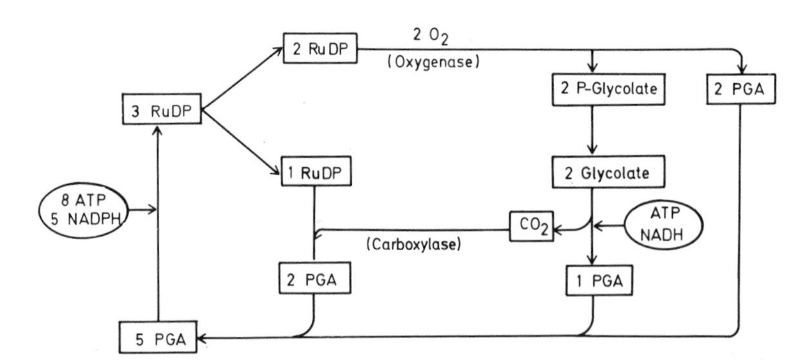

Fig. 5.13. Dissipation of photosynthetic energy at the CO_2 compensation point in a cycle composed of the carbon reduction and glycolate pathways (scheme after a suggestion by C.B. Osmond).

organic matter serves a physiological function. It has been suggested that photorespiration represents an evolutionary "hangover" and serves no useful purpose. On these grounds attempts by plant physiologists to block and eliminate photorespiration and thereby increase crop yields are understandable [171].

There may be, on the other hand, conditions, which require dissipation of energy trapped in the photosynthetic apparatus, if such dissipation is not possible by normal photosynthesis. Osmond and Björkman [128] have suggested that photorespiratory energy consumption may protect plants from photooxidative damage and is, therefore, of ecological advantage. Under water stress, leaves close their stomata and thereby prevent significant net photosynthesis even though the photosynthetic apparatus remains exposed to irradiation. Under these conditions energy-requiring photorespiratory processes leading to CO_2 evolution and refixation of the evolved CO_2 permit continued operation of the electron transport chain and may indeed represent one of the mechanisms to protect the chloroplast against photooxidation. Experimental evidence supporting this view is given in Fig. 5.14. It shows that intact chloroplasts preilluminated for only 7 min in the presence of catalase, but in the absence of a photosynthetic substrate such as CO_2, exhibit, after addition of bicarbonate, a much slower rate of photosynthesis than chloroplasts which had been permitted to photosynthesize from the beginning of the illumination period. The slower rate of photosynthesis in the chloroplasts exposed to illumination in the absence of substrate is interpreted to be an expression of damage which the photosynthetic apparatus suffered from the light.

If photorespiratory energy dissipation represents a protective mechanism, it should not be expected to operate only in a limited number of plants. In contrast to C3-plants, C4-plants such as maize or sugar cane exhibit the same rates of photosynthesis in the presence of low and of high partial pressures of oxygen [25]. Also, in contrast to C3-plants they are capable of depleting an atmosphere almost totally of CO_2 and do not show the phenomenon of photorespiratory CO_2 escape. However, their fluorescence and light scattering properties are qualitatively very similar to those of C3-plants. As in C3-plants, raising the oxygen content from 5 to 21% or more causes an increase in fluorescence and a decrease in light scattering indicating energy dissipation. It appears therefore, that at least qualitatively the same mechanisms of energy dissipation are operative in C4- and C3-plants. The different response of the former to CO_2 is caused by the different CO_2 trapping system. In C4-plants CO_2 is fixed primarily by PEP carboxylase instead of carboxydismutase [59].

If the view is correct that photorespiratory energy dissipation actually is a "safety valve" reaction, attempts to abolish photorespiration by selective blocking of its pathway may, if successful, produce disastrous results under conditions that would call for its operation.

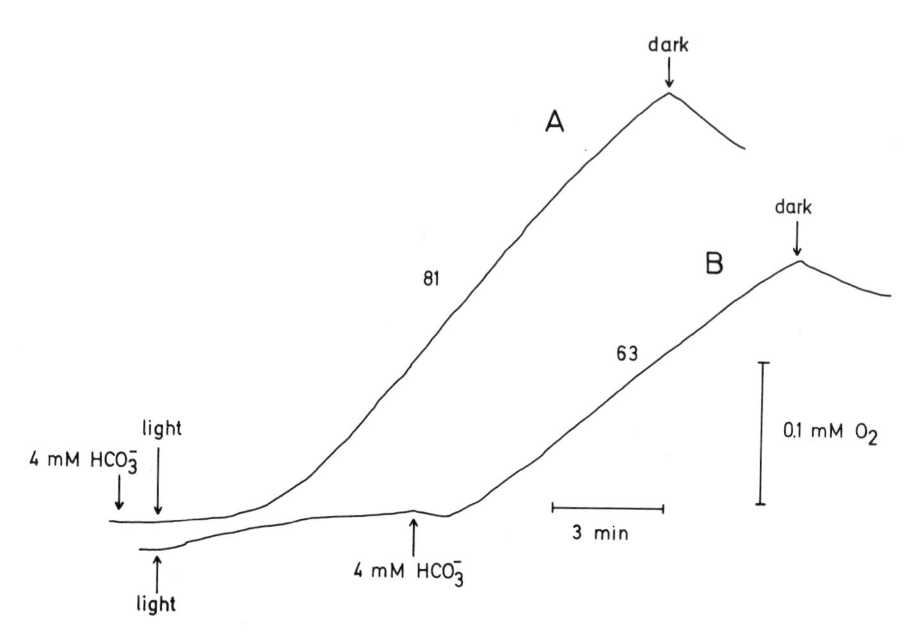

Fig. 5.14. CO_2-dependent oxygen evolution by intact spinach chloroplasts [71]. (A) Illumination in the presence of bicarbonate. (B) Preillumination for 7 min in the absence of substrate before bicarbonate was added. Numbers denote rates of O_2 evolution in μmoles /mg chlorophyll/h.

In conclusion, oxygen appears to affect the energy status of photosynthesizing chloroplasts in different ways, First, it can accept electrons under ATP deficiency providing extra ATP for CO_2 reduction and other energy-requiring cell activities. This reaction is saturated in vivo at a very low partial pressure of oxygen. Half saturation is seen at 0.1% oxygen in the atmosphere [61]. Second, by permitting the glycolate cycle to operate, oxygen causes dissipation of photosynthetic energy. The latter process is not completely saturated even at 100% oxygen [108] but assumes significant proportions in the natural atmosphere containing 300 ppm CO_2. Energy loss by photorespiration appears to be a protective device especially under high light intensities.

5.7. CONCLUDING REMARKS

In a way, chloroplasts are much more remarkable organelles than mitochondria. From a bioenergetic point of view, ATP formation by mitochondria is simple to understand. The large change in free energy of NADH oxidation is used to drive endergonic phosphorylation of ADP. Chloroplasts, on the other hand, achieve ATP formation while performing endergonic

NADP reduction. They are endowed with mechanisms to form additional ATP, if that provided during NADP reduction is insufficient. This requires regulatory devices permitting chloroplasts to "sense" ATP deficiency. If, on the other hand, no ATP is needed, the chloroplast electron transport chain can produce reductant without concomitant ATP formation. While producing reductant, chloroplasts also form oxidants. Highly reactive radicals are generated in the light especially under conditions that do not permit net carbon assimilation. Evolution has invented mechanisms either to destroy dangerous compounds such as the oxygen radical $O_2^{\cdot-}$ enzymically (for instance by superoxide dismutase), or to prevent excessive formation by keeping both the electron transport pathway and the carbon cycle busy with "normal" activity even under conditions that do not permit photosynthesis to proceed. To provide reducible substrates under such conditions appears to be the physiological function of photorespiration.

Chloroplasts are self-sufficient to the extent that after isolation from leaf cells they can reduce CO_2 in the light at normal rates without added cofactors. This clearly would not be possible if diffusion barriers did not prevent escape of ions and cofactors necessary for the operation of the enzymic machinery of photosynthesis. In spite of the presence of such barriers, chloroplasts export, via specific carriers, products of photosynthesis such as phosphoglycerate and triosephosphate which at the same time are intermediates of the carbon reduction cycle. Again it is obvious that delicate controls are required to prevent draining of necessary intermediates from the cycle which would result in the breakdown of photosynthesis. Although cofactors such as adenylates and pyridine nucleotides are effectively retained inside the chloroplasts, both phosphate energy and reducing equivalents can be either exported or imported. Through energy transfer, chloroplasts remain metabolically active in the dark and exert profound influence on cytoplasmic metabolism in the light.

Thus it appears that chloroplasts have adjusted well to contradictory situations. Although we seem to have gained some insight into the workings of a chloroplast, our knowledge is qualitative and fragmentary. Chloroplast metabolism and energetics remain a fascinating field for further study.

REFERENCES

1 Allen, J.F., Hall, D.O. (1973) Biochem. Biophys. Res. Commun., [52], 856–862.
2 Allen, J.F., Hall, D.O. (1974) Biochem. Biophys. Res. Commun., [58], 579–585.
3 Anderson, L.A., Ng, T.-C.L. and Park, K.-E.Y. (1974) Plant Physiol., [53], 835–839.
4 Andrews, T.J. Lorimer, G.H. and Tolbert, N.E. (1973) Biochemistry, [12], 11–17.
5 Arnon, D.I. (1956) Ann. Rev. Plant Physiol., [7], 325–354.
6 Arnon, D.I. (1961) Light and Life, Cell-free Photosynthesis and the Energy Conversion Process, Johns Hopkins, Baltimore.
7 Arnon, D.I., (1967) Physiol. Rev., [47], 317–358.

8 Arnon, D.I., Allen, M.B. and Whatley, F.R. (1954) Nature, [174], 394—396.
9 Arnon, D.I., Tsujimoto, H.Y. and McSwain, B.D. (1967) Nature, [214], 562—566.
10 Atkinson, D.E. (1968) Biochemistry, [7], 4030—4034.
11 Avron, M. and Gibbs, M. (1974) Plant Physiol., [53], 136—139.
12 Badger, M.R. and Andrews, T.J. (1974) Biochem. Biophys. Res. Commun., [60], 204—210.
13 Bahr, J.M. and Jensen, R.G. (1974) Arch. Biochem. Biophys., [164], 408—413.
14 Barber, J. and Telfer, A. (1974) in Membrane Transport in Plants (Zimmermann, E. and Dainty, J.,eds), pp. 281—288, Springer, Berlin.
15 Barber, J., Telfer, A., Mills, J. and Nicolson, J. (1974) in Proc. 3rd Int. Cong. Photosynthesis (Avron, M.,ed), pp. 53—63, Elsevier, Amsterdam.
16 Barber, J., Telfer, A. and Nicolson, J. (1974) Biochim. Biophys. Acta, [357], 161—165
17 Bassham, J.A. (1964) Ann. Rev. Plant Physiol., [15], 101—120.
18 Bassham, J.A. and Kirk, M. (1968) in Comparative Biochemistry and Biophysics of Photosynthesis (Shibata, K., Takamiya, A., Jagendorf, A.T. and Fuller, R.C., eds), p. 365, Univ. of Tokyo Press, Tokyo.
19 Bassham, J.A. and Kirk, M. (1973) Plant Physiol., [52], 407—411.
20 Bassham, J.A., Kirk, M. and Jensen, R.G. (1968) Biochim. Biophys. Acta, [153], 211—218.
21 Bassham, J.A. and Krause, G.H. (1969) Biochim. Biophys, Acta, [189], 207—221.
22 Bird, I.F., Cornelius, M.J., Dyer, T.A. and Keys, A.J. (1973) J. Exp. Bot., [24], 211—215.
23 Bird, I.F., Cornelius, M.J., Keys, A.J. and Whittingham, C.P. (1974) Phytochemistry, [13], 59—64.
24 Björkman, O. (1966) Physiol. Plant., [19], 618—633.
25 Björkman, O. (1971) in Photosynthesis and Photorespiration (Hatch, M.D., Osmond, C.B. and Slatyer, R.O., eds), pp. 18—32, Wiley Interscience, New York.
26 Böhme, H. and Trebst, A. (1969) Biochim. Biophys. Acta, [180], 137—148.
27 Bowes, G. and Ogren, W.L. (1972) J. Biol. Chem., [247], 2171—2176.
28 Bowes, G., Ogren, W.L. and Hagemann, R.H. (1971) Biochem. Biophys. Res. Commun., [45], 716—722.
29 Bradbeer, J.W. and Racker, E. (1961) Fed. Proc., [20], 88.
30 Chan, H.W.-S. and Bassham, J.A. (1967) Biochim. Biophys. Acta, [141], 426—429.
31 Chance, B. and Williams, G.R. (1956) Adv. Enzymol., [17], 65—98.
32 Cockburn, W., Baldry, C.W. and Walker, D.A. (1967) Biochim. Biophys, Acta, [143], 614—624.
33 Coombs, J. and Whittingham, C.P. (1966) Proc. Roy. Soc. Ser. B., [164], 511—520.
34 Douce, R., Holtz, R.B. and Benson, A.A. (1973) J. Biol. Chem., [248], 7215—7222.
35 Duysens, L.N.M. and Sweers, H.E. (1963) in Studies on Microalgae and Photosynthetic Bacteria (Japan. Soc. Plant Physiol., ed), pp. 353—372, University of Tokyo Press, Tokyo.
36 Egneus, H., Heber, U. and Mathiesen, U. (1976) Biochim. Biophys. Acta, [408], 252—268.
37 Eickenbusch, J.D. and Beck, E. (1973) FEBS Lett., [31], 225—228.
38 Ellyard, P.W. and Gibbs, M. (1969) Plant Physiol., [44], 1115—1121.
39 Elstner, E.F. and Heupel, A. (1973) Biochim. Biophys, Acta, [325], 182—188.
40 Elstner, E.F., Heupel, A. and Vaklinova, S. (1970) Z. Pflanzenphysiol., [62], 184—200.
41 Elstner, E.F. and Kramer, R. (1973) Biochim. Biophys. Acta, [314], 340—353.
42 Epel, B.L. and Neumann, J. (1972) in 6th Int. Cong. Photobiology, Bochum, (Schenk, G.O., ed), Abstract No. 237.
43 Epel, B.L. and Neumann, J. (1973) Biochim. Biophys.Acta, [325], 520—529.
44 Everson, R.G., Cockburn, W. and Gibbs, M. (1967) Plant Physiol., [42], 840—844.

45 Fork, D.C. (1963) Plant Physiol., [38], 323—332.
46 Forti, G. (1968) Biochem. Biophys. Res. Commun., [32], 1020—1024.
47 Forti, G., Rosa, L., Fuggi, A. and Garlaschi, F.M. (1974) in Proc. 3rd Int. Cong. Photosynthesis (Avron, M., ed), pp. 1499—1505, Elsevier, Amsterdam.
48 Frackowiak, B. and Kaniuga, Z. (1971) Biochim. Biophys. Acta, [226], 360—365.
49 Gibbs, M. (1971) in Structure and Function of Chloroplasts (Gibbs, M. ed), pp. 169—214, Springer, Berlin.
50 Gibbs, M., Bamberger, E.S., Ellyard, P.W. and Everson, R.G. (1967) in Biochemistry of Chloroplasts (Goodwin, T.W. ed), Vol. 3, pp. 3—38, Academic Press, London.
51 Gibbs, M., Latzko, E., Everson, R.G. and Cockburn, W. (1967) in Harvesting the Sun (San Pietro, A., Green, F.A. and Army, T.J., eds), pp. 111—130, Academic Press, New York.
52 Gibbs, M. and Plaut, Z. (1970) Plant Physiol., [45], 470—474.
53 Gibbs, M. and Shain, Y. (1971) Plant Physiol., [48], 325—330.
54 Gimmler, H. (1973) Z. Pflanzenphysiol., [68], 289—307.
55 Gimmler, H., Schäfer, G. and Heber, U. (1974) in Proc. 3rd Int. Cong. Photosynthesis (Avron, M., ed), pp. 1381—1392, Elsevier, Amsterdam.
56 Goldsworthy, A. (1970) Bot. Rev., [36], 321—340.
57 Gould, J.M. and Izawa, S. (1973) Biochim. Biophys. Acta, [314], 211—223.
58 Gromet-Elhanan, Z. (1968) Arch. Biochem. Biophys., [123], 447—456.
59 Hatch, M.D. (1971) in Photosynthesis and Photorespiration (Hatch, M.D., Osmond, C.B. and Slatyer, R.O., eds), pp. 139—152, Wiley Interscience, New York.
60 Heber, U. (1960) Z. Naturforsch., [15b], 95—109.
61 Heber, U. (1969) Biochim. Biophys. Acta, [180], 302—319.
62 Heber, U. (1973) Biochim. Biophys. Acta, [305], 140—152.
63 Heber, U. (1974) Ann. Rev. Plant Physiol., [25], 393—421.
64 Heber, U. (1974) in Proc. 3rd Int. Cong. Photosynthesis (Avron, M.,ed), pp. 1335—1348, Elsevier, Amsterdam.
65 Heber, U. (1976) J. Bioenergetics, Review Section, in press.
66 Heber, U., Andrews, T.J. and Boardman, N.K. (1975) Submitted to Plant Physiol.
67 Heber, U. and French, C.S. (1968) Planta, [79], 99—112.
68 Heber, U., Hallier, U.W. and Hudson, M.A. (1967) Z. Naturforsch., [22b], 1200—1215.
69 Heber, U. and Kirk, M.R. (1974) in Proc. 3rd Int. Cong. Photosynthesis (Avron, M., ed), pp. 1041—1046, Elsevier, Amsterdam.
70 Heber, U. and Kirk, M.R. (1975) Biochim. Biophys. Acta, [376], 136—150.
71 Heber, U. and Kirk, M.R. unpublished.
72 Heber, U., Kirk, M.R., Gimmler, H. and Schäfer, G. (1974) Planta, [120], 31—46.
73 Heber, U. and Krause, G.H. (1971) in Photosynthesis and Photorespiration (Hatch, M.D., Osmond, C.B. and Slatyer, R.O., eds), pp.218—225, Wiley Interscience, New York.
74 Heber, U. and Krause, G.H. (1972) in Proc. 2nd Int. Cong. Photosynthesis (Forti, G., Avron, M. and Melandri, A., eds), pp. 1023—1033, Junk, The Hague.
75 Heber, U. and Santarius, K.A. (1965) Biochim. Biophys. Acta, [109], 390—408.
76 Heber, U. and Santarius, K.A. (1970) Z. Naturforsch., [25b], 718—728.
77 Heber, U. and Willenbrink, J. (1964) Biochim. Biophys. Acta, [82], 313—324.
78 Heldt, H.W. (1969) FEBS Lett., [5], 11—14.
79 Heldt, H.W., Geller, G. and Werdan, K. (1972) Biochim. Biophys. Acta, [283], 430—441.
80 Heldt, H.W. and Rapley, L. (1970) FEBS Lett., [7], 139—142.
81 Heldt, H.W. and Rapley, L. (1970) FEBS Lett., [10], 143—148.
82 Heldt, H.W. and Sauer, F. (1971) Biochim. Biophys. Acta, [234], 83—91.

83 Heldt, H.W., Sauer, F. and Rapley, L. (1972) in Proc. 2nd Int. Cong. Photosynthesis (Forti, G., Avron, M. and Melandri, A., eds), pp.1345—1355, Junk, The Hague.
84 Heldt, H.W., Werdan, K., Milovancev, M. and Geller, G. (1973) Biochim. Biophys. Acta, [314], 224—241.
85 Hind, G. and McCarthy, R.E. (1973) in Photophysiology (Giese, A.C., ed), Vol. 8, p. 114, Academic Press, New York.
86 Hoch, G., Owens, O.v.H. and Kok, B. (1963) Arch. Biochem. Biophys., [101], 171—180.
87 Hohorst, H.J. (1963) in Funktionelle und morphologische Organisation der Zelle, (Karlson, P., ed), pp. 194—208, Springer, Berlin.
88 Horton, A.A. and Hall, D.O. (1968) Nature, [218], 386—388.
89 Izawa, S. and Good, N.E. (1968) Biochim. Biophys. Acta, [162], 380—391.
90 Izawa, S. and Hind, G. (1967) Biochim. Biophys. Acta, [143], 377—390.
91 Jackson, W.A. and Volk, R.J. (1970) Ann. Rev. Plant Physiol., [21], 385—432.
92 Jagendorf, A.T. and Hind, G. (1963) in Photosynthetic Mechanisms in Green Plants, Natl. Acad. Sci. Natl. Res. Council Publ. [1145], pp. 599—607.
93 Jensen, R.G. and Bassham, J.A. (1968) Biochim. Biophys. Acta, [153], 227—234.
94 Johnson, H.S. and Hatch, M.D. (1970) Biochem. J., [119], 273—280.
95 Kagawa, T. and Hatch, M.D. (1974) Aust. J. Plant Physiol., [1], 51—64.
96 Kandler, O. and Haberer-Liesenkötter, I. (1963) Z. Naturforsch., [18b], 718—730.
97 Kearney, P.C. and Tolbert, N.E. (1962) Arch. Biochem. Biophys., [98], 164—171.
98 Kelly, G.J. and Gibbs, M. (1973) Plant Physiol., [52], 674—676.
99 Keys, A.J. (1968) Biochem. J., [108], 1—8.
100 Keys, A.J. and Whittingham, C.P. (1969) in Progress in Photosynthesis Research, (Metzner, H., ed), Vol. I, pp. 352—358, Int. Union of Biol. Sci., Tübingen.
101 Klingenberg, M., Heldt, H.W. and Pfaff, E. (1969) in The Energy Level and Metabolic Control in Mitochondria (Papa, S., Tager, J.M., Quagliariello, E. and Slater, E.C., eds), pp. 237—255, Adriatica Editriche, Bari.
102 Kraayenhof, R. (1969) Biochim. Biophys. Acta, [180], 213—215.
103 Krause, G.H. (1971) Z. Pflanzenphysiol., [65], 13—23.
104 Krause, G.H. (1973) Biochim. Biophys. Acta, [292], 715—728.
105 Krause, G.H. (1974) Biochim. Biophys. Acta, [333], 301—313.
106 Krause, G.H. (1974) in Membrane Transport in Plants (Zimmermann, E. and Dainty, J., eds), pp. 274—280, Springer, Berlin.
107 Krause, G.H. (1974) in Proc. 3rd Int. Cong. Photosynthesis (Avron, M., ed), pp. 1021—1030, Elsevier, Amsterdam.
108 Krause, G.H., unpublished.
109 Krause, G.H. and Bassham, J.A. (1969) Biochim. Biophys. Acta, [172], 553—565.
110 Krause, G.H. and Heber, U. (1971) in 1st Eur. Biophysics Cong. Proc. (Broda, E., Locker, A. and Springer-Lederer, H., eds), Vol. IV, pp. 79—84, Verlag Wiener Medizin. Akad., Vienna.
111 Krause, G.H. and Heber, U. (1972) in 4th Int. Biophysics Cong. Vol. I, Abstracts, p. 323, I.U.P.A.B. and Acad. Sci. USSR, Moscow.
112 Lendzian, K. and Ziegler, H. (1972) in Proc. 2nd Int. Cong. Photosynthesis (Forti, G., Avron, M. and Melandri, A., eds), Vol. 3, pp. 1831—1838, Junk, The Hague.
113 Lilley, R., McCormick, A.V., Fitzgerald, M.P., Rienits, K.G. and Walker, D.A. (1975) New Phytol., in press.
114 Mathieu, Y. (1967) Photosynthetica, [1], 57—63.
115 Mayne, B.C., Edwards, G.E. and Black, C.C. (1971) in Photosynthesis and Photorespiration (Hatch, M.D., Osmond, C.B. and Slatyer, R.O., eds), pp. 361—371. Wiley Interscience, New York.
116 McGowan, R.E. and Gibbs, M. (1974) Plant Physiol., [64], 312—319.

117 McRobbie, E.A.C. (1970) Quart. Rev. Biophys., [3], 251—294.
118 Mehler, A.H. (1951) Arch. Biochem. Biophys., [33], 65—77.
119 Mehler, A.H. (1951) Arch. Biochem. Biophys., [34], 339—351.
120 Menke, W. (1962) Ann. Rev. Plant. Physiol., [13], 27—44.
121 Miginiac-Maslow, M. and Moyse, A. (1969) in Progress in Photosynthesis Research, (Metzner, H., ed), pp. 1203—1212, Laupp, Tübingen.
122 Mitchell, P. (1966) Biol. Rev., [41], 445—502.
123 Nobel, P.S. (1969) Biochim. Biophys. Acta, [172], 134—143.
124 Nobel, P.S., Chang, D.C., Wang, C., Smith, S. and Barcus, D.E. (1969) Plant Physiol., [44], 655—661.
125 Oh-Hama, T. and Miyachi, S. (1959) Biochim. Biophys. Acta, [34], 202—210.
126 Oh-Hama, T., Miyachi, S. and Tamiya, H. (1963) Colloq. Int. Centre Natl. Rech. Sci. No. 119, pp. 439—448, Editions du Centre National de la Recherche Scientifique, Paris.
127 Ongun, A. and Stocking, C.R. (1965) Plant Physiol., [40], 825—831.
128 Osmond, C.B. and Björkman, O. (1972) Carnegie Institution Yearbook, [71], 141—148.
129 Pflüger, R. (1973) Z. Naturforsch., [28c], 779—780.
130 Quitrakul, R. and Izawa, S. (1973) Biochim. Biophys. Acta, [305], 105—118.
131 Rabinowitch, E.I. (1945) Photosynthesis and Related Processes, Interscience, New York.
132 Reeves, S.G. and Hall, D.O. (1973) Biochim. Biophys. Acta, [314], 66—78.
133 Ried, A. (1968), Biochim. Biophys. Acta, [153], 653—663.
134 Robinson, M. and Gibbs, M. (1974) Plant Physiol., [53], 790—797.
135 Robinson, J.M. and Stocking, C.R. (1968) Plant Physiol., [42], 1597—1604.
136 Santarius, K.A. and Heber, U. (1965) Biochim. Biophys. Acta, [102], 39—54.
137 Santarius, K.A., Heber, U., Ullrich, W. and Urbach, W. (1964) Biochem. Biophys. Res. Commun., [15], 139—146.
138 Santarius, K.A. and Stocking, C.R. (1969) Z. Naturforsch., [24b], 1170—1179.
139 Schnarrenberger, C., Oeser, A. and Tolbert, N.E. (1973) Arch. Biochem. Biophys., [154], 438—448.
140 Shain, Y. and Gibbs, M. (1971) Plant Physiol., [48], 325—330.
141 Shavit, N. and Avron, M. (1967) Biochim. Biophys. Acta, [131], 516—525.
142 Shin, N. (1971) in Methods in Enzymology (San Pietro, A., ed), Vol. 23, pp. 440—447, Academic Press, New York.
143 Shin, M. and Arnon, D.I. (1965) J. Biol. Chem., [240], 1405—1411.
144 Simonis, W. and Urbach, W. (1973) Ann. Rev. Plant Physiol., [24], 89—114.
145 Stinson Jr. H.T. and Zucker, M. (1962) Arch. Biochem. Biophys., [96], 637—644.
146 Stocking, C.R. (1971) in Methods in Enzymology (San Pietro, A., ed), Vol. 23 pp. 221—228, Academic Press, New York.
147 Stocking, C.R. and Larson, S. (1969) Biochem. Biophys. Res. Commun., [37], 278—282.
148 Stokes, D.M. and Walker, D.A. (1971) in Photosynthesis and Photorespiration (Hatch, M.D., Osmond, C.B., and Slater, R.O., eds), pp.226—231, Wiley Interscience, New York.
149 Strotmann, H. and Berger, S. (1969) Biochem. Biophys. Res. Commun., [35], 20—26.
150 Strotmann, H. and Heldt, H.W. (1969) in Progress in Photosynthesis Research, (Metzner, H., ed), Vol. 3, pp. 1131—1140, Int. Union of Biol. Sci., Tübingen.
151 Tanner, W., Dächsel, L. and Kandler, O. (1965) Plant Physiol., [40], 1151—1156.
152 Tanner, W., Löffler, M. and Kandler, O. (1969) Plant Physiol., [44], 422—428.

153 Ting, I.P., Rocha, K., Mikerji, S.K. and Curry, R. (1971) in Photosynthesis and Photorespiration (Hatch, M.D., Osmond, C.B. and Slatyer, R.O., eds), pp. 534—540, Wiley Interscience, New York.
154 Tolbert, N.E. (1971) Annu. Rev. Plant Physiol., [22], 45—74.
155 Tolbert, N.E. (1971) in Photosynthesis and Photorespiration (Hatch, M.D., Osmond, C.B. and Slatyer, R.O., eds), pp. 458—471, Wiley Interscience, New York.
156 Tolbert, N.E. (1973) in Current Topics in Cellular Regulation (Horecker, B.L. and Stadtman, E.R., eds), Vol. 7, pp. 21—49, Academic Press, New York.
157 Trebst, A. and Reimer, S. (1973) Biochim. Biophys. Acta, [305], 129—139.
158 Walker, D.A. (1971) in Methods in Enzymology (San Pietro, A., ed), Vol. 23, pp. 211—220, Academic Press, New York.
159 Walker, D.A. (1973) New Phytol., [72], 209—235.
160 Walker, D.A. (1974) in Med. Techn. Publ. Int. Rev. Sci. Biochem. Ser. I (Northcote, D.H., ed), Vol. 11, pp. 1—49, Butterworth, London.
161 Walker, D.A. and Crofts, A.R. (1970) Ann. Rev. Biochem., [39], 389—428.
162 Walker, D.A., Kosciukiewicz, K. and Case, C. (1973) New Phytol., [72], 237—247.
163 Wang, C.T. and Nobel, P.S. (1971) Biochim. Biophys. Acta, [241], 200—212.
164 Werdan, K. and Heldt, H.W. (1972) in Proc. 2nd Int. Cong. Photosynthesis (Forti, G., Avron, M. and Melandri, A., eds), pp. 1337—1344, Junk, The Hague.
165 Werdan, K., Heldt, H.W. and Milovancev, M., (1976) Biochim. Biophys. Acta, [396], 276—292.
166 West, K.R. and Wiskich, J.T. (1968) Biochem. J., [109], 527—532.
167 Whitehouse, D.G., Ludwig, L.J. and Walker, D.A. (1971) J. Exp. Bot., [22], 772—791.
168 Wiesemann, R. (1972) Staatsexamensarbeit, University of Düsseldorf.
169 Wilson, A.T. and Calvin, M. (1955) J. Amer. Chem. Soc., [77], 5948—5957.
170 Winget, G.D., Izawa, S. and Good, N.E. (1965) Biochem. Biophys. Res. Commun., [21], 438—443.
171 Zelitch, I. (1971) Photosynthesis, Photorespiration and Plant Productivity. Academic Press, New York.

The Intact Chloroplast — edited by J. Barber

Chapter 6

Metabolite Transport in Intact Spinach Chloroplasts

HANS WALTER HELDT

Institut für Physiologische Chemie und Physikalische Biochemie der Universität München, 8 München 2, Goethestrasse 33 (W. Germany)

CONTENTS

Abbreviations: DAP, dihydroxyacetone phosphate; 1,3-DiPGA, 1,3-diphosphoglyceric acid; DMO, 5,5-dimethyloxazolidine-2,4-dione; Fum, fumarate; GAP, glyceraldehyde-3-phosphate; Mal, malate; P, phosphate; PGA, 3-phosphoglyceric acid.

6.1. INTRODUCTION

Chloroplasts are a constituent part of all eukaryotic photosynthesizing cells. They contain some DNA and a protein synthesizing system similar to that found in prokaryotic organisms (see Chapter 10 of this volume), and they multiply by division. It has been speculated that the chloroplasts are derived from photosynthesizing prokaryotic cells which entered into symbiosis with a non-photosynthesizing cell [31]. Regardless whether this is true or not, it illustrates the role of the chloroplast as a separate metabolic unit within the plant cell.

It is the metabolic function of the chloroplast to fix CO_2 in order to provide the plant cell with substances required for cell growth. The main substances produced by the chloroplast appear to be triosephosphates ([46] and Chapter 7 of this volume). This requires a transfer of H_2O, CO_2, and inorganic phosphate from the cytoplasm to the chloroplast stroma, which is the site of CO_2 fixation, and a transfer of triosephosphates in the other direction (Fig. 6.1). Besides the delivery of fixed carbon, the chloroplasts may also provide the cell with reducing equivalents, forming NADH or NADPH, or phosphorylating equivalents in the form of ATP (see Chapters 5 and 12). For this to occur there is also a requirement for the transfer of other substances between the chloroplast stroma and the cytoplasm. In order to understand the interrelationship between the two metabolic compartments, the stroma and the cytoplasm, it is necessary to know the permeability properties of the separating membranes. It is this aspect of chloroplast function which is the subject of this chapter.

6.2. PERMEABILITY MEASUREMENTS

The present report will mainly deal with direct measurements of the uptake of substances into chloroplasts. A very important method for such measurements is the silicone layer filtering centrifugation [22, 29, 49]. In this method, a centrifugation tube is filled at the bottom with a denaturing

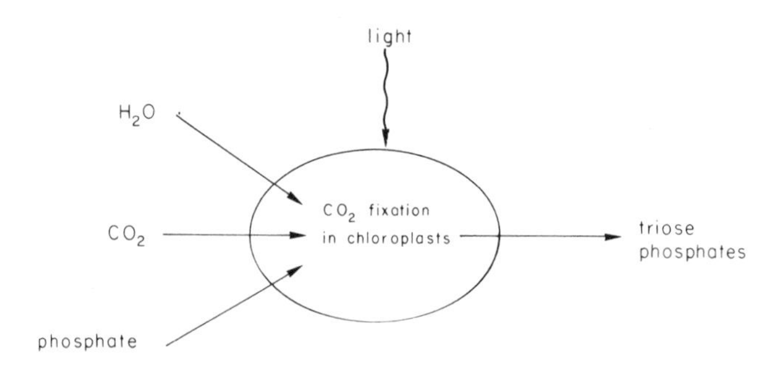

Fig. 6.1. Schematic diagram of metabolite fluxes during CO_2 fixation.

agent, e.g. perchloric acid, followed by a layer of silicone oil, and the chloroplast suspension on top of this (Fig. 6.2). A substance is added to the suspension on a stirrer, and after a certain time the chloroplasts are separated from the medium by sedimentation through the silicone layer into the perchloric acid. The uptake of the substance into the chloroplasts is determined from assay in the sediment fraction. For this the tip containing the perchloric acid is separated by cutting through the silicone layer and the substance is assayed in the acid-soluble extract. For convenience, radioactively labelled compounds are usually used, and the assay is simplified to radioactivity measurement. The small amount of medium adhering to the surface of the chloroplasts, as they migrate through the silicone layer, can be corrected for by adding radioactively labelled macromolecules (e.g. dextran) which are known not to permeate membranes.

When using rapid microcentrifuges, the sedimentation time of intact chloroplasts can be reduced to below 2 seconds. Therefore, silicone layer filtering centrifugation is a very useful method for kinetic measurements.

6.3. UNSPECIFIC PERMEABILITY OF THE OUTER MEMBRANE OF THE CHLOROPLAST ENVELOPE

The envelope of the chloroplast, separating the chloroplast stroma from the cytoplasm consists of two membranes, the outer and the inner membrane. The question arose, which of these two membranes is the functional partition between the stroma and the cytoplasm. For answering this question the permeability of these two membranes was studied by

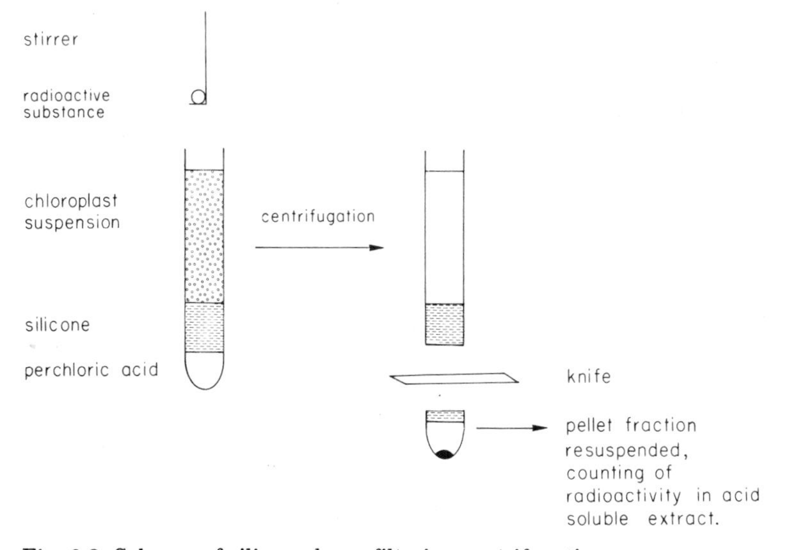

Fig. 6.2. Scheme of silicone layer filtering centrifugation.

silicone layer filtering centrifugation. It was found that part of the chloroplast volume was very rapidly penetrated by all compounds of low molecular weight which have been tested, such as nucleotides, inorganic phosphate, phosphate derivatives, carboxylic acids and sucrose [22]. On the other hand it was not penetrated by dextran. Since the size of the permeable space is usually determined with radioactively labelled sucrose, or sorbitol, it has been called the sucrose — or sorbitol-permeable space. The size of the sucrose-permeable space depends on the tonicity of the medium (Fig. 6.3A). It is relatively small when the chloroplasts are kept in a hypotonic medium (0.16 M sorbitol), whereas in a hypertonic medium (0.66 M sorbitol) almost half of the chloroplast space is permeated by sucrose.

To correlate these findings with the morphology of chloroplasts, electron microscopy studies were carried out at the same time. In a hypertonic medium the stroma of the chloroplasts appeared to be shrunken, the outer membrane being very loosely attached to the inner membrane with large empty spaces in between (intermembrane space). In a hypotonic medium the stroma was expanded with only little space between the two membranes. For more quantitative information, the relative sizes of the intermembrane spaces in the chloroplasts were evaluated by planimetry of electron micrographs. The diagrams of Fig. 6.3B show average values from a large number of electron micrographs.

There is a striking parallel between the relative sizes of the space within the outer and the inner membrane, as observed with electron microscopy and the sucrose-permeable space measured in our experiments. It is therefore concluded that the sucrose-permeable space is identical with

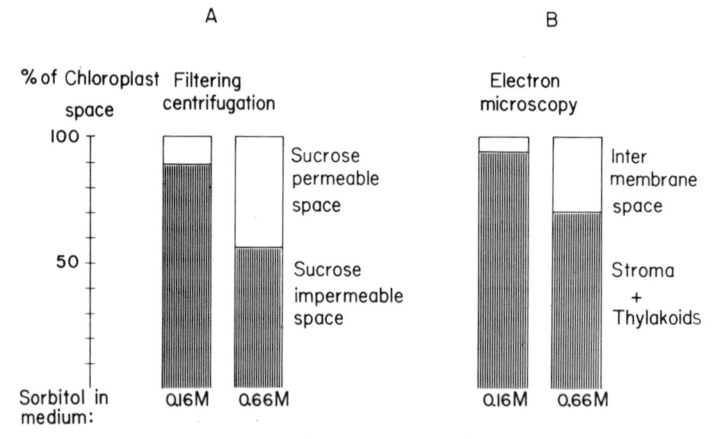

Fig. 6.3. Comparison of the relative sizes of the sucrose-permeable spaces measured by filtering centrifugation with the sizes of the inter-membrane spaces obtained by planimetry of electron micrographs of spinach chloroplasts. (From ref. 23.) For details see ref. 1.

the intermembrane space and that the outer membrane is unspecifically permeable to sucrose and other molecules of similar size (Fig. 6.4).

The osmotic response of the stroma space clearly indicates that the inner membrane is the osmotic barrier of the envelope, whereas the intermembrane space is freely accessible to the metabolites in the cytoplasm. Since it has been shown that the thylakoid membranes of the chloroplasts are formed from invagination of the inner membrane during development of the plastid [32,33] it might be possible that the thylakoid space is connected to the intermembrane space. In this case the thylakoid space should also be freely accessible to small molecules in the medium. This possibility is ruled out from pH measurements in the thylakoid space of intact chloroplasts [24]. When chloroplasts, kept in a medium of pH 7.6, are illuminated, the pH in the thylakoid space is found to be around pH 5.4. The existence of such a large pH gradient between the thylakoid space and the medium excludes any free connection between these two spaces.

6.3.1. *Comparison between the structure of chloroplasts and of mitochondria*

There are similarities between the structure of chloroplasts and of mitochondria. As with chloroplasts, the outer mitochondrial membrane was

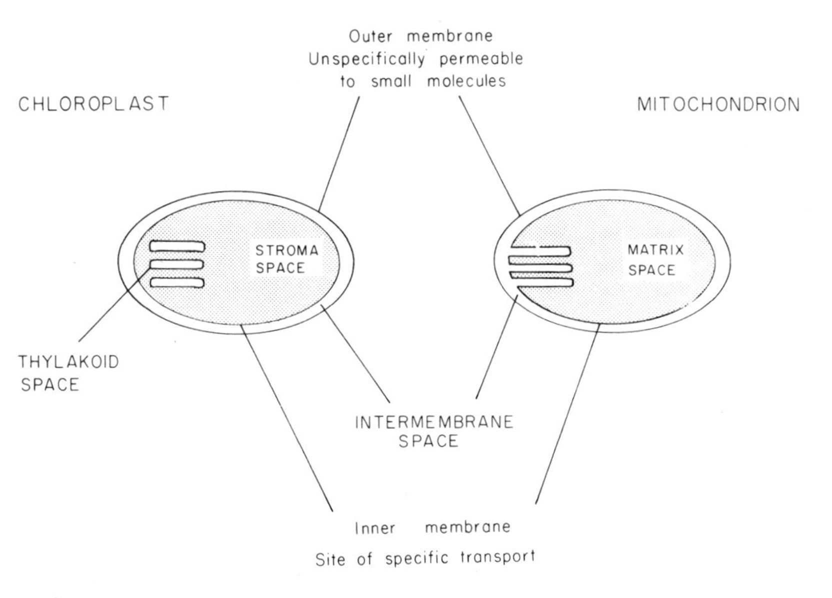

Fig. 6.4. Schematic diagram of the structure of chloroplasts and of mitochondria. (From ref. 22.)

found to be unspecifically permeable to solutes of low molecular weight, with the inner membrane being the site of specific transport [35]. The main difference appears that in the mitochondria the intracristae space is part of the intermembrane space, whereas in the chloroplasts the thylakoid space is separated from the intermembrane space. Thus in relation to the creation of ion gradients during photosynthesis or oxidative electron flow, the thylakoid space corresponds to the mitochondrial intermembrane space and the stroma to the matrix.

In membrane preparations of the chloroplast envelope, only small amounts of chlorophyll have been found, which may be due to contamination by thylakoid membranes [7,30,37]. Although it has not yet been proven definitely it is almost certain that these envelope preparations contain the inner membrane. It therefore seems that there is no photosynthetic electron transport in the inner membrane of the envelope. In the mitochondria the inner membrane has a dual function. It is the site of electron transport and also the site of metabolite transport, whereas in the chloroplasts these two functions are separated in two different membranes, the thylakoid membrane and the inner membrane of the envelope.

6.4. PERMEABILITY PROPERTIES OF THE INNER MEMBRANE OF THE CHLOROPLAST ENVELOPE

From the discussions above it has been concluded that the inner membrane of the chloroplast envelope is the functional barrier between the chloroplast stroma and the cytoplasm. The inner membrane is permeable to carbon dioxide [48] and certain monocarboxylic acids, e.g. acetic acid ([26] and Werdan and Heldt, unpubl.), glyceric acid [17] and glycolic acid (Werdan and Heldt, unpubl.). It is less permeable to amino acids, as will be discussed later, and it appears to be impermeable to sucrose, sorbitol, and various anions, e.g. di- and tricarboxylates, phosphate and phosphorylated compounds like nucleotides and sugar phosphates [22]. This impermeability of the inner membrane is overcome by specific carriers for transport of phosphate, phosphoglycerate and dihydroxyacetone phosphate, for transport of dicarboxylates and for transport of ATP, which will be dealt with later.

6.4.1. Uptake of bicarbonate into the chloroplast stroma

Bicarbonate is taken up very rapidly and a certain concentration is reached in the stroma [48] (Fig. 6.5). Even at low temperatures the kinetics of uptake cannot be resolved. From Fig. 6.5 a minimal rate of bicarbonate (0.5 mM) uptake is estimated as 100 μmoles/mg chlorophyll/h (9°C), but the actual value is likely to be much higher. The bicarbonate concentration

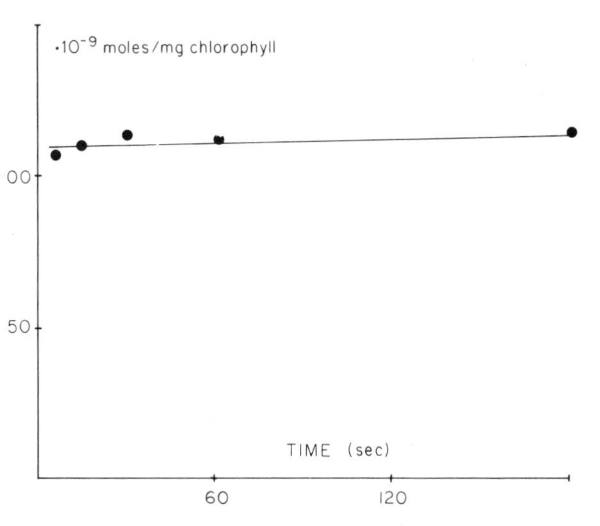

Fig. 6.5. Bicarbonate uptake into the sucrose-impermeable space of illuminated spinach chloroplasts. Temp. 9°C, bicarbonate 0.5 mM. The fixed carbon has been substrated. (Data from ref. 48.)

in the stroma is linearly dependent on the bicarbonate concentration in the medium. CO_2 has been shown to diffuse through artificial lipid bilayers very rapidly [3], indicating that biological membranes probably do not act as permeability barriers to CO_2. It is feasible that bicarbonate in the medium equilibrates with CO_2, the latter diffuses rapidly across the inner membrane and equilibrates with HCO_3^- in the stroma.

$$HCO_3^- + H^+ \rightleftharpoons \overset{\text{external}}{H_2O + CO_2} \;\; \rightleftharpoons\!\!\!\!\!\!\!\;\; \| \;\; \overset{\text{internal}}{CO_2 + H_2O} \rightleftharpoons HCO_3^- + H^+$$

If the rate of diffusion is not a limiting step, the CO_2 concentration on each side of the membrane should be equal. Consequently the logarithm of the ratio of bicarbonate concentrations in the stroma and in the medium should be equal to the ΔpH across the inner membrane:

$$\log \frac{[HCO_3^-]_{int.}}{[HCO_3^-]_{ext.}} = pH_{int.} - pH_{ext.}$$

The validity of this mechanism can be tested by plotting the logarithm of the ratio of bicarbonate concentrations against the ΔpH across the inner membrane. In the experiment of Fig. 6.6 the external pH was varied and the pH in the stroma ($pH_{int.}$) was assayed from the distribution of [^{14}C]-dimethyloxazolidinedione (DMO). There is indeed a linear function observed with a slope of 1.

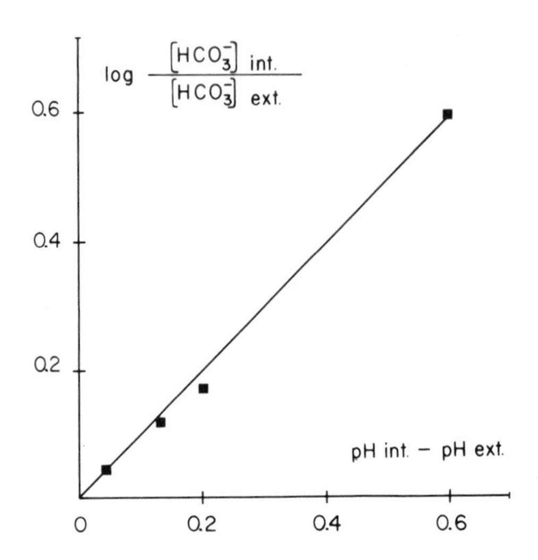

Fig. 6.6. The logarithm of bicarbonate distribution between the stroma and the medium depending on the ΔpH between the two spaces. Spinach chloroplasts, illuminated. (Data from ref. 48.)

Thus the experimental data seem to agree rather well with the above mechanism. Rapid equilibration of bicarbonate with CO_2 requires the activity of carbonic anhydrase. A very high activity of this enzyme has been located in the chloroplast stroma [9,10,37,48]. With isolated chloroplasts there will be always sufficient stroma carbonic anhydrase in the medium, due to some broken chloroplasts. In vivo CO_2 and not bicarbonate is the source of carbon taken up by terrestrial plants.

From studies in our laboratory with isolated chloroplasts there was no indication for a direct transport of bicarbonate as postulated by Poincelot [38]. Such transport of bicarbonate could not be reconciled with the data of Fig. 6.6 and other results [48]. Thus at present, the experimental evidence for the uptake of bicarbonate via diffusion of CO_2 across the inner membrane may be regarded as being much stronger than the evidence presented by Poincelot [38] for a direct transport of bicarbonate. Illumination of chloroplasts, which is known to cause an alkalization in the stroma [24], increases the bicarbonate concentration in this compartment [48]. This is in accordance with the data of Fig. 6.6. The increase of bicarbonate concentration, however, does not alter the CO_2 concentration in the stroma, which is important in regard to the observation that CO_2 rather than bicarbonate is the substrate of ribulosediphosphate carboxylase [6]. Therefore, a light-dependent increase of bicarbonate concentration does not imply a "CO_2 pump".

Since CO_2 enters the chloroplast stroma, and CO_2 is also utilized for CO_2 fixation, one may ask, what is the physiological role of the high carbonic anhydrase activity in the stroma? It has been speculated that carbonic anhydrase may cause a local rise of CO_2 concentration at the site of ribulosediphosphate carboxylase [48]. It may also be that rapid equilibration of CO_2 with bicarbonate enhances CO_2 diffusion across the chloroplast stroma [8].

6.4.2. Uptake of amino acids into the chloroplast stroma

From studies of the distribution of metabolites in plant cells it has been suggested that there is rapid movement of certain amino acids, e.g. alanine, glycine and serine between the cytoplasm and the chloroplast stroma [1,40]. Table I shows the rates of uptake of various neutral amino acids. In comparison to the rates of phosphate or dicarboxylate transport (see later) these rates are very low. Those amino acids having a large hydrophobic moiety like phenylalanine or isoleucine are taken up the most rapidly, whereas short amino acids like serine and alanine are taken up very slowly. This concurs with similar results by Gimmler et al. [12], but is in contrast to a relatively high permeability for glycine and serine reported in pea chloroplasts [34]. The concentration of neutral amino acids found in the chloroplast stroma is about equal to the concentration in the medium. The rate of uptake depends linearly on the concentration in the medium (Fig. 6.7), indicating that the uptake of neutral amino acids proceeds via diffusion across the inner membrane. Studies on the amino acid permeability of liposomes showed, that the rate of permeation was dependent on the solubility of the amino acid in the lipid phase of the membrane [50]. Thus phenylalanine was shown to pass an artificial membrane comparatively rapidly, whereas serine moved very slowly across it. Essentially the same results are obtained in our experiments. It has been proposed by Nobel and Cheung [34] that there are specific carriers for

TABLE I

Amino acid uptake into the sucrose-impermeable space of spinach chloroplasts
Temp. 20° C. (From ref. 25.)

L-Amino acid (1 mM)	Uptake (μmoles/mg chlor./h)
Glycine	0.70
Alanine	0.41
Serine	0.56
Isoleucine	3.04
Phenylalanine	3.56

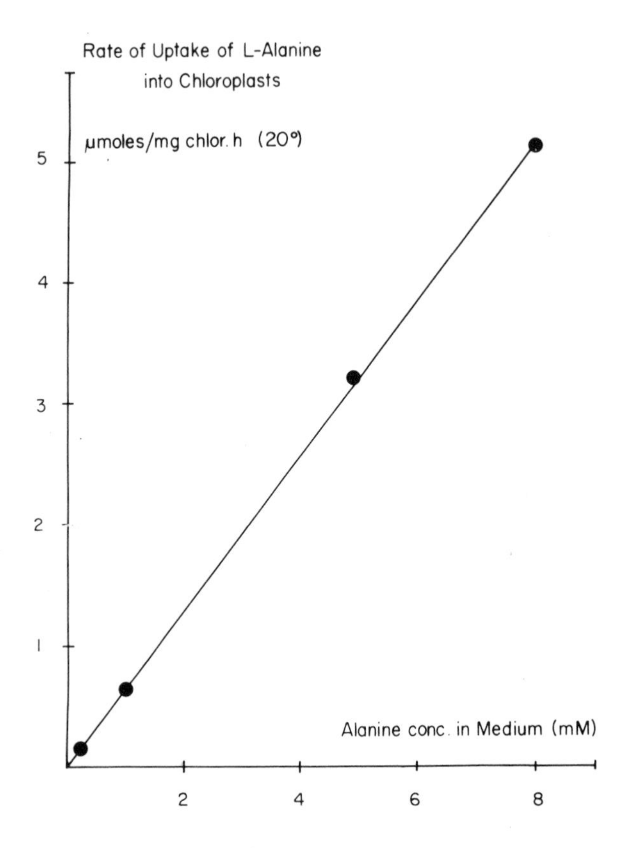

Fig. 6.7. The rate of alanine uptake into the sucrose-impermeable space of spinach chloroplasts depending on the concentration in the medium. Temp. 20° C (Werdan, Milovancev and Heldt, unpublished).

amino acids in pea chloroplasts with K_m in the range of 100 mM. The data shown here do not contradict such a possibility, but it is unlikely that there are carriers requiring completely unphysiological substrate concentrations for half-saturation of transport.

6.4.3. *Transport of ATP*

As in mitochondria, a counter exchange of adenine nucleotides was also found in chloroplasts [19,43]. The specificity of this transport for external nucleotides is shown in Table II. There are basic differences, however, between the translocation of adenine nucleotides in chloroplasts and in mitochondria.

Atractyloside, a strong inhibitor of the mitochondrial system [20] does not inhibit the transport of adenine nucleotide in chloroplasts. Whereas in mitochondria the transport is specific for external ADP [36], in chloro-

TABLE II

Adenine nucleotide translocation in spinach chloroplasts
(From ref. 19.)

Nucleotide (0.25 mM)	Activity of translocation (20° C) (μmoles/mg chlor./h)
ATP	2.04
ADP	0.24
AMP	0.07
CTP	< 0.03
UTP	< 0.01
GTP	< 0.01
ITP	< 0.01

plasts it is specific for external ATP. In spinach chloroplasts, the maximal rates obtained were 5 μmoles/mg chlorophyll/h, measured at 20°C. When this is compared with the transport rate for the substrates of the phosphate and dicarboxylate translocator, the activity of the ATP translocator appears to be about two orders of magnitude lower than the activity of the other two transport systems mentioned.

6.4.4. The phosphate translocator

Functional studies (e.g. measurement of ^{14}C-labelled compounds released from the chloroplasts during CO_2 fixation, or the effect of added substances on CO_2 fixation and measurements of metabolite levels in the intact plant tissue) suggested that 3 phosphoglycerate, triosephosphates and inorganic phosphate are able to permeate the chloroplast envelope ([2,5,18,41,45,46] and Chapter 5 of this volume). Direct measurements, which shall be summarized here, revealed that these compounds are specifically transported across the inner membrane. Fig. 6.8 shows rapid uptake of 3-phosphoglycerate and inorganic phosphate into the chloroplast stroma, whereas hexosephosphates are taken up only very slowly. The rapid uptake can be saturated. A double reciprocal plot of the rate of transport depending on the concentration in the medium yields a linear curve (Fig. 6.9), enabling the determination of K_m and V_{max}. 3-Phosphoglycerate, which is also rapidly taken up, competitively inhibits the uptake of phosphate. Similarly phosphate, 3-phosphoglycerate and dihydroxyacetone phosphate were shown to compete with each other for transport [47]. In all cases the K_m for transport, and the K_i for inhibition of transport were identical. From this it was concluded that there is one specific carrier catalyzing the transport of dihydroxyacetone phosphate, 3-phosphoglycerate and inorganic phosphate, which has been named "phosphate translocator" [21]. The specificity of the transport is shown in Table III. The carrier accepts

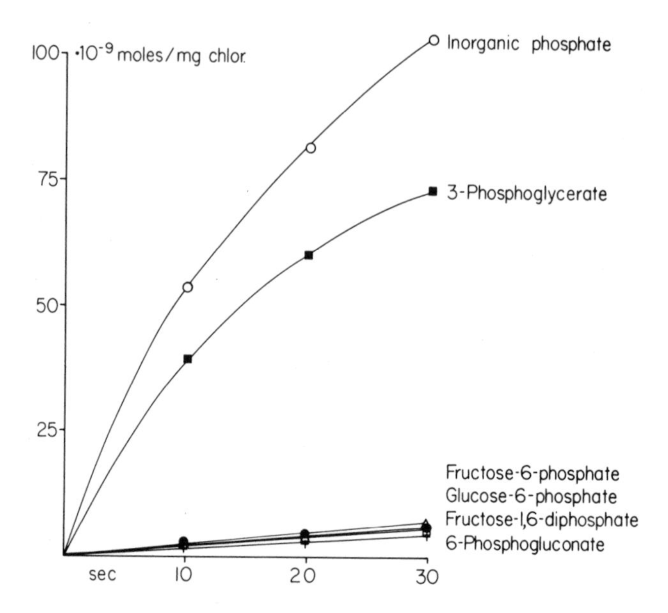

Fig. 6.8. Transport of radioactively labelled metabolites into the sucrose-impermeable space of spinach chloroplasts. Metabolite concentration 1 mM, temp. 4°C (From ref. 23.)

TABLE III

Specificity of the phosphate translocator in spinach chloroplasts (4°C) (From ref. 25.)

Phosphate	K_m	0.20 [mM]
3-Phosphoglycerate	K_m	0.15
Dihydroxyacetone phosphate	K_i	0.13
Arsenate	K_i	0.35
Pyrophosphate	K_i	1.8
2-Phosphoglycerate	K_i	6.5
1-Glycerophosphate	K_i	1.3
2-Glycerophosphate	K_i	7.7
Phosphoenolpyruvate	K_i	4.7
Erythrose-4-phosphate	K_i	2.30
Ribose-5-phosphate	K_i	10.0
Fructosediphosphate	K_i	8.5
Fructose-6-phosphate	K_i	12.5
Glucose-6-phosphate	K_i	40.0
6-Phosphogluconate	K_i	20.0

either inorganic phosphate (or arsenate) or a phosphate molecule attached to the end of a 3-carbon chain, like 3-phosphoglycerate and triosephosphates. Thus 1 glycerophosphate is also transported to some extent. 4-carbon compounds like erythrose-4-phosphate are only poorly transported and hexosephosphates not at all. 3-carbon compounds, in which the phosphate is attached to carbon atom 2, like 2-phosphoglycerate, phosphoenolpyruvate and 2-glycerophosphate, are not accepted by the carrier, which demonstrates its strong specificity.

The activity of this transport is very high. At 4°C the V_{max} for uptake of phosphate, 3-phosphoglycerate and dihydroxyacetone phosphate was found to be in the range of 50 μmoles/mg chlorophyll/h and at 20°C it may be about 5 times higher (Fliege and Heldt, unpubl.). The phosphate translocator is inhibited by low concentrations of mercurials, e.g. p-chloromercuriphenylsulphonic acid, a reagent which interacts with SH groups [47], indicating that SH groups are involved in the functioning of the translocator.

The phosphate translocator catalyzes a counter-exchange, as shown from the experiment in Fig. 6.10. The chloroplasts were preincubated with ^{32}P-labelled phosphate and washed afterwards. ^{14}C-labelled 3-phosphoglycerate was added and the uptake and release of these compounds was measured simultaneously. The amount of phosphate released is equivalent to the amount of 3-phosphoglycerate taken up, indicating strict coupling of inward and outward transport. This concurs with the observation that inorganic phosphate does not leak out from intact isolated chloroplasts.

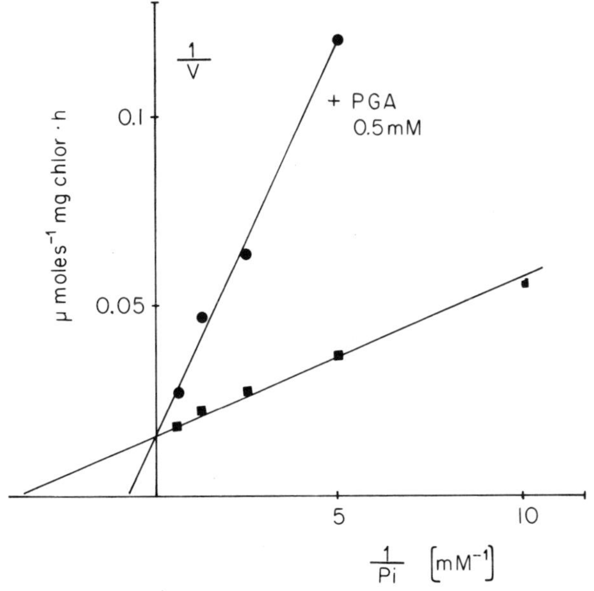

Fig. 6.9. Concentration dependence of the transport of inorganic phosphate (P_i), inhibition by 3-phosphoglycerate (PGA). Dark, temp. 4°C (Fliege and Heldt, unpublished).

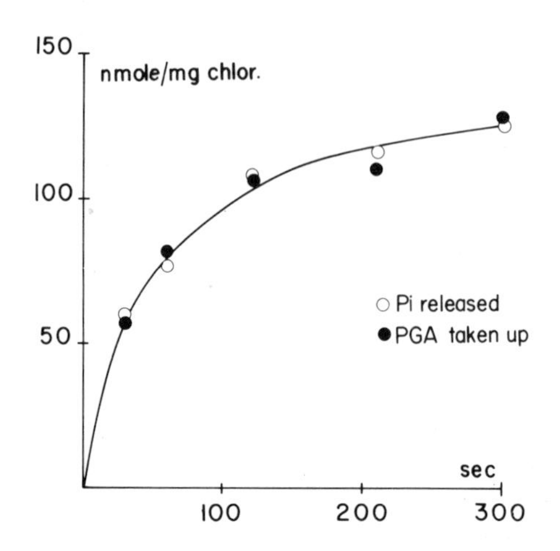

Fig. 6.10. Simultaneous measurement of the uptake of [^{14}C]3-phosphoglycerate and the release of [^{32}P]phosphate. The chloroplasts were preincubated with phosphate in the dark. Temp. 4°C (From ref. 25.)

6.4.5. Transport of dicarboxylates

Several dicarboxylates are transported across the inner membrane of the chloroplast envelope by a specific transporter, which has been named dicarboxylate translocator [21]. Also this transport shows substrate saturation (Fig. 6.11). All dicarboxylates transported compete with each other for transportation into the stroma. Thus transport of fumarate is inhibited by malate and vice versa (Fig. 6.11).

The K_m of fumarate transport equals the K_i of fumarate for inhibition of malate transport within the experimental error and the same is true for the K_m of the malate transport. Similar results have been obtained with the other dicarboxylates transported, indicating that all these compounds are transported by the same carrier. Therefore the specificity of the carrier for a dicarboxylate can be either expressed by the K_m or K_i.

Table IV shows the specificity of dicarboxylate transport expressed as K_i. Because of the very high activity of the transport, the measurements were carried out at 4°C. There appears to be a wide specificity for various C_4 and C_5 dicarboxylates. Maleate is not transported in contrast to its transisomer fumarate, and the C_3 compound malonate does not react with the carrier. Furthermore, the carrier does not transport tricarboxylates nor phosphate. Thus the dicarboxylate transport in chloroplasts appears to be entirely different from the dicarboxylate transport in mitochondria. For dicarboxylate transport in chloroplasts the following values for V_{max} (μmoles/mg chlorophyll/h) have been obtained at 4°C: malate, 19; L-aspartate, 31; L-glutamate, 16 [25].

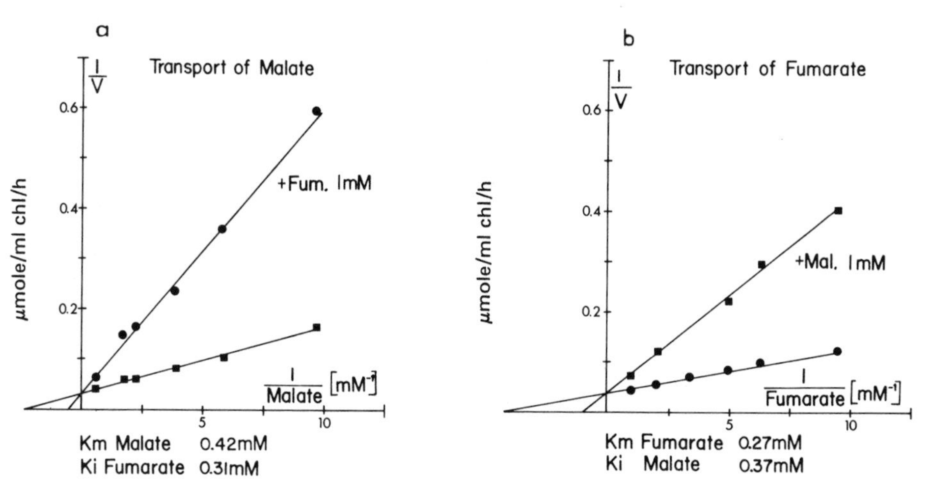

Fig. 6.11. Concentration dependence of the transport of malate, and its inhibition by fumarate (a) and of the transport of fumarate and its inhibition by malate (b). Temp. 4°C (Lehner and Heldt, unpublished).

The dicarboxylate carrier facilitates a counter exchange of anions, but this counter exchange is not as strictly coupled as transport by the phosphate translocator [25]. Net transport in one direction only is possible, but the rate is about one order of magnitude lower than the rate of counter transport. This is in agreement with the observation that dicarboxylates leak out even from highly intact chloroplasts.

TABLE IV

Specificity of the dicarboxylate translocator in spinach chloroplasts
(Competitive inhibition of the transport of L-malate (K_m 0.34). Temp. 4°C.
(From ref. 25.)

Inhibitor	K_i (mM)
Oxaloacetate	0.26
Fumarate	0.32
Succinate	0.35
α-Ketoglutarate	0.40
L-Aspartate	0.58
L-Glutamate	1.00
L-Tartrate	1.08
meso-Tartrate	0.83
D-Tartrate	2.71
Thiomalate	0.91
Glutarate	1.97
ortho-Phthalate	2.43
Malonate	no inhibition
Maleate	no inhibition

6.5. PHYSIOLOGICAL SIGNIFICANCE OF TRANSFER PROCESSES ACROSS THE INNER MEMBRANE OF THE ENVELOPE

6.5.1. Diffusion

The diffusion of CO_2 across the inner membrane enables the supply of inorganic carbon for CO_2 fixation occurring in the stroma. Transfer of glycolate and glycerate, which might occur as diffusion of the undissociated acid across the inner membrane, is involved in the overall reaction of photo-respiration (see Chapters 5 and 8). Glycolate, which may be formed from the oxidation of ribulosediphosphate and hydrolysis of phosphoglycolate in the stroma, is released from the chloroplast. In a series of reactions occurring in the peroxysomes and in the mitochondria it is transformed to glycerate. This is transferred back to the chloroplast stroma, where it is further metabolized [44].

6.5.2. Specific transport

The inner membrane of the envelope is in principle impermeable to nucleotides, as has been mentioned before. Therefore it is not possible to transport reducing equivalents between the stroma and the cytoplasm by transfer across the inner membrane. However, as already mentioned in Chapter 5, there is a possibility of transferring reducing equivalents via metabolite shuttles. The dicarboxylate translocator has been shown to trans-port oxaloacetate and malate. Since there is malic dehydrogenase in the stroma as well as in the cytoplasm [13,27], a malate oxaloacetate shuttle may transfer reducing equivalents between the two compartments (Fig. 6.12A). Such a shuttle has been demonstrated to operate in vitro [15,25]. However, doubts have been raised whether this shuttle may be of physiologi-cal significance, since the concentration of oxaloacetate in the plant cell is rather low. It has been suggested as shown in Fig. 6.12B, that oxaloacetate might undergo transamination with glutamate leading to the formation of α-ketoglutarate and aspartate, the latter two compounds being transported across the inner membrane in exchange with glutamate and malate [16]. This cycle has been known to transfer reducing equivalents into the mitochondria [4]. The dicarboxylate translocator in chloroplasts is able to transport all these substances with high velocity, which indicates that this cycle may indeed have a physiological role. This shuttle would be suited to reducing cytoplasmic NAD.

However, cytoplasmic NADPH is required for biosynthetic processes. A reduction of cytoplasmic NADP is possible through a metabolite shuttle, which is facilitated by the phosphate translocator (Fig. 6.13). In this cycle 3-phosphoglycerate is reduced in the stroma to dihydroxacetone phosphate at the expense of ATP and NADPH. Dihydroxyacetone phosphate is trans-

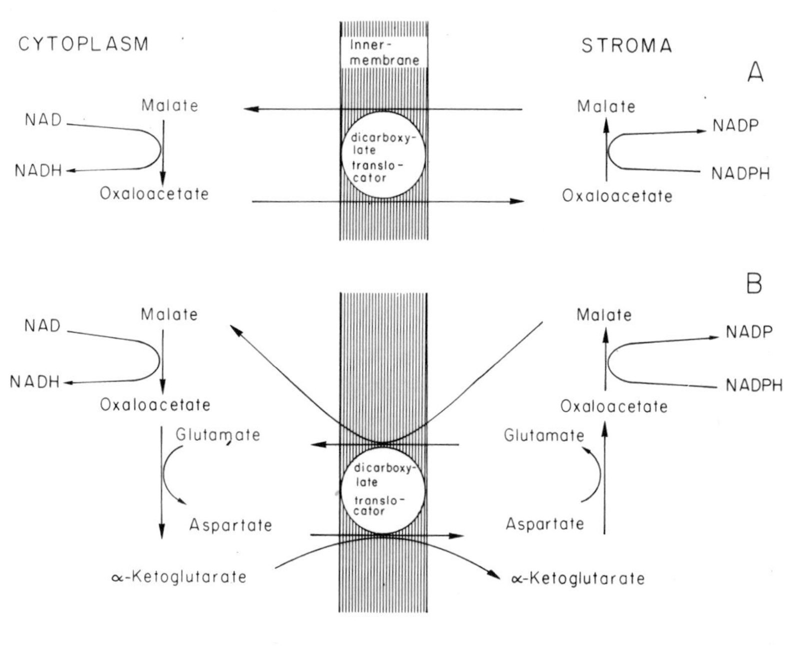

Fig. 6.12. Schematic diagram of metabolite shuttle facilitated by the dicarboxylate translocator.

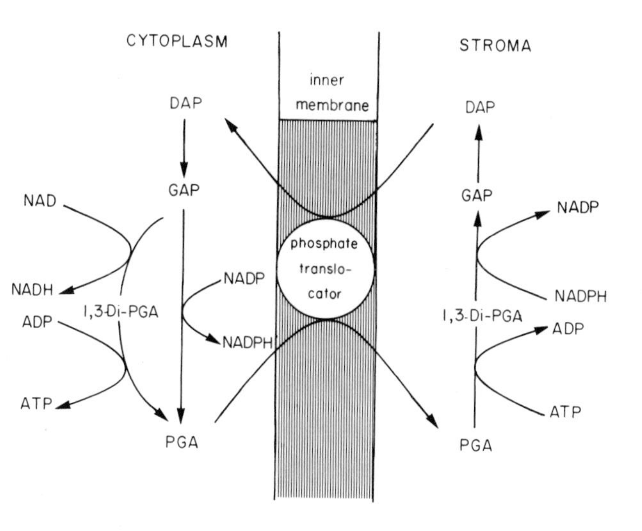

Fig. 6.13. Schematic diagram of metabolite shuttle facilitated by the phosphate translocator.

ported into the cytoplasm in exchange for 3-phosphoglycerate. In principle, also glyceraldehyde phosphate could be transported. Since the concentration of this compound in equilibrium with dihydroxyacetone phosphate is very low, and the activities of trioseisomerase are very high, transport of dihydroxyacetone phosphate is expected to be much higher than that of glyceraldehyde phosphate (see also ref. 14). The dihydroxyacetone phosphate can be reoxidized in the cytoplasm by the non-reversible, non-phosphorylating triosephosphate dehydrogenase, yielding the reduction of NADP [11]. Alternatively the dihydroxyacetone phosphate in the cytoplasm can be converted to 3-phosphoglycerate by the glycolytic pathway yielding ATP and NADH. In this way a transfer of reducing equivalents across the inner membrane is accompanied by an indirect transfer of ATP. Both variants of this cycle have been shown to operate with isolated chloroplasts [11,14,42] but it has been found that the non-phosphorylating triosephosphate dehydrogenase has a lower K_m for glyceraldehyde phosphate than the glycolytic enzyme [28]. Apparently the export of reducing equivalents from the stroma to the cytoplasm in the form of NADPH, as required for biosynthetic processes, appears to have priority over the supply of ATP together with NADH to cytoplasm.

Direct transport of ATP across the inner membrane is very low. It appears that the adenine nucleotide translocation of chloroplasts does not contribute to an export of ATP from the stroma. The high specificity for external ATP suggests that the ATP translocator may act in the opposite direction, trans-

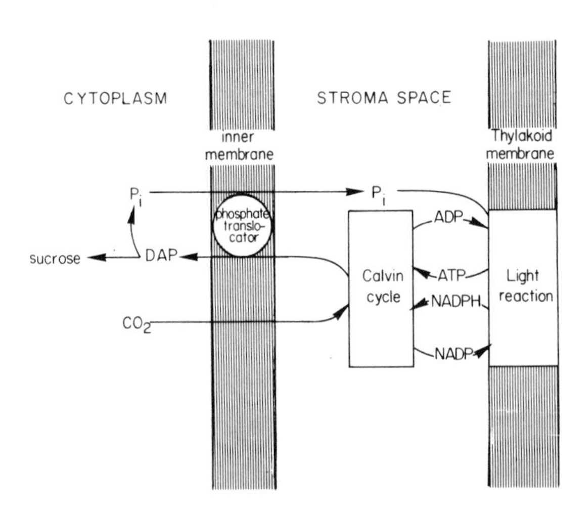

Fig. 6.14. Schematic diagram of the participation of the phosphate translocator in the overall reaction of photosynthesis.

porting ATP from the cytoplasm into the stroma. This could play a role for supplying the chloroplasts during the night phase with ATP generated by glycolysis or respiration.

The main function of the phosphate translocator is to enable the export of reduced carbon from the chloroplasts (Fig. 6.14). An export of triosephosphates requires an uptake of inorganic phosphate. The strict counter-exchange between release of dihydroxyacetone phosphate and phosphate may be important to maintain a constant level of phosphate in the chloroplasts in order to insure CO_2 fixation.

6.6. OUTLOOK

It has been shown that there are metabolite carriers in the inner membrane of the chloroplast envelope, and some of their basic properties have been established. Since these investigations have all been carried out with spinach chloroplasts, it will be necessary to extend these studies to other plants in order to draw general conclusions.

6.7. ACKNOWLEDGEMENT

This work has been supported by a grant of the Deutsche Forschungsgemeinschaft.

REFERENCES

1 Aach, H.G. and Heber, U. (1967) Z. Pflanzenphysiol., [57], 317—322.
2 Bassham, J.A., Kirk, M., and Jensen, R.G. (1968) Biochim. Biophys. Acta, [153], 211—218.
3 Blank, M. and Roughton, F.J.W. (1960) Trans. Faraday Soc., [56], 1832—1836.
4 Borst, P. (1963) in Funktionelle und morphologische Organisation der Zelle (Karlson, P., ed), pp. 137—158, Springer, Berlin.
5 Cockburn, W., Walker, D.A. and Baldry, C.W. (1968) Biochem. J., [107], 89—95.
6 Cooper, T.G., Filmer, D., Wishnick, M. and Lane, M.D. (1969) J. Biol. Chem., [244], 1081—1083.
7 Douce, R., Holtz, R.B., Benson, A.A. (1973) J. Biol. Chem., [248], 7215—7222.
8 Enns, T. (1967) Science, [155], 44—47.
9 Everson, R.G. and Slack, C.R. (1968) Phytochemistry, [7], 581—586.
10 Everson, R.G. (1970) Phytochemistry, [9], 25—32.
11 Bamberger, E.S., Ehrlich, B.A., Gibbs, M. (1974) in Proc. 3rd Int. Cong. Photosynthesis (Avron, M., ed), pp. 1349—1362, Elsevier, Amsterdam.
12 Gimmler, H., Schäfer, G., Krammer, M. and Heber, U. (1974) Planta, [120], 47—61.
13 Heber, U. (1960) Z. Naturforsch., [15b], 100—109.
14 Heber, U. and Santarius, K.A. (1970) Z. Naturforsch., [25b], 718—728.
15 Heber, U. and Krause, G. (1972) in Proc. 2nd Int. Cong. Photosynthesis (Forti, G., Avron, M. and Melandri, A., eds), Vol. 2, pp. 1023—1033, Junk, The Hague.

16 Heber, U. (1974) in Proc. 3rd Int. Cong. Photosynthesis (Avron, M., ed), pp. 1335—1348, Elsevier, Amsterdam.
17 Heber, U., Kirk, M.R., Gimmler, H. and Schäfer, G. (1974) Planta, [120], 34—46.
18 Heber, U. (1974) Ann. Rev. Plant Physiol., [25], 393—421.
19 Heldt, H.W. (1969) FEBS Lett., [5], 11—14.
20 Heldt, H.W. (1969) in 20th Mosbacher Colloquium (Bucher, Th. and Sies, H., eds), pp. 301—317, Springer, Heidelberg.
21 Heldt, H.W. and Rapley, L. (1970) FEBS Lett., [10], 143—148.
22 Heldt, H.W. and Sauer, F. (1971) Biochim. Biophys. Acta, [234], 83—91.
23 Heldt, H.W., Sauer, F. and Rapley, L. (1972) in Proc. 2nd Int. Cong. Photosynthesis (Forti, G., Avron, M. and Melandri, A., eds), Vol. 2, pp. 1345—1355, Junk, The Hague.
24 Heldt, H.W., Werdan, K., Milovancev, M. and Geller, G. (1973) Biochim. Biophys. Acta, [314], 224—241.
25 Heldt, H.W., Fliege, R., Lehner, K., Milovancev, M. and Werdan, K. (1974) in Proc. 3rd Int. Cong. Photosynthesis (Avron, M., ed), pp. 1369—1379, Elsevier, Amsterdam.
26 Jacobson, B.S. and Stumpf, P.K. (1972) Arch. Biochem. Biophys., [153], 656—663.
27 Johnson, H.S. and Hatch, M.D. (1970) Biochem. J., [119], 272—280.
28 Kelly, G.J. and Gibbs, M. (1973) Plant Physiol., [52], 111—118.
29 Klingenberg, M. and Pfaff, E. (1967) Methods in Enzymology, [10], 680—684.
30 Mackender, R.O. and Leech, R.M. (1970) Nature, [228], 1347—1349.
31 Margulis, L. (1970) Origin of Eucaryotic Cells, Yale University Press, New Haven, Conn.
32 Menke, W. (1964) Ber. Deut. Bot. Ges., [77], 340—354.
33 Mühlethaler, K. and Frey-Wyssling, A. (1959) J. Biophys. Biochem. Cytol., [6], 507—619.
34 Nobel, P.S. and Cheung, Y.N.S. (1972) Nature New Biol., [237], 207—208.
35 Pfaff, E., Klingenberg, M., Ritt, E. and Vogell, W. (1968) Eur. J. Biochem., [5], 222—230.
36 Pfaff, E., Heldt, H.W. and Klingenberg, M. (1969) Eur. J. Biochem., [10], 484—493.
37 Poincelot, R.P. (1972) Biochim. Biophys. Acta, [258], 637—642.
38 Poincelot, R.P. (1974) Plant Physiol., [54], 520—525.
39 Poincelot, R.P. (1973) Arch. Biochem. Biophys., [159], 134—142.
40 Roberts, G.R., Keys, A.J. and Whittingham, C.P. (1970), J. Expt. Bot., [21], 683—692.
41 Robinson, J.M. and Stocking, C.R. (1968) Plant Physiol., [43], 1597—1603.
42 Stocking, C.R. and Larson, S. (1969) Biochem. Biophys. Res. Commun., [37], 278—282.
43 Strotmann, H. and Heldt, H.W. (1969) in Progress in Photosynthesis Research, (Metzner, H., ed), pp. 1131—1140. I.U.B.S., Tübingen.
44 Tolbert, N.E. (1971) in Photosynthesis and Photorespiration (Hatch, M.D., Osmond, C.B. and Slatyer, R.O., eds), pp. 458—471, Wiley Interscience, New York.
45 Urbach, W., Hudson, M.A., Santarius, K.A. and Heber, U. (1965) Z. Naturforsch., [20b], 890—898.
46 Walker, D.A. (1975) in Med. Techn. Publ. Int. Rev. Sci. Biochem. Ser. I, (Northcote, D.H., ed), Vol. II, pp. 1—49, Butterworth, London.
47 Werdan, K. and Heldt, H.W. (1972) in Proc. 2nd Int. Cong. Photosynthesis (Forti, G., Avron, M. and Melandri, A., eds), Vol. 3, pp. 1337—1344, Junk, The Hague.
48 Werdan, K., Heldt, H.W. and Geller, G. (1972) Biochim. Biophys. Acta, [283], 430—441.
49 Werkheiser, W.C. and Bartley, W. (1957) Biochem. J., [66], 79—85.
50 Wilson, P.D. and Wheeler, K.P. (1973) Biochem. Soc. Transact., [1], 369—372.

The Intact Chloroplast — edited by J. Barber

Chapter 7

CO_2 Fixation by Intact Chloroplasts: Photosynthetic Induction and its Relation to Transport Phenomena and Control Mechanisms

D.A. WALKER

Department of Botany, The University of Sheffield, Sheffield (Great Britain)

CONTENTS

Abbreviations: DHAP, Dihydroxyacetone phosphate; G3P, glyceraldehyde-3-phosphate; OAA, oxaloacetate; P_i, inorganic orthophosphate; PP_i, inorganic pyrophosphate; PGA, 3-phosphoglycerate; RBP, ribulose-1,5-bisphosphate; RBPCase, ribulose-1,5-bisphosphate carboxylase; R5P, ribose-5-phosphate; Ru5P, ribulose-5-phosphate; TP, triosephosphate. phate.

7.1. SIMPLE INDUCTION IN WHOLE PLANTS

7.1.1. Introduction

When green leaves are strongly illuminated after a period of darkness, photosynthetic oxygen evolution and CO_2 fixation does not normally reach a maximum for several minutes. This initial lag is called "induction" or sometimes "simple induction" in order to distinguish it from a variety of transient fluctuations in gaseous exchange ("gulps" and "bursts") which occur under some conditions [90]. Simple induction is not an artefact of measurement. It occurs in species which lack stomata and it can be readily observed in experiments with isolated chloroplasts (for reviews, see e.g. refs. 45,106,112). It is an intrinsic and fundamental feature of photosynthesis and it is the aim of this article to examine its significance and its relation to transport between the chloroplast and its immediate environment. Many of the facts and arguments presented here were first published in the New Phytologist [106].

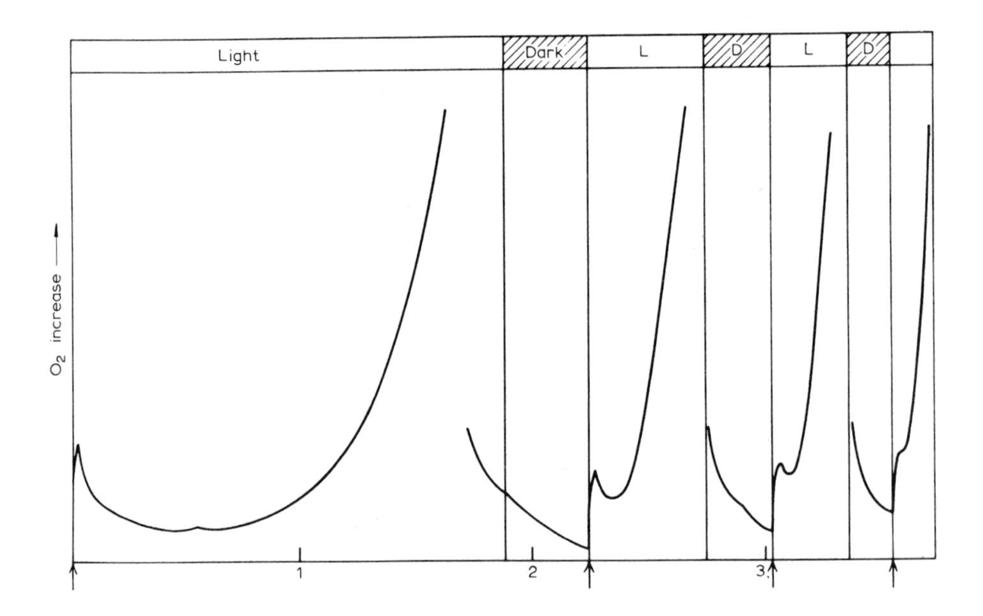

Fig. 7.1. Induction in *Ricinus* leaf (after ref. 23). An O_2 electrode was held against the leaf surface with agar. Before the onset of measurement the leaf was kept in the dark for 10 min. Note the shortening of the lag after increasingly brief dark intervals, the initial burst of O_2 evolution and the progressive increase in post-illumination O_2 uptake.

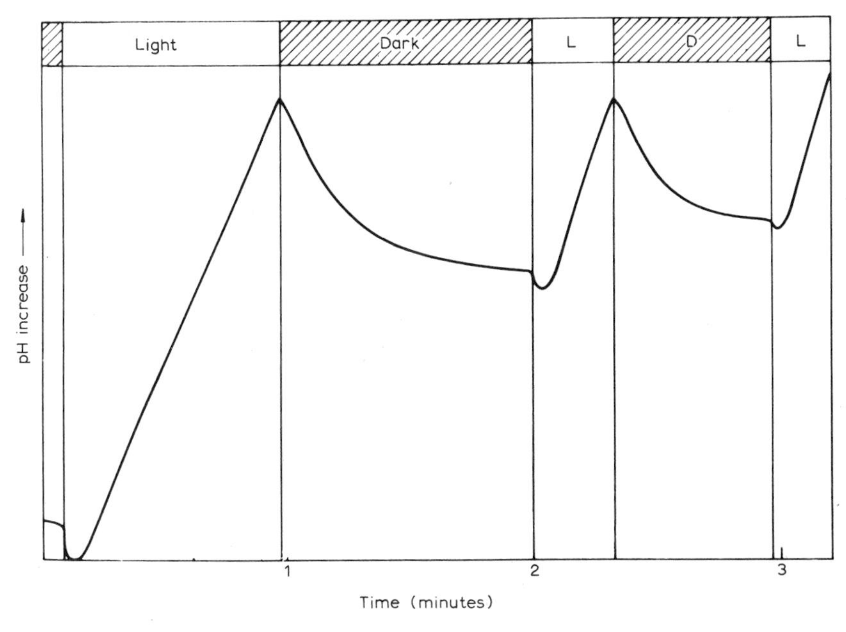

Fig. 7.2. Induction in pond lily (after ref. 23). Uptake and release of CO_2 caused the pH changes in the medium recorded with a glass electrode. Again there is a discernible shortening in induction after a brief dark interval. The increased release of CO_2 immediately after illumination is similar to the increased oxygen seen in Fig. 7.1.

7.1.2. Historical aspects

Induction periods which lasted several minutes at room temperature were first reported by Osterhout and Haas [82] in experiments with *Ulva* in 1918 and by Warburg [117] (using *Chlorella*) in 1920. In 1938 Blinks and Skow [23] made continuous records of lags in oxygen evolution from *Ricinus* leaves and in pH changes associated with the onset of photosynthesis in waterlily (see Figs. 7.1, 7.2). Steeman-Nielsen [94] also observed induction in photosynthetic oxygen evolution from *Fucus*. Probably the most extensive and useful measurements during this period were made by McAlister and Myers [76] who followed CO_2 uptake by wheat leaves using a spectrographic (infra-red) method (Figs. 7.3, 7.4). A comprehensive account of these and other observations appears in Rabinowitch's monograph [90]. Some contemporary results similar to those of Blinks and Skow [23] and McAlister [76] are illustrated in Figs. 7.5 to 7.9.

7.1.3. The role of stomata

When higher plants are taken from the dark their stomata will normally be

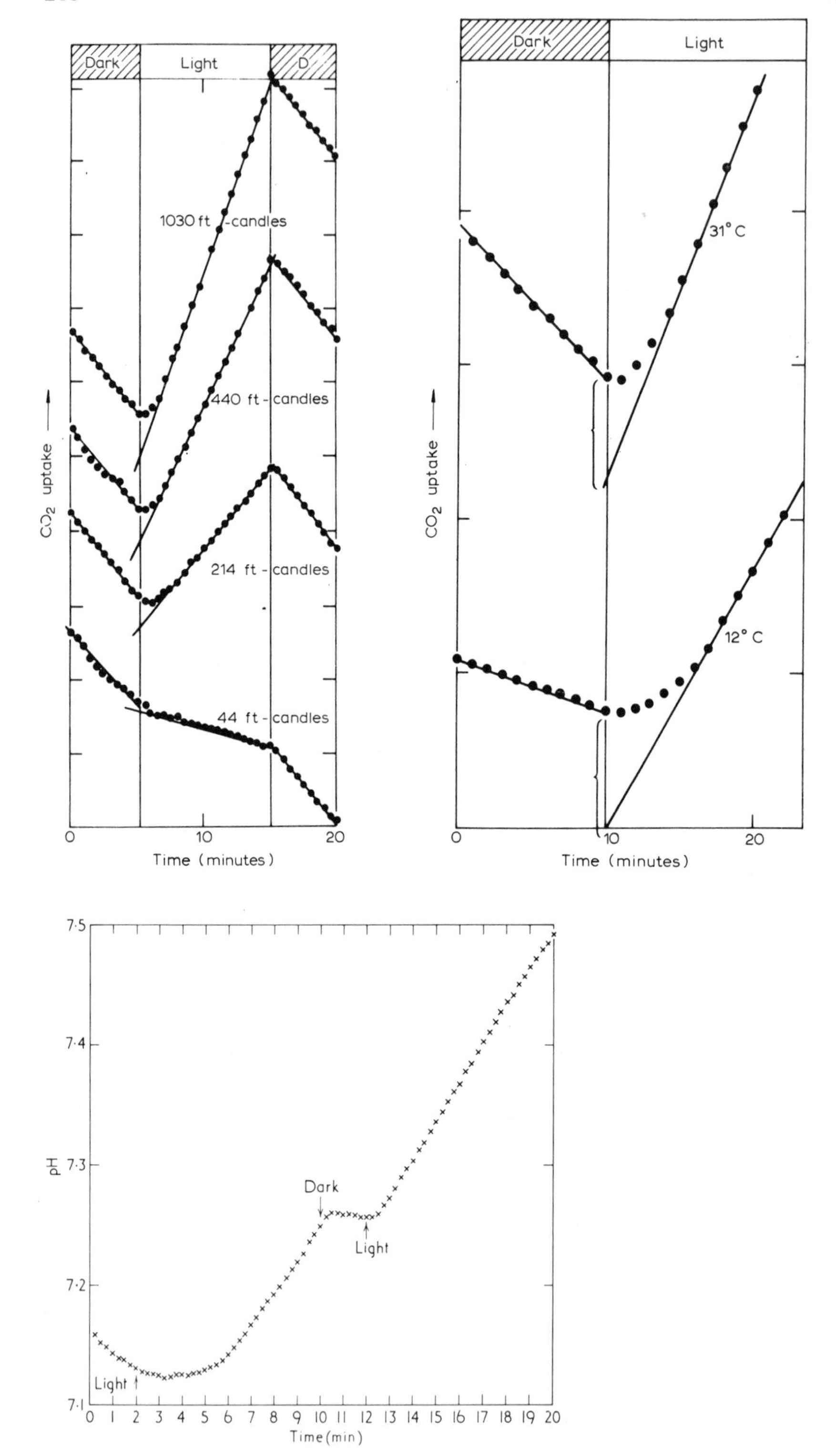

Dark
Light
D
1030 ft -candles
440 ft - candles
214 ft - candles
44 ft - candles
CO2 uptake
Time (minutes)
Dark
Light
31° C
12° C
CO2 uptake
Time (minutes)
7·5
7·4
7·3
7·2
7·1
pH
Dark
Light
Light
Time (min)

closed and will commence to open upon illumination (see e.g. refs. 77,128, 100). The time course of opening is often very similar to the time course of induction and is similarly affected by temperature and (to some extent) by light intensity. Particularly when infra-red gas analysis is used to follow photosynthesis at air-levels of CO_2 (Figs. 7.3, 7.4 and 7.9) it is clearly difficult to distinguish between inductive effects which are independent of stomata and those occasioned by stomatal opening. Moreover, there seems little doubt that the rate of diffusion of CO_2 from the external atmosphere to the site of carboxylation will affect the rate of photosynthesis or that endo-diffusion of CO_2 will be affected by stomatal resistance. It is therefore not unreasonable to ask if there is an inductive lag in higher plant photosynthesis other than that brought about by stomatal opening. On the basis of comparative metabolism it would have to be concluded that induction is a normal feature of photosynthesis. Certainly, many studies of induction have been made with unicellular algae or aquatic plants which do not possess stomata (e.g. refs. 117,94,57). Moreover, stomata open well in moderate light [126,127,77] and yet some induction can still be observed in passing from moderate to saturating light [76,94]. Induction has also been observed in leaf discs from which the epidermis has been stripped (Fig. 7.8). Similarly, induction is consistently observed in photosynthesis by chloroplasts isolated from the leaves of higher plants (see section 7.3).

7.1.4. Effects of light and temperature on induction in leaves

Induction in intact leaves is largely independent of light intensity (Fig. 7.3) but strongly dependent on temperature (Fig. 7.4). This strongly suggests that it is more directly related to the so-called "dark biochemistry" of photosynthesis (i.e. carbon assimilation and associated reactions) rather than to the primary photochemical events, and this is borne out by in vitro experiments (see e.g. Figs. 7.14 to 7.16). Induction is shortened by pre-illumination (Figs. 7.1, 7.2, 7.6 to 7.9) and is therefore minimal when a period of strong illumination is briefly interrupted by a brief dark interval (Figs. 7.5 to 7.9). Again it must be inferred that some constraint is lifted during the first minutes of photosynthesis and that this is not immediately reimposed in the dark.

Fig. 7.3. Induction in wheat leaves (after ref. 76). Changes in CO_2 followed by an infra-red method. In these experiments the initial lag was independent of light intensity.

Fig. 7.4. As for Fig. 7.3 showing the extension of the lag at a lower temperature.

Fig. 7.5. Induction in Lemna. Whole plants illuminated at 20°C in bicarbonate solution (approx. 10^{-3} M). As in the original experiments of Blinks and Skow [23] the pH increases as CO_2 is taken up from solution. Note long initial lag which is not repeated after a short dark interval.

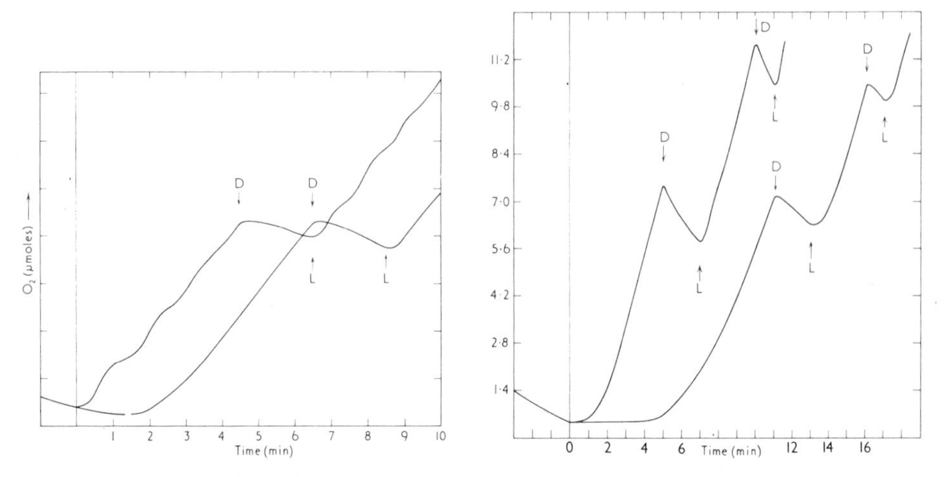

Fig. 7.6. Sunflower. Left, from light; right, from dark.

Fig. 7.7. Spinach. Left, from light; right, from dark. The oscillations in A are not always seen but have been observed in several species on a number of occasions, usually following illumination and are reminiscent of the results of Van der Veen [129] and others.

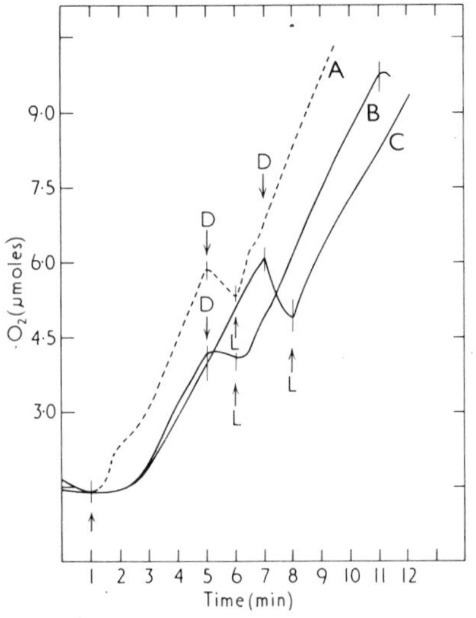

Fig. 7.8. Tobacco. (A) from light, (C) from dark, (B) as for C but with upper epidermis stripped from the leaf.

Figs. 7.6, 7.7 and 7.8. Examples of induction in leaves. Leaf discs (10 cm^2) were illuminated in a small (3 ml) chamber containing an oxygen electrode and 0.2 ml of 1.0 M bicarbonate on a filter paper disc. The concentration of CO_2 in the gas phase during the course of the experiment is indeterminate (balanced as it is between uptake by the leaf and release from the unbuffering bicarbonate solution) but is believed to be saturating. Certainly it is sufficient to maintain rapid rates of O_2 evolution (250-350 μmoles.mg^{-1} chlorophyll.h^{-1}). Silicone rubber impressions of the leaf surface made according to the method of Zelitch [126] immediately after illumination showed no evidence of stomatal opening but it is assumed that at these concentrations closed stomata would offer only minimal resistance to gaseous exchange (cf. ref. 128). All of the species examined showed the characteristic initial lag which is longer in leaves taken from the dark rather than from the light. Other consistent features include the greatly decreased lag following a brief dark interval despite increased post-illumination oxygen uptake (Walker, unpublished).

7.1.5. *The molecular basis of induction*

From the outset [82], induction has been explained in terms of the activation of catalysts and the building up of intermediates. Rabinowitch [90] restated and extended the original Osterhout-Haas [82] hypothesis and talked of the "deactivation in the dark of some enzymes needed for photosynthesis" and of "a deficiency of this acceptor (the CO_2-acceptor) at the beginning of a light period and of an autocatalytic adjustment of its concentration in light to the level needed to maintain photosynthesis at the rate corresponding to the prevailing light intensity". There continues to be increasing support for both of these proposals.

7.2. THE BENSON—CALVIN CYCLE AS AN AUTOCATALYTIC SEQUENCE

A primary carboxylating mechanism [107] ought, ideally, to meet the following criteria:

(*i*) Because of the low concentration of CO_2 in the atmosphere the actual carboxylation should have a favourable equilibrium position;

(*ii*) For the same reason, the enzyme which catalyses the carboxylation should have a high affinity for CO_2 or bicarbonate;

(*iii*) So that the process can continue there must be regeneration of the CO_2-acceptor;

(*iv*) So that the organism can grow and so that the process can respond positively to transient changes in light intensity etc., the sequence must be autocatalytic.

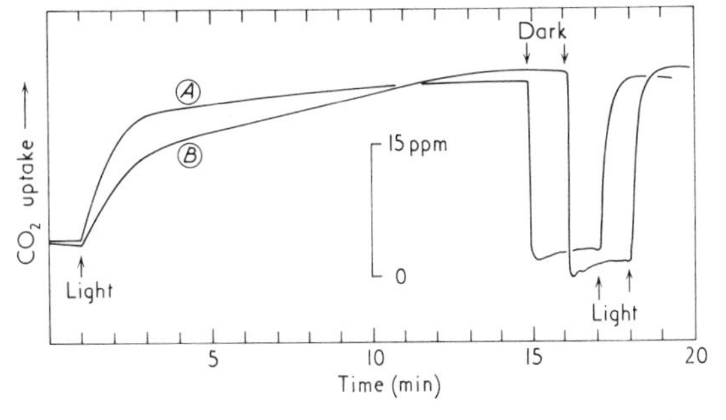

Fig. 7.9. CO_2 uptake by spinach, followed by infra-red gas analysis (Delaney, unpublished). Leaf discs (10 cm^2) were strongly illuminated (ca.6000 ft candles) at 20° C in air containing 300 ppm CO_2 flowing at 1 l/min. Curve A, disc taken from leaf previously illuminated in bright light for 1 h. Curve B, disc taken from plant in room light (ca. 150 ft candles). As in Figs. 7.6 and 7.7 there is a lag of some minutes before the maximum rate is reached and the rate increases more rapidly in the disc which has been pre-illuminated in strong, rather than weak light. After a brief dark interval, the lag is greatly diminished. Under these conditions part of the initial lag may be attributed to the relative slowness of the stomatal response.

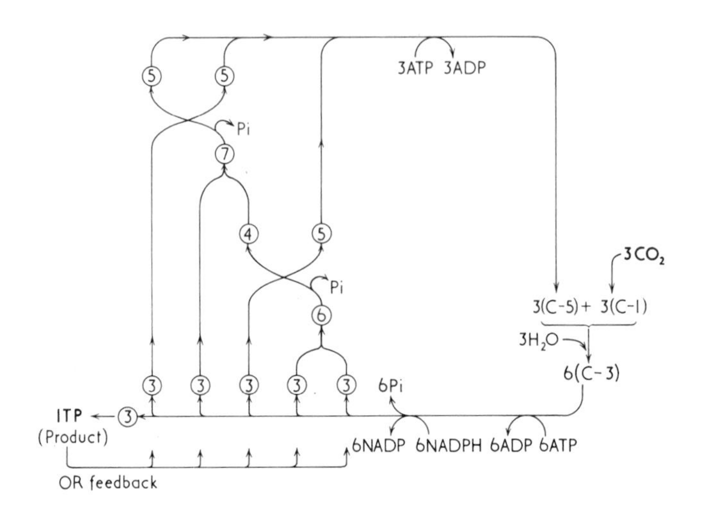

Fig. 7.10. The Benson—Calvin cycle as an autocatalytic sequence or breeder reaction. Represented in triplicate so that the condensation of three molecules of RBP with three molecules of CO_2 leads to the eventual formation of six molecules of TP. Normally five molecules of TP would then be rearranged in the reactions of the sugar phosphate "shuffle" to regenerate three molecules of CO_2 -acceptor whilst the remaining TP would constitute net product. The circled numbers represent TPs (3), erythrose-4-phosphate (4), pentose phosphates (5), fructose-1,6-bisphosphate (6) and sedoheptulose-1,7-bisphosphate (7). Similarly C-1 is CO_2 , C-3 is PGA and C-5 is RBP. The sequence starts at the right with the hydrolytic carboxylation (dismutation) of three molecules of RBP to yield six molecules of PGA. These are phosphorylated at the expense of ATP to yield six molecules of glycerate-1,3-bisphosphate and these, in turn, are reduced by NADPH to G3P. Two molecules of this TP (one as G3P the other as DHAP) then undergo an aldol condensation to yield fructose-1,6-bisphosphate which is hydrolysed to fructose-6-phosphate. This is followed by the first of two reactions catalysed by transketolase with G3P as the acceptor of 2-carbon unit. Thus G3P gaining a 2-carbon ketose unit become R5P and fructose-6-phosphate losing 2 carbons becomes erythrose-4-phosphate. This sequence of condensation and transfer is then repeated; but because the condensing partners are now erythrose-4-phosphate and G3P the first product is sedoheptulose-1,7-bisphosphate. Similarly with sedoheptulose bisphosphate as the 2-carbon donor the products of the transketolase reaction are R5P and xylulose-5-phosphate. These, together with the previously formed molecule of R5P, are all converted to Ru5P prior to the second phosphorylation in which the 3 molecules of CO_2 -acceptor are finally regenerated. Overall, this may be summarised as

$$3CO_2 + 5H_2O + \rightarrow 1\ TP + 9ADP + 8P_i$$

In an autocatalytic phase, it is assumed that the triose phosphate product could feed back into the cycle. Disregarding the loss of triose phosphate in export from the chloroplast, etc., this would permit a theoretical doubling of the concentration of the acceptor for every fifteen molecules of CO_2 fixed. In the steady state TP will be released from the cycle to accumulate as an end-product such as starch or to be exported unchanged from the chloroplast to the cytoplasm.

The Benson—Calvin cycle [17] has all of these features although the affinity of ribulose-1,5-bisphosphate carboxylase (RBPCase) for CO_2 has been in question for some years and surprisingly little attention has been paid to the autocatalytic requirement [107]. At least, in retrospect, the latter is self evident. The vast majority of metabolic processes draw on preformed substrates, often synthesised by other organisms. At the end of the line stands the primary carboxylating mechanism which must not only replenish its own substrate but also function as a breeder reaction, increasing the amount of substrate within the system. As set out in Fig. 7.10 the Benson—Calvin cycle is seen as a sequence in which the input is comprised of three molecules of CO_2 and one molecule of orthophosphate and the output is one molecule of triosephosphate (TP). The TP can, of course, re-enter the cycle and if it does then the concentration of ribulose bisphosphate (RBP) will increase. The extent of the increase will depend upon the extent to which TP and other intermediates escape from the cycle and from the chloroplast. In theory, one additional molecule of RBP can be formed for every 5 molecules of CO_2 fixed and this would lead to an exponential increase in rate until such time as other rate-limiting processes imposed a ceiling [106].

7.3. INDUCTION IN VITRO

7.3.1. Introduction

Experiments with isolated chloroplasts virtually started with the experiments of Hill in the 1930s. At first (e.g. refs. 55,56) these were largely concerned with O_2 evolution and the concomitant reduction of exogenous electron acceptors and neither the Hill reaction nor associated photophosphorylation show any sign of induction of the type under discussion. When photosynthetic CO_2 fixation was first demonstrated with isolated chloroplasts lags were observed but this might have been an artefact resulting from the procedure then employed. The first unambiguous records of induction in vitro must therefore be attributed to Gibbs and his colleagues [12,13,101]. This was confirmed by Walker [102] in 1965 (Fig. 7.11) and corresponding lags in CO_2-dependent O_2 evolution were reported by Walker and Hill [111] in 1967. In 1964, Heber and Willenbrink [49] had shown a lag in CO_2 fixation by chloroplasts within the intact leaf (by following exposure to $^{14}CO_2$ with isolation of chloroplasts in non-aqueous medium).

7.3.2. The causes of in vitro induction

Except that there can be no stomatal contribution, or effects resulting from gaseous diffusion through the leaf tissues, an explanation for in vitro induction must also be sought (cf. section 7.1.5) in terms of the Osterhout—

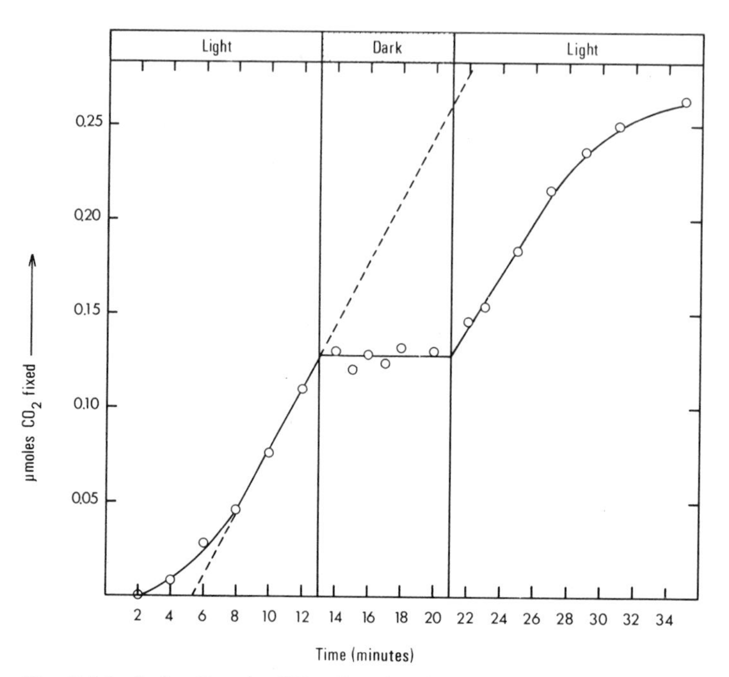

Fig. 7.11. Induction in CO_2 fixation by isolated chloroplasts (after ref. 109). Fixation only reached its maximum after several minutes' illumination. It then ceased abruptly in the dark and was resumed without discernible lag after a brief dark interval.

Haas hypothesis [82] and its elaboration by Rabinowitch [90]. However, although induction occurs in intact leaves, and in chloroplasts within intact leaves [49], there is also the possibility that some aspects of in vitro induction might relate to the conditions which have been employed in many experiments. The most important of these is that the chloroplasts are normally incubated in mixtures such that the chlorophyll concentration does not exceed 100 μg/ml of suspending medium. These proportions have been chosen because they permit easily recorded rates and something approaching light saturation. Nevertheless the ratio of chloroplast volume to suspending medium volume will be in the range of 1/500 to 1/1000, whereas the corresponding ratio of chloroplast volume to volume of cytoplasm within the parent tissue will be nearer to unity. The practical consequence is that substances lost from isolated chloroplasts will not accumulate in the medium as rapidly as they might in the cytoplasm. Conversely, the further metabolism of these substances will be different because they will no longer be exposed to a full complement of cytoplasmic enzymes or to organelles such as microbodies. Some aspects of the changes occasioned by differences in suspending volume etc. have been investigated by Coombs and Baldry [42] and they must always be borne in mind in attempting to interpret other observations.

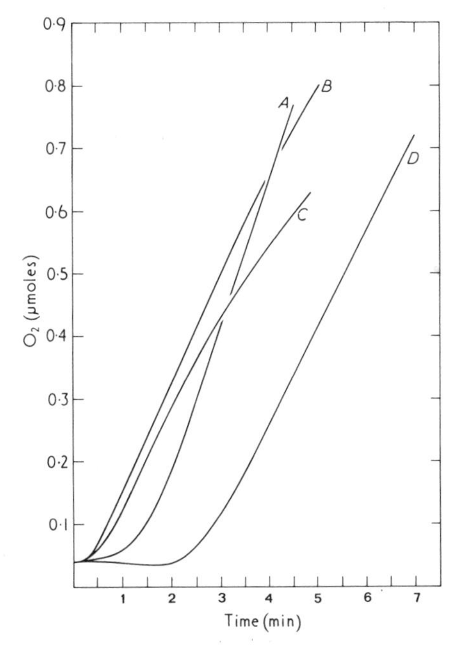

Fig. 7.12. Shortening of induction by compounds released from illuminated chloroplasts and by the addition of cycle intermediates. Curve D, CO_2-dependent O_2 evolution by chloroplasts prepared from leaves harvested after 12 h darkness. Curve C, as for D but the chloroplasts resuspended in medium in which other chloroplasts from the same preparation had been illuminated for 30 min and then removed by centrifugation. Curves A and B, as for D but with R5P added to A, PGA added to B (Walker, unpublished but cf. ref.10). Note that the lag is shortened by compounds released from illuminated chloroplasts in much the same way as it is by added PGA, indicating a major contribution by this compound. R5P also shortens the lag but the kinetics are then more like those exhibited by chloroplasts from pre-illuminated leaves (cf. Fig. 7.13A). The lower and declining rate which results from the addition of PGA at the outset is characteristic.

7.3.3. The contribution of autocatalysis

It seems clear that many observations which have been made are consistent with the proposal that the concentration of intermediates of the Benson—Calvin cycle within the isolated chloroplasts is initially less than that required to support maximal photosynthesis and that the ensuing lag represents the time which must elapse before the full steady state concentration is achieved (see e.g. refs. 45,106,112).

The evidence is as follows:

(*i*) Although there is an initial lag [13,101—103] this is not repeated (or is greatly abbreviated) following the introduction of a short dark interval once the maximal (or near-maximal) rate has been reached (see e.g. Fig. 7.11 and refs. 102,111,114). This suggests that intermediates have accumulated which abolish the lag.

(*ii*) As in Fig. 7.12, if chloroplasts are added to a medium from which other chloroplasts have been removed, after photosynthesis, then the initial lag is decreased [10].

(*iii*) Induction may often be shortened (Fig. 7.12) by the addition of those compounds which are believed to penetrate the chloroplast envelope (e.g. refs. 10,13,29,102,108,110,111) and in doing this they can act catalytically, i.e. the response which they produce is greater than that which would follow stoichiometric conversion to other cycle intermediates [10].

(*iv*) Induction may be lengthened by washing isolated chloroplasts in appropriate media [10,37].

(*v*) Induction may be lengthened (see section 7.3.5) by incubation with orthophosphate which is believed to facilitate the export of certain intermediates [38,108].

(*vi*) As in Fig. 7.13, induction is more prolonged in chloroplasts isolated from dark-stored leaves (in which intermediates may be depleted) than in chloroplasts from leaves which have been undergoing photosynthesis until the moment of extraction [37,102,111].

(*vii*) As in intact tissues, induction is largely independent of light intensity but is prolonged by low temperatures, implying that it is more directly related to "dark" events with temperature coefficients of about 2 rather than to photochemical processes such as electron transport [10,114].

(*viii*) Certain intermediates, particularly pentose monophosphates, are present at lower concentration during induction than in steady state photosynthesis [64].

It should be noted, of course, that while these observations strengthen the proposed involvement of autocatalysis they do not necessarily imply that steady state photosynthesis is governed entirely by the concentration of intermediates within the functioning cycle. Obviously, at this level of control, there must also be almost endless possibilities of feedback activation and inhibition.

Nevertheless it seems clear that induction is not concerned with the photochemistry as such but with the generation and reduction of glycerate-1,3-bisphosphate (the final electron acceptor in the Benson—Calvin cycle). Results which support this proposal are illustrated in Figs. 7.14 to 7.16. It will be seen that electron transport and its associated reactions start immediately (the only lags are those imposed by the relatively slow response of the recording instruments). True induction is only observed when the oxidant has to be formed by the operation of the photosynthetic cycle and something approaching induction if this synthesis is simulated by slow addition of phosphoglycerate (PGA) (Fig. 7.17).

7.3.4. The effect of cycle intermediates

Envelope-free chloroplasts immediately start to evolve oxygen at their full

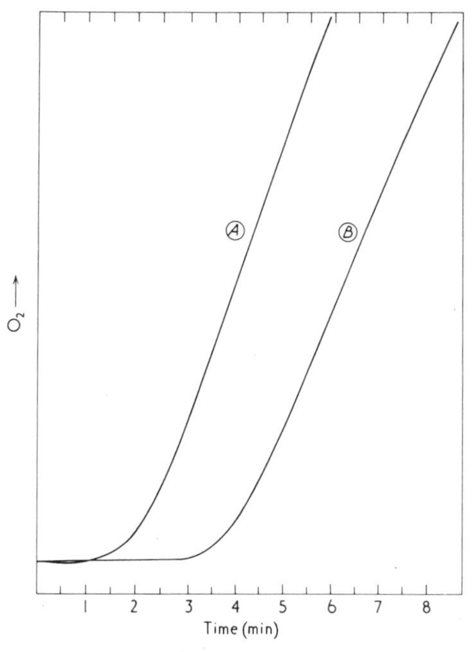

Fig. 7.13. Shortening of induction by pre-illumination of the parent tissue. Curve A, CO_2-dependent O_2 evolution by spinach chloroplasts isolated from leaves pre-illuminated for 30 min in white light at 2000 ft candles. Curve B, as for A but chloroplasts isolated at the same time from non-illuminated leaves. All of the leaves were harvested after 12 h darkness and randomised prior to extraction. (Walker, unpublished but cf. ref. 10.)

rate if they are illuminated in the presence of an artificial electron acceptor or Hill oxidant such as ferricyanide. Similar kinetics are observed with NADP as the acceptor with ruptured chloroplasts supplemented with exogenous ferredoxin (Fig. 7.15). Intact chloroplasts will also evolve oxygen if provided with PGA (Fig. 7.16A) and again there is no initial lag of any considerable size [111]. The two TPs, glyceraldehyde-3-phosphate (G3P) and dihydroxyacetone phosphate (DHAP) also shorten induction in CO_2 fixation and CO_2-dependent O_2 evolution (Fig. 7.12) by intact isolated chloroplasts [13,110]. Similar effects are sometimes seen in the presence of other cycle intermediates such as pentose monophosphates (Fig. 7.12) and fructose-1,6-bisphosphate [13,102]. The magnitude of the observed effect is very variable. With spinach chloroplasts prepared in sorbitol-pyrophosphate or sorbitol-MES media, the initial lag is usually not more than 1 or 2 min, particularly if the chloroplasts are separated from previously pre-illuminated tissue. For this reason the extent to which the lag can be shortened is correspondingly small although the effect of PGA on the time course of O_2 evolution is always pronounced (Fig. 7.12). Conversely, if pea chloroplasts are prepared in a medium containing 0.1 M orthophosphate, lags of 20—30 min may be observed and com-

Figs. 7.14 to 7.16. Induction as a consequence of the need to generate the natural oxidant. (Note that the recordings in these figures were made at successively lower speeds.)

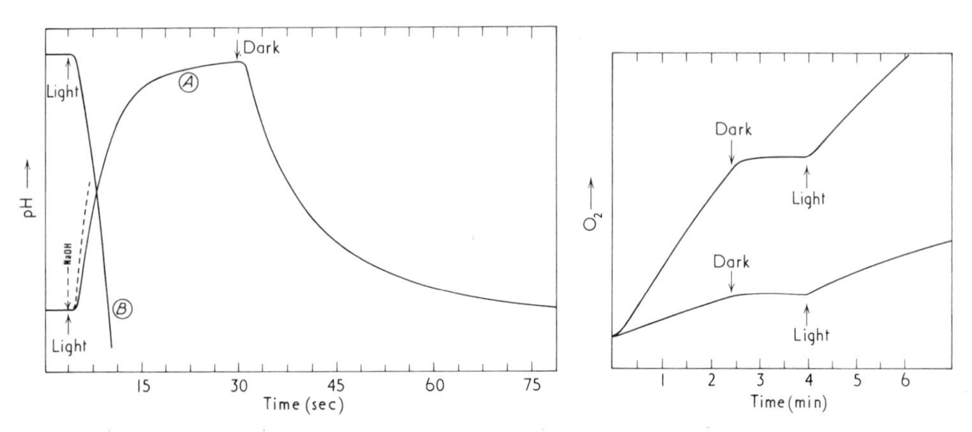

Fig. 7.14. (A) Light-generated pH shift. The broken line records the change in pH which followed the addition of NaOH to osmotically shocked chloroplasts in the dark and is included to show the response time of the apparatus. The lower trace shows the response of the same chloroplasts to illumination in the presence of pyocyanine. This reagent is believed to return electrons from the "top" (reducing) end of photosystem II to some point between photosystems I and II. The pH shift is associated with electron transport and this figure shows that it can be discerned as rapidly as the pH meter can respond. (B) Ferricyanide-dependent pH shift. Ferricyanide accepts electrons but not protons so that the solution becomes more acid as it is reduced. The reaction mixture contained nigericin to inhibit the converse pH shift directly associated with electron transport. Again it is seen that ferricyanide reduction commences as rapidly as the instrument can record.

Fig. 7.15. NADP-dependent O_2 evolution. As in Fig. 7.14B this is a Hill reaction except that the Hill oxidant is natural rather than artificial and (unlike ferricyanide) NADP does not accept electrons directly from the photochemical apparatus. Instead electrons have to traverse both photosystems and then pass to NADP via soluble ferredoxin and a flavoprotein. Again there is no discernible lag. The upper trace was obtained in the presence of the uncoupler nigericin.

pounds such as ribose-5-phosphate (R5P) will bring about a marked decrease in induction (see e.g. ref. 111). Because of the interaction between orthophosphate (P_i), inorganic pyrophosphate (PP_i) and metabolites (see e.g. refs. 68,109), no statement concerning the effect of a particular additive should be taken out of the particular experimental context in which it was first applied. For example, if chloroplasts are illuminated in mixtures containing concentrations of orthophosphate greater than optimal they will normally exhibit a relatively long induction period which can be shortened or virtually eliminated by PGA (see e.g. ref. 38). The final rate will then usually equal or exceed that seen in the presence of CO_2 alone or in the presence of CO_2 plus R5P. Conversely, if the P_i concentration is near optimal the lag will be much

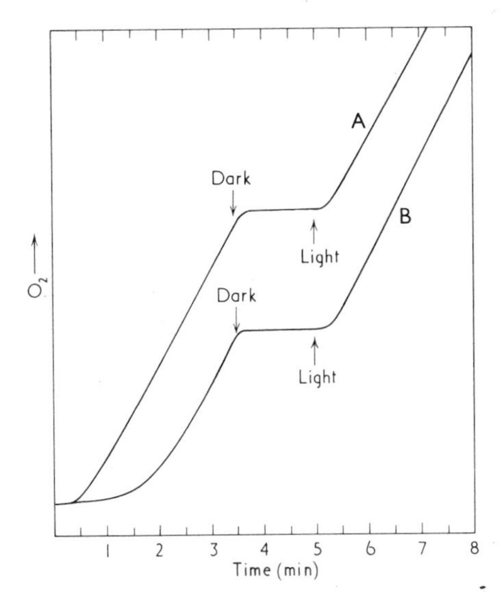

Fig. 7.16. (A) PGA-dependent O_2 evolution. This is similar to that in Fig. 7.15 except that this experiment was carried out with intact chloroplasts. These do not evolve O_2 when first illuminated in the presence of CO_2 (B) but with PGA the delay is minimal. During this period the PGA must enter and be converted first to glycerate-1,3-bisphosphate and then to G3P. The second of these reactions reoxidises NADPH which has been formed by the reductive sequence followed in Fig. 7.15. (B) CO_2 -dependent O_2 evolution. As in Fig. 7.15 O_2 evolution involves the transfer of electrons from water to NADP but in intact chloroplasts this can be seen only if steps are taken to permit the reoxidation of the small quantity of this coenzyme which is present. When the natural Hill oxidant, glycerate-1, 3-bisphosphate is formed through the autocatalytic operation of the Benson—Calvin cycle, the lag is longer.

shorter, and although PGA will shorten it still further, the final rate will be usually less than that seen in the presence of CO_2 alone or in the presence of CO_2 + R5P (Fig. 7.12). In these circumstances PGA will also usually give a faster rate if its addition is delayed. R5P will stimulate in high P_i and inhibit in low [38]. The precise nature of all of these responses is still largely a matter for conjecture but many can be traced to a disturbance of the normal metabolic balance. Thus in high P_i, PGA and R5P can overcome the loss of cycle intermediates brought about by enforced export (see section 7.3.5). In low P_i they will act as sinks for P_i, lowering its concentration still further and increasing ADP to inhibitory levels.

7.3.5. *The effect of orthophosphate*

Prolongation of induction in CO_2-dependent O_2 evolution by P_i in excess of 10^{-5} M [38] is shown in Fig. 7.18. It can be seen that the lag is progressively increased with increasing P_i concentration until, eventually, photosyn-

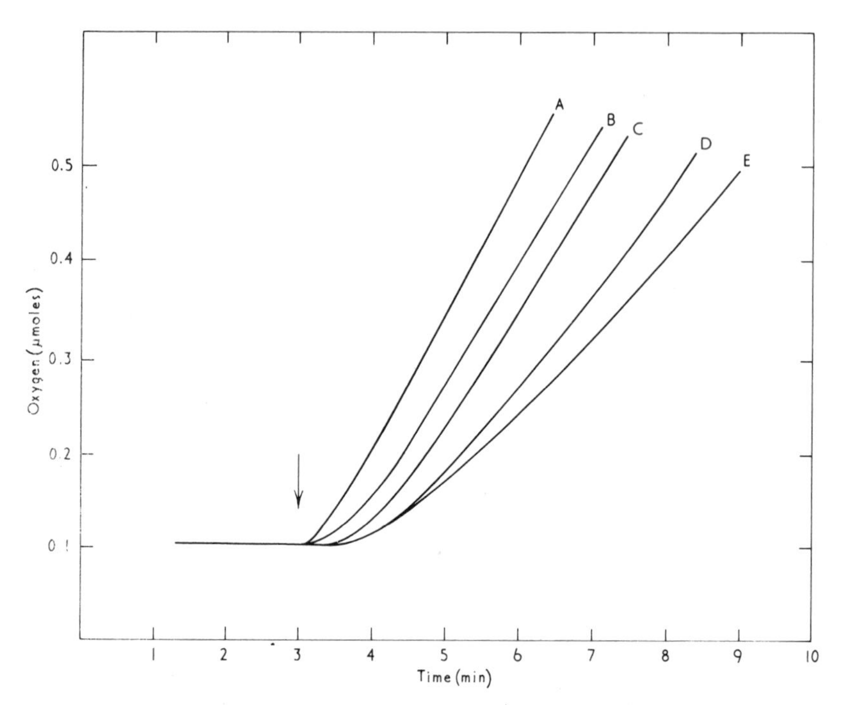

Fig. 7.17. Simulation of induction by varying the rate of addition of PGA to isolated chloroplasts (Fitzgerald, unpublished). CO_2-dependent O_2 evolution was inhibited by high P_i (cf. Fig. 7.18), PGA was then added (arrow) at the following rates: (A) 10, (B) 1.28, (C) 0.625, (D) 0.31 and (E) 0.16 μmoles.min^{-1}. At the lower rates of addition there is an increasingly longer interval before the attainment of maximal rate. There seems little doubt that in many circumstances the length of induction will be similarly related to the accumulation of PGA and glycerate-1,3-bisphosphate. In the endogenous process, however, PGA formation will be autocatalytic, thus terminating the lag more abruptly.

thesis is almost entirely suppressed within the period of measurement. The degree of lag extension is variable. It is most marked in chloroplasts prepared from tissues which have not been pre-illuminated. To some extent it is related to the concentration of magnesium ion in the external medium (cf. Figs. 7.18 and 7.22 with 7.24, 7.26 and 7.27), possibly for this reason, P_i inhibition can be partially overcome by chelating agents such as PP_i [68]. PP_i may, however, reverse P_i inhibition in an entirely different fashion by interfering with the action of the phosphate translocator. It should be noted that isolated spinach chloroplasts do not normally produce free sugars (for a recent review, see ref. 109) and that in most circumstances polysaccharide synthesis is relatively slow. In consequence these chloroplasts are phosphate-consuming organelles and in the absence of exogenous P_i net O_2 evolution or CO_2 fixation will virtually cease [39] as the endogenous supply is incorporated into sugar phosphates [9]. Photosynthesis can then be restarted by the addition of P_i and, if the aliquots of added P_i are small, there is a rough stoichiometry of 3 molecules of CO_2 fixed or O_2 evolved for each molecule of P_i added [39]. PP_i will also restore photosynthesis in these circumstances (provided that Mg is also present in the external medium) and the stoichiometry is then 6 to 1 [39,93]. PP_i does not penetrate the intact envelope and in this instance its effect is clearly related to external hydrolysis by pyrophosphatase released from ruptured chloroplasts [68,93].

7.3.6. The inter-relationship between induction, orthophosphate and cycle intermediates

The prolongation of induction by P_i may be reversed by various cycle intermediates (Fig. 7.19) including those which shorten the normal induction period observed in the presence of optimal P_i [40]. All of the compounds which reverse P_i inhibition are believed to cross the chloroplast envelope or to be capable of ready conversion to compounds for which there is independent evidence that they are capable of ready penetration [108,109]. Conversely, the ability of intermediates to reverse P_i inhibition has been taken as an indication of penetration. Compounds such as PGA, G3P and DHAP act more or less immediately, whereas the pentose monophosphates produce a response only after a lag [40]. Fructose-1,6-bisphosphate reverses P_i inhibition but probably only after its external conversion to TP. Compounds such as RBP are ineffective.

The reversal of P_i inhibition by PGA is remarkable to the extent that if added at the outset there is a marked difference between its effect on O_2 evolution, which follows immediately and its effect on CO_2 fixation which picks up only slowly [40]. By contrast, if added some time after the commencement of illumination its effect is more pronounced (cf. ref. 54) and O_2 evolution and CO_2 fixation are initiated simultaneously [40]. So far as is known, none of the other intermediates can bring about separate responses in O_2 evolution and CO_2 fixation and the kinetics for O_2 and CO_2 are therefore quantitatively and qualitatively similar. Related sugar phosphates which are not intermediates (such as 6-phosphogluconate) are without effect [40].

7.3.7. Induction and the phosphate translocator

The inhibition of photosynthesis by P_i and the reversal of this inhibition by cycle intermediates could be explained, at least in part, by the operation of the phosphate translocator which has been formulated by Hans Heldt and his colleagues [51,52,54,119] (and see also Chapter 6 of this volume). This mechanism, for which there is ample direct evidence, would permit a direct exchange of external or cytoplasmic P_i for intermediates such as PGA and DHAP which are formed in the stroma during the operation of the Benson—Calvin cycle. It now seems extremely probable that the major imports into

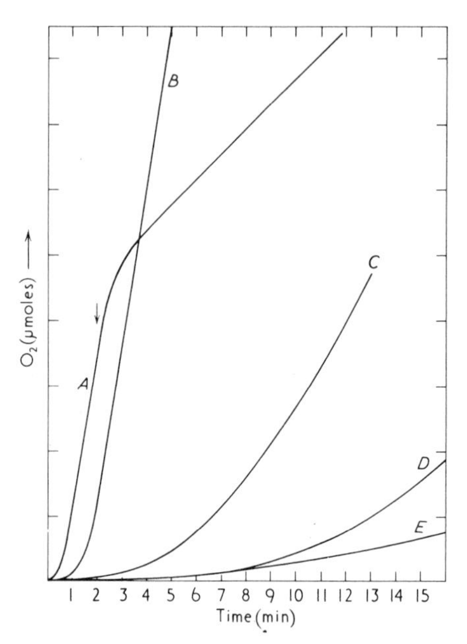

Fig. 7.18. Some effects of P_i on induction exhibited by intact isolated spinach chloroplasts (Walker, previously unpublished but cf. ref. 38). P_i always brings about lag extension but the response with respect to concentration varies according to the preparation and the Mg concentration in the medium. This figure illustrates the most simple situation in which the lag is extended by increasing P_i but the kinetics are not otherwise complex. Curve A. Zero added P_i; Curve B, 0.5 mM P_i; Curve C, 5 mM P_i; Curve D, 7.5 mM P_i; Curve E, 10 mM P_i. Each reaction mixture contained chloroplasts in sorbitol medium containing 1 mM $MgCl_2$. For full experimental details see Cockburn et al. [38]. Relatively large quantities of P_i were required for maximal inhibition indicating that the chloroplasts (prepared from pre-illuminated leaves) retained relatively high levels of cycle intermediates and enough endogenous P_i to allow photosynthesis to start with minimal lag in the absence of added P_i (Curve A). The same point is made more clearly by the addition of 10 mM P_i at the arrow in A. The inhibition then produced (i.e. after 2 min accumulation of intermediates) was then considerably less than that observed when this concentration of P_i was present from the start.

the chloroplast are CO_2 and P_i and that the major exports are TPs [48,108, 109]. This traffic would be facilitated by the translocator and P_i inhibition would be seen as an exaggeration of the normal exchange. Triosephosphate newly formed in photosynthesis must always be subject to several alternative uses. In some species it may be converted to starch which is stored within the stroma. In all species a proportion must go to the regeneration of the CO_2 acceptor and at some stage a proportion (or a derivative) must be exported to the cytoplasm. Normally, during induction, TP is likely to be retained within the stroma because its concentration will be relatively low and that of P_i relatively high (see Fig. 7.20). As P_i is consumed according to the overall equation

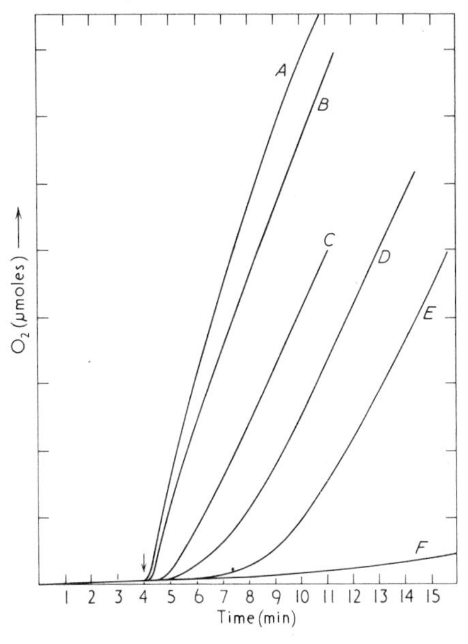

Fig. 7.19. Reversal of P_i inhibition by cycle intermediates (Walker, previously unpublished but cf. ref. 40). The chloroplast preparation etc. was the same as that in Fig. 7.18 except that all of the mixtures used contained inhibitory (10 mM) P_i. Cycle intermediates (to a final concentration of 1.0 mM) were added at the arrow. A, PGA; B, DHAP; C, G3P; D, fructose-1,6-bisphosphate; E, R5P; F, fructose-6-phosphate. It is believed (see e.g. refs. 48,108,109) that fructose bisphosphate may undergo external lysis (by aldolase released from ruptured chloroplasts) prior to entry as TP.

$$3\ CO_2 + P_i + 6\ NADPH_2 \rightarrow TP + 3\ H_2O + 6\ NADP$$

its concentration within the stroma will fall and that of the TP will rise. In the steady state a balance will be achieved in which TP (in excess of that required for the regeneration of RBP) will be exported to the cytoplasm in exchange for P_i. For isolated chloroplasts it is proposed that if an excess of P_i is present in the external medium from the outset, there will be obligatory export such that the normal autocatalytic build up of intermediates is unable to take place (as shown in Fig. 7.20). Induction will therefore be prolonged indefinitely unless terminated by the further application of an intermediate capable of penetrating the inner envelope.

7.3.8. The relationship between starch synthesis and orthophosphate concentration

If the movement of metabolites between the chloroplast and its surrounding medium is regulated by the concentration of P_i (section 7.3.7) then low

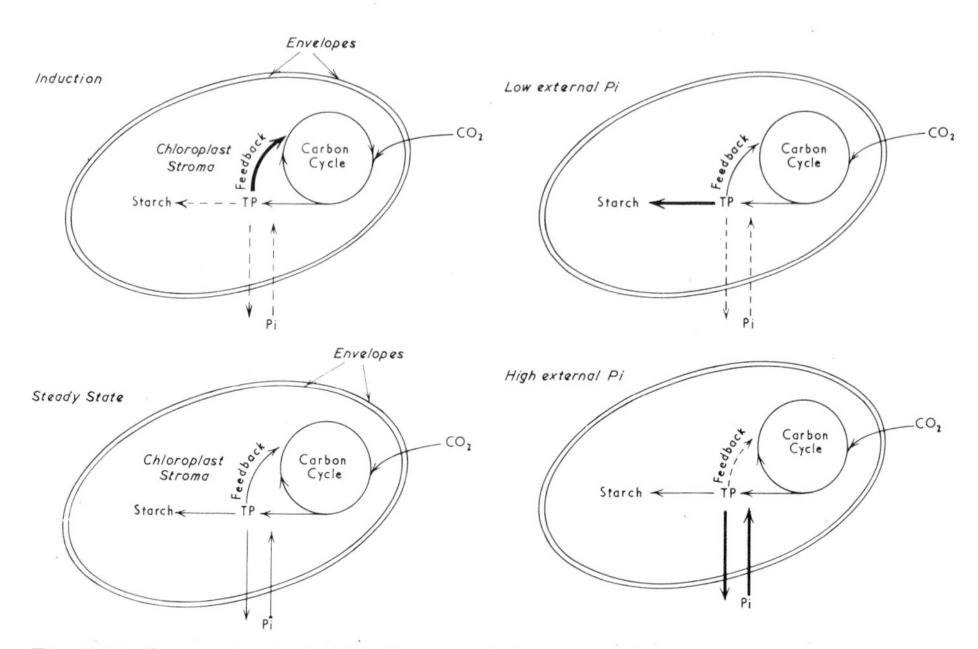

Fig. 7.20. Proposed relationship between induction, P_i and metabolite movement. In normal concentrations of external P_i, TP formed in the photosynthetic carbon cycle will either (*a*) feed back into the cycle, (*b*) undergo further conversion to starch etc. on the stroma or (*c*) be exported. At the onset of illumination the lag which is observed will reflect the time taken for autocatalytic feedback to raise the concentration of cycle intermediates to the steady-state level dictated by the prevailing light intensity etc. During this period feedback will be favoured by rapid utilisation within the cycle. As the steady state is approached relative surplus will increasingly permit export and internal storage. In low external P_i (which it is believed can be achieved experimentally in some leaf tissue by feeding mannose [33] export will be diminished and surplus TP diverted into starch synthesis etc. In high external P_i, obligatory export will diminish the proportion of TP available for feedback and prolong induction beyond the normal period of 1—3 min (see e.g. Fig. 7.18). In vivo less extreme fluctuations in P_i than those illustrated will influence the balance between storage within the chloroplast and export to the cytoplasm.

P_i should favour retention just as high P_i imposes excessive export (Fig. 7.20). Preliminary work (Heldt, unpublished; Chen-She Sheu-Hwa and Fitzgerald, unpublished) indicates that polysaccharide synthesis by isolated chloroplasts is indeed favoured by low P_i. It has also been argued that starch synthesis in vivo should be enhanced by sequestration of cytoplasmic P_i and again this prediction has been borne out by experiment [33]. Conversion of cytoplasmic P_i to a form in which it was not readily available for further metabolism was achieved by feeding leaf discs with mannose (cf. refs. 47,75). Mannose is readily phosphorylated by ATP in the presence of hexokinase but in some species (such as spinach beet) there is no appreciable further conversion of mannose phosphate mannose ⟶ mannose-6-P; ATP ⟶ ADP; P_i ⟶ H_2O

As Fig. 7.21 shows there was a 10-fold increase in starch synthesis when mannose was supplied to spinach beet leaf discs at 10^{-2} M in the light. There was no starch synthesis in the dark nor, when labelled mannose was supplied in the light, was there any significant incorporation of radioactivity into the starch fraction. Instead, the radioactivity was recovered in fractions corresponding to the free sugar and to mannose phosphate. On the face of it these results are clearly in accord with the operation of the phosphate translocator as illustrated in Fig. 7.20 and it is evident that the stimulation by mannose is indirect and does not result from incorporation of mannose, or a mannose derivative, into the polysaccharide skeleton. Sequestration of cytoplasmic P_i could, of course, produce its effects in several ways. The enzyme ADP glucose pyrophosphorylase is inhibited by P_i and stimulated by PGA (see e.g. ref. 88). Phosphorolytic degradation of starch would also be decreased if the P_i concentration were lowered. Whether one, two, all or any of these factors is involved remains to be certainly established but there is certainly a strong implication that a disturbance in the normal course of phosphate metabolism can produce profound effects on the movement of metabolites from the chloroplast into the cytoplasm.

Starch accumulation has also been observed as a symptom of P_i deficiency in spinach grown in water culture (Herold, unpublished).

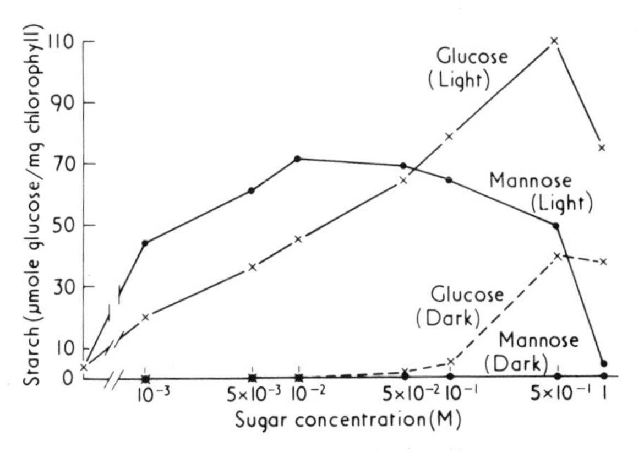

Fig. 7.21. Stimulation of photosynthetic starch formation in leaf discs by exogenous sugars (after ref. 33). In the light a 10—15-fold increase in starch synthesis occurred in leaf discs (spinach beet) floated on 10^{-2} M mannose even though this sugar did not promote starch synthesis in the dark. Little or no label from [^{14}C]mannose finds its way into starch and the stimulation is seen as an indirect effect related to the sequestration of cytoplasmic P_i.

7.3.9. The relationship between induction, orthophosphate concentration and Antimycin A

In some circumstances Antimycin A can materially affect the course of photosynthesis by isolated chloroplasts (see e.g. refs. 30—32,78,91,92).

Considerable stimulation of CO_2 fixation and O_2 evolution can be observed in concentrations (1 μM) which are probably too low to have significant direct effects on photophosphorylation or electron transport. A variety of proposals have been put forward to account for these observations and many of these are quite outside the scope of this chapter. Two, however, seem highly relevant to induction and the relationship between induction and the proposed role of orthophosphate in metabolite transport. These are (*a*) that Antimycin A is an antagonist of phosphate inhibition [31] and (*b*) that it activates hexose bisphosphatase [92]. In considering these proposals it must be emphasised that the conditions employed are crucial and that stimulatory effects have often been recorded in mixtures in which (in the absence of inorganic pyrophosphate) there is an inhibitory concentration of P_i (see e.g. ref. 78).

7.3.9.1. Antimycin A and orthophosphate at low concentrations

At relatively low concentrations of P_i, Antimycin A has been observed to stimulate CO_2 fixation. Schacter and Bassham [92] have published time-courses showing the changes in various cycle intermediates which follow the addition of Antimycin A and there are several more or less consistent features which have also been noted (sometimes as transients) by Shacter et al. [91] and Miginiac-Maslow and Champigny [78]. These include an increase in RBP, a decrease in fructose and sedoheptulose bisphosphates, a decrease in DHAP and an increase in PGA. Many of these features could be explained by increased fructose bisphosphatase activity and this has been suggested by Schacter and Bassham [92] and elaborated by Miginiac-Maslow and Champigny [78]. The activation is assumed to be a consequence of increased Mg and increased pH. Fructose bisphosphatase activity is known to be affected by a variety of factors, including Mg and pH (see e.g. refs. 26,89), and the magnitude of the changes in activity which can be brought about by simultaneous alteration in Mg and pH have recently been effectively illustrated by Baier and Latzko [8]. In this laboratory Antimycin A has also been observed to offset, to some extent, the decline in CO_2-dependent O_2 evolution which occurs in phosphate-deficient mixtures. In all of these regards the argument would run as follows. Antimycin A would increase the stromal Mg concentration and pH. Increased fructose bisphosphatase activity would liberate P_i and increase the flow of fructose-6-phosphate into the transketolase reaction. More pentose phosphate would be formed and consumed and the level of RBP would rise. The carboxylase, similarly activated, would give rise to increased levels of PGA but with the formation of TP, now rate-limiting, no stimulation of O_2-evolution would be seen. Declining O_2 production occasioned by low P_i and high ADP could however be temporarily offset by P_i released from the hexose and heptose bisphosphate pool. Although they have been sought, changes in pH consequent upon Antimycin A have not yet been demonstrated (Heldt, private communication) but it is conceivable that Antimycin A might still activate the bisphosphatases by influencing the Mg concentration or the reduction status of the stroma (cf. ref. 8). Preliminary experiments with the reconstituted system (Walker and Fitzgerald, unpublished) indicate that Antimycin A may affect the bisphosphatases in much the same way as dithiothreitol.

7.3.9.2. Antimycin A and orthophosphate at high concentrations

Antimycin A also has profound effects on the kinetics of inhibition by high concentrations of P_i and, in these circumstances therefore, on the nature of the lag. If high P_i were to act solely by draining TP from the chloroplast at an unchanging rate (see section 7.3.7.) then photosynthesis in its presence would eventually increase to the same rate as that observed at lower concentrations. However, P_i not only produces a lag extension but, at higher concentrations, the final rate of photosynthesis is also depressed. This implies the operation of secondary effects. If, for whatever reason, the rate of export increased gradually with time at high P_i then this would result not only in a longer lag but also a lower steady-state rate. On the other hand the lower steady state could result from a variety of additional effects. For example, P_i is a competitive inhibitor of RBPCase and it seems plausible that this inhibition could become increasingly important with the probable coincidence of high P_i and low pentose phosphate (cf. ref. 64). In the presence of small quantities of Antimycin A the kinetics of P_i inhibition approximate much more closely to those which would be predicted solely on the basis of a drain on intermediates (Figs. 7.22 to 7.25). Again this could be seen as a response to increased bisphosphatase activity which would, in turn, produce the effects listed in section 7.3.9.1. In the presence of high P_i there would, however, be additional consequences. Under normal conditions TP will be utilised in a number of ways. It may undergo aldol condensations leading to fructose and sedoheptulose bisphosphates. It may enter the transketolase reaction or it may move out into the reaction mixture. In high P_i the latter will be favoured. Increased bisphosphatase activity brought about by Antimycin A would tend to decrease the TP loss by increasing the opportunity for entry into the transketolase reaction. Any consequent increases in the pentose phosphate pool would also tend to alleviate competitive inhibition of RBPCase by P_i.

It is also conceivable that the depression of the final rate in the presence of high P_i is caused, in part, by ATP inhibition of electron transport and indeed at some concentrations biphasic kinetics suggest that there may be a more or less sudden change in the nature of the inhibition with time. If

Figs. 7.22 to 7.25. Some effects of Antimycin A on induction (Walker, previously unpublished).

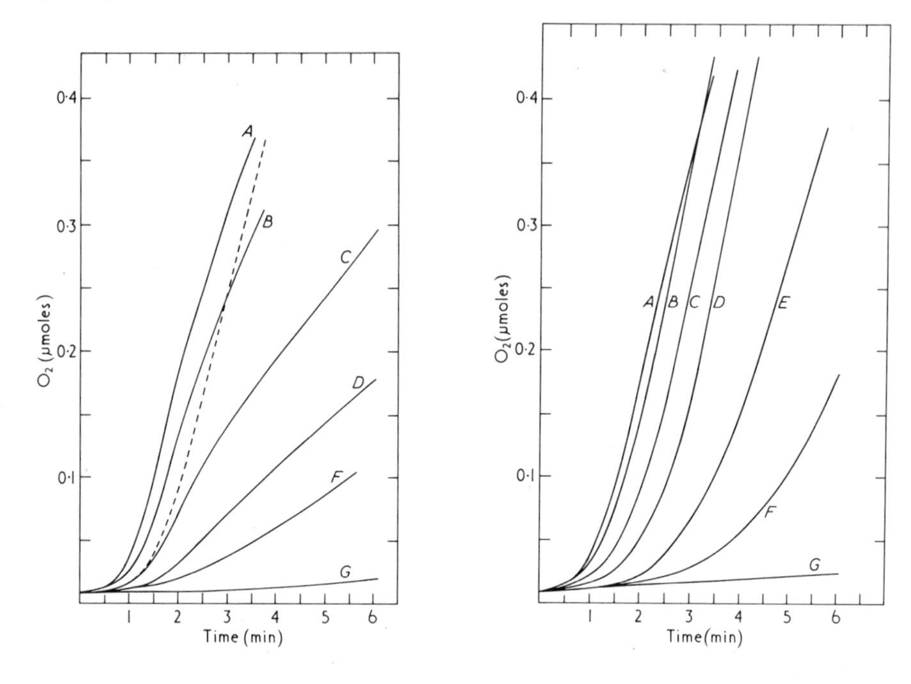

Fig. 7.22 is basically similar to Fig. 7.19 and shows the effect of P_i on induction exhibited by isolated spinach chloroplasts. The concentrations required for maximal inhibition were however smaller than in Fig. 7.19 (reflecting a lower level of endogenous intermediates). Curve A, 0.1 mM P_i; Curve B, 0.25 P_i; Curve C, 0.5 P_i; Curve D, 1.0 P_i; Curve E, 1.5 P_i; Curve F, 2.5 P_i; Curve G, 5.0 P_i. Broken line 0.5 mM P_i + 5 mM PP_i. It may be noted that C is distinctly biphasic, D suggests a progressive inhibition and only in F is there a suggestion of the partial recovery with time which is clearly seen in Fig. 7.19. The falling off at the lowest concentrations may be partly attributed to P_i deficiency and, for example, with a P_i to O_2 ratio of 1 to 3 [39] the added P_i in A would be exhausted after the evolution of 0.3 μmoles of O_2. In this concentration range addition of further P_i at this point will produce an acceleration whereas at higher concentrations it would be without effect or inhibitory. All reaction mixtures (1 ml) contained 5 μl of absolute ethanol to permit direct comparison with Fig. 7.23 in which Antimycin A was added in 5 μl of ethanol. Experimental details otherwise very similar to those described by Cockburn et al. [38]. The broken line illustrates the way in which PP_i can function as a non-inhibitory P_i source (cf. refs. 68,93,105).

Fig. 7.23. As for Fig. 7.22 but with 1 μM Antimycin A present from the outset. The effect of increasing P_i is now clearly seen as a lag extension. In short, Antimycin A seems to reverse some secondary inhibition by high P_i so that the kinetic behaviour approaches that which would be predicted if the sole response to increasing P_i was increased export of cycle intermediates.

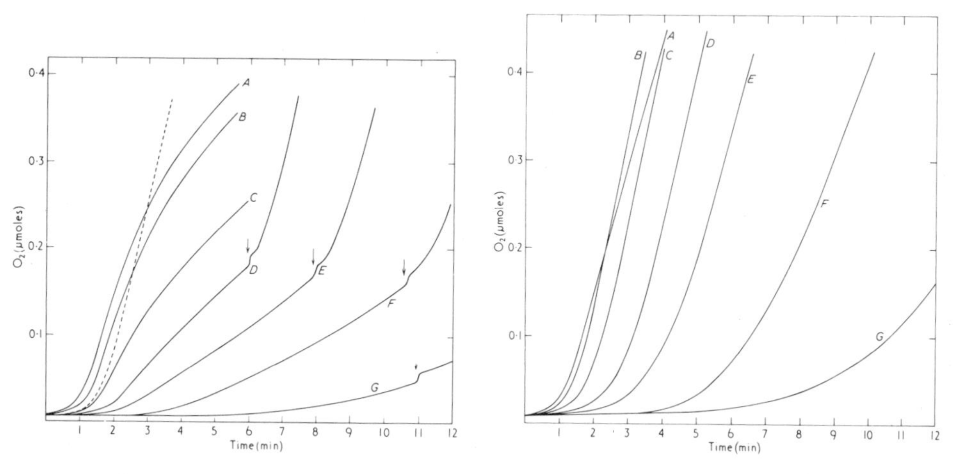

Fig. 7.24. As for Fig. 7.22 but with different chloroplasts and solutions containing 5 mM $MgCl_2$ rather than 1.0 mM $MgCl_2$. The higher Mg exaggerates the falling off in activity which occurs in the lower P_i concentration range, but as in Fig. 7.22 it should be noted that at concentrations of 0.25 mM (B) and above (C and D) this decline cannot be attributed to P_i deficiency occasioned by consumption with a 1:3 ratio of P_i:O_2. At still higher concentrations (E, F, G) there is again a trend towards recovery. The addition of (1 μM) Antimycin A (arrows) brought about a marked acceleration (the initial surge is an artefact caused by the ethanol in which the Antimycin A was dissolved).

Fig. 7.25. As for Fig. 7.24 but with Antimycin A present from the outset. Again (cf. Fig. 7.23) the combination of increasing P_i and constant Antimycin A produces kinetics approaching those of a simple lag extension rather than a lag extension combined with secondary inhibition.

Antimycin A increased the pentose phosphate concentration this could, in turn, lead to a more favourable ATP/ADP ratio. The complex, variable and varying nature of the P_i inhibition is itself illustrated by Figs. 7.26 and 7.27. Thus in the absence of Antimycin A, P_i inhibition may be progressive or, particularly at lower concentrations it may decrease with time. In the presence of Antimycin A it always decreases with time (at very low P_i the rate will always eventually fall off with time) but this is merely a deficiency symptom which can be abolished by the addition of more P_i.

If Antimycin A does not act in the manner suggested the only readily apparent and feasible alternative at this time is that it acts directly on the phosphate translocator. Again, however, the existing evidence for direct action is negative (Heldt, private communication). Like PP_i (see e.g. ref. 105) (Figs. 7.26 and 7.27) Antimycin A behaves as a phosphate "antagonist" [31] but it seems unlikely that both Antimycin A and PP_i will alleviate P_i inhibition in the same way.

Figs. 7.26 and 7.27. The effect of PP_i on lag extension by P_i (Walker, previously unpublished but cf. ref. 68). These figures are included for comparison with those (Figs. 7.22 to 7.25) which illustrate the effects of Antimycin A in similar circumstances. The general effect of PP_i is relatively independent of Mg concentration but these particular experiments were carried out in the absence of added Mg to eliminate any effects caused by chelation of Mg with PP_i.

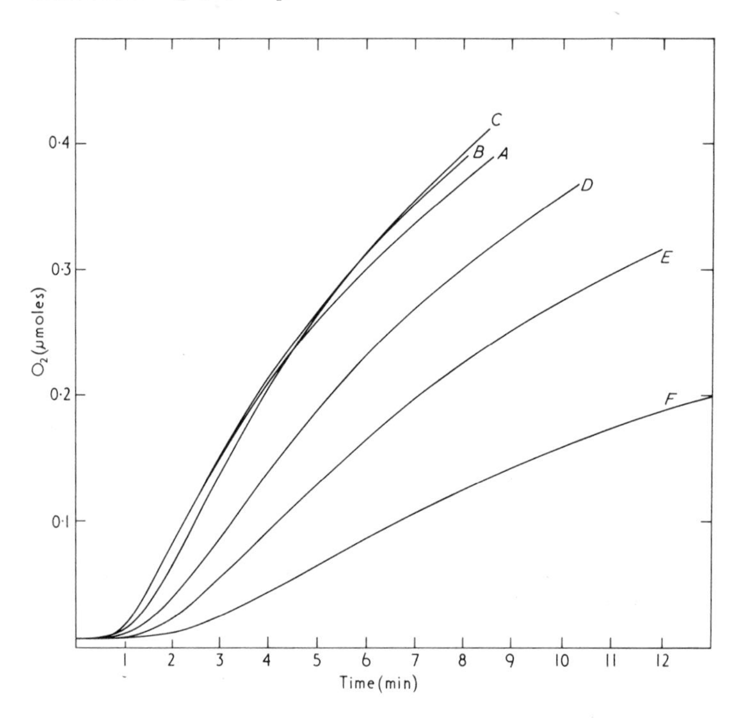

Fig. 7.26. In low Mg, the lag extension caused by increasing P_i is still apparent but the inhibition is progressive at all concentrations whereas, at higher Mg, progression is either absent (Fig. 7.18) or present only in the lower P_i range (Fig. 7.22 and 7.24).

Even if these two compounds act at entirely different sites, however, the end result may be the same. The possibility that Antimycin A may reduce loss of TP has been considered in this section and is regarded as entirely feasible. Recent evidence [11] that export of G3P is lessened by PP_i supports the proposal [108] that PP_i may interfere with the normal function of the phosphate translocator and therefore decrease the drain on PGA and G3P imposed by high external P_i.

7.4. INDUCTION AND THE LIGHT ACTIVATION OF PHOTOSYNTHETIC ENZYMES

As previously noted (section 7.1.5) the Osterhout–Haas hypothesis

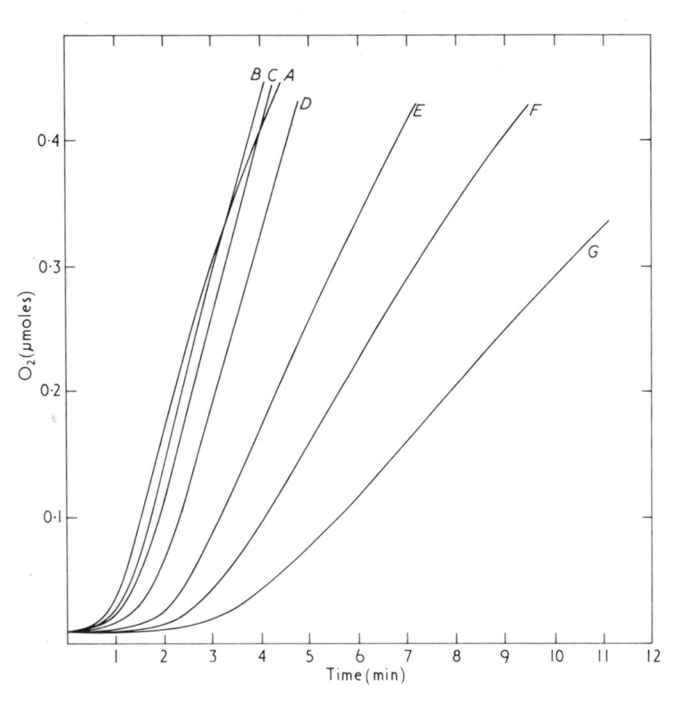

Fig. 7.27. As for Fig. 7.26 but with constant 5 mM PP_i. Like Antimycin A this has an ameliorating effect on the P_i inhibition and largely eliminates its progressive aspect.

proposes that induction is related to the building up of intermediates and the light activation of catalysts. Evidence for the former is listed in section 7.3. The light activation of enzymes has, on the whole, been investigated more recently and is not well understood. At some future date it may be possible to divide light activation into several categories including "direct" activation in which modification of a catalytic site is closely linked to the absorption of light by a pigment (cf. ref. 124) and "indirect" activation in which the concentration of a cofactor, coenzyme or metabolite changes as a consequence of photosynthesis and in so doing exerts a regulatory effect. At present such a rigid classification is precluded by ignorance of the mechanisms involved. The remainder of this section will therefore be devoted to a few specific examples and is not intended to be a comprehensive catalogue or to be a discussion of mechanisms at anything other than a superficial level. It will also be immediately obvious that the "building up of intermediates" can not be divorced from the "activation of catalysts". Similarly induction itself is inextricably linked to the interaction between the chloroplast and its immediate environment.

7.4.1. Activation and deactivation of ribulose bisphosphate carboxylase

It has been known since some of the early time-course experiments on $^{14}CO_2$ fixation by isolated chloroplasts that incorporation of radioactivity ceases abruptly in the dark (see e.g. Fig. 7.11). Bassham and Jensen [18] and Jensen and Bassham [61], showed that fixation ceased despite the fact that there was still ample RBP apparently available for carboxylation (cf. ref. 64) and concluded that there must be a dark deactivation of the carboxylase. Some form of activation was also implied by the marked discrepancy between the apparent activity of the enzyme and that of the parent tissue. In 1956 Weissback et al. [118] put the K_m (bicarbonate) of the carboxylase at about 11 mM (K_m is used here in the conventional and restricted sense of the substrate concentration required to give half maximal velocity) and Peterkofsky and Racker [84] reported rates of 150 μmoles CO_2 fixed.mg^{-1} chlorophyll.h^{-1} at 20°C (using spinach carboxylase and 20 mM bicarbonate at pH 7.8). On this basis, and supposing a concentration of CO_2 at the carboxylation site of approx. 0.01%, the enzyme was entirely inadequate for its postulated role. More than 500 times as much enzyme would have been needed to catalyse the rates of photosynthesis which the intact leaf could achieve under reasonably favourable conditions. In the absence of activation, or some mechanism which would have increased the concentration of CO_2 at the enzyme surface to a value several hundred times that in the external atmosphere, photosynthesis would have been impossible (cf. ref. 46). This problem has now been partially if not entirely resolved. A succession of workers, building on the early observations of Pon [85] and the more recent work of Sugiyama and others (see e.g. refs. 7,60,96,98,99,104,113,115) reported carboxylase preparations with increasingly higher affinities for CO_2. Attention was also given to extraction and other details of assay [7,60,69] and at the present time it is possible [72,115] to extract from spinach chloroplasts an extract which is equal or almost equal to its task in vivo. This last conclusion is based on the values given in Table I in which the concentration of CO_2 at the carboxylation site is calculated for an enzyme with a K_m (CO_2) of approx. 45 μM and a V_{max} of 940 (these being the values actually observed by Walker and Lilley [115] and Lilley and Walker [72]. The table also shows corresponding values for higher values of the V_{max} because it is believed that the V_{max} could be underestimated by a factor of 2 or more [73]. These values for the V_{max}, particularly the less favourable, are still only barely adequate to account for the performance of the parent tissue and fall short of the biochemist's maxim that it is necessary to demonstrate ten times as much enzyme as the reaction rate demands. If, however, the affinity of the enzyme has remained unchanged during evolution the present bare adequacy might simply reflect the extent to which plants have been able to offset decreasing atmospheric CO_2 by synthesising more and more carboxylase. Certainly the fact that the carboxy-

TABLE I

The relationship between carboxylation rate and CO_2 concentration at the site of carboxylation

This table shows the rates of CO_2 fixation which would be obtained on reaction mixtures in equilibrium with partial pressures of CO_2 between 50—200 ppm assuming a K_m of 46.5 μM CO_2 and maximal velocities in the range of 800—2000 μmoles.mg^{-1} chl.h^{-1}. In addition, in the last column, the CO_2 concentration required to give a rate of 100 μmoles.mg^{-1} chl.h^{-1} is listed. It will be seen (figures in italics) that, for the highest value of V_{max} actually recorded [115], a partial pressure of 100 ppm would have yielded a rate of 61. For a rate of 100 (which is taken as that of the average plant in its natural environment) a partial pressure of 172 ppm CO_2 would have been required at the carboxylation site.

Maximal velocity μmoles.mg^{-1} chl.h^{-1}	Rates in CO_2 concentration				CO_2 conc (ppm) to give fixation rate of 100 μmoles.mg^{-1} chl.h^{-1}
	50 ppm	100 ppm	150 ppm	200 ppm	
800	27	52	75	98	206
940	*32*	*61*	*88*	*114*	*172*
1000	33	64	93	120	161
1250	42	80	116	150	127
1600	53	102	148	192	98
2000	67	130	188	244	82

lase may constitute 50% or more of the soluble leaf protein [1] is an otherwise surprising aspect of plant physiology. It should also be noted that these values (and those derived for intact chloroplasts which are discussed in section 7.4.2) neglect any contribution which the oxygenase function of the carboxylase may have made to the results. The literature at the time indicated that this would be negligible (see e.g. ref. 4) but more recent work (see e.g. ref. 6) suggests that the contribution by the oxygenase may be somewhat larger. If any inhibition by oxygen were fully competitive [6] the V_{max} would remain unchanged and the apparent K_m (CO_2) would decrease so that, especially at low CO_2 concentrations, the values in Table I might again under-estimate the real situation. On the other hand, the stroma will not be anaerobic during photosynthesis and for this reason the present conservative estimates are probably more realistic than any which might be derived in relation to a lower concentration of O_2.

7.4.2. Activation of ribulose bisphosphate carboxylase by increases in magnesium and pH

A mechanism of activation based on light-induced Mg movements has also been proposed by several workers (see e.g. refs. 1,18,20,60,61,74,106,107) and a composite scheme incorporating many of these proposals is given in Fig. 7.28. The light-induced movement of Mg from the thylakoid compartment to the stromal compartment would appear to be sufficient to bring about the changes in concentration which this scheme calls for (see e.g. refs. 14,15,44,59,74 and Chapter 3 of this volume).

Magnesium activation of RBPCase has also been simulated in a reconstituted chloroplast system (see e.g. refs. 71,106). The actual Mg requirement for carboxylation in such mixtures will vary according to the concentration and nature of chelating agents such as ATP which also are present but dramatic changes can be brought about by increases in the range of 2.5—5 mM. For example, Fig. 7.29 shows that electron transport and the steps between PGA and G3P will proceed rapidly at Mg concentrations which permit only very slow rates of carboxylation. A subsequent increase in Mg then increases CO_2-dependent O_2 evolution to the same level as PGA-dependent O_2 evolution. The response of the carboxylase to increased Mg is immediate [71] and the kinetics of activation by this mechanism are consistent with the rapid resumption of CO_2 fixation by intact chloroplasts after a dark interval. The relationship between RBPCase activation and induction is much more complex. Magnesium fluxes are extremely rapid and if activation depended entirely on movement between the thylakoid compartment and the stroma it would have to be assumed that the carboxylase assumes its active configuration as rapidly after first illumination as after a brief dark interval. The position is complicated, however, by the fact that with the extracted enzyme the carboxylation never reaches the same rate if the reaction is initiated by

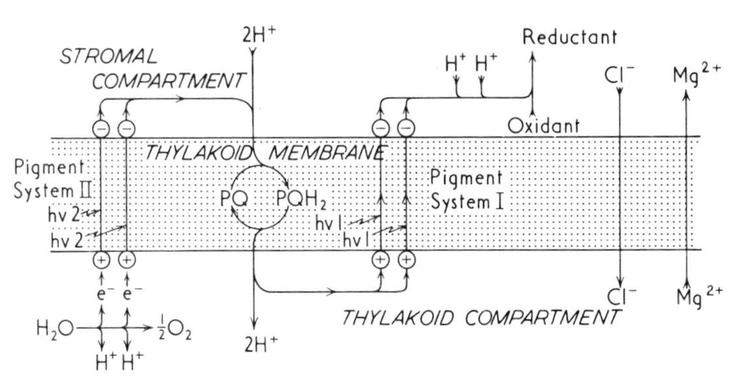

Fig. 7.28. Factors leading to the activation of RBP carboxylase, fructose-1,6-bisphosphatase and other enzymes of the cycle. This greatly simplified figure incorporates a number of contemporary views concerning photosynthetic electron transport and associated ion movements (see e.g. 58,79,80,125 and Chapters 2, 3 and 4 of this volume) and the consequences of these events as enumerated in the text. In the light, for every 4 photons which are absorbed 2 electrons are transported from water to an oxidant which becomes reduced. O_2 is evolved. Plastoquinone accepts electrons from pigment system II and protons from the stromal compartment. It then donates electrons to pigment system I and protons to the thylakoid compartment. The resultant charge separation is offset by inward movement of negatively charged ions and outward movement of positively charged ions. The precise nature of the oxidant within the stroma is not specified but in this context it will include soluble ferredoxin and NADP. The consequence of these events with regard to light activation and dark deactivation are as follows:

Light	Dark
Electron transport starts	Electron transport ceases
Stromal pH increases	Stromal pH decreases
Stromal Mg^{2+} increases	Stromal Mg decreases
ATP/ADP increases	ATP/ADP decreases
Stromal reducing potential increases	Stroma becomes less reducing

The in vitro activity of a number of cycle enzymes is increased by sulphydryl reagents (which may mimic the effect of light generated reductants). Other enzymes are activated by increased Mg^{2+} and by more alkaline pH. Some, such as fructose bisphosphatase may be simultaneously affected by all of these changes. The reduction of PGA to TP is particularly responsive to the ATP/ADP ratio.

Mg as it does if it is initiated by RBP (see e.g. ref. 71). Presented with RBP, in the absence of optimal Mg, the carboxylase appears to lock into a relatively inactive configuration. Whether or not this same state of configuration is produced by withdrawal of Mg from the fully active state is not known. If it is, then the observer would be driven to the unlikely conclusion that full activity would never be again realised. While RBPCase activity can also be affected in some circumstances by certain metabolites (see section 7.4.3) there is no suggestion at present that any of these compounds can convert the enzyme from the relatively inactive state which results from preincubation with RBP.

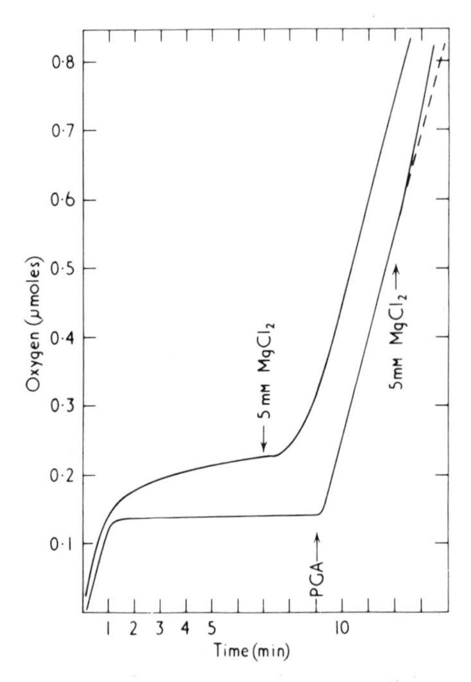

Fig. 7.29. Simulation of light activation of RBPCase by magnesium. Both traces show oxygen evolution in the reconstituted system. Both reaction mixtures contained catalytic NADP and 1 mM $MgCl_2$. In the lower trace O_2 evolution ceased when all the NADP was reduced. It was then restarted by the addition of PGA (which reoxidises NADPH as it is reduced to TP). The second mixture also contained RBP and CO_2 and the slow rate of O_2 evolution observed after the first minute is related to the slow formation of PGA by carboxylation. Following the addition of $MgCl_2$ to 5 mM the carboxylation was able to supply PGA at a greatly increased rate. In short, all of the reactions but the carboxylation were able to proceed at rapid rates in the presence of 1 mM $MgCl_2$, whereas the carboxylation required 5 mM $MgCl_2$ (from ref. 71). It should be noted that the relatively slow response to added Mg reflects the time taken to increase the reduction of newly formed PGA to its full capacity — the effect of Mg on the formation of PGA is immediate [71].

RBPCase is obviously a remarkably complex enzyme and at this stage speculation about its behaviour within the chloroplast is not simplified by the fact that pH optimum, V_{max} and K_m all apparently change as a function of Mg concentration. Alkalisation of the stromal compartment, which is believed to occur at the same time as Mg influx was at one stage thought to be unhelpful or possibly even counter-productive. Alkaline pH would favour bicarbonate accumulation but this would not increase the rate of an enzyne using *dissolved* CO_2 [43] unless the acid-producing carboxylation could establish a micro-environment of low pH at the carboxylase surface as proposed by Werdan et al. [120] and already mentioned in Chapter 6 of this volume. Similarly the pH optimum of the *purified* enzyme is known to shift

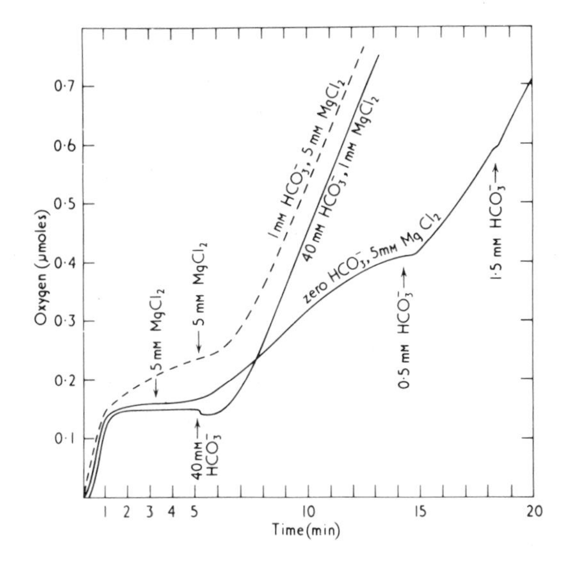

Fig. 7.30. RBP-dependent O_2 evolution in the reconstituted system as a function of [Mg] and [CO_2]. The broken curve is essentially the same as the upper curve in Fig. 7.29 and shows that maximal rate could be achieved in the presence of 5 mM $MgCl_2$ and 1 mM $NaHCO_3$. The lower curve records O_2 evolution from a similar mixture initially containing no added bicarbonate and shows that the same rate could be achieved by adding extra bicarbonate rather than extra Mg (i.e. 40 mM bicarbonate + 1 mM $MgCl_2$ was equivalent to 1 mM bicarbonate + 5 mM $MgCl_2$). The middle curve also shows that the Mg is affecting the affinity of the enzyme for CO_2. Thus when the Mg is increased the endogenous CO_2 is sufficient to support some O_2 evolution before it is exhausted (from ref. 71).

towards neutrality in the presence of high Mg (see e.g. refs. 81,98). More recently, Lilley and Walker [72] obtained the highest values yet recorded for RBPCase in a mixture containing 15 mM $MgCl_2$ and 10 mM bicarbonate at pH 7.9 and 20°C. The rate at pH 7.6. was substantially *lower* but above pH 7.9 the profile was very flat and the behaviour evidently somewhat similar to the broad pH optimum in the 8.2—8.6 range (for 25 mM $MgCl_2$?) cited by Badger and Andrews [6]. Similarly there is good evidence that carboxylation is favoured by the alkaline shift from about 7.6 to 8.0 which occurs in the stroma as a consequence of illumination [53,121].

The picture which is presently emerging is therefore one in which *both* proton efflux and Mg influx contribute to the activation of the carboxylase.

It should also be noted that the increase in V_{max} in response to increased Mg is at least as important as the decrease in K_m (i.e. the increase in affinity for CO_2 which is illustrated in Fig. 7.30). For example, Bahr and Jensen [7] have described an unstable form of the enzyme which works at half maximal velocity in air levels of CO_2 but because this form also has a relatively low V_{max}, it is somewhat inferior in terms of overall activity to an "intermediate" form (see section 7.4.1) in which a relatively low K_m (about 4 × air-level of

CO_2) is allied to an exceptionally high V_{max} [72,115]. Although the carboxylase used by Badger and Andrews [6] was unstable during assay, its apparent K_m (CO_2) *in air* and relatively high activity (about 340 μmoles mg^{-1} chlorophyll.h^{-1} at 25°C) suggest that it was nearer to the "intermediate" form than the "low K_m" form of Bahr and Jensen [7].

Evidence has also been obtained which suggests that the characteristics of the "intermediate" form, correspond closely to the characteristics of the enzyme within the intact functioning chloroplast. The CO_2 concentration required for half maximal rate in chloroplasts is very low (about 0.03%) as it is for the intact tissue but the new evidence [72,115] shows that at saturation the rate of fixation is limited by electron transport rather than carboxylation. The CO_2 required by the chloroplast for half maximal velocity is therefore substantially less than that required by the carboxylase within it. If the departure from linearity imposed by the electron transport ceiling is disregarded in the double reciprocal plot of rate against substrate the "K_m" derived by extrapolation corresponds to that of the extracted carboxylase (Fig. 7.31).

In one respect RBPCase activation must clearly affect induction in respect to the building up of intermediates. Several independent observations (for reviews, see refs. 48,108,109) indicate that RBP does not move freely across the chloroplast envelope. If neither carboxylation nor export occurred in the dark to any appreciable extent, then this key component of the Benson—Calvin cycle would persist at a higher level than would otherwise be possible and would shorten the lag in the next light period. If carboxylation continued unchecked the ΔF of the reaction (about −8 kcal [21]) would ensure virtually complete utilisation. It is conceivable of course that RBP may be degraded by other routes (see e.g. ref. 5) and will undoubtedly fall to comparatively low levels in the dark [18,64,65] but whether or not it will fall to zero (cf. ref. 36) remains to be established. In this respect, even the persistence of very low concentrations may be of crucial importance. The Benson—Calvin cycle may be able to lift itself, autocatalytically, by its own bootstrings but the operation of the compound interest law will not lead to an increase in capital without an initial investment. In the steady state the stroma may well be saturated with RBP (cf. ref. 92). During induction the build up of RBP from the low values implied by the ΔF of the carboxylation is unlikely to be less important than that of the other intermediates and indeed such a build up has been reported [92].

7.4.3. Activation of ribulose bisphosphate carboxylase by other agents

Buchanan and Schürmann [27,28] have reported activation of RBPCase by fructose-6-phosphate and inhibition by fructose-1,6-bisphosphate. The activation by fructose-6-phosphate has not been confirmed by other workers

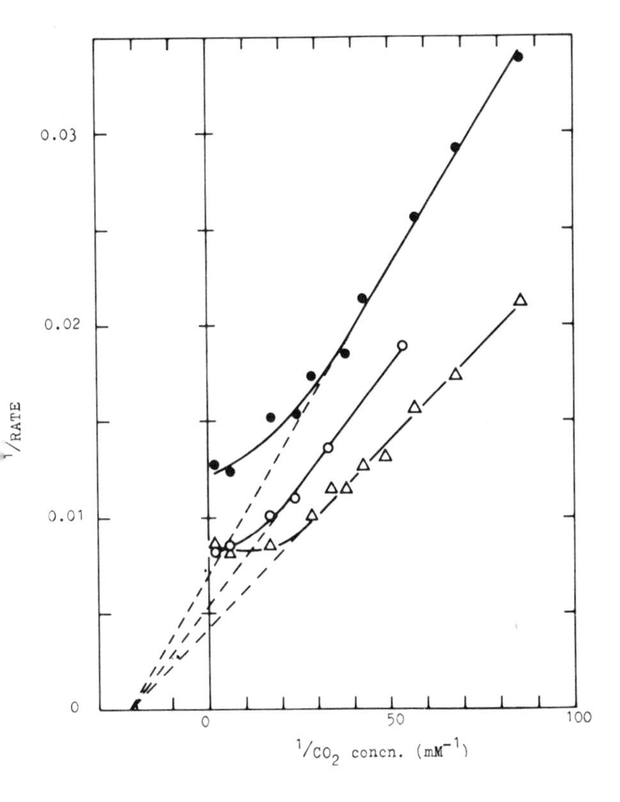

Fig. 7.31. CO_2-dependent O_2 evolution by intact chloroplasts as a function of CO_2 concentration. In this double reciprocal plot there is a marked departure from linearity at high concentrations of CO_2. If this departure is ignored the curves extrapolate to give a value of CO_2 required for half maximal activity equal to the K_m (CO_2) for the isolated carboxylase. The departure from linearity is attributed to a ceiling imposed by electron transport. Thus electron transport limits photosynthesis in saturating CO_2. Because the true potential rate of carboxylation is not realised the concentration of CO_2 required for half maximal velocity is similarly depressed and the affinity of the chloroplast for CO_2 is exaggerated.

(see e.g. refs. 5,35,36) but this may be related to the quality of the fructose-6-phosphate used and to differences in procedure. Fructose-1,6-bisphosphate inhibition was not observed in the system used by Avron and Gibbs [5] but has also been reported by Bowes and Ogren [24] and Chu and Bassham [36]. Apart from any direct effect fructose bisphosphate will alter the concentration of free magnesium in a reaction mixture and although it is perhaps unlikely that this could account for all of the observed inhibitions, some of which have been seen in the presence of 5 and 10 mM [24,36] it is an aspect of this work which must always be borne in mind. ATP can also affect the rate of carboxylation, according to the timing of its addition relative to other reactants and there seems little doubt that in these particular circumstances, this is an indirect effect brought about by chelation of Mg (Walker and Holborow, unpublished).

RBPCase is activated and inhibited (according to concentration) by 6-phosphogluconate [34—36] and it has been suggested (see e.g. ref. 34) that this might assist a switch from the reductive pentose phosphate cycle in the light to the oxidative pentose phosphate cycle in the dark. The same authors [36], have recently reported a similar activation by NADPH and propose that this compound might take over the role of regulator as the 6-phosphogluconate concentration decreases in the light. Bicarbonate, which is also believed to increase in concentration within the stromal compartment of the illuminated chloroplast may also function as an allosteric activator [99].

The relevance of these effects to induction is as yet uncertain but any hypothesis would need to be compatible with the kinetics illustrated in Fig. 7.11.

Electron transport and Mg movements are rapid processes which would be completed within seconds of illumination, whereas metabolites such as 6-phosphoglucanate would change in concentration relatively slowly, like the intermediates of the Benson—Calvin cycle. It would be reasonable to suppose that soon after first illumination the concentration of Mg ions, protons, bicarbonate and NADPH would all favour activation of RBPCase and therefore that the delay in the onset of carboxylation must be related to the operation of other factors. If the oxidative pentose phosphate pathway operates in the chloroplast in the dark [19,50,63] and if the operation leads to an inhibitory concentration of 6-phosphogluconate then this could clearly account for some part of the observed induction. However, the fact that induction in chloroplasts from dark-stored leaves is largely, though not entirely eliminated, by the addition of exogenous TPs (see section 7.3.4) suggests that the low level of RBP is more likely to be the over-riding factor.

The extent to which 6-phosphogluconate is involved in light deactivation of RBPCase is also a matter for speculation at this time but if fixation ceases as rapidly as Fig. 7.11 suggests (see also ref. 61) it seems probable that control is extended by something which itself alters very rapidly such as Mg or pH. If 6-phosphogluconate increases rapidly in the dark to an inhibitory level at the expense of glucose-6-phosphate this might also be expected to delay or offset to some extent the (deactivating?) decline in NADPH associated with the reduction of pre-formed PGA. The rapid resumption in photosynthesis which occurs after a brief dark interval (during which time RBP persists at a measurable level [18]) is again consistent with a very rapid process (such as Mg transport) rather than one in which control is exerted by relatively slow changes in the concentration of a metabolite.

7.4.4. *Light activation of other enzymes of the cycle*

In the above discussions, attention has been focused on the light activation of RBPCase but (as noted in the legend to Fig. 7.28) other enzymes of the Benson—Calvin cycle are similarly affected and at the time of writing it is by no means certain that the activation of the carboxylation is of more importance than that of the other reactions. Ru5P kinase is thought to be activated by a photosynthetically produced reductant [2,3,5,64—66]. Fructose-1,6-bisphosphatase activity can be increased in many ways and a number of these could constitute forms of light activation (see e.g. refs. 16,25,26,83, 87,89) and it seems likely that increased reducing power, increased Mg and increased pH may combine (Fig. 7.28) to bring this and other enzymes into full operation in the light [8]. Ru5P from spinach is reported to be sensitive to ADP [62,67] although, surprisingly, this sensitivity does not extend to the same enzyme from pea leaves [3]. The reduction of PGA is also extremely sensitive to ADP [70] and again it has been established that the reaction involved is that catalysed by PGA kinase [67] (and Slabas and Walker, unpublished). Enzymes present in the chloroplast but not immediately associated with the cycle may also be indirectly activated by light-triggered events (see e.g. ref. 88,89).

Clearly all of these factors could be important in the realisation of the full photosynthetic capacity of the illuminated chloroplast and in its general regulation of photosynthesis. Moreover the product of one reaction will very often affect the rate of another. Nevertheless, as discussed elsewhere in this article (e.g. section 7.4.2.) many of these parameters (such as pH, Mg concentration etc.) presumably change too rapidly upon illumination to account for the characteristically slow advent of photosynthesis upon first illumination.

7.5. THE RELEVANCE OF ATP-ADP REGULATION TO INDUCTION

Photosynthetic electron transport is normally accelerated by ADP and slows when all of the available adenylate has been converted to ATP. This is the basis of the well known phenomenon of photosynthetic control, analogous to respiratory control, which was first described by West and Wiskich [122]. Conversely, at least two reactions in the Benson—Calvin cycle are particularly susceptible to inhibition by ADP (see sections 7.4.3 and 7.5.1). In the steady state, the consumption of ATP must permit enough electron transport to sustain photophosphorylation at an unchanging level. If, for any reason, ATP consumption is curtailed, photosynthetic control will sooner or later impose a restriction on electron transport and associated phosphorylation. If, for any reason, ADP formation outstrips ATP regeneration, there will be a corresponding decline in ATP consumption. In the steady state, therefore, these controls will tend to maintain the status quo and ensure that demand does not exceed production. During induction it must be assumed that the ATP/ADP ration will quickly rise to its maximum value and subsequently fall as ATP utilisation increases with the onset of rapid carboxylation. If there were a pronounced suppression of electron transport during the initial phase, the proton and Mg gradient might also be lowered and consequently the activation and RBPCase and fructose bisphosphatase. This argument has, in fact, been used by Miginiac-Maslow and Champigny [78] to account for the stimulation of CO_2 fixation by Antimycin A (see also refs. 31,91,92).

7.5.1. The effect of ADP and ATP on O_2 evolution in the reconstituted system

As noted (section 7.4.3) the reduction of PGA to TP and its associated O_2 evolution is particularly susceptible to ADP. For example (in systems which rely on endogenous photophosphorylation) PGA-dependent O_2 evolution is interrupted by the addition of R5P, which serves as a sink for ATP, thus lowering the ATP/ADP ratio (Slabas and Walker, unpublished). Similarly R5P-dependent O_2 evolution will not start until all or most of the R5P has undergone conversion to RBP, and once started may be interrupted by the addition of small quantities of ADP (Slabas and Walker, unpublished). If an additional ATP-generating system is added to reinforce endogenous photophosphorylation these effects are abolished or diminished but if the ATP/ADP ratio is kept at a markedly high value electron transport is slowed because of photosynthetic control. High ATP could contribute to induction (section 7.5) but it is difficult to envisage circumstances in which the onset of photosynthesis could be delayed for several minutes by a relative excess of ADP.

7.6. INDUCTION IN THE RECONSTITUTED CHLOROPLAST SYSTEM

During the last five years renewed attention has been paid to the reconstituted chloroplast system in which ruptured chloroplasts are supplemented with stromal protein. This system was first used by Whatley et al. [123]. The rates of photosynthesis which were then observed were low but of the same order as those recorded with "intact" chloroplasts. More recently it has been possible to achieve the same level of photosynthetic performance with isolated chloroplasts as with the parent tissue and this, in turn, has led to comparable rates with the reconstituted system. At first rapid rates of O_2 evolution were seen only with PGA [95]. In essence this is simply an elaborated Hill reaction in which NADP reduced via ferredoxin is reoxidised by glycerate-1,3-bisphosphate formed by phosphorylation of PGA. It has proved possible to move successively backwards through the Benson—Calvin cycle from this step (see e.g. refs. 71,96,97,113), replacing

PGA by RBP and CO_2, RBP by R5P, R5P by fructose-1,6-bisphosphate and so on. Although the rates observed are only about one third of those seen with PGA it is even possible to demonstrate O_2 evolution with TP as substrate, thus embracing the complete cycle (Walker, unpublished). Similarly Bassham et al. [22] have demonstrated *CO_2 fixation with PGA as substrate.*

In principle the full reconstituted system should exhibit induction in the same way as the intact chloroplast. In practice there are obvious difficulties in securing an appreciable build-up of intermediates in the relatively large volume of the reaction mixture compared with the relatively minute volume of the stromal compartment. With excess stromal protein, however, increased rates of catalysis off-set low concentrations of substrate and autocatalysis can be detected [116]. This supports the view that induction is an intrinsic feature of the Benson—Calvin cycle and does not depend on the physical structure of the intact chloroplast for its demonstration.

7.7. CONCLUDING REMARKS

The initial lag in O_2 evolution and CO_2 uptake which is normally observed when plants are first illuminated after a period of darkness is attributed to an initial deficiency in Benson—Calvin cycle intermediates which is then overcome by the autocatalytic action of the cycle. The light activation of enzymes and transient imbalance between the production and consumption of ATP and NADPH may well contribute to induction but the fact that carbon assimilation ceases very rapidly in the dark and is then resumed almost as quickly after a brief dark interval suggests that these factors can not in themselves account for delays of several minutes.

Induction is of interest to the biochemist because it can be regarded as a demonstration of the autocatalytic function of the Benson—Calvin cycle. From the outset it has been recognised that any primary carboxylating mechanism must incorporate a facility for the regeneration of the carbon dioxide acceptor. Less attention has been paid to the fact that the cycle must function as a breeder reaction, producing more substrate than it utilises. If this were not so the cycle could not accelerate to accommodate increases in light intensity, nor could the chloroplast afford to export metabolites to the cytoplasm. The immediate end-products of photosynthesis are used in various ways. They may be stored as polysaccharides, fed back into the cycle or exported to the cytoplasm. The loss of starch (from the chloroplasts of those plants which produce it) during relatively short periods in the dark is visual evidence of the operation of sinks within the cytoplasm. The free energy change of reactions (such as that of RBP carboxylation) point to depletion in the dark even if catalysis is substantially slowed by dark deactivation of enzymes. All of these events are natural consequences of the metabolism of the chloroplast. Autocatalysis, and its manifestation as induction, is essential in order that anabolism in the light can offset catabolism in the dark, and replenish metabolites consumed by catabolism in the dark.

7.8. SUMMARY

The initial lag in photosynthesis is interpreted as a period in which intermediates of the Benson—Calvin cycle are built up to a steady-state working concentration. The light activation of enzymes, the movement of metabolites to the immediate environment and various aspects of photosynthetic control may affect induction but are seen as secondary effects. Once the steady state has been achieved, induction is negligible after a short dark interval even though this may have been of sufficiently long duration to deactivate several key enzymes.

7.9. ACKNOWLEDGEMENTS

Sooner than ask permission to reproduce a number of figures I was prompted by the requirements of this article to repeat some old work and even to undertake some new, I therefore gladly and gratefully acknowledge (*a*) the original contributions made to the facts and theories outlined above by my past colleagues C.W. Baldry, C. Bucke, W. Cockburn, T. Delieu, R. McC. Lilley, J. Ludwig, A.V. McCormick, J.D. Schwenn and D.M. Stokes; (*b*) the invaluable advice and criticism which we have all received and continue to receive from R. Hill and (*c*) the skilled and unflagging labours of Krystyna Holborow and June Devereux.

REFERENCES

1 Akazawa, T. (1970) in Progress in Phytochemistry (Reinhold, L. and Liwschitz, Y., eds), Vol. 2. pp. 107—141, Wiley Interscience, London.
2 Anderson, L.E. (1973) Biochim. Biophys. Acta, [321], 484—488.
3 Anderson, L.E. (1973) Plant Sci. Lett., [1], 331—334.
4 Andrews, T.J., Lorimer, G.H. and Tolbert, N.E. (1973) Biochemistry, [12], 11—18.
5 Avron, M. and Gibbs, M. (1974) Plant Physiol., [53], 140—143.
6 Badger, M.R. and Andrews, T.J. (1974) Biochem. Biophys. Res. Commun., [60], 204—210.
7 Bahr, J.T. and Jensen, R.G. (1974) Plant Physiol., [53], 39—44.
8 Baier, D. and Latzko, E. (1975) Biochim. Biophys. Acta, [396], 141—148.
9 Baldry, C.W., Bucke, C. and Walker, D.A. (1966) Nature, [210], 793—796.
10 Baldry, C.W., Walker, D.A. and Bucke, C. (1966) Biochem. J., [101], 641—646.
11 Bamberger, E.S., Ehrlich, B.A. and Gibbs, M. (1974) in Proc. 3rd Int. Cong. Photosynthesis (Avron, M., ed), pp. 1349—1362. Elsevier, Amsterdam.
12 Bamberger, E.S. and Gibbs, M. (1963) Plant Physiol. Suppl., 38, X.
13 Bamberger, E.S. and Gibbs, M. (1965) Plant Physiol., [40], 919—926.

14 Barber, J., Mills, J. and Nicolson, J. (1974) FEBS Lett., [49], 106—110.
15 Barber, J., Telfer, A, and Nicolson, J. (1974) Biochim. Biophys. Acta, [357], 161—165.
16 Bassham, J.A. (1971) Science, [172], 526—534.
17 Bassham, J.A. and Calvin, M. (1957) The Path of Carbon in Photosynthesis, pp. 1—107, Prentice-Hall, Englewood Cliffs, N.J.
18 Bassham, J.A. and Jensen, R.G. (1967) in Harvesting the Sun (San Pietro, A., Greer, F.A. and Army, T.J. eds), pp. 79—110, Academic Press, New York.
19 Bassham, J.A. and Kirk, M. (1968) in Comparative Biochemistry and Biophysics of Photosynthesis (Shibata, K., Takamiya, A., Jagendorf, A.T. and Fuller, R.C., eds), pp. 365—378, University of Tokyo Press, Tokyo.
20 Bassham, J.A., Sharp, P. and Morris, I. (1968) Biochim. Biophys. Acta, [153], 898—900.
21 Bassham, J.A. and Krause, G.H. (1969) Biochim. Biophys. Acta, [189], 207—221.
22 Bassham, J.A., Levine, G. and Forger, J. III (1974) Plant Sci. Lett., [2], 15—21.
23 Blinks, L.R. and Skow, R.K. (1938) Proc. Natl. Acad. Sci. USA, [24], 413—419.
24 Bowes, G. and Ogren, W.L. (1972) J. Biol. Chem., [247], 2171—2176.
25 Buchanan, B.B., Kalberer, P.P. and Arnon, D.I. (1967) Biochem. Biophys. Res. Commun., [29], 74—79.
26 Buchanan, B.B., Schurmann, P. and Kalberer, P.P. (1971) J. Biol. Chem., [246], 5952—5959.
27 Buchanan, B.B. and Schurmann, P. (1972) FEBS Lett., [23], 157—159.
28 Buchanan, B.B. and Schurmann, P. (1973) J. Biol. Chem., [248], 4956—4964.
29 Bucke, C., Walker, D.A. and Baldry, C.W. (1966) Biochem. J. [101], 636—641.
30 Champigny, M.L. and Gibbs, M. (1969) in Proc. 1st Int. Cong. Photosynthesis (Metzner, H., ed), Vol. III, pp. 1534—1537, Intern. Union. Biol. Sci., Tübingen.
31 Champigny, M.L. and Miginiac-Maslow, M. (1971) Biochim. Biophys. Acta, [234], 335—343.
32 Champigny, M.L., Mathieu, Y. and Miginiac-Maslow, M. (1972) in Proc. 2nd Int. Cong. Photosynthesis (Forti, G., Avron, M. and Melandri, A., eds), Vol. III, pp. 1909—1916, Junk, The Hague.
33 Chen-She Sheu-Hwa, Lewis, D.H. and Walker, D.A. (1975) New Phytol., [74], 381—390.
34 Chu, D.K. and Bassham, J.A. (1972) Plant Physiol., [50], 224—227.
35 Chu, D.K. and Bassham, J.A. (1973) Plant Physiol., [52], 373—379.
36 Chu, D.K. and Bassham, J.A. (1974) Plant Physiol., [54], 556—559.
37 Cockburn, W., Baldry, C.W. and Walker, D.A. (1967) Biochim. Biophys. Acta, [143], 603—613.
38 Cockburn, W., Baldry, C.W. and Walker, D.A. (1967) Biochim. Biophys. Acta, [143], 614—624.
39 Cockburn, W., Baldry, C.W. and Walker, D.A. (1967) Biochim. Biophys. Acta, [131], 594—596.
40 Cockburn, W., Walker, D.A. and Baldry, C.W. (1968) Biochem. J., [107], 89—95.
41 Cockburn, W., Walker, D.A. and Baldry, C.W. (1968) Plant Physiol., [43], 1415—1418.
42 Coombs, J. and Baldry, C.W. (1971) Plant Physiol., [48], 379—381.
43 Cooper, T.G., Filmer, D., Wishnick, M. and Lane, M.D. (1969) J. Biol. Chem., [244], 1081—1083.
44 Dilley, R.A. and Vernon, L.P. (1965) Arch. Biochem., [111], 365—375.
45 Gibbs, M. (1971) in Structure and Function of Chloroplasts (Gibbs, M., ed), pp. 169—214, Springer, Berlin.

46 Gibbs, M., Latzko, E., Everson, R.G. and Cockburn, W. (1967) in Harvesting the Sun (San Pietro, A., Green, F.A. and Army, T.J., eds), pp. 111—130, Academic Press, New York.
47 Goldsworthy, A. and Street, H.E. (1965) Ann. Bot., N.S., [29], 45—58.
48 Heber, U. (1974) Ann. Rev. Plant Physiol., [25], 393—421.
49 Heber, U. and Willenbrink, J. (1964) Biochim. Biophys. Acta, [82], 313—324.
50 Heber, U., Hallier, U.W. and Hudson, M.A. (1967), Z. Naturforsch., [22b], 1200—1215.
51 Heldt, H.W. and Rapley, L. (1970) FEBS Lett. [10], 143—148.
52 Heldt, H.W. and Sauer, F. (1971) Biochim. Biophys. Acta, [234], 83—91.
53 Heldt, H.W., Werdan, K., Milovancev, M. and Geller, G. (1973) Biochim. Biophys. Acta, [314], 224—241.
54 Heldt, H.W., Sauer, F. and Rapley, L. (1972) in Proc. 2nd Int. Cong. Photosynthesis (Forti, G., Avron, M. and Melandri, A., eds), Vol. II, pp. 1345—1355, Junk, The Hague.
55 Hill, R. (1939) Proc. Roy. Soc. (London), Ser. B., [127], 192—210.
56 Hill, R. (1965) Essays Biochem., [1], 121—151.
57 Hill, R. and Whittingham, C.P. (1953) New Phytol., [52], 133—148.
58 Hill, R. and Bendall, F. (1960) Nature, [186], 136—137.
59 Hind, G., Nakatani, H.Y. and Izawa, S. (1974) Proc. Natl. Acad. Sci. USA, [71], 1484—1488.
60 Jensen, R.G. (1971) Biochim. Biophys. Acta, [234], 360—370.
61 Jensen, R.G. and Bassham, J.A. (1968) Biochim. Biophys. Acta, [153], 227—234.
62 Johnson, E.J. (1966) Arch. Biochem. Biophys., [114], 178—183.
63 Krause, G.H. and Bassham, J.A. (1969) Biochim. Biophys. Acta, [172], 553—565.
64 Latzko, E. and Gibbs, M. (1969) Plant Physiol., [44], 396—402.
65 Latzko, E. and Gibbs, M. (1969) Prog. Photosynthesis Res., Vol. III, 1624—1630.
66 Latzko, E., Garnier, R.V. and Gibbs, M. (1970) Biochem. Biophys. Res. Commun., [39], 1140—1144.
67 Lavergne, D., Bismuth, E. and Champigny, M.L. (1974) Plant Sci. Lett., [3], 391—397.
68 Lilley, R. McC., Schwenn, J.D. and Walker, D.A. (1973) Biochim. Biophys. Acta, [325], 596—604.
69 Lilley, R. McC. and Walker, D.A. (1974) Biochim. Biophys. Acta, [358], 226—229.
70 Lilley, R. McC. and Walker, D.A. (1974) Biochim. Biophys. Acta, [368], 269—278.
71 Lilley, R. McC., Holborow, K. and Walker, D.A. (1974) New Phytol., [73], 659—664.
72 Lilley, R. McC. and Walker, D.A. (1975) Plant Physiol., [55], 1087—1092.
73 Lilley, R. McC., Fitzgerald, M.P., Rienits, K.G. and Walker, D.A. (1975) New Phytol., [75], 1—10.
74 Lin, D.C. and Nobel, P.S. (1971) Arch. Biochem. Biophys. [145], 622—632.
75 Loughman, B.C. (1966) New Phytol., [65], 388—397.
76 McAlister, E.D. (1937) Smithson. Misc. Coll., [95], 1—17.
77 Meidner, H. and Mansfield, T.A. (1968) Physiology of Stomata, pp. 1—179, McGraw-Hill, London.
78 Miginiac-Maslow, M. and Champigny, M.L. (1974) Plant Physiol., [53], 856—862.
79 Mitchell, P. (1961) Nature, [191], 144—148.
80 Mitchell, P. (1966) Biol. Rev., [41], 445—502.
81 Nishimura, M. and Akazawa, T. (1974) Biochemistry, [13], 2277—2287.
82 Osterhout, W.J.V. and Haas, A.R.C. (1918) J. Gen. Physiol., [1], 1—16.
83 Pedersen, T.A., Kirk, M. and Bassham, J.A. (1966) Physiol. Plant., [19], 219—231.
84 Peterkofsky, A. and Racker, E. (1961) Plant Physiol., [36], 409—414.
85 Pon, N.G. (1959) Ph.D. thesis, University of California, Berkeley.
86 Pon, N.G., Rabin, B.R. and Calvin, M. (1963) Biochem. Z., [338], 7—19.

87 Preiss, J., Biggs, M.L. and Greenberg, E. (1967) J. Biol. Chem., [242], 2292—2294.
88 Preiss, J., Ghosh, H.P. and Wittkop, J. (1967) in The Biochemistry of Chloroplasts (Goodwin, T.W., ed), Vol. 2, pp. 131—153, Academic Press, New York.
89 Preiss, J. and Kosuge, T. (1970) Ann. Rev. Plant Physiol., [21], 433—466.
90 Rabinowitch, E.I. (1956) Photosynthesis and Related Processes Vol. 2, Part 2, pp. 1211—2088, Wiley Interscience, New York.
91 Schacter, B.Z., Gibbs, M. and Champigny, M.L. (1971) Plant Physiol., [48], 443—466.
92 Schacter, B.Z. and Bassham, J.A. (1972) Plant Physiol., [49], 411—416.
93 Schwenn, J.D., Lilley, R. McC. and Walker, D.A. (1973) Biochim. Biophys. Acta, [325], 586—595.
94 Steeman-Nielsen, E. (1942) Dansk. Bot. Ark., [11], 2.
95 Stokes, D.M. and Walker, D.A. (1971) Plant Physiol., [48], 163—165.
96 Stokes, D.M., Walker, D.A. and McCormick, A.V. (1972) in Proc. 2nd Int. Cong. Photosynthesis (Forti, G., Avron, M. and Melandri, A., eds), pp. 1779—1785, Junk, The Hague.
97 Stokes, D.M. and Walker, D.A. (1972) Biochem. J., [128], 1147—1157.
98 Sigiyama, T., Matsumoto, C. and Akazawa, T. (1969) Arch. Biochem. Biophys., [129], 597—602.
99 Sugiyama, T., Nakayama, N. and Akazawa, T. (1968) Arch. Biochem. Biophys., [126], 737—745.
100 Thomas, M., Ranson, S.L. and Richardson, J.A. (1973) Plant Physiology, pp. 1—1062, Longman, London.
101 Turner, J.F., Black, C.C. and Gibbs, M. (1962) J. Biol. Chem., [237], 577—579.
102 Walker, D.A. (1965) in Proc. NATO Adv. Study Inst. Biochemistry of Chloroplasts, Aberystwyth, 1965, Photosynthetic Activity of Isolated Pea Chloroplasts (Goodwin, T.W., ed), Vol. 2, pp. 53—69, Academic Press, New York.
103 Walker, D.A. (1965) Plant Physiol., [40], 1157—1161.
104 Walker, D.A. (1972) in Proc. 2nd Int. Cong. Photosynthesis (Forti, G., Avron, M. and Melandri, A., eds), Vol. III, pp. 1773—1778, Junk, The Hague.
105 Walker, D.A. (1971) in Methods in Enzymology (San Pietro, A., ed), Vol. 23, pp. 211—220, Academic Press, London.
106 Walker, D.A. (1973) New Phytol., [72], 209—235.
107 Walker, D.A. (1974) in Proc. Phytochem. Soc. Symp. Edinburgh, 1973 (Pridham, J.B., ed), pp. 7—26, Academic Press, London.
108 Walker, D.A. (1974) in Med. Tech. Publ. Int. Rev. Sci. Biochem. (Northcote, D.H., ed), Ser. I, Vol. II, pp. 1—49, Butterworth, London.
109 Walker, D.A. (1975) in Encyclopedia of Plant Physiology (New Series) (Pirson, A., and Zimmermann, M., eds), pp. 000—000, Springer, Berlin. (In Press).
110 Walker, D.A., Cockburn, W. and Baldry, C.W. (1967) Nature, [216], 597—599.
111 Walker, D.A. and Hill, R. (1967) Biochim. Biophys. Acta, [131], 330—338.
112 Walker, D.A. and Crofts, A.R. (1970) Ann. Rev. Biochem., [39], 389—428.
113 Walker, D.A., McCormick, A.V. and Stokes, D.M. (1971) Nature, [233], 346—347.
114 Walker, D.A., Kosciukiewicz, K. and Case C. (1973) New Phytol., [72], 237—247.
115 Walker, D.A. and Lilley, R. McC. (1976) Proc. 50th Ann. Meeting of the Soc. for Exp. Biol. Cambridge, 1974 (Sunderland, N., ed), pp. 189—198, Pergamon, London.
116 Walker, D.A. and Lilley, R. McC. (1974) Plant Physiol., [54], 950—952.
117 Warburg, O. (1920) Biochem. Z., [103], 188—217.

118 Weissbach, A., Horecker, B.L. and Hurwitz, J. (1956) J. Biol. Chem., [218], 795—810.
119 Werdan, K. and Heldt, H.W. (1972) in Proc. 2nd Int. Conf. Photosynthesis (Forti, G., Avron, M. and Melandri, A., eds), Vol. II, pp. 1337—1344, Junk, The Hague.
120 Werdan, K., Heldt, H.W. and Geller, G. (1972) Biochim. Biophys. Acta, [283], 430—441.
121 Werdan, K., Heldt, H.W. and Milovancev, M. (1975) Biochim. Biophys. Acta, [396], 276—292.
122 West, K.R. and Wiskich, J.T. (1968) Biochem. J., [109], 527—532.
123 Whatley, F.R., Allen, M.B., Rosenberg, L.L., Capindale, J.B. and Arnon, D.I. (1956) Biochim. Biophys. Acta, [20], 462—468.
124 Wildner, G.F. and Criddle, R.S. (1969) Biochem. Biophys. Res. Commun., [37], 952—960.
125 Witt, H.W. (1971) Quart. Rev. Biophys., [4], 365—477.
126 Zelitch, I. (1961) Proc. Natl. Acad. Sci. USA, [47], 1423—1433.
127 Zelitch, I. (1965) Biol. Rev., [40], 463—482.
128 Zelitch, I. (1971) Photosynthesis, Photorespiration and Plant Productivity, pp. 1—347, Academic Press, New York.
129 Veen, R. van der (1949) Physiol. Plant., [2], 217—234.

The Intact Chloroplast — edited by J. Barber

Chapter 8

Interactions Between Chloroplasts and Cytoplasm in C4 Plants

J. COOMBS

Tate and Lyle Ltd., Group Research and Development, P.O. Box 68, Reading (Great Britain)

CONTENTS

Abbreviations: DCMU, 3-(3,4-dichlorophenyl)-1,1-dimethylurea; DHAP, dihydroxyactone phosphate; G3P, glyceraldehyde-3-phosphate; G6P, glucose-6-phosphate; OAA, oxalo-acetate; P_i, inorganic orthophosphate; PP_i, inorganic pyrophosphate; P_{700}, photosystem I reaction centre; PCR, photosynthetic carbon reduction; PEP, phosphoenol pyruvate; PGA, 3-phosphoglycerate; RBP, ribulose-1,5-bisphosphate; TCA, tricarboxylic acid.

8.1. INTRODUCTION

It is generally recognised that in the eukaryotic cell opposing chemical reactions, or series of reactions, such as photosynthesis and respiration are separated and usually confined within specific membrane-bound subcellular compartments or organelles. Several years ago it would have been universally accepted that the process of photosynthesis is restricted to the chloroplast. During the early 1960s considerable effort was expended [168,170] in demonstrating that chloroplasts isolated from temperate plants, such as *Pisum sativum* and *Spinacia oleracea*, were capable of carrying out the complete process of photosynthesis in an integrated manner, with the stoichiometry required by the basic equation, at rates commensurate with those observed in intact leaves. As discussed in detail in Chapter 7 of this book these studies have resulted in a clear picture of the photosynthetic process in what are now known as C3 plants. This may be summarised briefly as follows.

Chloroplasts are small, generally saucer-shaped, organelles 5 to 10 μm in diameter and several μm thick bounded by a selectively permeable double membrane or envelope. Within this, two compartments, stromal and thylakoid, are separated by the complex internal membrane system comprising granal and intergranal lamellae (see Chapter 1). The light reactions of photosynthesis (primary quantum conversion, oxygen production, electron transport and photophosphorylation) are largely associated with the internal membrane system whereas the dark reactions (fixation and reduction of carbon dioxide) are associated with the soluble proteins of the stroma. As explained in Chapter 7 carbon dioxide is assimilated into organic matter in the carboxylation reaction catalysed by ribulose-1,5-bisphosphate carboxylase (RBP carboxylase, EC 4.1.1.39; otherwise ribulose-1,5-diphosphate carboxylase, carboxydismutase or 3-D-phosphoglycerate carboxylase). This enzyme, which represents the bulk of the soluble, fraction I, protein in leaves of C3 plants catalyses the condensation of one molecule of ribulose bisphosphate (RBP) with one molecule of CO_2 [53], in a reaction which requires Mg^{2+} ions as cofactor. The reaction product, the three carbon (C3) D-3 phosphoglyceric acid (PGA) is then converted to fructose-1,6-bisphosphate in a series of reactions which are essentially the reverse of those which occur during the respiratory breakdown of carbohydrate in glycolysis. This requires the initial phosphorylation of PGA, to give diphosphorylglyceric acid, which is then reduced by the NADP-specific triose dehydrogenase to form glyceraldehyde-3-phosphate (G3P). Isomerization of this compound produces dihydroxyacetone phosphate (DHAP) which is then condensed with G3P. The resultant sugar phosphate represents the net product of photosynthesis, although part is recycled to be used in the regeneration of the primary CO_2 acceptor molecule RBP. The complete sequence of reactions comprise the well known Calvin, or photosynthetic carbon reduction (PCR), cycle [13] and is given in detail in Chapter 7 (Fig. 7.10). This is driven in the direction of net synthesis using the energy and reducing power of ATP and NADPH produced in the light reactions.

Since all the reactions of photosynthesis are confined within the chloroplasts all that is required to maintain the process is light, and the influx of CO_2 and water. Oxygen will diffuse from the chloroplasts. The organic products may be converted to starch and stored in situ, or pass to the cytoplasm as sugar phosphates. In addition, other compounds, with regulatory function, or associated with the synthesis of other metabolites such as amino acids and lipids, may also pass in and out of the chloroplast (see Chapters 6,9,10,11 and 12 of this volume). The movement of such compounds through the outer envelope of chloroplasts from C3 plants has been studied in some detail. Hence, it is possible to provide a comprehensive picture of the relationship between the chloroplasts and cytoplasm in these species, as has been done in various chapters of this book and in two recent reviews [92,169]. On the other hand, little if any similar work has been carried out using chloroplasts isolated from C4 plants.

This lack of information on the subject, which prevents a similar treatment of chloroplastic/cytoplasmic relationships in C4 plants is a pity, since the present widely accepted concept of C4 photosynthesis places an increased importance on the movement of fixed-carbon within the leaf. According to these views the fixation of an individual molecule of atmospheric carbon dioxide into an end product of photosynthesis such as sucrose requires the co-operative effort, not only of cytoplasmic and chloroplastic enzymes, but also of chloroplasts and other organelles situated in two separate and specific layers of photosynthetic tissue. At the same time it is suggested that the chloroplasts of C4 plants have evolved in such a way that some, at least, perform a function which is more specialised but at the same time less complex than that of the chloroplast from a C3 plant.

Since information on metabolite transfer is not available for C4 plants this chapter will be concerned mainly with a consideration of the evidence on which the various schemes of C4 photosynthesis are based. In general the subjects discussed relate to the flow of carbon from the atmosphere, through photosynthetic intermediates, to sucrose. However, at a more practical level, the interest in C4 photosynthesis arises from the fact that it is amongst the C4 species that plants capable of producing the highest annual yields of organic material are found. Therefore, some consideration will also be given to the extent to which this productivity can be explained in terms of the basic biochemistry of photosynthesis in these plants.

8.2. C4 PLANTS

So far the terms C3 and C4 have been used without definition. The distinction has its origin in the number of carbon atoms in the first product of photosynthesis to become radioactive when leaves of a given species of higher plant are exposed to $^{14}CO_2$ in the light. In the one case this initial product is the three-carbon acid PGA. In the C4 plants the initial product is the four-carbon oxaloacetic acid (OAA). However, the distinction does not rest on this point alone. There is now considerable evidence which suggests that C4 plants can be distinguished by a number of specific physiological, anatomical and biochemical features which together make up what has been termed the C4 syndrome. Many aspects of the early work which led to this concept are discussed in the proceedings of a conference held in Canberra in 1970 [85].

High rates of photosynthesis (over 60 mg of CO_2 assimilated per dm^2 per h) and production of dry matter in the field (over 50 g dry weight per m^2 ground area per day) can result in an annual yield of over 120 tons per acre — this compares with figures of the order of 20 mg CO_2 per dm^2 per h or 30 g dry weight per m^2 for C3 plants. However, to reach this level of productivity the C4 plants require high temperatures (optimum between 30 and 40°C) and high light intensities.

The high efficiency, in assimilating CO_2 from the atmosphere, shown by C4 plants reflects a low compensation point and a resistance towards inhibition by atmospheric concentrations of oxygen. With C4 plants values recorded for the CO_2-compensation point, which represents the concentration of CO_2 reached when a leaf or plant is placed in an illuminated chamber and the atmosphere within allowed to reach an equilibrium, falls below 10 ppm [33]. In a similar manner a light-compensation point can be reached by decreasing the light intensity until assimilation and respiration again balance one another with no net gas exchange. Under these conditions light-compensation points of about 10 ft c. are obtained for C4 plants. These values compare with CO_2-compensation points of between 20 and 100 ppm, and light-compensation points of between 100 and 500 ft c. for C3 plants.

The low compensation points reached with C4 plants reflect the fact that leaves of these species do not show a measurable efflux of CO_2 in the light as found in C3 plants. This light-dependent production of CO_2 and concomitant consumption of O_2 is termed photorespiration. The exact magnitude of this phenomenon is difficult to assess, due not only to competition between respiration and photosynthesis but also to effects of illumination on dark respiration. As a result a wide range of techniques has been devised in order to quantify the "true" rate of photorespiration [101]. These include the measurement of *(a)* CO_2-compensation point at atmospheric concentrations of O_2; *(b)* effects of varying O_2 concentration on net photosynthesis or growth rate; *(c)* release of CO_2 into CO_2-free air, O_2 or N_2, (*d*) post-illumination CO_2 burst; (*e*) rates of gas exchange as heavy or isotopic flux in and out of illuminated leaves. All such techniques provide positive indications of photorespiration in C3 plants. However, similar experiments indicate that an efflux of CO_2 does not occur in the light from C4 plants [58,61,131,160].

A further physiological peculiarity of C4 plants is that they discriminate to a lesser extent against the heavy isotope of carbon (^{13}C) than do C3 plants. This is reflected in the value recorded for $\delta\,^{13}C$, which represents the difference per mille of the $^{13}C/^{12}C$ ratio of the sample of plant material relative to a standard such as a specific limestone. Values obtained for C4 plants are usually of the order of —10 to —20, whereas values recorded for C3 plants lie closer to —33 [16,154]. This effect has been used as a diagnostic feature in the identification of species of C4 plants [28,153].

In the C4 plants the photosynthetic tissue is located in two distinct concentric layers arranged around the vascular bundles. The significance of this Kranz-type (wreath-like) anatomical peculiarity has been discussed in detail in relation to the C4 syndrome in a recent review [119] which also considers many aspects of the ultrastructure of these plants. The two distinct cell layers, outer mesophyll tissue and inner bundle sheath, are found in most C4 plants — both monocotyledons and dicotyledons. However, some variations in this structure do occur [111].

In the more highly evolved C4 plants as typified by panicoid grasses such as sugar cane, sorghum or maize, the chloroplasts of the two cell layers differ. In the mesophyll layer the chloroplasts appear normal, with many well developed grana. In contrast grana are absent, or only poorly developed in the plastids of the bundle sheath. These chloroplasts, which are larger than those in the mesophyll cells, preferentially accumulate starch. The location of the chloroplasts, within the bundle sheath cells, may vary in different groups of C4 plants [75,82]. In the monocotyledons the plastids are generally arranged in a centrifugal position — with the exception of members of the *Chlorideae*, *Eragrostis*, *Sporobolus* and *Panicum*. In these latter species and in the dicotyledons such as *Atriplex*, *Amaranthus* and *Froelichia* the chloroplasts of the bundle sheath are arranged centripetally.

In general, the chloroplasts of C4 plants are characterised by the possession of a series of anastomosing tubules located in the peripheral region of the stroma. These tubules, which are apparently contiguous with the inner membrane of the chloroplast envelope, have been termed the peripheral reticulum. There have been some suggestions that a similar structure also occurs in C3 plants [72,97] although this interpretation has been questioned [119].

Various other ultrastructural features have been suggested as typical of C4 plants. These include considerations of chloroplast frequency, presence or absence of grana, presence of starch grains in the bundle sheath, and frequency of other organelles such as mitochondria and peroxisomes [25].

However, it would appear that the most constant anatomical or ultra-structural feature is the Kranz anatomy.

Many of the original observations which led to the concept of C4 plants as a separate group were made on panicoid grasses. It is now clear that C4 plants occur in many families of both monocotyledons and dicotyledons (these include *Amaranthaceae*, *Aizoaceae*, *Chenopodiaceae*, *Compositeae*, *Cyperaceae*, *Euphorbiaceae*, *Graminae*, *Nyctaginaceae*, *Portulacaceae* and *Zygophyllaceae*) and must have had a polyphyletic origin. More detailed lists of C4 species are available in a number of publications including refs. 25,75,85,102 and 153.

8.3. $^{14}CO_2$ ASSIMILATION

The occurrence of four carbon organic acids as initial products of $^{14}CO_2$ assimilation in the light was first noted during attempts to demonstrate the presence of the PCR cycle in leaves of sugar cane. These observations, and similar ones on maize, were made in the late 1950s but did not receive wide publicity prior to the publication of Kortschak et al., in 1965 [114]. Further reference to this early work can be found in refs. 22,85,89,109 and 119.

The incorporation of radioactivity into C4 compounds was not of itself a new observation. However, as reported by Kortschak [113], "Finally the facts became too clear to be denied. When radioactive CO_2 is fed to a sugar cane leaf the first stable compounds to become radioactive are malic and aspartic acids. PGA is an intermediate product appearing after these and before hexose phosphate . . . essentially all of the carbon gained by the leaf passed through malic acid then through PGA." This description pin-points the important facts. First, the C4 acids are early products of photo-synthesis and not end products as found in C3 plants. Second, this carbon fixed into C4 acids is rapidly transferred to sugar phosphates. Third, most of the carbon fixed apparently passed through this route.

Extensive subsequent investigations of $^{14}CO_2$ incorporation, largely by Hatch, Slack and their co-workers have amply confirmed these initial impressions [77,86,91,103]. In particular it was established that the first formed compound is in fact OAA, which is rapidly reduced to malate or transaminated to aspartate. After very brief periods of illumination (seconds) essentially all the fixed ^{14}C is located in C4 acids. The incorporation of radioactivity into these acids is linear with time and comparable to the initial rate of photosynthesis. On the other hand a significant lag occurs before PGA and sugar phosphates attain a steady rate of labelling. Similar results were obtained irrespective of the CO_2 concentration, light intensity or age of leaves used.

When the radioactive products of photosynthesis were degraded and the intramolecular distribution of radioactivity studied it was found that ^{14}C entered the fourth carbon of the C4 acids and the first carbon of PGA, but that the rate of incorporation into malate was faster than that into PGA. This observation suggested that carbon four of malate was transferred to the carboxyl group of PGA. This sequence was supported by radiotracer pulse-chase experiments in which leaves were transferred to $^{12}CO_2$ following a brief period of photosynthesis in $^{14}CO_2$. Under these conditions it could be shown that radioactivity moved from malate to PGA and thence through sugar phosphates to sucrose and starch. The transfer of ^{14}C from C4 acids to PGA is strictly light-dependent but independent of the concentration of $^{12}CO_2$ in the outer atmosphere. A quantitative relationship was established between the loss of ^{14}C from malate and the gain in radioactivity of sugar phosphates. This has been taken as an indication that this represents the sole route for entry of external CO_2 into PGA in C4 plants. Subsequent studies established that the fourth carbon of aspartate could also be transferred in a similar quantitative manner to PGA in some species of C4 plant [34,143]. In general more recent studies of $^{14}CO_2$ assimilation by C4 plants support these early conclusions [127]. A comparison of the size and nature of the "carbon donor pool" in *Pisum sativum* and *Zea mays* [69] led to the conclusion that PGA is derived from a small (50 nmoles per g fresh weight) volatile CO_2 pool in equilibrium with the outer atmosphere in the C3 plant but from a larger (540 nmoles per g fresh weight) organic acid pool in the C4 plant. On the other hand it has been concluded that there is an intermediate pool of carbon dioxide in leaves of C4 plants [77]. Furthermore, there is some evidence which suggests that not all CO_2 assimilated into PGA must pass through the C4 acids. In *Pennisetum purpureum* illuminated under high concentrations of $^{14}CO_2$, kinetics suggesting initial labelling of PGA were obtained [46]. Results of experiments in which light-stimulated dark fixation was studied using leaves of *Z.mays* [118] have also been interpreted as evidence that CO_2 may be fixed directly into PGA. The importance of enzyme-catalysed exchange reactions, without net fixation of CO_2, in labelling of C4 acids has also been discussed [67]. In addition one or two peculiarities have been recorded.

The formation of ^{14}C-labelled alanine, derived from pyruvate, has been observed in *Portulaca oleracea* [110]. Experiments indicated that the normal precursor-product relationship, PGA to pyruvate to alanine, did not exist since the labelling of alanine preceded that of PGA. Furthermore, pulse-chase experiments indicate that alanine was rapidly metabolised further to sugar phosphates and end products of photosynthesis. Another C4 plant *(Eleusine coracana)* has been reported [143] in which there is practically no detectable incorporation of $^{14}CO_2$ into malate during short periods of illumination, but a high degree of incorporation of label into aspartate. In this plant there was a light-dependent transfer of this radioactivity from aspartate to PGA and sugar phosphates.

The general conclusion from these radioactive labelling studies is that the four carbon acids which are labelled prior to intermediates of the PCR cycle donate their carboxyl carbon to form the carboxyl group of PGA. Detailed studies of the enzymes involved in these reactions have enabled these results to be interpreted at the biochemical level.

8.4. ENZYMES OF C4 PHOTOSYNTHESIS

8.4.1. PEP carboxylase

Phosphoenol pyruvate carboxylase (EC 4.1.1.31) catalyses the essentially irreversible reaction which results in the incorporation of inorganic carbon into OAA. Although it has been suggested that the enzyme from *Zea mays* uses CO_2 [171] it would appear that the true substrate is in fact bicarbonate [51,54].

It is now clear that a number of different forms of PEP carboxylase exist in higher plants. These can be distinguished on the basis of their elution profiles from ion-exchange resins, kinetic behaviour and response to allosteric effectors. The properties of the enzyme purified from leaves of a C4 species of *Atriplex (A.spongiosa)* differed significantly from those of the enzyme extracted from leaves of a C3 species, *A.hastata* [162]. The C4 form of the enzyme had a higher maximum reaction velocity (V_{max}) and a higher Michaelis constant (K_m), indicating a lower substrate affinity, for both PEP and the metal-ion cofactor Mg^{2+}. In other studies [71] it was shown that a form of enzyme similar to the C3 form occurred in etiolated leaves of sugar cane. This contrasted with the enzyme found in greened tissue which resembled the high K_m C4 form. As a result of these and further observations [163] a number of distinct forms of the enzyme are now recognised. These include *(i)* The C4 photosynthetic form with a high K_m for PEP (about 0.6 mM) and Mg^{2+} (0.5 mM) and a high V_{max} (about 30 μmoles min^{-1} . mg^{-1} chlorophyll); *(ii)* a C3 photosynthetic form with low K_m for PEP (0.14 mM) and Mg^{2+} (0.09 mM) and a low V_{max} (about 5% of that of the C4 form); *(iii)* a crassulacean form with a low K_m for PEP and a high V_{max}; *(iv)* a non-green form with low K_m and low V_{max}.

It is of particular interest that it has now been shown [172] that the low discrimination against the heavy (^{13}C) isotope of carbon can be demonstrated at the enzyme level. In these particular experiments it was found that the carbon atoms of glucose and malate were enriched by 2 to 3 parts per thousand in ^{12}C with respect to the $^{12}CO_2/^{13}CO_2$ fed. Enzymic synthesis of malate from PEP and HCO_3^- using enzyme from *Sorghum* resulted in fractionation of a similar magnitude. In contrast the enzymic synthesis of PGA from RBP and CO_2 in preparations from the same plants resulted in an enrichment of between 18 and 34 parts per thousand. It was concluded that the small enrichment in ^{12}C, observed with C4 plants in vivo, takes place at the PEP carboxylase step. This result supports the idea that most CO_2 enters the PCR cycle indirectly.

There is now considerable evidence to suggest that PEP carboxylase is a regulatory enzyme in C4 photosynthesis and has allosteric properties [44]. This subject is discussed in more detail in section 8.

8.4.2. Malate dehydrogenase

The NADP-specific malate dehydrogenase (EC 1.1.1.82) found in C4 plants [104] catalyses the reversible reaction:

$$L\text{-malate} + \text{NADP} \rightleftharpoons \text{OAA} + \text{NADPH}$$

Using enzyme purified from leaves of *Zea mays* it was found that the velocity in the direction of malate synthesis (pH optimum about 8.0) was ten times faster than that in the reverse direction, where the pH optimum was closer to 9.0. The enzyme has a high affinity for both OAA and NADP with K_m values for both substrates in the region of 0.01 to 0.1 mM. The enzyme has an absolute requirement for a reducing compound such as a thiol in order to demonstrate its activity. With green leaves the enzyme is inactivated by a period in the dark, but may be reactivated by illumination of the intact leaf prior to extraction. The inactive enzyme extracted from darkened leaves may be reactivated in vitro by addition of thiols to reaction mixtures.

Leaves of C4 plants also contain high levels of the NAD-specific form of the enzyme (EC 1.1.1.37) which catalyses a similar reaction. The properties of this form have also been studied in detail using enzyme purified from *Pennisetum purpureum* [48]. This enzyme also has a high affinity for both NADH and OAA with K_m values similar in magnitude to those of the NADP-specific enzyme. The kinetic characteristics and product inhibition patterns of the NAD-specific enzyme are such that malate production is favoured if an adequate supply of NADH is available.

Studies of a light-dependent carboxylation of PEP in chloroplasts of sugar cane [7] suggest that OAA may also be reduced directly to malate by reduced ferredoxin or other primary photoreductant.

8.4.3. Malic enzyme

Malic enzyme, or L-malate:NADP oxidoreductase (decarboxylating) (EC 1.1.1.40) catalyses the reaction:

$$\text{L-malate} + \text{NADP} \overset{K}{\rightleftharpoons} \text{pyruvate} + CO_2 + \text{NADPH}$$

This enzyme is highly specific for NADPH, but will also catalyse the decarboxylation of OAA. The kinetics and other properties have been studied in some detail [48,104]. The K_m for malate varies with the pH of the reaction mixture and the pH optimum varies with malate concentration. Although it has been suggested that the preferred metal cofactor is Mg^{2+} [104] other workers have found greater activity with Mn^{2+} [48,180]. Although the enzyme from C3 plants may have allosteric properties this is not true for the enzyme isolated from members of the *Graminae* [56].

Not all C4 plants contain high levels of this enzyme. Activity is higher in extracts of C4 plants which belong to the *Andropogoneae*, *Maydeae*, *Paniceae* and *Aristideae* amongst monocotyledons, and to species of *Froelichia* amongst the dicotyledons [75].

In other species of C4 plants such as *Amaranthus*, *Atriplex* and some *Panicums*, higher levels of the NAD-specific form of malic enzyme (EC 1.1.1.39) are found [79,80]. This enzyme requires NAD and Mn^{2+} ions and may be stimulated by addition of CoA or acetyl-CoA to assay mixtures. The kinetics of activation have been studied in detail [84].

8.4.4. Pyruvate P_i dikinase

This enzyme (EC 2.7.9.1) which catalyses the regeneration of PEP from pyruvate was first reported in higher plants from leaves of C4 species [87]. The reaction catalysed is as follows:

$$\text{ATP} + \text{pyruvate} + \text{P}_i \rightleftharpoons \text{AMP} + \text{PEP} + \text{PP}_i$$

The suggested reaction mechanism [3] is more complex, involving the formation of an intermediate phosphorylated enzyme which then reacts with pyruvate to form PEP. When plants are placed in the dark prior to extraction the levels of dikinase drop, the activity is restored by subsequent illumination [88,147].

The enzyme isolated from sugar cane was found to be stable at pH 8.3 only if stored at 20° C in the presence of Mg^{2+} and a thiol. At this optimum pH the initial velocity in the direction of PEP synthesis was about six times that of the reverse reaction. The other reaction products (AMP and PP_i) are removed by the activities of pyrophosphatase and adenylate kinase. Both of these enzymes have been shown to be active in extracts of leaves of C4 plants.

8.4.5. PEP carboxykinase

This enzyme (EC 4.1.1.49), which only occurs at significant levels in some species of C4 plants including *Panicum*, *Chloris* and *Sporobolus* [65,75,82], catalyses the decarboxylation of OAA with the formation of PEP:

$$OAA + ATP \rightleftharpoons PEP + CO_2 + ADP$$

8.4.6. Amino transferase

Two amino transferases, aspartate (EC 2.6.1.1) and alanine (EC 2.6.1.2) are believed to be of importance in some C4 species. The reactions catalysed are:

L-aspartate + 2 oxoglutarate ⇌ OAA + L-glutamate
L-alanine + 2 oxoglutarate ⇌ pyruvate + L-glutamate

Isoenzymes of both activities have been partially purified from leaves of *Atriplex spongiosa* [78,83]. The identities of these enzymes, separated by ammonium sulphate fractionation and ion-exchange chromatography, was established by comparing mobilities of various fractions on acrylamide gels. Two isoenzymes could be distinguished for both aminotransferases. The reaction velocities, activities and apparent locality of these suggested that the enzymes in the mesophyll layer convert OAA to aspartate and alanine to pyruvate, whereas the isoenzymes in the bundle sheath convert aspartate to OAA and pyruvate to alanine.

8.4.7. Carbonic anhydrase

This enzyme (EC 4.2.1.1) catalyses the reversible hydration of carbon dioxide:

$$H^+ + HCO_3^- \rightleftharpoons CO_2 + H_2O$$

Levels in C4 plants are similar to those in C3 plants [141] and apparently associated with the chloroplasts in both mesophyll and bundle sheath cells [166].

8.4.8. RBP carboxylase and PCR cycle activity

Original estimates of the levels of RBP carboxylase suggested that C4 plants contained low levels of this enzyme [149]. Subsequently, better extraction procedures have established that C4 plants do in fact contain sufficient levels of this, and other enzymes of the PCR cycle, to support the observed rates of photosynthesis [4,21]. In general the properties of these enzymes resemble those from C3 plants, although it has been suggested that some PCR cycle enzymes (G3P-kinase and ribose-5-P kinase) differ in their response on light activation [157].

8.5. CELLULAR DISTRIBUTION STUDIES

8.5.1. Techniques and enzymes

All the more detailed schemes concerning the movement of carbon compounds in C4 photosynthesis have been devised on the basis of the recorded activities of certain enzymes which can be extracted from the leaf tissue, and attributed to a particular cellular or sub-cellular location. These studies can be divided into three areas. First, those investigations carried out before it had been demonstrated that C4 plants contained levels of RBP carboxylase sufficient to support photosynthesis [148,150]. Second, experiments in which differential or progressive grinding techniques were used to selectively release the cell contents of specific tissues [9,17,21,29]. Third, experiments in which mesophyll cells or protoplasts were separated from bundle sheath cells by mechanical means, with or without prior enzyme digestion, and the resulting cell fractions separated and purified by filtration through graded sieves, or on density gradients [63,64,66,107,108].

Results of experiments using this last technique will be considered first. Particular emphasis has been placed on such results since it is claimed that by these means very pure fractions of the contents of the two types of photosynthetic cell can be obtained with little or no cross contamination [22]. This contrasts with preparations obtained using differential grinding techniques in which each fraction is merely enriched in the contents of a given cell type.

Enzymic digestion of leaves with 2% cellulase plus pectinase for several hours yields intact protoplasts from mesophyll cells mixed with epidermal tissue, vascular strands, broken protoplasts and chloroplasts [107]. The cell fragments can be removed using filters of different porosity, and intact protoplasts harvested at the interphase of a dextran polyethylene-glycol liquid/liquid density system [108]. When this technique was applied to leaves of *Zea mays* both light and electron microscopy, and measurement of the ratio of chlorophyll *a* to chlorophyll *b* (about 3.1:1 in the mesophyll protoplasts and around 6.3:1 in the bundle sheath strands) suggested a clear separation of the two tissues. High levels of PEP carboxylase and NADP-malate dehydrogenase were observed in the mesophyll preparations, whereas the bundle sheath preparations contained high activities of RBP carboxylase, phosphoribulosekinase and NADP-malic enzyme. G3P-dehydrogenase was recorded in both fractions. These results were consistent with a number of previous observations obtained using non-aqueous [148,150], differential grinding [17,21,79,99] or cell separation [35,36] methods and have been supported by more recent observations [76,116]. On the other hand, results of experiments using the differential grinding procedure have not always been so clear-cut. In particular results from our laboratory [9,29,30] have been interpreted [42,43] as evidence that RBP carboxylase is not rigidly confined to the bundle sheath cells, a conclusion supported by other work of a similar nature [141].

Results of studies on enzyme distribution, coupled to observations on $^{14}CO_2$ assimilation patterns, have led to suggestions [58] that C4 plants can be divided into two groups — malate and aspartate formers. The first group consists of those species in which malate is labelled preferentially on exposure to $^{14}CO_2$ and in which NADP-malic enzyme is the most active of the decarboxylating enzymes. The occurrence of extreme dimorphic chloroplasts, in the two cell layers, is restricted to C4 plants from this group. It is generally accepted that these plants possess high levels of PEP carboxylase, pyruvate P_i dikinase, NADP-malate dehydrogenase, and G3P dehydrogenase in the mesophyll cells. Malic enzyme, and all the enzymes of the PCR cycle occur in the bundle sheath layer. Enzymes of starch and sucrose synthesis can be demonstrated in both cell layers [30,60].

It was originally suggested that PEP carboxylase was located within mesophyll chloroplasts. However, initial suggestions [9,29,44] that this enzyme is cytoplasmic are now generally accepted [79,159]. Although the C4 PEP carboxylase is cytoplasmic it can be regarded as a photosynthetic enzyme since levels are low in etiolated tissue and increase in parallel with chloroplast enzymes on illumination [38,73,90].

8.5.2. *Photochemical activity of chloroplasts with reduced grana*

The photochemical activity of chloroplasts isolated from C4 plants has also been investigated in some detail. Particular attention has been paid to the properties of the bundle sheath chloroplasts from those species in which these plastids contain reduced grana. These chloroplasts show higher ratios of chlorophyll *a* (chl_a) to chlorophyll *b* (chl_b) and of P_{700} to chlorophyll than those found in the mesophyll cells [24,31,98,175]. They may also have a higher capacity for cyclic photophosphorylation [142]. In general results indicate that bundle sheath chloroplasts are deficient in photosystem II activity [2,19,59,174] and that this deficiency may develop during the ontogeny of the plastids [62]. Estimates of the actual level of photosystem II activity in bundle sheath chloroplasts vary from about 50% of that in mesophyll chloroplasts [15,20] to less than 30% [129] or as low as 4% [117]. In a detailed survey of a number of C4 plants [128] it was concluded that the mesophyll cells contained 95%, 90% and 80% of the total leaf photosystem II activity in sorghum, maize and *Digitaria* respectively, although these cells contained only about 60% of the total leaf chlorophyll. The ratios of chl_a to chl_b were close to 4 for the whole leaf, this represented the average between values of 3 to 4 for mesophyll protoplasts and values close to 6 for bundle sheath cells. The level of P_{700} was also higher. Rates of Hill reaction were faster with mesophyll chloroplasts studied using a variety of non-physiological electron acceptors such as *p*-benzoquinone, dichlorphenolindophenol or ferricyanide [117]. In general 20 mM HCO_3^- was not an effective electron acceptor, although rates of about 14 μmoles . mg chl^{-1}. h^{-1} O_2 evolution were obtained with mesophyll protoplasts from *Panicum capillare*. This compared with rates of 53 μmoles . mg chl^{-1}. h^{-1} for bundle sheath cells from the same species. However, it should be noted that the chloroplasts of this species possess normal grana.

With bundle sheath chloroplasts isolated from maize, results may be complicated by the loss of soluble components of the electron transport system during preparation. In a series of studies [1,19,152] using bundle sheath cells or chloroplasts isolated from leaves of *Zea mays* at various stages of development it was found that the complete electron transport chain could be demonstrated.

The bundle sheath chloroplasts were found to be capable of reducing ferricyanide, dichlorophenol, cytochrome *c*, plastocyanin and cytochrome b_{552} in the light. These reductions were inhibited by the specific inhibitor of photosystem II 3-(3,4-dichlorophenyl)-1,1-dimethyl urea (DCMU). Rates of photoreduction of NADP from water were low, except with bundle sheath chloroplasts isolated from very young leaves. On the other hand, photosystem I activity was observed in chloroplasts from leaves at all stages of development. Further investigations on intact bundle sheath cells indicated that electron flow could occur between system II and cytochrome *f* resulting in reduction of the cytochrome. This cytochrome could also be oxidised by system I, suggesting that the chloroplasts should be capable of the complete non-cyclic electron transport. This could in fact be demonstrated by the addition of soluble components of the electron transport chain (plastocyanin and ferredoxin) and ferredoxin/NADP reductase to reaction mixtures. On a chlorophyll basis the photosystem I activity of these chloroplasts was about twice that of the mesophyll chloroplasts.

8.5.3. Partial reactions in mesophyll and bundle sheath cells

Complete photosynthesis, that is stoichiometric CO_2 fixation with production of O_2 and sugar phosphates, has not generally been demonstrated using chloroplasts, protoplasts, or isolated cells from either type of photosynthetic tissue. Inconsistent with this statement are results of studies with mesophyll chloroplasts isolated from young leaves of *Z.mays* [70,95,136] which indicate that they do not differ significantly from those of C3 plants such as spinach. Chloroplasts isolated from 4- to 6-day-old seedlings of *Z.mays* were capable of a light-dependent assimilation of CO_2 at rates of up to 45 μmoles . mg chl^{-1} . h^{-1}. These rates were consistent with the observed levels of RBP carboxylase present in the preparations. The nature of the products of photosynthesis and the response to added co-factors, intermediates of the PCR cycle, O_2 and inhibitors of photosynthesis were all similar to those previously observed with spinach chloroplasts under the same conditions. These chloroplasts were not capable of metabolising malate. Normal C3 type photosynthesis has also been observed in sugar cane stalk parenchyma tissue culture [115].

A number of other significant partial reactions of C4 photosynthesis have been demonstrated. Addition of PEP or pyruvate will stimulate CO_2 fixation by chloroplasts [7] or protoplasts [76] in vitro. High rates of conversion of pyruvate to PEP have been recorded using maize mesophyll chloroplasts [79,106]. Rates were increased when PGA, OAA or methyl viologen were included in reaction mixtures. PGA and OAA were converted to DHAP and malate with a concomitant evolution of oxygen. The rate of reaction was not altered by the addition of ADP or NADP. However, when photophosphorylation was uncoupled by addition of methylamine the reduction of OAA was stimulated and the conversion of PGA to DHAP inhibited. Again, no evolution of O_2 could be demonstrated on addition of bicarbonate, nor was there evidence for CO_2 fixation by these chloroplasts. The photoreduction of PGA has also been demonstrated using mesophyll cells isolated from *Digitaria sanguinalis* [146]. The production of O_2 in the coupled reaction was inhibited by DCMU and uncouplers of photophosphorylation. These results suggest that PGA can act as a Hill oxidant, using NADPH in the reduction of 1,3 diPGA in the reaction catalysed by G3P dehydrogenase. Photophosphorylation is needed for the production of ATP used in the conversion of the fed PGA to diPGA.

Carbon dioxide assimilation has been recorded using bundle sheath strands from maize. This CO_2 assimilation can be stimulated by addition of ribose-5-P and ATP under which conditions rates of dark CO_2 fixation of about 70 μmoles . mg chl^{-1}. h^{-1} are observed [37]. This compares with light-driven rates of about 8 μmoles . mg chl^{-1}. h^{-1} [41].

A number of studies also indicate that the bundle sheath cells can catalyse the decarboxylation of malate. Preparations of cells isolated by grinding and filtration from leaves of *Z.mays*, *Sorghum bicolor* and *Digitaria sanguinalis* were found to catalyse the decarboxylation of malate from the C4 carboxyl position [57,100,105]. These cells, which have a high content of malic enzyme, require the addition of Mg^{2+} or Mn^{2+} and NADP for maximum activity. Evidence has also been produced which suggests that the reducing equivalents formed during this reaction can be used in the assimilation of CO_2 [138].

Indirect evidence which suggests that some of the enzymes of the C4 cycle and of the PCR cycle are spatially separated comes from studies of the selective inhibition of mesophyll chloroplast development by low night temperatures [151]. When plants of *Sorghum*, *Digitaria* or *Paspalum* were exposed to temperatures of below 4° C for a single night a transverse irreversibly chlorotic band of tissue developed within 36 h at the base of the leaf. This chlorosis was associated with the presence of chlorophyll-deficient, structurally abnormal plastids in the mesophyll cells. On the other hand the chloroplasts in the adjacent bundle sheath cells were green and structurally normal. The ultrastructure of other organelles in the chlorotic cells also appeared normal. Levels of cytoplasmic rRNA, of non-plastid lipids and of isocitrate dehydrogenase were similar to those in the normal portion of the leaf. The extractable activities of the enzymes of the C4 pathway (PEP carboxylase, NADP malate-dehydrogenase and adenylate kinase) were reduced to a much greater extent than those of the PCR cycle (RBP carboxylase and fructose bisphosphate aldolase) and that of the C4 cycle believed to be located in bundle sheath cells (NADP-malic enzyme). Further studies [158] indicated that the light-activated enzymes of the C4 pathway (NADP-malate dehydrogenase and pyruvate P_i dikinase) are most sensitive to low temperatures. Mesophyll cells have also been shown to be more prone to damage by water stress than are bundle sheath cells [74].

The general conclusion from these studies of malate-forming C4 plants is that neither the mesophyll cells, nor the bundle-sheath cells are capable of the complete process of photosynthesis. Chloroplasts in the mesophyll layer are deficient in carboxylation capacity (lack of RBP carboxylase), whereas chloroplasts in the bundle sheath cells are deficient in photochemical capacity (low photosystem II).

8.5.4. Aspartate formers

In the second group of C4 plants to be considered, the aspartate formers, the chloroplasts have well developed grana, irrespective of their origin, in bundle sheath or mesophyll cells [75,82]. The photochemical characteristics of these plastids have been studied in detail [117,128]. In some species the bundle sheath cells contain over 50% of the total leaf chlorophyll. However, there is little difference in the Chl_a to Chl_b ratio found in the mesophyll cells as compared with that determined for bundle sheath chloroplasts. The bundle sheath chloroplasts apparently resemble plastids from C3 plants, showing high rates of oxygen evolution in the Hill reaction with benzoquinone as electron acceptor. Over 50% of the total leaf capacity for photochemical reduction of NADP is located in the bundle sheath cells, which also contain high levels of RBP carboxylase and other enzymes of the PCR cycle.

Hence, it can be concluded that the bundle sheath cells from aspartate formers are essentially normal in respect to both electron transport capacity and levels of carboxylation enzymes and should thus be capable of complete photosynthesis. This has been indicated in experiments using bundle sheath isolates from leaves of *Panicum capillare* [76]. These were capable of assimilating CO_2 in a light-dependent reaction at rates of over 100 μmoles . mg chl^{-1}. h^{-1}.

It is generally accepted that in the aspartate formers PEP carboxylase is located in the cytoplasm of the mesophyll cells. The main points of interest in enzymes of these C4 plants concern the levels and distribution of aminotransferases, and the nature of the most active enzyme-catalysing decarboxylation of a C4 acid. In particular these plants have high levels of alanine and aspartate aminotransferase [5,64], low levels of NADP-malic enzyme and high levels of the NAD-specific form [80] or of PEP carboxykinase [65].

Detailed investigations have been carried out using the dicotyledons *Amaranthus edulis* and *Atriplex spongiosa*, and the monocotyledon *Panicum miliacium* [64,83]. Levels of the two aminotransferases recorded in extracts from leaves of these species are about 20 times those found in leaves of malate-forming C4 plants or of C3 plants. The total activity of both enzymes is distributed equally between the mesophyll cells and the bundle sheath layer. However, the activity in the two cell layers is apparently catalysed by different isoenzymes [78]. In the mesophyll layer these enzymes are located in the cytoplasm, whereas in the bundle sheath the aspartate aminotransferase isoenzyme is located in mitochondria, and the alanine aminotransferase is again cytoplasmic. Investigations of other enzymes in these plants suggested that the mesophyll cells contained all of the leaf pyruvate P_i dikinase and PEP carboxylase activity and the major proportion of the adenylate kinase and pyrophosphatase. Enzymes of the PCR cycle (PGA kinase, NADP, G3P dehydrogenase and triose Pi isomerase) were distributed equally between the two cell layers or confined to the bundle sheath cells (RBP carboxylase, ribose-5-P kinase, ribose-5-P isomerase, fructose bisphosphate aldolase and alkaline fructose bisphosphatase).

A number of C4 aspartate formers, such as *Panicum maximum*, *P.texanum*, *Sporobolus poirettii*, *S.fimbriatus*, and *Chloris gayana* contain high levels of PEP carboxykinase which is located predominantly in the bundle sheath cytoplasm [65,75,82]. Decarboxylation of aspartate by this enzyme of course requires the prior deamination of the C4 amino acid to form OAA. In other C4 plants it is suggested that OAA is reduced to malate in the bundle sheath mitochondria and then decarboxylated in the reaction catalysed by a NAD-specific malic enzyme [80]. Species with high levels of this enzyme include *Atriplex spongiosa*, *Amaranthus edulis*, *Panicum miliaceum*, *Eleusine indica* and *Portulaca oleracea*. The most detailed studies of the properties, activity, cellular and subcellular distribution of NAD-specific malic enzyme have been carried out using the first three of these species [81,84]. The enzyme is most active with NAD and Mn^{2+} and can be stimulated from 5- to 15-fold by low concentrations of CoA or acetyl-CoA. Practically all the activity, which was low in dark-grown leaves and increased on illumination, was located in the bundle sheath cells. During illumination the increase in this activity paralleled that of the photosynthetic enzyme phosphoribulokinase, whereas the levels of the non-photosynthetic NAD-malate dehydrogenase remained fairly constant. The activity sedimented in a particulate fraction which also contained the aspartate aminotransferase and mitochondrial enzymes. The conclusion that the NAD-malic enzyme is located in mitochondria was correlated with the fact that the bundle sheath cells of these species contain a large number of prominent mitochondria as compared with C3 and other types of C4 plants [80—82].

Further studies with *A.spongiosa* relate to the functional capacities of various fractions from mesophyll cells and bundle sheath strands. When mesophyll chloroplasts were illuminated with $[^{14}C]$-PGA [106] label rapidly appeared in DHAP but was not detected in hexose phosphate, sucrose or starch. However, low rates of starch formation could be demonstrated if these plastids were provided with glucose-1-phosphate. Both PGA and OAA stimulated the production of O_2 by these plastids. Photoreduction of OAA was apparently mediated by the NAD-dependent malate dehydrogenase since levels of the NADP-specific form were low in this plant. Addition of PGA to these plastids also stimulated the light dependent conversion of pyruvate to PEP while the addition of pyruvate stimulated the conversion of PGA to DHAP and the evolution of O_2. The stimulation of PGA and OAA-dependent evolution of O_2 by pyruvate can be explained in terms of increased utilisation of ATP in the pyruvate P_i dikinase reaction stimulating the rate of coupled non-cyclic electron transport.

With bundle sheath cells of *A.spongiosa*, capable of assimilating ^{14}C from bicarbonate into intermediates of the PCR cycle at rates of over 100 μmoles . mg chl^{-1} . h^{-1}, addition of malate stimulated similar rates of evolution of O_2 [105]. Both bundle sheath cells and isolated mitochondrial particles were capable of catalysing the decarboxylation of malate; at the same time the C4 carboxyl group of malate was rapidly assimilated into intermediates of the PCR cycle. The highest rates of decarboxylation occurred when malate was added in the presence of aspartate and oxoglutarate. Assimilation of radioactivity from aspartate, into products of photosynthesis, by bundle sheath chloroplasts isolated from leaves of *Eleusine coracana* has also been reported [143].

8.5.5. Sub-groups of C4 plants

On the basis of these studies coupled to observations of ultrastructure, it has been suggested that C4 plants should be divided into three groups, rather than into malate and aspartate formers [75,82]. These groups are based on the identity of the most active extractable decarboxylating enzyme present in leaves of a given C4 species. The groups, and suggested abbreviated designation, are as follows:

Group I. NADP-ME — high levels of NADP-malic enzyme
II. NAD-ME — high levels of NAD-malic enzyme
III. PCK — high levels of PEP carboxykinase

As will be apparent from the above discussion of enzyme levels malate formers represent group I, whereas aspartate formers are found in groups II and III.

8.6. PATHWAYS OF CARBON FLOW

The above studies of the distribution of enzyme activities and partial reactions of photosynthesis in C4 plants have resulted in the formulation of a number of complex descriptions of the pathways of carbon assimilation. However, the essential reactions are in fact very simple. The complete cycle consists of the four steps carboxylation, reduction, decarboxylation and substrate regeneration. This sequence of reactions does not result in a net assimilation of CO_2 into organic matter. Net CO_2 fixation is achieved through the refixation of CO_2 in the reaction catalysed by RBP carboxylase. Hence, a common feature of the proposed schemes is that they consider the C4 pathway as a mechanism for concentrating atmospheric CO_2 at the site of reductive re-assimilation through the PCR cycle.

The primary carboxylation reaction is always that catalysed by a cytoplasmic PEP carboxylase located in the mesophyll cells, and the secondary carboxylation, that catalysed by RBP carboxylase, is located in the bundle sheath chloroplasts. A further reaction common to all C4 plants is the regeneration of PEP by pyruvate P_i dikinase in the mesophyll chloroplasts. The differences lie in the nature of the carbon compounds translocated from one cell layer to the other, and in the nature of the most active decarboxylation reactions.

8.6.1. Malate formers

Since much of the early work on photosynthesis in C4 plants was carried out using the large economically important tropical grasses, such as sugar cane and maize, proposals for these plants are best known and have at times been regarded as "the pathway of C4 photosynthesis". However, using the classification of C4 plants as detailed above such schemes would apply to group I (NADP-ME) malate formers. The suggested flow of carbon in these species is the least complex, so it is both simplest and historically valid to consider this first.

For this group of C4 plants the overall sequence of reactions required for the synthesis of a molecule of starch or sucrose from atmospheric CO_2 will be as follows:

(i) CO_2 is assimilated by PEP carboxylase in the cytoplasm of the mesophyll cells with the production of OAA; *(ii)* OAA is reduced to malate in the NADP-dependent reaction in the mesophyll chloroplasts; *(iii)* Malate is

transported to the bundle sheath cells or chloroplasts where it is decarboxylated; *(iv)* The resultant CO_2 is re-assimilated by RBP carboxylase in the bundle sheath chloroplasts; *(v)* The resultant PGA is reduced in part to triose-P, using either photochemically produced NADPH or NADPH produced during the decarboxylation of malate; *(vi)* The rest of the PGA formed in the reaction catalysed by RBP carboxylase is transported back to the mesophyll layer and reduced to triose-P using photochemically generated NADPH produced in mesophyll chloroplasts; *(vii)* Pyruvate, the second product of the NADP-malic enzyme activity in the bundle sheath, is also transported back to the mesophyll chloroplasts where it is used in the regeneration of PEP in the reaction catalysed by pyruvate P_i dikinase; *(viii)* Triose-P produced in the mesophyll chloroplasts is transported back to the bundle sheath cells where it will contribute to the formation of hexose-P; *(ix)* The hexose-P may be used for the synthesis of starch in the bundle sheath chloroplasts, or the synthesis of sucrose in the bundle sheath cytoplasm; *(x)* Alternatively, the hexose-P may be transported back to the mesophyll cells where starch and sucrose are again formed in chloroplasts and cytoplasm respectively.

8.6.2. Function of mesophyll cells

The suggested reaction sequences for group II and group III C4 plants are essentially similar except that carbon is transported to the bundle sheath in the form of aspartate, and returned as alanine. Hence the overall reactions which are required in the mesophyll cells are as shown in Fig. 8.1.

This requires the intercellular transport of considerable quantities of aspartate, alanine, pyruvate, malate, PGA, triose-P, hexose-P and sucrose at high rates, in opposing directions through the cell wall between the mesophyll and bundle sheath cell. Transport of pyruvate, malate, PGA, triose-P and hexose-P, in and out of the mesophyll chloroplast, is also required.

The return of PGA to the mesophyll chloroplasts is suggested for C4 plants from group I since the deficiency in photosystem II in the bundle sheath chloroplasts prevents reduction of all the PGA at the site of production. The return of PGA is required in the aspartate formers (group II and III) since the cytoplasmic aminotransferase reactions of the mesophyll cell do not require photochemical reductant. Hence, in the absence of PGA derived from the bundle sheath cells these chloroplasts would have no function other than regeneration of PEP from pyruvate, which could of course be accomplished using ATP generated in cyclic photophosphorylation. Such a limited function is inconsistent with the demonstrated photochemical activities of these plastids. Transport back of hexose-P is required since the mesophyll cells can accumulate high concentrations of starch in

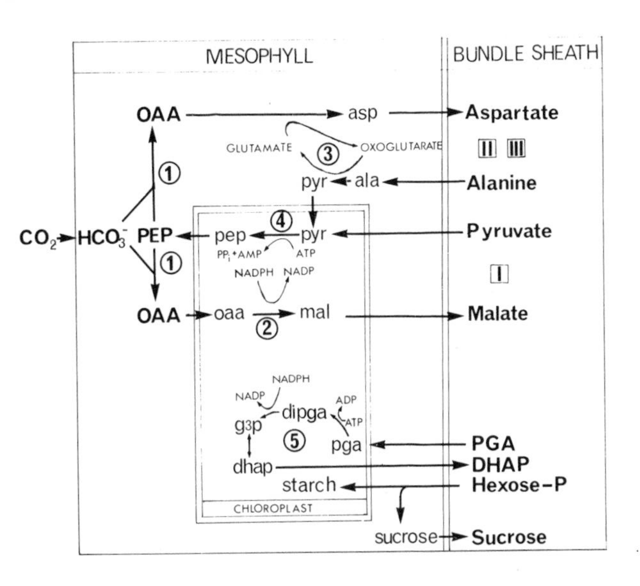

Fig. 8.1. Partial reactions of C4 photosynthesis located in the mesophyll cells of malate (group I) and aspartate (group II and III) formers. Reactions are catalysed by (1) PEP carboxylase; (2) NADP-specific malate dehydrogenase; (3) aspartate and alanine amino-transferases; (4) pyruvate P_i dikinase; (5) triose-P dehydrogenase.

their chloroplasts, and furthermore contain over 50% of the enzymes associated with the synthesis of sucrose.

Although there is no information available concerning movement of metabolites between either cellular or sub-cellular compartments in C4 plants, the requirements for transport of materials in and out of the chloroplasts does not present a problem if these chloroplasts have properties similar to those of C3 plants. In the C3 plants both dicarboxylate and phosphate translocator sites can be recognised in the chloroplast envelope. A detailed account of these and related phenomena can be found in Chapter 6 of this volume and also in recent reviews [92,169]. Briefly, the decarboxylate translocator is capable of moving malate, OAA, aspartate and glutamate across the plastid membrane at rates of about 100 μmoles . mg chl^{-1} . h^{-1}. The phosphate translocator moves P_i, PGA or DHAP at similar rates. However, it would appear that hexose-P moves at a lower rate. Traditionally carbon moves between cells as sucrose, or a similar inert carbohydrate. There is, as yet, no significant demonstration of the transport of other metabolites between cell layers in C4 plants.

8.6.3. *Function of bundle sheath cells*

Further complications arise when the pathways of carbon flow in the bundle sheath cells are considered for C4 plants belonging to groups II and III. The reactions are shown schematically in Fig. 8.2, which also shows the details of the more direct route of malate decarboxylation suggested for the C4 plants of group I. In both types of aspartate formers the C4 amino acid is transaminated to give OAA — this reaction occurs in mitochondria. In the group II plants this OAA is then reduced to malate by the mitochondrial NAD-malate dehydrogenase using reductant (NADH) generated during the subsequent decarboxylation reaction catalysed by the mitochondrial NAD-malic enzyme. In the group III plants OAA is decarboxylated directly by a cytoplasmic PEP carboxykinase. The fate of the resultant PEP is not clear. However, it has been assumed that this is converted to pyruvate and then transaminated to alanine. No clear suggestions have been made concerning reasons why malate should not be transported from the mesophyll layer to the bundle sheath cells in these plants of group II and III. In the same way it is not clear why PEP generated in the carboxykinase reaction should not pass back to the mesophyll layer, except that this would leave the mesophyll plastids with no obvious function.

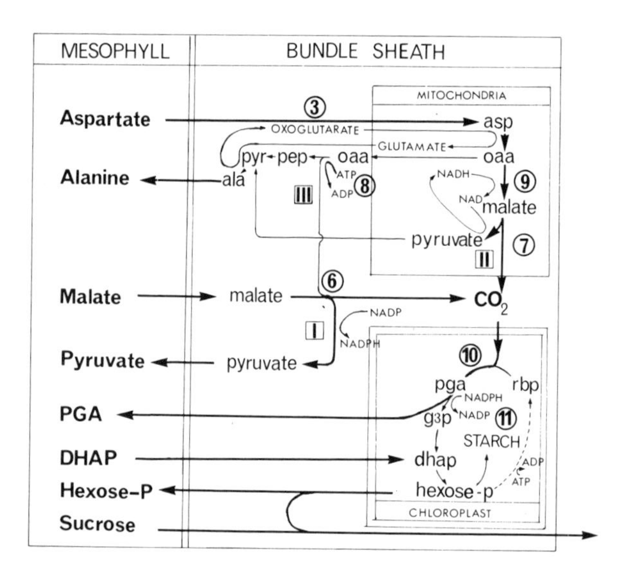

Fig. 8.2. Partial reactions of C4 photosynthesis located in the bundle sheath cells of malate (group I) and aspartate (groups II and III) formers. Reactions are catalysed by (3) aspartate and alanine aminotransferases; (6) NADP-specific malic enzyme; (7) NAD-specific malic enzyme; (8) PEP carboxykinase; (9) NAD-malate dehydrogenase; (10) RBP carboxylase; (11) PCR cycle.

These suggestions again require considerable movement of carbon compounds between chloroplasts, cytoplasm and additionally mitochondria. The remarks above concerning such movement in mesophyll cells apply equally well here. However, there is some evidence for movement of carbon from mitochondria to the chloroplasts [32]. It was found that a large light-dependent transfer of label from intermediates of the tricarboxylic acid (TCA) cycle, to photosynthetic products, was a feature of C4 plants. This transfer was inhibited by TCA cycle inhibitors such as malonate and fluoroacetate. Using specific labelled [^{14}C]succinate it was confirmed that malate and aspartate, used in photosynthesis, were derived from the TCA cycle and not from re-assimilated respired CO_2. It has also been suggested that the light-dependent movement of Cl^- ions observed in C4 plants [126] may be associated with transport of intermediates of the C4 pathway.

8.7. AN ALTERNATIVE HYPOTHESIS

In general the pathways of carbon flow discussed above have been proposed on the basis of the levels of activities, of a small number of enzymes, which have been recorded in vitro using disrupted leaf tissue. Above all these schemes rest on the conclusion that mesophyll chloroplasts do not contain RBP carboxylase. Acceptance of the demonstration of significant levels of this enzyme in the mesophyll chloroplasts makes the above complex pathways unnecessary.

The possibilities that endogenous inhibitors present in leaves of C4 plants can lower the apparent activities of extractable enzymes have been considered in some detail in the past [8,10] and so will not be reconsidered here. On this basis we have suggested [9,29,30] that mesophyll chloroplasts do contain RBP carboxylase. A similar conclusion has been reached on the basis of other studies [70,115,120,141]. Such conclusions have been subject to detailed criticism [22,116] on the basis of the low observed rates and the possibilities of cross contamination in the differential grinding procedures used. However, if the mesophyll cells do contain RBP carboxylase then the C4 pathway would act as a shuttle mechanism between cytoplasm and chloroplasts of the same cell, as shown in Fig. 8.3.

This figure follows the initial suggestions along these lines [9,42] which assumed that the PCR cycle activity was mainly associated with the mesophyll cells and that bundle sheath chloroplasts functioned mainly as starch storage organs. This conclusion has again been criticised [22] on the basis of the demonstration of RBP carboxylase activity in the bundle sheath of C4 plants from all three groups. On the other hand, as discussed in detail in a recent review [119], a similar conclusion has been reached previously on the basis of physiological observations. Hence this scheme can represent an alternative hypothesis for the pathway of carbon assimilation in malate formers.

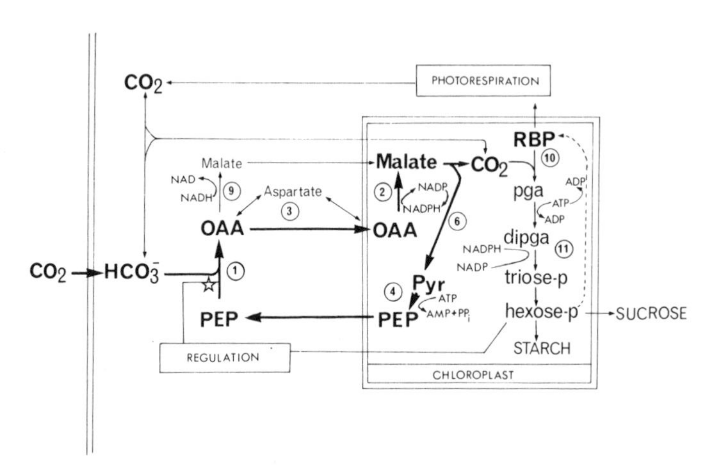

Fig. 8.3. An alternative hypothesis for the pathway of carbon in C4 photosynthesis. Reactions are catalysed by (1) PEP carboxylase; (2) NADP-specific malate dehydrogenase; (3) aminotransferases; (4) pyruvate P_i di-kinase; (6) NADP-malic enzyme; (9) NAD-malate dehydrogenase; (10) RBP carboxylase; (11) PCR cycle. ☆ Feedback regulation of PEP carboxylase by the end product of the PCR cycle G6P.

It is now clear that in many species of C4 plants, particularly those of group II and III, the bundle sheath cells do have a high capacity for CO_2 assimilation in vitro. It is possible that the extent of this activity, in vivo, reflects the distribution of the other photosynthetic activities.

These suggestions do not require a stoichiometric flow of carbon from the C4 cycle to the PCR cycle, but assume some direct fixation of CO_2 by RBP carboxylase. It is possible that the main function of the C4 cycle is not as a CO_2-pump, but rather as a CO_2-trap preventing the loss of photorespiratory CO_2 from the leaf. As such, the operation of the C4 pathway would be of much greater importance at low concentrations of atmospheric CO_2. Under these conditions the relative activities of the two carboxylating enzymes (PEP carboxylase and RBP carboxylase) would be controlled by a feedback mechanism mediated through the end products of photosynthesis such as glucose-6-phosphate (G6P). There is now considerable evidence which suggests that such control mechanisms do exist.

8.8. METABOLIC REGULATION

The initial observations which suggested that PEP carboxylase is a regulatory enzyme were made independently in three laboratories [44,161, 173]. In general these, and other reports [47,124,134] indicate that the C4 form of PEP carboxylase differs kinetically from the C3 form as found in plants such as spinach [130]. In general it would appear that the C4 form is activated by sugar phosphates but inhibited by organic acids. The pattern of inhibition by $SO_3{}^{2-}$ ions observed with the enzyme from *Z.mays* [179] also differs from that observed with the C3 form from spinach [133].

In particular the effects of G6P, on PEP carboxylase from the C4 plant *Pennisetum purpureum*, have been studied in detail [47]. The initial velocity of incorporation of ^{14}C from bicarbonate into OAA could be modified over a wide range by inclusion of G6P in reaction mixtures. When the concentration of G6P exceeded that of PEP the initial velocity of reaction tended towards zero. Activity reached a maximum under conditions where the concentration of G6P equalled that of PEP, and declined to a level similar to that of control samples when the concentration of PEP exceeded that of G6P. These results were interpreted in terms of the allosteric activation of the enzyme by G6P.

Although a number of other sugar phosphates and phosphorylated compounds have been reported as modifying the activity of PEP carboxylase in vitro it has been concluded that this may be caused in part by the interaction between the apparent effectors and Mg^{2+} ions [45]. Similar conclusions have been reached [50] concerning the effects observed with adenylates and energy charge, where energy charge = ATP + 0.5 ADP / AMP + ADP + ATP. On the other hand it has been suggested that regulation by energy charge is of importance in vivo [173]. This suggestion was based on results of experiments in which it was found that the activity of PEP carboxylase decreased as charge was increased. As the energy charge will in fact increase on illumination these results are inconsistent with the postulated role, and light activation of the enzyme in C4 plants. Effects of adenylates were therefore re-investigated in detail [50].

In general, addition of adenylates decreased the observed enzyme activity in vitro which was consistent with other observations [134,161] using the enzyme extracted from C4 plants. The inhibitory effects could be overcome by increasing the concentration of Mg^{2+} in reaction mixtures, or by the addition of the allosteric effector G6P. Response to energy charge was investigated in detail, using either calculated concentrations of all adenylates, or mixtures of AMP and ATP pre-incubated with adenylate kinase. Inhibition at high charge could be demonstrated, but this was reduced in magnitude when the concentration of Mg^{2+} in the reaction mixture was increased. The inhibition could also be overcome by the addition of G6P. However, when a buffered system was used to keep the concentration of Mg^{2+} constant over the full range of energy charge no significant changes were observed in PEP carboxylase activity.

Addition of pyridine nucleotides to reaction mixtures also cause apparent changes in the activity of PEP carboxylase [47]. In general the observed initial velocities of the reaction are increased by NADH, but decreased by NADP, NAD and NADPH. The increased rates observed in the presence of NADH obviously resulted from the reduction of OAA (an end product inhibitor) to the less inhibitory malate. This reaction is catalysed by NAD-specific malate dehydrogenase present as an impurity in the preparation of PEP carboxylase.

The magnitude and direction of a number of these effects depend on both the pH of the reaction mixture, and the availability of Mg^{2+}. In particular changes in the concentration of available Mg^{2+} ions affect the kinetics of the reaction catalysed by PEP carboxylase [45,50]. In a more detailed study of these kinetics, using an enzyme partially purified from leaves of *Z. mays*, it was found that a negative cooperativity with respect to binding of Mg^{2+} to the enzyme, was combined with a positive cooperativity in respect to catalysis [132].

Illumination is known to affect the concentration of both hydrogen and Mg^{2+} ions within the various metabolic compartments of a green leaf (see Chapters 3 and 7 of this volume). Hence, such light-dependent changes coupled to feedback activation by G6P (the end product of the PCR cycle) or inhibition by malate (the end product of the C4 pathway) would permit close regulation of the amount of CO_2 assimilated through the C4 pathway [49].

8.9. CARBOXYLATION EFFICIENCY AND PHOTORESPIRATION

As a result of the studies discussed so far it can be concluded that C4 plants show higher rates of photosynthesis than C3 plants either because CO_2 is pumped to the PCR cycle, or because photorespiratory CO_2 is refixed before it is lost from the leaf. In the past it has been suggested that a CO_2 pumping mechanism is necessary, even in C3 plants, to achieve the recorded rates of photosynthesis under the present normal low concentrations of atmospheric CO_2 (see Chapters 6 and 7 and also ref. 169). Briefly, the argument is as follows (for more details of gas exchange in photosynthesis see ref. 177).

8.9.1. CO_2 uptake

Under conditions of low CO_2 concentration (0.03%) and high O_2 partial pressure (21%) photosynthesis in C3 plants will be limited by their ability to assimilate CO_2. During steady-state photosynthesis CO_2 will enter the leaf by diffusion, hence a gradient develops between the chloroplast and the atmosphere. The rate of photosynthesis depends on the size of this gradient and the magnitude of the various resistances to diffusion (boundary layer, stomatal pore, and mesophyll cell wall) encountered. In addition to these various true physical resistances apparent chemical and photochemical resistances may also be determined. However, if the total resistance is constant under steady-state conditions, the rate of CO_2 uptake will depend on the size of the gradient. This presents conceptual problems, since the rate of diffusion into the leaf will be greatest with a low internal concentration of CO_2, whereas the enzyme RBP carboxylase requires a high concentration of CO_2 for maximum activity. The exact concentration required depends, of course, on the affinity of RBP carboxylase for CO_2 i.e. K_m for CO_2.

In the past the experimentally determined value of the K_m (CO_2) for purified RBP carboxylase has been regarded as too high to permit the observed rates of CO_2 fixation. On the other hand, values determined for isolated chloroplasts or intact leaves were more favourable. This could indicate that a CO_2-concentrating mechanism, possibly dependent on carbonic anhydrase, operates in the chloroplasts of C3 plants. However, more recent results indicate that a form of RBP carboxylase, with a high affinity (low K_m) for CO_2, can be demonstrated in chloroplasts isolated from leaves of C3 plants [12]. In order to obtain the high affinity form (K_m (HCO_3^-) of 0.5 to 0.8 mM at pH 7.8) it was necessary to assay the enzyme immediately following rupture of the chloroplasts. If this was not done the low K_m form decayed rapidly to a high K_m form which resembled the purified enzyme. At the experimental pH used the CO_2/bicarbonate equilibrium is such that the value of the K_m can be recalculated as 18 μM, in respect of CO_2, for the high affinity form. The total activity of this form of the enzyme is sufficient to account for the observed rates of photosynthesis without evoking the need for a CO_2 pump in C3 plants (for further discussion on this see Chapter 7 of this volume). Hence, it is unlikely that the higher rates of photosynthesis observed in C4 plants are due simply to the C4 cycle acting as a CO_2-concentrating mechanism. The alternative hypothesis, that C4 photosynthesis overcomes the deleterious effects of photorespiration by preventing CO_2 efflux from the leaf will therefore be considered.

8.9.2. Photorespiration in C3 plants

Photorespiration consists of a light-dependent consumption of O_2 with a concomitant production of CO_2. The biochemistry of this process is now well established [164,165] and shown to differ from dark or mitochondrial respiration, depending on the production and further metabolism of glycollic acid. Glycollate is derived from intermediates of the PCR cycle in the chloroplasts, in reactions which also require the participation of molecular O_2. In the first proposed mechanism [52] the interaction between sugar phosphates and O_2 is indirect. The initial reaction involves the formation of a peroxide in a Mehler-type reaction between O_2 and reduced ferredoxin. The peroxide, in turn, can oxidise a two-carbon fragment from the transketolase intermediate or from RBP, forming glycollic acid. This type of reaction can be reproduced in vitro using isolated chloroplasts [140].

More recently it has been suggested that the main route of glycollate production is through an oxygenase reaction catalysed by fraction I protein [26,27]. It would appear that the same protein can catalyse both the RBP carboxylase reaction resulting in the formation of two molecules of PGA and an oxygenase reaction which results in the production of one molecule of PGA and one of P-glycollate. The actual mechanism of production of P-glycollate has been studied in some detail [6,11]. Chloroplasts also contain a specific P-glycollate phosphatase which releases free glycollate which then passes from the chloroplast.

The importance of the oxygenase reaction in photorespiration was first suggested on the basis of the observation that O_2 is a competitive inhibitor of the isolated carboxylase [135]. In a more detailed study [122] on carboxylase and oxygenase from soybean these activities were correlated with observed net CO_2 assimilation at various temperatures, CO_2 concentrations and O_2 partial pressures. It was found that the inhibition of net CO_2 fixation by O_2 increased as the ambient temperature was increased. This inhibition reflected a decrease in the affinity of the leaf for CO_2 and an increase in the affinity for O_2. With the purified enzyme, O_2 inhibition of CO_2 fixation and the ratio of oxygenase to carboxylase activity also increased with increased temperature, the similarity in response suggesting that the results in vivo reflected the increased oxygenase activity. With isolated chloroplasts it has been found [11] that the concentration of CO_2 necessary to cause half inhibition of the oxygenase activity is similar to the K_m for CO_2 of the carboxylase, suggesting that the two activities may involve similar binding sites.

Although considerable emphasis has been placed on the role of oxygenase activity in the production of glycollate, some results indicate that this is not the only source of glycollate. For instance it has been shown that the formation of glycine and serine from glycollate continues under conditions of high CO_2 when oxygenase activity would be inhibited [155]. Similarly, it has been concluded, on the basis of experiments with isolated chloroplasts incubated with ribose-5-P and fructose-1,6-bisphosphate, that the inhibition of carboxylation activity is not of necessity allied to glycollate synthesis [145]. More recent studies on algae [14] suggest that glycollate may be produced by both oxygenase reaction and oxidation of sugar phosphates.

Glycollate, lost from the chloroplasts, is metabolised further through glyoxylate, glycine, serine and hydroxypyruvate back to sugars in what is termed the C_2 pathway [165]. Many of the partial reactions of this pathway occur in specific microbodies, the leaf peroxisomes [164]. These contain high activities of catalase, glycollate oxidase, glutamate/glyoxylate aminotransferase, serine/glyoxylate amino transferase, hydroxypyruvate reductase, NAD-specific malate dehydrogenase and NADP-specific isocitrate dehydrogenase.

The CO_2 produced in photorespiration is liberated during the conversion of glycine to serine in mitochondria [18]. The reaction requires ADP and O_2 and results in the formation of ATP.

These results suggest that rates of photosynthesis, in C3 plants, are reduced by a number of related effects. There can be a net loss of carbon from the plant as the photorespiratory efflux of CO_2. This release of CO_2 will also decrease the rate of photosynthesis by decreasing the size of the diffusion gradient into the leaf. Loss of carbon from the PCR cycle, in the form of glycollate, will deplete the pools of carboxylation substrate (RBP) and other intermediates of the PCR cycle. Finally, the inhibition of RBP carboxylase by O_2 will further reduce the carboxylation capacity of the plant (see Chapter 5 for a possible physiological role of photorespiration). The operation of the C4 cycle in a higher plant leaf apparently overcomes these effects.

8.9.3. Photorespiration in C4 plants

Illuminated leaves of C4 plants do not show an efflux of CO_2 to the atmosphere, although the initial O_2-consuming reaction of photorespiration may occur [55]. The magnitude of this O_2 consumption, which may be detected by use of the heavy isotope ^{18}O, increases at the CO_2 compensation point. It has been concluded from concurrent measurements of O_2 evolution, O_2 uptake and assimilation of CO_2, that photorespiratory processes do occur in leaves of *Z.mays* but that any CO_2 generated in this process was recycled internally [167].

A post-illumination respiration, measured as a CO_2 efflux into CO_2-free air, which can use glycollate as a substrate has also been demonstrated in maize leaves [93]. This efflux was eliminated when the partial pressure of O_2 was decreased to less than 0.04% and increased by feeding various metabolites, including glycollate to excised leaves. This glycollate-stimulated respiration was inhibited by the specific inhibitor α-hydroxysulphonate. Maize leaf segments will also catalyse the production of CO_2 from fed glycine in an O_2-stimulated reaction [112].

These results suggest that maize leaves can metabolise glycollate, a conclusion consistent with the demonstration of peroxisomes [25,68,96] and enzymes of photorespiration [125,137,139,144] in leaves of C4 plants, although it has been suggested that the route of synthesis of glycollate in C4 plants differs from that which occurs in C3 plants [178]. Furthermore, the results of more recent studies of the post-illumination CO_2 burst in C4 plants [176] has led to the conclusion that this was derived from the decarboxylation of C4 acids rather than from glycollate. A CO_2 burst, of similar origin, has also been demonstrated [159] using leaves of maize illuminated after a period of darkness in an atmosphere of N_2.

High concentrations of O_2 do in fact affect C4 plants. In particular oxygen inhibits photosynthesis by isolated bundle sheath cells [39,40]; inhibits light-stimulated dark fixation of $^{14}CO_2$ [156]; increases the amount of carbon assimilated into malate and aspartate during short-term photosynthesis [123] and suppresses the germination of seeds of C4 plants [94].

There would seem little doubt that C4 plants have the capacity for photorespiration, but this does not result in CO_2 production or inhibition of photosynthesis. A simple explanation for these effects is that CO_2 released in the photosynthetic tissue is refixed before it can be lost from the leaf [121]. Hence, irrespective of the CO_2 or O_2 concentration of the outer atmosphere the CO_2 concentration in the substomatal cavity will remain low, and CO_2 will continue to diffuse into the leaf. It is also possible that the inhibitory effects of O_2 are overcome in part by the C4 cycle increasing the local concentration of CO_2 in the vicinity of the RBP carboxylase reaction, thus decreasing the extent of the competing oxygenase reaction.

8.10. SUMMARY

The title of this volume — *The Intact Chloroplast* — reflects an opinion that chloroplasts are self-sufficient organelles capable of complete photosynthesis. This belief is now challenged by a widely held concept of the pathways of photosynthesis in C4 plants. This requires the co-operation between chloroplasts and cytoplasmic organelles in two separate cell layers, and depends on rapid transport of compounds in and out of chloroplasts and between the mesophyll layer and bundle sheath cells. Although there is nothing in these suggestions which is incompatible with results of experiments on metabolite transfer in C3 plants there is as yet little evidence for such transport in C4 plants.

The complex schemes proposed for mechanisms of C4 photosynthesis rest, not on the elegant demonstrations of enzyme activities and partial reactions within the two types of photosynthetic tissue, but rather on a single assumption of low levels of RBP carboxylase in mesophyll cells. Clear demonstrations of such activity would make many of these suggestions unnecessary.

REFERENCES

1 Andersen, K.S., Bain, J.M., Bishop, D.G. and Smillie, R.M. (1972) Plant Physiol., [49], 461—466.
2 Anderson, J.M., Woo, K.C. and Boardman, N.K. (1971) Biochim. Biophys. Acta, [245], 398—408.
3 Andrews, T.J. and Hatch, M.D. (1969) Biochem. J., [114], 117—125.

4 Andrews, T.J. and Hatch, M.D. (1971) Phytochemistry, [10], 9—16.
5 Andrews, T.J., Johnson, H.S., Slack, C.R. and Hatch, M.D. (1971) Phytochemistry, [10], 2005—2014.
6 Andrews, T.J., Lorimer, G.H. and Tolbert, N.E. (1973) Biochemistry, [12], 11—18.
7 Baldry, C.W., Bucke, C. and Coombs, J. (1969) Biochem. Biophys. Res. Commun., [37], 828—832.
8 Baldry, C.W., Bucke, C. and Coombs, J. (1970) Planta (Berl.), [94], 124—133.
9 Baldry, C.W., Bucke, C. and Coombs, J. (1971) Planta (Berl.), [97], 310—319.
10 Baldry, C.W., Bucke, C., Coombs, J. and Gross, D. (1970) Planta (Berl.), [94], 107—123.
11 Bahr, J.T. and Jensen, R.G. (1974) Arch. Biochem. Biophys., [164], 408—413.
12 Bahr, J.T. and Jensen, R.G. (1974) Plant Physiol., [53], 39—44.
13 Bassham, J.A. and Calvin, M. (1957) The Path of Carbon in Photosynthesis, Prentice Hall, Englewood Cliffs, N.J.
14 Bassham, J.A. and Kirk, M. (1973) Plant Physiol., [52], 407—411.
15 Bazzaz, M.B. and Govindjee (1973) Plant Physiol., [52], 257—262.
16 Bender, M.M. (1971) Phytochemistry, [10], 1239—1244.
17 Berry, J.A., Downton, W.H.S. and Tregunna, E.B. (1970) Can. J. Bot., [48], 777—786.
18 Bird, I.F., Cornelius, M.J., Keys, A.J. and Whittingham, C.P. (1972) Phytochemistry, [11], 1587—1594.
19 Bishop, D.G., Andersen, K.S. and Smillie, R.M. (1972) Plant Physiol., [49], 467—470.
20 Bishop, D.G., Andersen, K.S. and Smillie, R.M. (1972) Plant Physiol., [50], 774—777.
21 Bjorkman, O. and Gauhl, E. (1969) Planta (Berl.), [88], 197—203.
22 Black, C.C. (1973) Annu. Rev. Plant Physiol., [24], 253—286.
23 Black, C.C., Edwards, G.E., Kanai, R. and Mollenhauer, H.H. (1972) in Proc. 2nd Int. Cong. Photosynthesis (Forti, G., Avron, M. and Melandri, A., eds), pp. 1745—1757, Junk, The Hague.
24 Black, C.C. and Mayne, B.C. (1970) Plant Physiol., [45], 738—741.
25 Black, C.C. and Mollenhauer, H.H. (1971) Plant Physiol., [47], 15—23.
26 Bowes, G. and Ogren, W.L. (1972) J. Biol. Chem., [247], 2171—2176.
27 Bowes, G.W., Ogren, W.L. and Hageman, R.H. (1971) Biochem. Biophys. Res. Commun., [45], 716—722.
28 Brown, W.V. and Smith, B.N. (1972) Nature, [239], 345—346.
29 Bucke, C. and Long, S.P. (1971) Planta (Berl.), [99], 199—210.
30 Bucke, C. and Oliver, I.R. (1975) Planta (Berl.), [122], 45—52.
31 Chang, F.H. and Troughton, J.H. (1972) Photosynthetica, [6], 57—65.
32 Chapman, E.A. and Osmond, C.B. (1974) Plant Physiol., [53], 893—898.
33 Chen, T.M., Brown, R.H. and Black, C.C. (1970) Weed Sci., [18], 399—403.
34 Chen, T.M., Brown, R.H. and Black, C.C. (1971) Plant Physiol., [47], 199—203.
35 Chen, T.M., Campbell, W.H., Dittrich, P. and Black, C.C. (1973) Biochem. Biophys. Res. Commun., [51], 461—467.
36 Chen, T.M., Dittrich, P., Campbell, W.H. and Black, C.C. (1974) Arch. Biochem. Biophys., [163], 246—342.
37 Chollet, R. (1974) Arch. Biochem. Biophys., [163], 521—551.
38 Chollet, R. and Ogren, W.L. (1972) Z. Pflanzenphysiol., [68], 45—54.
39 Chollet, R. and Ogren, W.L. (1972) Biochem. Biophys. Res. Commun., [46], 2062—2066.
40 Chollet, R. and Ogren, W.L. (1972) Biochem. Biophys. Res. Commun., [48], 684—688.
41 Chollet, R. and Ogren, W.L. (1973) Plant Physiol., [51], 789—792.

42 Coombs, J. (1971) Proc. Roy. Soc. Lond. B, [179], 221—235.
43 Coombs, J. (1973) Curr. Adv. Plant Sci., [1], 1—10.
44 Coombs, J. and Baldry, C.W. (1972) Nature New Biol., [238], 268—270.
45 Coombs, J. and Baldry, C.W. (1975) Planta (Berl.), [124], 153—158.
46 Coombs, J., Baldry, C.W. and Brown, J.E. (1973) Planta (Berl.), [110], 121—129.
47 Coombs, J., Baldry, C.W. and Bucke, C. (1973) Planta (Berl.), [110], 95—108.
48 Coombs, J., Baldry, C.W. and Bucke, C. (1973) Planta (Berl.), [110], 109—120.
49 Coombs, J., Baldry, C.W. and Bucke, C. (1976) in Perspectives in Experimental Biology, Vol. 2. Botany (Sunderland, N., ed), pp. 177—188, Pergamon, Oxford.
50 Coombs, J., Maw, S.L. and Baldry, C.W. (1974) Planta (Berl.), [117], 279—292.
51 Coombs, J., Maw, S.L. and Baldry, C.W. (1975) Plant. Sci. Lett., [4], 97—102.
52 Coombs, J. and Whittingham, C.P. (1966) Proc. Roy. Soc. Lond. B., [164], 511—520.
53 Cooper, T.G., Filmer, D., Wishnick, M. and Lane, M.D. (1969) J. Biol. Chem., [244], 1081—1083.
54 Cooper, T.G. and Wood, H.G. (1971) J. Biol. Chem., [246], 5488—5490.
55 D'Aoust, A.L. and Canvin, D.T. (1973) Can. J. Bot., [51], 457—464.
56 Davies, D.D., Nascimento, K.H. and Patil, K.D. (1974) Phytochemistry, [13], 2417—2425.
57 Dittrich, P., Salin, M.L. and Black, C.C. (1973) Biochem. Biophys. Res. Commun., [55], 104—110.
58 Downton, W.J.S. (1970) Can. J. Bot., [48], 1795—1800.
59 Downton, W.J.S., Berry, J.A. and Tregunna, E.B. (1970) Z. Pfanzenphysiol., [63], 194—198.
60 Downton, W.J.S. and Hawker, J.S. (1973) Phytochemistry, [12], 1551—1556.
61 Downton, W.J.S. and Tregunna, E.B. (1968) Can. J. Bot., [46], 207—215.
62 Downton, W.J.S. and Pyliotis N.H. (1971) Can. J. Bot., [49], 179—180.
63 Edwards, G.E. and Black, C.C. (1971) Plant Physiol., [47], 149—156.
64 Edwards, G.E. and Gutierrez, M. (1972) Plant Physiol., [50], 728—732.
65 Edwards, G.E., Kanai, R. and Black, C.C. (1971) Biochem. Biophys. Res. Commun., [45], 278—285.
66 Edwards, G.E., Lee, S.S., Chen, T.M. and Black, C.C. (1970) Biochem. Biophys. Res. Commun., [39], 389—395.
67 Farineau, J. (1972) in Proc. 2nd Int. Cong. Photosynthesis (Forti, G., Avron, M. and Melandri, A., eds), pp. 1971—1979, Junk, The Hague.
68 Frederick, S.E. and Newcomb, E.H. (1971) Planta (Berl.), [96], 152—174.
69 Galmiche, J.M. (1973) Plant Physiol., [51], 512—519.
70 Gibbs, M., Latzko, E., O'Neal, D. and Hew, C.S. (1970) Biochem. Biophys. Res. Commun., [40], 1356—1361.
71 Goatly, M.B. and Smith, H. (1974) Planta (Berl.), [117], 67—73.
72 Gracen, V.E., Hilliard, J.H., Brown, R.H. and West, S.H. (1972) Planta (Berl.), [107], 189—204.
73 Graham, D., Hatch, M.D., Slack, C.R. and Smillie, R.M. (1970) Phytochemistry, [9], 521—532.
74 Giles, K.L., Beardsell, M.F. and Cohen, D. (1974) Plant Physiol., [54], 208—212.
75 Gutierrez, M., Gracen, V.E. and Edwards, G.E. (1974) Planta (Berl.), [119], 279—300.
76 Gutierrez, M., Kanai, R., Huber, S.C., Ku, S.B. and Edwards, G.E. (1974) Z. Pflanzenphysiol., [72], 305—319.
77 Hatch, M.D. (1971) Biochem. J., [125], 425—432.
78 Hatch, M.D. (1973) Arch. Biochem. Biophys., [156], 207—214.
79 Hatch, M.D. and Kagawa, T. (1973) Arch. Biochem. Biophys., [159], 842—853.

80 Hatch, M.D. and Kagawa, T. (1974) Arch. Biochem. Biophys., [160], 346—349.
81 Hatch, M.D. and Kagawa, T. (1974) Aust. J. Plant Physiol., [1], 357—369.
82 Hatch, M.D., Kagawa, T. and Craig, S. (1975) Aust. J. Plant Physiol., [2], 111—128.
83 Hatch, M.D. and Mau, S. (1973) Arch. Biochem. Biophys., [156], 195—206.
84 Hatch, M.D., Mau, S. and Kagawa, T. (1974) Arch. Biochem. Biophys., [165], 188—200.
85 Hatch, M.D., Osmond, C.B. and Slatyer, R.O. (eds) (1971) Photosynthesis and Photorespiration, Wiley Interscience, New York.
86 Hatch, M.D. and Slack, C.R. (1966) Biochem. J., [101], 103—111.
87 Hatch, M.D. and Slack, C.R. (1968) Biochem. J., [106], 141—146.
88 Hatch, M.D. and Slack, C.R. (1969) Biochem. J., [112], 549—558.
89 Hatch, M.D. and Slack, C.R. (1970) Ann. Rev. Plant Physiol., [21], 141—162.
90 Hatch, M.D., Slack, C.R. and Bull, T.A. (1969) Phytochemistry, [8], 697—706.
91 Hatch, M.D., Slack, C.R. and Johnson, H.S. (1967) Biochem. J., [102], 417—422.
92 Heber, U. (1974) Annu. Rev. Plant Physiol., [25], 393—421.
93 Heichel, G.H. (1970) Plant Physiol., [46], 359—362.
94 Heichel, G.H. and Day, P.R. (1972) Plant Physiol., [49], 280—283.
95 Hew, C.S. and Gibbs, M. (1970) Can. J. Bot., [48], 1265—1269.
96 Hilliard, J.H., Gracen, V.E. and West, S.H. (1971) Planta (Berl.), [97], 93—105.
97 Hilliard, J.H. and West, S.H. (1971) Planta (Berl.), [99], 352—356.
98 Holden, M. (1973) Photosynthetica, [7], 41—49.
99 Huang, A.H.C. and Beevers, H. (1972) Plant Physiol., [50], 242—248.
100 Huber, S.C., Kanai, R. and Edwards, G.E. (1973) Planta (Berl.), [113], 53—66.
101 Jackson, W.A. and Volk, R.J. (1970) Ann. Rev. Plant Physiol., [21], 385—432.
102 Johnson, H.S. and Hatch, M.D. (1968) Phytochemistry, [7], 375—380.
103 Johnson, H.S. and Hatch, M.D. (1969) Biochem. J., [114], 127—134.
104 Johnson, H.S. and Hatch, M.D. (1970) Biochem. J., [119], 273—280.
105 Kagawa, T. and Hatch, M.D. (1974) Biochem. Biophys. Res. Commun., [59], 1326—1332.
106 Kagawa, T. and Hatch, M.D. (1974) Aust. J. Plant Physiol., [1], 51—64.
107 Kanai, R. and Edwards, G.E. (1973) Plant Physiol., [51], 1133—1137.
108 Kanai, R. and Edwards, G.E. (1973) Plant Physiol., [52], 484—490.
109 Karpilov, Y.S. (ed.) (1974) Photosynthesis in Maize, Academy of Sciences of the USSR, Puschino-on-oka.
110 Kennedy, R.A. and Laetsch, W.M. (1974) Plant Physiol., [54], 608—611.
111 Kent Crookston, R. and Moss, D.N. (1973) Plant Physiol., [52], 397—402.
112 Kisaki, T., Yaio, N. and Hirabayashi, S. (1972) Plant Cell. Physiol. Tokyo, [13], 581—584.
113 Kortschak, H.P. (1968) in Photosynthesis in Sugar Cane (Coombs, J., ed), pp. 18—26, Tate and Lyle, London.
114 Kortschak, H.P., Hartt, C.H. and Burr, G.O. (1965) Plant Physiol., [40], 209—213.
115 Kortschak, H.P. and Nickell, L.G. (1970) Plant Physiol., [45], 515—516.
116 Ku, S.B., Gutierrez, M. and Edwards, G.E. (1974) Planta (Berl.), [119], 267—278.
117 Ku, S.B., Gutierrez, M., Kanai, R. and Edwards, G.E. (1974) Z. Pflanzenphysiol., [72], 320—337.
118 Laber, L.J., Latzko, E. and Gibbs, M. (1974) J. Biol. Chem., [249], 3436—3439.
119 Laetsch, W.M. (1974) Ann. Rev. Plant Physiol., [25], 27—52.
120 Laetsch, W.M. and Kortschak, H.P. (1972) Plant Physiol., [49], 1021—1023.
121 Laing, W.A. and Forde, B.J. (1971) Planta (Berl.), [98], 221—231.
122 Laing, W.A., Ogren, W.L. and Hageman, R.H. (1974) Plant Physiol., [54], 678—685.

123 Lewanty, Z., Maleszewski, S. and Poskuta, J. (1971) Z. Pflanzenphysiol., [65], 469—472.
124 Lowe, J. and Slack, C.R. (1971) Biochim. Biophys. Acta, [235], 207—209.
125 Lui, A.Y. and Black, C.C. (1972) Arch. Biochem. Biophys. [149], 269—280.
126 Luttge, U., Ball, E. and von Willert, K. (1971) Z. Pflanzenphysiol., [65], 336—350.
127 Mahon, J.D., Fock, H., Holer, T. and Canvin, D.T. (1974) Planta (Berl.), [120], 113—123.
128 Mayne, B.C., Dee, A.M. and Edwards, G.E. (1975) Z. Pflanzenphysiol., [74], 275—291.
129 Mayne, B.C., Edwards, G.E. and Black, C.C. (1971) Plant Physiol., [47], 600—605.
130 Miziorko, H.M., Nowak, T. and Mildvan, A.S. (1974) Arch. Biochem. Biophys., [163], 378—389.
131 Moss, D.N., Willmer, C.M. and Kent Crookson, R. (1971) Plant. Physiol., [47], 847—849.
132 Mukerji, S.K. (1974) Plant Sci. Lett., [2], 243—248.
133 Mukerji, S.K. and Yang, S.F. (1974) Plant Physiol., [53], 829—834.
134 Nishikido, T. and Takanashi, H. (1973) Biochem. Biophys. Res. Commun., [45], 716—722.
135 Ogren, W.L. and Bowes, G. (1971) Nature New Biol., [230], 159—160.
136 O'Neal, D., Hew, C.S., Latzko, E., and Gibbs, M. (1972) Plant Physiol., [49], 607—614.
137 Osmond, C.B. (1969) Biochim. Biophys. Acta, [172], 144—149.
138 Osmond, C.B. (1974) Aust. J. Plant Physiol., [1], 41—50.
139 Osmond, C.B. and Harries, B. (1971) Biochim. Biophys. Acta, [234], 270—282.
140 Plaut, Z. and Gibbs, M. (1970) Plant Physiol., [45], 470—474.
141 Poincelot, R.P. (1972) Plant Physiol., [50], 336—340.
142 Polya, G.M. and Osmond, C.B. (1972) Plant Physiol., [49], 267—269.
143 Rathnam, C.K.M. and Das, V.S.R. (1975) Z. Pflanzenphysiol., [74], 377—393.
144 Rehfeld, D.W., Randall, D.D. and Tolbert, N.E. (1970) Can. J. Bot., [48], 1219—1226.
145 Robinson, J.M. and Gibbs, M. (1974) Plant Physiol., [53], 790—797.
146 Salin, M.L. and Black, C.C. (1974) Plant. Sci. Lett., [2], 303—308.
147 Slack, C.R. (1968) Biochem. Biophys. Res. Commun., [30], 483—488.
148 Slack, C.R. (1969) Phytochemistry, [8], 1387—1391.
149 Slack, C.R. and Hatch, M.D. (1967) Biochem. J., [103], 660—665.
150 Slack, C.R., Hatch, M.D. and Goodchild, D.J. (1969) Biochem. J., [114], 489—498.
151 Slack, C.R., Roughan, P.G. and Bassett, H.C.M. (1974) Planta (Berl.), [118], 57—73.
152 Smillie, R.M., Andersen, K.S., Tobin, N.F., Entsch, B. and Bishop, D.G. (1972) Plant Physiol., [49], 471—475.
153 Smith, B.N. and Brown, W.V. (1973) Am. J. Bot., [60], 505—513.
154 Smith, B.N. and Epstein, S. (1971) Plant Physiol., [47], 380—384.
155 Snyder, F.W. and Tolbert, N.E. (1974) Plant Physiol., [53], 514—515.
156 Stamieszkin, I., Maleszewski, S. and Poskuta, J. (1972) Z. Pflanzenphysiol., [67], 180—182.
157 Steiger, E. Ziegler, I. and Ziegler, H. (1971) Planta (Berl.), [96], 109—118.
158 Taylor, A.O., Slack, C.R. and McPherson, H.G. (1974) Plant Physiol., [54], 696—701.
159 Thibault, P. (1973) Planta (Berl.), [114], 109—118.

160 Throughton, J.H. (1971) Planta (Berl.), [100], 87—92.
161 Ting, I.P. and Osmond, C.B. (1973) Plant Sci. Lett., [1], 123—128.
162 Ting, I.P. and Osmond, C.B. (1973) Plant Physiol., [51], 439—447.
163 Ting, I.P. and Osmond, C.B. (1973) Plant Physiol., [51], 448—453.
164 Tolbert, N.E. (1971) Ann. Rev. Plant Physiol., [22], 45—74.
165 Tolbert, N.E. and Yamazaki, R.K. (1969) Ann. N.Y. Acad. Sci., [168], 325—341.
166 Triolo, L., Bagnara, D., Anselini, L. and Basenelli, C. (1974) Physiol. Plant., [31], 86—91.
167 Volk, R.J. and Jackson, W.A. (1972) Plant Physiol., [49], 218—223.
168 Walker, D.A. (1970) Nature, [226], 1204—1208.
169 Walker, D.A. (1974) M.T.I. Int. Rev. Sci., [11], 3—44.
170 Walker, D.A. and Crofts, A.R. (1970) Ann. Rev. Biochem., [39], 389—428.
171 Waygood, E.R., Mache, R. and Tan, C.K. (1969) Can. J. Bot., [47], 1455—1458.
172 Whelen, T., Sackett, W.N. and Benedict, C.R. (1973) Plant Physiol., [51], 1051—1054.
173 Wong, K.F. and Davies, D.D. (1973) Biochem. J., [131], 451—458.
174 Woo, K.C., Andersen, J.M., Boardman, N.K., Downton, W.J.S., Osmond, C.B. and Thorne, S.W. (1970) Proc. Natl. Acad. Sci. USA, [67], 18—25.
175 Woo, K.C., Pyliotis, N.A. and Downton, W.J.S. (1971) Z. Pflanzenphysiol., [64], 400—413.
176 Wynn, T., Brown, H., Campbell, W.H. and Black, C.C. (1973) Plant Physiol., [52], 288—291.
177 Zelitch, I. (1971) Photosynthesis, Photorespiration and Plant Productivity, Academic Press, New York.
178 Zelitch, I. (1973) Plant Physiol., [51], 299—305.
179 Ziegler, I. (1973) Pytochemistry, [12], 1027—1030.
180 Ziegler, I. (1974) Biochim. Biophys. Acta, [364], 28—37.

The Intact Chloroplast – edited by J. Barber

Chapter 9

Photosynthetic Sulfate Reduction by Chloroplasts

J.D. SCHWENN and A. TREBST

Department of Biology, Ruhr University, Bochum (G.F.R.)

CONTENTS

Abbreviations: APS, adenosine-5′-phosphosulfate; Fd, ferredoxin; PAP, 3′-phosphoadenosine-5′-phosphate; PAPS, 3′-phosphoadenosine-5′-phosphosulfate; X-S : SH, bound sulfide; X-S : SO_3H, bound sulfite.

9.1. THE ESSENTIAL ROLE OF SULFUR AS NUTRIENT

Sulfur as nutrient for plants and autotrophic algae appears in the soil as sulfate, i.e. in the highest positive valency state of sulfur (6+). Yet the plant needs sulfur in the lowest valency state (2—) for the biosynthesis of essential sulfur-containing compounds. The sulfur amino acids cysteine and methionine are obligatory constituents of proteins, cysteamine plays an important part in the function of coenzyme A, and various other coenzymes contain sulfhydryl groups such as lipoic acid or glutathion [49,72]. Far less compounds are found in plants in which the sulfur remains in the valency state of the sulfate. They are found in the sulfate esters of the sulfated polysaccharides predominantly occurring in marine algae, choline sulfate and sulfonic acids. Besides these, sulfate in an unusual C-S bond is found in the plant sulfolipids [12,13]. Extensive reviews on the sulfur compounds of the plants are given by Kjaer [33] and Thompson [72].

The occurrence of compounds containing reduced sulfur requires a mechanism by which the nutrient sulfate is reduced and then incorporated as reduced sulfur into the various substances by autotrophic plants and algae. This process is defined as the assimilatory sulfate reduction.

9.2. SULFATE REDUCTION BY WHOLE PLANTS

Most of our knowledge about the physiology of sulfate assimilation in plants is based on early nutritional studies with whole plants. These results have been reviewed most elaborately by Thomas [71]. However, a few details will be given here with emphasis on the uptake and incorporation of the sulfur in plants, primarily because of their relevance to the location of the sulfate-reducing process in the plant and secondly to its mechanism as a light driven process.

9.2.1. Uptake and transport of sulfate in plants

Studies with micro-autoradiographs [37] demonstrated that the sulfate taken up by the root appears in the outer layers of the xylem. Lüttge found evidence that the smaller xylem cells are preferred to the wider ones [38]. Whether this is an active secretion into the xylem cells is still a matter of discussion [39]. When the sulfate has reached the vascular system of the leaves it was found to spread evenly into the interveinal areas.

Sulfur in the valency state of 4+, e.g. sulfur dioxide, also is taken up by the leaf. Plants practically inhale the gaseous sulfur dioxide. From tracer studies it was observed that sulfur dioxide was accumulated in the stomata [80] and in the phloem of the bundle sheaths [70a]. After prolonged periods of growth the labelled sulfur initially applied to the leaf as sulfur dioxide is found in the phloem and particularly in the cambium of the whole plant.

A transport of sulfur from the leaves to the roots has been postulated from data with fumigated leaves [21]. The sulfite formed from this sulfur dioxide in the leaf cell (cf. ref. 80) is rapidly converted into sulfate [6] even under conditions which clearly favour assimilatory sulfate reduction.

Although there is definite evidence that the sulfur in the valency state of sulfate is transported from the root to the leaves there is no direct evidence that the reduced sulfur in the valency state of sulfide is transported back to the root.

9.2.2. Incorporation of reduced sulfur in amino acids and proteins

In tracer studies with ^{35}S-labelled sulfate [70a,71] it was shown that whole wheat plants and sugar beet incorporate sulfur into proteins. Similar results were obtained by Kylin [34] studying the metabolism of sulfate in young wheat seedlings. Labelled sulfur was predominantly found in the amino acids cysteine and methionine and in proteins. More recently this was confirmed by Willenbrink for tomato plants [81]. Sinha and Cossins [52] found that the leaves, hypocotyls and cotyledons of radish seedlings incorporated the label into the amino acids with cysteine being the first S-amino acid labelled.

9.2.3. Localisation of the sulfate-reducing process

It was not until the work of Fromageot and Asahi that the site of sulfate reduction in plants was localised and intermediates between sulfate and cysteine came into discussion [20,32]. These authors reported that excised leaves of mung bean and of tobacco could reduce sulfate to sulfite in a light-driven step. Willenbrink [82] found that excised leaves of tomato plants also reduce sulfate to cysteine, methionine and glutathion upon illumination. Fig. 9.1 compares the amount of reduced labelled sulfur in light- and dark-grown plants and leaves according to Willenbrink. Within 48 h the label of light-grown plants increased drastically whereas the label of dark-grown plants remained constant after the first half hour.

The unanimous observation that light plays an essential part in the formation of sulfite [20,32] and the formation of cysteine and methionine [52,82] as the final products of assimilatory sulfate reduction in the leaf led to a proposal of a relationship between the sulfate-reducing process and the photochemical activity of the leaf. Finally Asahi [7] assumed that the reduction of sulfate is connected to the photoreducing capacity of the chloroplast.

Studies of the sulfate reduction in algae were in line with the idea of a light-driven assimilation of sulfate, hence the photosynthetic sulfate reduction seems to be a general feature of green autotrophic plants.

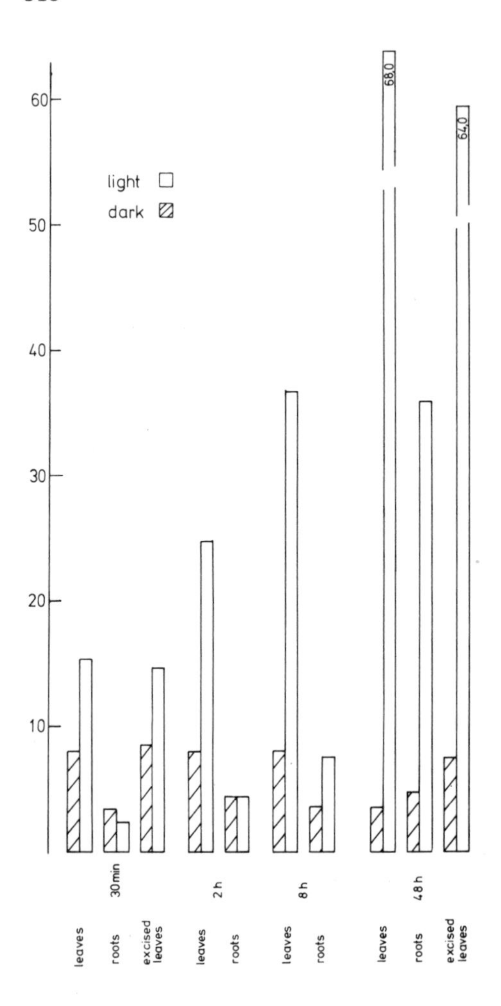

Fig. 9.1. Light-dependent incorporation of ^{35}S into the soluble organic fraction (including proteins) in young tomato plants [81]. Coordinate: fixed ^{35}S compounds in % of the total label incorporated; abscissa: time (as indicated). Dark (▨)- and light (□)-grown plants.

Although there is positive evidence that the leaves are able to reduce sulfate in the light, this does not exclude that the root, heterotrophically grown algae or tissue cultures may also be capable of sulfate reduction [19, 24,53].

9.3. PHOTOSYNTHETIC SULFATE REDUCTION IN CHLOROPLASTS

9.3.1. First evidence for the sulfate-reducing process in chloroplasts

The very first results on sulfate reduction in a cell-free system from plants were obtained with "broken" chloroplasts [7]. Sulfite formation from sulfate in this type of chloroplast could only be demonstrated after complementation with sulfate activating enzymes together with the sulfite carrier protein, both being obtained from yeast [7,10,83]. Yet, no reduction further than sulfite was achieved. ATP sulfurylase was detected in these chloroplast particles [7,9,50], but the APS kinase seemed to be absent from the chloroplast [19]. This in contrast to the yeast system where both enzymes were proven to be necessary for the activation of sulfate before further reduction.

It was shown in 1968 by Schmidt that intact chloroplasts do indeed contain the complete sulfate-reducing process and that assimilatory sulfate reduction can be demonstrated in this cell-free system from plants [59,75]. They found that intact chloroplasts are capable of reducing sulfate to the valency state of sulfide in a light-driven reaction. The illuminated chloroplasts reduced sulfate to "acid volatile sulfur", that is sulfite and sulfide, at a rate of 30 nmoles.mg $chl^{-1}.h^{-1}$. In the dark practically no reduction of sulfate was found. The experiments were carried out with chloroplasts prepared according to Jensen and Bassham [30] and ^{35}S-labelled sulfate as substrate. Labelled sulfite and sulfide were trapped in alkali. Among other products formed from the labelled sulfate, cysteine was identified.

Some parameters controlling photosynthetic sulfate reduction by chloroplasts have been published [59,75]. Adenosine nucleotides (ATP, ADP) were found to stimulate the formation of acid volatile sulfur in the light. In the presence of 4 mM ADP the optimal concentration for sulfate was found to be 2 mM. It was assumed that the stimulatory effect of ATP and ADP indicated sulfate activation processes in the chloroplast similar to those found in the yeast system [11]. These results have to be revalued since more recently it has been shown that the nucleotides do not rapidly permeate the chloroplast envelope (cf. ref. 54 and Chapters 5 and 6 of this volume). From this view it is more likely that the stimulation by ATP or ADP as observed by Trebst and Schmidt could only occur because the chloroplasts used were partly damaged explaining the relatively poor rates of sulfate reduction by these preparations.

However, these early results made it clear that a light-dependent sulfate-reducing capacity of the leaf is associated with the chloroplasts. Secondly it became apparent that isolated chloroplasts represent an in vitro system which bears the possibility of investigating further aspects of sulfate reduction in green plants.

9.3.2. Kinetics of the photosynthetic sulfate reduction in intact chloroplasts

In further experiments advantage was taken of the improved isolation procedures for intact chloroplasts. This type of chloroplast also reopened the study of photosynthetic carbon fixation and other chloroplasmic processes as emphasised by the various chapters of this book. Such intact chloroplasts were shown to reduce sulfate in the light to cysteine without further addition of any cofactors. In kinetic experiments on the sulfate reduction with good intact chloroplasts the identity and the sequence of reaction products from labelled sulfate were investigated [66]. The experiments were carried out under saturating white light. The addition of unlabelled carrier sulfite or sulfide to the chloroplast preparation was omitted for reasons that will be discussed below. The separation of the labelled reaction products was performed by ion exchange chromatography according to Iguchi [29]. The radioactivity was monitored continuously in a flow analyzer. An elution profile obtained by this procedure is shown in Fig. 9.2. It represents a typical diagram from an experiment with illuminated chloroplasts. Under the given experimental conditions six labelled substances have been isolated by column chromatography from intact chloroplasts fed with labelled sulfate in the light. They were identified as PAPS, APS, an unknown substance in

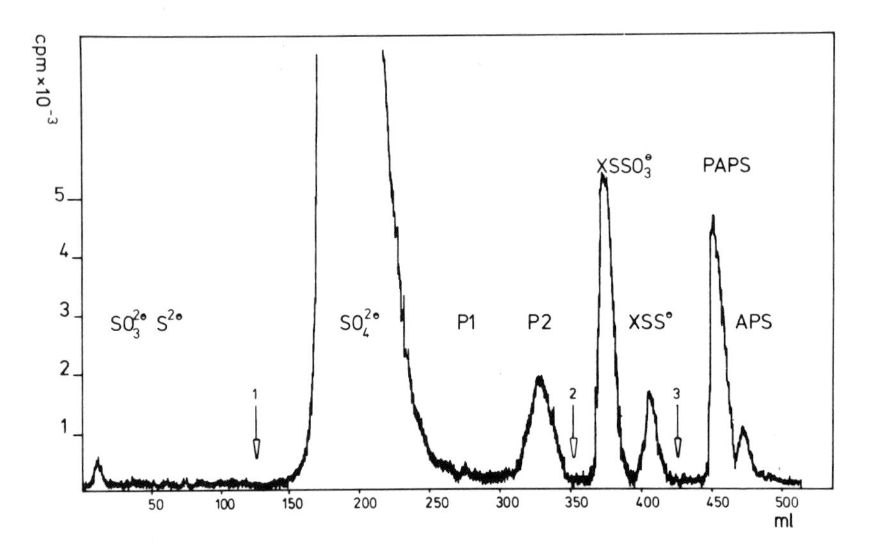

Fig. 9.2. A representive separation of labelled substances formed by intact chloroplasts from [^{35}S]sulfate in the light. The ion exchange chromatography was performed according to Iguchi [29], arrows indicate the buffer changes. Coordinate: radioactivity in cpm; abscissa: elution volume in ml.

peak two, bound sulfite (X-S:SO_3H) and bound sulfide (X-S:SH); substances formed non-enzymatically like peak one have been neglected. Significantly no free sulfite, thiosulfate or sulfide is found in chloroplasts. APS and PAPS are well known substances, identified early in sulfate activation by Lipmann's group. Thus the established identification procedures could be followed for the verification of the nature of these substances.

The substance in peak two has not yet been identified. The liberation of sulfate upon acidic hydrolysis points towards an O-sulfo compound rather than a C-sulfo compound. Its function in assimilatory sulfate reduction remains to be elucidated.

An intermediate bound sulfite has been postulated already very early in sulfate reduction particularly in the yeast system [26a,74]. The bound sulfite X-S:SO_3^- found in the chloroplast system was identified by isotope exchange reactions. The bound sulfite is very likely the source of sulfite found in the leaves [20,32] and in the chloroplasts [59] because it exchanges its labelled sulfite moiety with the free unlabelled sulfite, according to:

$$\text{X-S:}^{35}SO_3^- + SO_3^{2-} \rightleftharpoons \text{X-S:}SO_3^- + {}^{35}SO_3^{2-} \quad (1)$$

Similarly it exchanges the labelled sulfite of the bound sulfite with additional thiosulfate according to:

$$\text{X-S:}^{35}SO_3^- + \text{S:}SO_3^{2-} \rightleftharpoons \text{X-S:}SO_3^- + \text{S:}^{35}SO_3^{2-} \quad (2)$$

This exchange produces labelled thiosulfate [2,78,79]. Schiff and colleagues recently pointed out that thiosulfate also is generated by the action of dithiols used in the assay as antioxidants [76].

In previous work excess unlabelled sulfite always had to be added for methodical reasons when small amounts of sulfite were to be determined as "acid volatile sulfur" by employing acidic distillation and trapping into alkali. The bound sulfite in Fig. 9.2, however, is detected by means of column chromatography only in the absence of additional sulfite.

The occurrence of bound sulfide has been a new proposal. Bound sulfide is found in the absence of added hydrogen sulfide only. Its characterisation is based on its exchange with the unlabelled free sulfide:

$$\text{X-S:}^{35}S^- + S^{2-} \rightleftharpoons \text{X-S:}S^- + {}^{35}S^{2-} \quad (3)$$

according to a mechanism proposed by Gurjanowa [23,55,56]. Furthermore thiolysis with thiols leads to the liberation of labelled sulfide:

$$\text{X-S:}^{35}S^- + RS^- \rightleftharpoons \text{X-S:SR} + {}^{35}S^{2-} \quad (4)$$

Upon gel filtration no separation of bound sulfite and bound sulfide can be seen. This is taken as presumptive evidence for similarity in size between the carrier molecule of the bound sulfite and sulfide. Its molecular weight, determined by gel filtration, ranges from 4000 to 6000 daltons [61]. From the isotope exchange and the behaviour against thiols it was deduced that the binding site for the sulfite and sulfide is a sulfhydryl group. Although the nature of the carrier is not yet identified evidence has accumulated establishing its protein character [68,69].

Among these substances which are regarded as the intermediates in assimilatory sulfate reduction by intact chloroplasts the final product, the amino acid cysteine, is also found. The rates of its formation from sulfate in the light are found to be in the range of 0.5 nmoles.mg $chl^{-1}.h^{-1}$ [66]. For identification it has been converted into cysteine sulfonic acid according to Moore [44].

Relying on the above identification of the compounds formed by intact chloroplasts during the assimilatory sulfate reduction a time course of their formation has been established [66,61] as depicted in Fig. 9.3. The data suggested the following reaction sequence:

$$SO_4^{2-} \rightarrow APS \rightarrow (\dot{P}APS) \rightarrow P(2) \rightarrow X\text{-}S{:}SO_3^- \rightarrow X\text{-}S{:}S^- \rightarrow \text{cysteine}$$

Fig. 9.3 was obtained by plotting the percentage of all compounds formed from sulfate vs. the time of illumination. Accordingly it has been assumed as a working hypothesis that APS is the first labelled compound participating in assimilatory sulfate reduction. APS metabolised to PAPS, as the second substance labelled. At the expense of the activated sulfate, PAPS, a third substance is formed concomitantly: the unknown substance in peak two. There is some evidence from the kinetic data obtained with intact chloroplasts that it is PAPS which transfers its sulfate group onto the sulfhydryl group of the endogenous carrier "X" leading thereby to the formation of the bound sulfite $X\text{-}S{:}SO_3H$ (this reaction will be discussed explicitly on page 326. This S-sulfo compound is reduced to a persulfide (X-S:SH) in the light. The reductive step from the bound sulfite $X\text{-}S{:}SO_3H$ to the bound sulfide X-S:SH consumes six electrons generated at the necessary redox potential level (reduced ferredoxin) via the electron transport chain of the chloroplasts. The formation of cysteine is regarded as the final step in the reaction sequence. The reduced sulfur for the cysteine formation is very likely derived from the bound sulfide (see page 331 for the detailed mechanism). Another two electrons are needed during this process, possibly at a different redox potential level than that required for the reduction of bound sulfite (e.g. NADPH). Thus a total of eight electrons are used up in the reduction of one sulfate to the valency state of sulfur in cysteine.

Overall the data on the assimilatory sulfate reduction by intact chloroplasts discussed here are basically consistent with the studies on the assimilatory

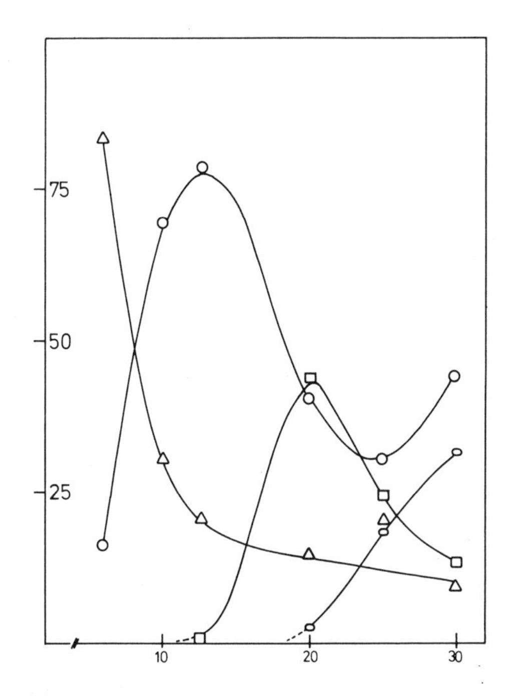

Fig. 9.3. The reaction kinetics of photosynthetic sulfate reduction by intact chloroplasts. △, APS + PAPS; ○, substances in peak two; □, bound sulfite; o, bound sulfide. Coordinate: percent (%) of fixed ^{35}S-labelled compounds in chloroplasts; abscissa: time (min).

sulfate reduction by yeast [11,73,74,83]. Except for the unknown substance in peak two and for the postulation of the bound sulfide (X-S:SH) the reaction sequences in both systems are analogous.

Schiff and his colleagues have also recently found that in the green alga, *Chlorella* a pathway with bound intermediates exists [57] together with an optional pathway of minor importance using free sulfite [65]. Although at the beginning of their studies thiosulfate had been identified as main intermediate [28,35,36] a bound sulfite (CarS:SO_3H) was later found as key intermediate [1] for the reduction of APS to acid volatile sulfur. The molecular weight of this intermediate was estimated to be around 1200 daltons. It differs from the bound sulfite of the chloroplasts by its tight binding to the enzyme APS sulfo transferase as a prosthetic group [1]. For further enzymatic details see page 328.

9.3.3. *The mechanism of photosynthetic sulfate reduction in chloroplasts*

Studies on the mechanism have been carried out with freshly ruptured chloroplasts [14], reconstituted chloroplasts [68,69], enzyme preparations from chloroplasts, leaf homogenates and algae [26,62—64] and with mutants of the green alga *Chlorella* (see ref. 57).

An updated scheme of the reactions involved, based on an earlier proposal by Schmidt and Schwenn [61], is presented in Fig. 9.4.

Following the change of valency of the sulfur during assimilation, the mechanism can be subdivided into four phases: (*i*) uptake of sulfate, (*ii*) activation and transfer of the sulfo group, (*iii*) reduction of the bound sulfite and (*iv*) the formation of cysteine.

9.3.3.1. *The uptake of sulfate*

The mechanism of sulfate uptake through the envelope of intact chloroplasts remains to be solved. In cell cultures of tobacco [24,53] the uptake of sulfate is controlled by the internal level of the sulfur amino acids. This finding essentially agrees with data arising from the work with microorganisms. In microorganisms sulfate uptake occurs in a highly controlled fashion via specific sulfate binding proteins [18,47]. However, specific permeases for sulfate in plant cells have not yet been detected. Mitochondria which show numerous similarities to chloroplasts, including membrane transport mechanisms, exchange dicarboxylate ions like malate, malonate and succinate for sulfate and sulfite [15,16]. With mitochondria the phosphate translocator seems not to be involved in sulfate transport but is indirectly influenced by the exchange of sulfite [15,16].

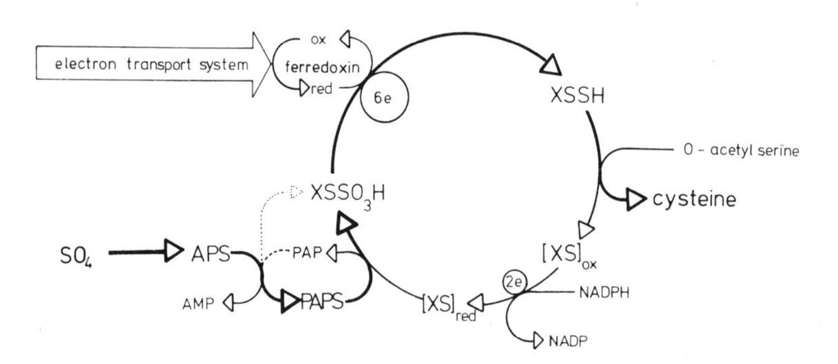

Fig. 9.4. A scheme of the assimilatory sulfate reduction by intact chloroplasts updated from an earlier proposal by Schmidt and Schwenn [61]. The heavy line follows the path of sulfur in the intact chloroplast, the dashed line shows as a hypothesis the action of a sulfokinase with PAP as recycling cofactor. The dotted arrow indicates the enzymic function of an APS-sulfo transferase as it is proposed for sulfate assimilation in the green alga *Chlorella*.

Although highly speculative, the inhibition by sulfate of carbon dioxide fixation and its reversal by orthophosphate, as shown by Baldry et al. [8] may be discussed on this basis of metabolite efflux from the chloroplast in exchange for sulfate uptake.

9.3.3.2. The activation of sulfate and transfer of the sulfo group

After sulfate entry into the chloroplast stroma it is metabolised to APS according to:

$$SO_4^{2-} + ATP \rightleftharpoons APS + PP_i \quad (5)$$

The equilibrium of the reaction given in Eqn. (5) is highly unfavorable for the forward reaction, thus APS does not accumulate in appreciable amounts. The reaction is catalysed by the enzyme ATP: sulfate adenylyl transferase, i.e. "sulfurylase". As mentioned already, in plants the enzyme was originally detected by Asahi [7] in broken chloroplasts. Since then several purification procedures have been described for leaf homogenates [9,50,59], employing one of the following assay methods; the molybdate (substituting for sulfate)-dependent hydrolysis of ATP, the back reaction in the presence of APS and PP_i and the sulfate dependent ATP—PP_i exchange. Using the latter assay Shaw and Anderson [50] reported a 1000-fold purification of the enzyme from spinach leaf tissue. The K_m of this reaction was found to be 3.1 mM for sulfate and 0.35 mM for ATP. The enzyme exhibited maximal activity at pHs ranging from 7.5 to 9.0. The enzyme was found insensitive to thiol group reagents and Mg^{2+} or CO^{2+} were required for maximal activity. The investigation of the sulfate-dependent ATP-[^{32}P]pyrophosphate exchange indicated a bireactant sequential mode of action, with $MgATP^{2-}$ as the first substrate to react with the enzyme [50]. Supporting evidence has recently been published by Marx et al. [41], demonstrating that in the yeast enzyme a complex of ATP with sulfurylase is formed in the forward reaction. However, binding of sulfate in the absence or sequentially in the presence of ATP has not yet been found.

The action of ATP-sulfurylase as the first enzyme of sulfate assimilation in chloroplasts can only be detected in the presence of pyrophosphatase which pulls the reaction in a forward direction by removing the pyrophosphate. A highly active pyrophosphatase has been reported in the stroma of chloroplasts [67].

The further activation of sulfate is accomplished by the action of the adenosine-5′-sulfatophosphate kinase which catalyses the irreversible reaction:

$$APS + ATP \rightarrow PAPS + ADP \quad (6)$$

Until recently the presence of APS kinase in chloroplasts was uncertain because no direct evidence could be obtained. The evidence for the existence of this enzyme was based on the occurrence of PAPS in plants and chloroplasts.

An explanation for the conflicting results on the existence and the physiological role of the enzyme could lie in the finding that the enzyme apparently is bound to the thylakoid membrane [14]. APS kinase is not extracted into aqueous chloroplast extracts but it is released into the extracting buffer only after ultrasonic treatment. This agrees well with the earlier observation by Mercer and Thomas [43] who showed that accumulation of PAPS occurred in extracts from sonicated maize and bean chloroplasts. Although APS-kinase has definitively been found in the chloroplast, the enzyme is not necessarily participating in sulfate reduction (see page 327).

The formation of bound sulfite X-S:SO_3H during assimilatory sulfate reduction is formally a transfer of an activated sulfate (mixed anhydride of sulfuric acid) from APS or PAPS onto an endogenous thiol, according to:

$$\text{adenosine—O—}\overset{\overset{\displaystyle O}{\|}}{\underset{\underset{\displaystyle O^-}{|}}{P}}\text{—O:}SO_3^- + \text{X-S:}^- \rightarrow \text{X-S:}SO_3^- + \text{adenosine—O—}\overset{\overset{\displaystyle O}{\|}}{\underset{\underset{\displaystyle O^-}{|}}{P}}\text{—O:}^- \quad (7)$$

where X-S:$^-$ represents the endogenous sulfite carrier protein in a reduced form. The enzyme catalysing the reaction given in Eqn. (7) is by definition a sulfo transferase. The transfer of the activated sulfate from an O-anhydride onto the thiol SH to form the bound sulfite is not yet a reduction of sulfate because its valency state in the thioester is not changed. The exchange of bound sulfite with free sulfite (see page 321) and the splitting of the bound sulfite with sufhydryl reagents to sulfite (Eqns. (1) and (4); the latter is a reduction) should not be confused with this.

The sulfate donor for the transfer reaction in chloroplasts of higher plants is assumed to be PAPS [14,61] which is readily formed in the intact and in the osmotically shocked chloroplasts. This assumption is supported indirectly by kinetic experiments with intact chloroplasts [66].

From the work with *Chlorella* extracts and mutants by Schiff and co-workers [57] and by Schmidt [63] it is proposed that the donor of the activated sulfate reduction is APS and not PAPS. Schiff and colleagues found that in the cell-free system of the alga, PAPS was only utilised after reconversion to APS, a degradation which is catalysed by a specific enzyme, 3′-nucleotidase, splitting PAPS to APS. Schmidt [63,64] found that APS serves as the sulfate donor in *Chlorella* extracts when reduced glutathione is used as acceptor in place of the endogenous carrier thiol "X". Binding experiments with mutants of *Chlorella* carried out by Abrams and Schiff [1] have been interpreted in favor of Schmidt's proposal. Abrams found that APS could be used in the formation of exchangeable sulfite (see also page 328). In recent communications [22,77] the role of the nucleotidase is seen in the generation of sufficient amounts of APS from stored PAPS. This is assumed to happen under conditions where the rate of APS formed by the sulfurylase would otherwise severely limit the complete sulfate reducing sequence. Moreover a regulating function by this enzyme (also called fraction "A") is discussed as a switch between sulfate ester formation or assimilatory sulfate reduction (for review see ref. 57). Strong but not conclusive evidence has been obtained that PAPS is used for the formation of sulfolipids in plants and algae [14a]. Of course, a nucleotidase as proposed by Schiff's group would have to be present also when PAPS alone would serve as sulfate donor (in sulfate reduction as well as in O-ester formation), because the resulting PAP would have to be degraded to AMP and orthophosphate to prevent its unwanted accumulation. However, to date this very specific enzyme is found in cell extracts from algae only [22] but it seems to be virtually absent from chloroplasts [14].

The sulfate acceptor is represented by the endogenous carrier X-SH and has recently been isolated, as bound sulfite X-S:SO_3H from chloroplasts [68,69]. A heat-stable non-dialysable protein fraction from the chloroplast extract is assumed to contain the endogenous low molecular weight sulfite carrier protein. This fraction was shown to increase the amount of bound sulfite (X-S:SO_3H) formed in the light by reconstituted chloroplasts [68]. Using the same type of chloroplasts evidence was found that the acceptor first is reduced to the equivalent thiol by NADPH if in the disulfide state [69]. See also page 331.

In the *Chlorella* system (as discussed above) Abrams and Schiff [1] found that the exchangeable sulfite is bound to a small acceptor molecule of an APS-sulfo transferase complex forming the CarS:SO_3H. This CarS:SO_3H substance is assumed to serve as the physiologically true substrate of a bound sulfite reducing enzyme. According to this proposal it would be the APS-sulfo transferase which forms the key intermediate bound sulfite, in the form of an enzyme-bound prosthetic group. A reductase which reduces this enzyme-bound sulfite to enzyme bound sulfide would then have to act upon the APS-sulfo transferase complex. The consequences of this model will be discussed in section 9.3.3.3(*i*).

9.3.3.3. The reduction of bound sulfite

(*i*) *The thiosulfonate reductase ("bound sulfite reductase").* The evidence presented in the course of the above discussions led to the proposal of an enzyme which reduces the bound sulfite to bound sulfide [61] according to:

$$\text{X-S:SO}_3\text{H} + 6\,e \rightarrow \text{X-S:SH} \qquad (8)$$

The enzyme has been designated thiosulfonate reductase [64]. It catalyses the reductive step which consumes six electrons like nitrite or nitrogen reduction. No intermediate valency states of sulfur have been detected.

A soluble thiosulfonate reductase has also been reported from *Chlorella* [61,64]. In these studies analogous substrates like S-sulfo glutathione have been used, because no substrate amounts of the bound sulfite X-S:SO_3H were available. Schmidt found that this enzyme works with reduced ferredoxin as the physiologically relevant reductant for the reduction of S-sulfo glutathione to the glutathione persulfide [61].

Ferredoxin could be replaced by methyl viologen in the thiosulfonate reductase assay. Moreover Schmidt found that the enzyme could easily be enriched with an assay system in which dithionite serves as substrate. The reaction is followed by hydrogen sulfide formation showing extremely high rates. According to Schmidt, the purified enzyme still reduces S-sulfo glutathione in a ferredoxin-dependent reaction [64].

With an enzyme, purified according to the above method (by following the dithionite reduction) from spinach leaves, Hennies found that the plant enzyme reduces S-sulfo glutathione; however, only up to a certain grade of purity [25,68]. The enzyme purified to homogeneity did neither reduce S-sulfo glutathione, nor free sulfite nor the physiologically important bound sulfite prepared from the chloroplast. Therefore the relevance of the dithionite-reducing enzyme to the thiosulfonate reductase (Eqn. (8)) in the sulfate-reducing process becomes somewhat doubtful. It actually seems possible that a number of different enzymes may have dithionite-reducing activity (Hennies, pers. commun.). Thus, the reduction of dithionite is not at all a model reaction for the proposed thiosulfonate reductase.

Relying on data from studies with extracts from *Chlorella*, Schiff's group also discussed that a thiosulfonate-reducing enzyme supposingly acts upon a bound sulfite ("CarS:SO_3H") of the APS-sulfo transferase complex to yield a bound sulfide ("CarS:SH") [1,57]. The compounds CarS:SO_3H and CarS:SH are presumably analogues to the bound intermediates X-S:SO_3H and X-S:SH found in the chloroplast earlier, see page 327. With crude lyophilised extracts from *Chlorella* some evidence has recently been presented [65] that APS is reduced to cysteine by the thiosulfonate reductase prepared from *Chlorella* according to Schmidt [64]. From the previous data it remains unclear whether the bound sulfite is a prosthetic group of the thiosulfonate reductase [64,65] or the prosthetic group of the APS-sulfo transferase complex as proposed by Abrams [1]. In this respect, the binding experiments with the partially purfied enzymes from *Chlorella* are difficult to assess because both enzymes, the thiosulfonate reductase and the APS-sulfo transferase, are reported to exchange the labelled sulfite of the prosthetic group. Note that in the first case the sulfite derives from S-sulfo glutathione [64] and in the latter it derives from APS [1].

(*ii*) *The sulfite reductase (free sulfite reducing).* A ferredoxin-dependent sulfite reductase (reducing free sulfite to free hydrogen sulfide) from plants and microorganisms is well known. Sulfite reductases from microorganisms [48] and from plants [3,4,25,40,59,70] have been investigated in the past.

The plant-type sulfite reductase uses methyl viologen as artificial [4,5], and ferredoxin as physiological, reductant [25,59]. The absence of a flavin component in the enzyme [5,25] is possibly linked with the specificity towards the reductant, ferredoxin, because the plant enzyme does not use NADPH as compared to the sulfite reductases from microorganisms which contain the flavin. It should be mentioned in this context that the prosthetic group of the NADPH-linked enzyme from *E. coli* has recently been identified as a "siroheme" [45]. The previous results are carefully compiled in the monograph by Roy and Trudinger [48].

The physiological importance of a sulfite reductase acting upon free sulfite remains questionable. As discussed before in section 9.3.2, free sulfite does not occur as an intermediate of assimilatory sulfate reduction in plants and the sulfite reductase purified from spinach does not reduce the bound sulfite (cf. ref. 63). These conflicting results made it obvious that in plants a specific enzyme is required for the reduction of bound sulfite. However, it should not be ruled out completely that the sulfite reductases isolated from spinach possibly represent a subunit of a complex or a denatured part of an enzyme capable of the reduction of bound sulfite in vivo. It is tempting to speculate that the sulfite reductase and the thiosulfonate reductase have an identical core. Schiff recently discussed the role of sulfite reductase as an optional enzyme removing accidentally produced sulfite (cf. ref. 65).

(iii) The thylakoid membrane-bound thiosulfonate reductase. Very recently it has been found that the bound sulfite is formed by the thylakoid membrane of the chloroplast. This activity was retained by the lamellae even after consecutive washes. On the other hand, the extract of the chloroplast stroma stimulated the amount of bound sulfite formed in and by the membrane fraction. Reducing equivalents in the form of NADPH were required for this stimulation. This is interpreted as the reduction of the sulfite-binding sites localized in the lamellae. The reduction which presumably is carried out by a stromal enzyme thereby increases the capacity for the binding of sulfite [69]. Furthermore, a reduction of bound sulfite to bound sulfide by the thylakoid membrane fraction has been observed in reconstituted chloroplasts [25,68]. Strong evidence for the proposal of a membrane-bound thiosulfonate reductase arises from experiments with antibodies which originally have been made against particles of the chloroplast lamellae [25]. Although the membrane-bound enzyme has not yet been purified to a homogeneous state, its spectral data resemble those of the soluble sulfite reductase from spinach [25,26].

More significant was its capability to equally catalyse the reduction of isolated bound sulfite from chloroplasts, free sulfite and dithionite. The reduction of bound sulfite to bound sulfide uses reduced ferredoxin as electron carrier. It could be shown that the partly purified enzyme fraction, if added to a reconstituted chloroplast system fed with labelled sulfate in the light, stimulated the reduction of bound sulfite to bound sulfide [25,68]. However, the possibility remains that other enzymatic activities of this fraction were responsible for the stimulation because of the lack of absolute purity.

The view of a membrane-bound thiosulfonate reductase is supported by results from experiments with reconstituted chloroplasts [69]. The data indicated that even in washed thylakoid membranes a sulfate-reducing activity is found but not in the soluble stromal protein of the chloroplasts.

9.3.3.4. *The formation of cysteine*

The terminal step in assimilatory sulfate reduction is the formation of cysteine according to:

$$H_2S + \text{L-serine} \rightarrow \text{L-cysteine} \quad (9)$$
$$H_2S + \text{O-acetyl L-serine} \rightarrow \text{L-cysteine} + \text{acetate} \quad (10)$$
$$\text{X-S:SH} + \text{O-acetyl L-serine} + 2e \rightarrow \text{L-cysteine} + \text{acetate} + \text{X-S:}^- \quad (11)$$

The reaction given in Eqn. (9) is catalysed by the L-serine sulfhydrylase which is found in microorganisms, plants like spinach and in animal tissue. For further details the reader is referred to the reviews of Thompson [72] and Roy and Trudinger [48].

In the reaction given in Eqn. (10) the formation of cysteine is catalysed by the O-acetyl L-serine sulfhydrylase which has been described for Salmonella and *E. coli* [11a,33a] and plant material [21a,b,46].

The proposal of a bound sulfide [61,66] requires a mechanism according to reaction (11). In this reaction bound sulfide donates the reduced sulfur in a reaction for which two electrons are necessary (see below). Reaction (11) also covers the proposal by Schmidt [64]. He postulated a mechanism in which an intermediate enzyme—product complex is formed which then has to be cleaved in a two electrons requiring process.

The formation of cysteine has been observed to occur in intact chloroplasts with sulfate as the source of sulfur [59,66]. This finding supports the proposal that the complete sulfate-reducing cycle is localised inside the chloroplast. However, the rates obtained with intact chloroplasts are comparatively low — in the range of 0.5 $nmoles.mg\ chl^{-1}.h^{-1}$. Higher rates of cysteine formation were found recently with reconstituted chloroplasts [69]. This type of chloroplast formed cysteine at rates in the range of 12 $nmoles.mg\ chl^{-1}.h^{-1}$. It was shown that O-acetyl L-serine is probably used very likely as carbon precursor of cysteine, confirming earlier results obtained with cell-free systems from plants [46] and from algae [64].

Enzymatic studies with partially purified enzymes from spinach or other plant material have been carried out previously. Two principal pathways are discussed: (*a*) the pyridoxal-dependent formation and (*b*) a pathway using O-acetyl L-serine independently of pyridoxal phosphate (see review ref. 72). The second pathway has recently been confirmed to exist in higher plants [46]. The plant enzymes have been studied with hydrogen sulfide as sulfur source and O-acetyl L-serine as the carbon skeleton. Serine proved rather ineffective and the enzyme from rape seedlings seemed to be specific for acetylserine, although it has not yet been isolated from a plant source. The general metabolism of cysteine is beyond the intention of this article and the reader is referred to the review by Thompson [72].

As mentioned above, the pathway with bound sulfide demands a reductive cleavage of the sulfur-sulfur bond in the persulfide according to the reaction given in Eqn. (11). Here an early proposal by Hilz et al. [26a] may be applicable. According to Eqn. (12) the sulfur-sulfur bond can be split by an intramolecular thiolysis with an adjacent sulfhydryl group:

$$X\langle\begin{matrix} S{:}SH \\ SH \end{matrix} + \text{O-acetyl-serine} \rightarrow X\langle\begin{matrix} S \\ \vdots \\ S \end{matrix} + \text{cysteine} + \text{acetate} \qquad (12)$$

The carrier "X" would be oxidised after this reaction, yet, it has to be reduced to serve as sulfo group acceptor in the formation of bound sulfite, as discussed on page 326, Eqn. (7). The regeneration of the carrier consumes two electrons. Schwenn et al. [69] recently suggested that the NADPH requirement of the formation of bound sulfite in reconstituted chloroplasts possibly reflects the reduction of the oxidised endogenous carrier disulfide.

The actual transfer of the sulfur which becomes necessary in the reaction given by Eqn. (12) could be catalysed by an enzyme of the rhodanese type. However, there is no final evidence that this enzyme exists in chloroplasts.

9.4. SUMMARY

Physiological evidence has led to the proposal that photosynthetic sulfate reduction occurs in the leaf and is possibly localized in the chloroplasts. Isolated intact chloroplasts were shown to contain the complete sulfate-reduction mechanism from sulfate to cysteine which is apparently coupled to the light reactions of photosynthesis.

Intermediates of the reduction process were identified as activated sulfate (APS and PAPS), a thiol-bound sulfite (X-S:SO_3H) as well as a thiol-bound sulfide (X-S:SH). A ferredoxin-dependent thiosulfonate reductase is shown to catalyse the six electron requiring reduction of X-S:SO_3H to X-S:SH. Two more electrons are consumed for the formation of cysteine including the regeneration of the carrier XSH. The electron transport system of photosynthesis therefore provides energy for sulfate reduction, that is ATP, reduced ferredoxin and NADPH (see Fig. 9.5).

Sulfate reduction shows significant similarities to nitrate reduction, in so far as the six electron reduction of nitrite is also directly coupled to the electron transport system of chloroplasts via the electron carrier

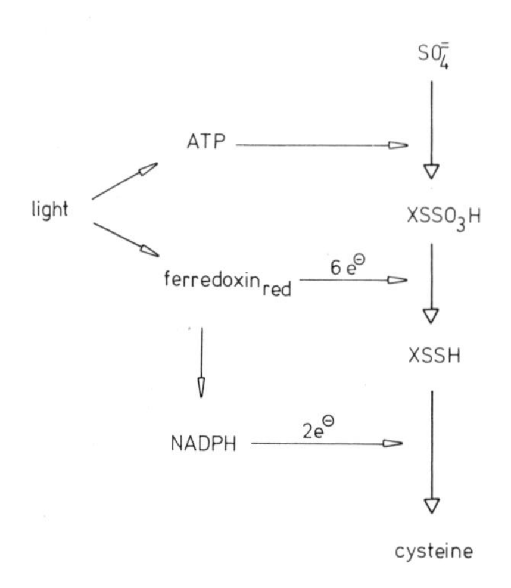

Fig. 9.5. The coupling of the light reaction to assimilatory sulfate reduction in chloroplasts.

ferredoxin (see Chapter 11 of this volume). Furthermore, the prosthetic group of both nitrite and sulfite reductase has been identified as "siroheme" [45].

Chloroplasts then generate in the light the reducing power for three important reduction processes in which inorganic material like carbon dioxide, nitrate and sulfate are converted into organic substances and thus enables the plant to perform autotrophic growth.

REFERENCES

1 Abrams, W.R. and Schiff, J.A. (1973) Arch. Mikrobiol., [94], 1—10.
2 Ames, D.P. and Willard, I.E. (1951) J. Am. Chem. Soc., [73], 164—172.
3 Asada, K. (1967) J. Biol. Chem., [242], 3646—3654.
4 Asada, K., Tamura, G. and Bandurski, R.S. (1968) Biochem. Biophys. Res. Commun., [30], 554—559.
5 Asada, K., Tamura, G. and Bandurski, R.S. (1969) J. Biol. Chem., [244], 4904—4915.
6 Asada, K. and Kiso, K. (1973) Eur. J. Biochem., [33], 253—257.
7 Asahi, T. (1964) Biochim. Biophys. Acta, [82], 58—66.
8 Baldry, C.W., Cockburn, W. and Walker, D.A. (1968) Biochim. Biophys. Acta, [153], 476—483.
9 Balharry, G.J.E. and Nicholas, D.J.D. (1970) Biochim. Biophys. Acta, [220], 513—524.
10 Bandurski, R.S., Wilson, L.G. and Asahi, T. (1960) J. Am. Chem. Soc., [82], 3218—3219.
11 Bandurski, R.S. (1965) in Plant Biochemistry (Bonner, J. and Varner, J.E., eds), pp. 467—490, Academic Press, New York.
11a Becker, M.A., Kredich, N.M. and Tomkins, G.M. (1969) J. Biol. Chem., [244], 2418—2427.

12 Benson, A.A., Daniel, H. and Wiser, R. (1959) Proc. Natl. Acad. Sci. USA, [45], 1582—1587.
13 Benson, A.A. (1966) J. Am. Oil. Chem. Soc., [43], 265—270.
14 Burnell, J.M. and Anderson, J.W. (1973) Biochem. J., [134], 565—579.
14a Davis, W.H., Mercer, E.I., Goodwin, T.W. (1966) Biochem. J., [98], 369—373.
15 Crompton, M., Palmieri, F., Capano, M. and Quagliariello, E. (1974) Biochem. J., [142], 127—137.
16 Crompton, M., Palmieri, F., Capano, M. and Quagliariello, E. (1974) FEBS Lett., [46], 247—250.
17 Cupoletti, J. and Segel, I.H. (1974) J. Membrane Biol., [17], 239—252.
18 Dreyfus, J. and Pardee, A.B. (1966) J. Bacteriol., [91], 2275—2280.
19 Ellis, R.J. (1969) Planta, [88], 34—42.
20 Fromageot, P. and Perez-Milan, H. (1959) Biochim. Biophys. Acta, [32], 457—464.
21 Furrer, O.J. (1966) in Proc. Symp. Isotopes in Plant Nutrition and Physiology, pp. 403—407, International Atomic Energy Agency, Vienna.
21a Giovanelli, J. and Mudd, S.H. (1967) Biochem. Biophys. Res. Commun., [25], 366—371.
21b Giovanelli, J. and Mudd, S.H. (1968) Biochem. Biophys. Res. Commun., [27], 150—156.
22 Goldschmidt, E.E., Tsang, M.L.-S. and Schiff, J.A. (1975) Plant Sci. Lett., [4], 293—299.
23 Gurjanowa, J.N. and Nassiljewa, W.N. (1954) J. Physik. Chemie, [28], 576—578.
24 Hart, J.W. and Filner, P. (1969) Plant Physiol., [44], 1253—1259.
25 Hennies, H.-H. (1974) Thesis University Bochum.
26 Hennies, H.-H. (1975) Z. Naturforsch., [30c], 359—362.
26a Hilz, H., Kittler, M. and Knape, G. (1959) Biochem. Z. [332], 151—161.
27 Hodson, R.C., Schiff, J.A., Scarsella, A.J. and Levinthal, M. (1968) Plant Physiol., [43], 563—569.
28 Hodson, R.C. and Schiff, J.A. (1971) Plant Physiol., [47], 296—299.
29 Iguchi, A. (1958) Bull. Chem. Soc. Japan, [31], 600—605.
30 Jensen, R.G. and Bassham, J.A. (1966) Proc. Natl. Acad. Sci. USA, [56], 1095—1101.
31 Kaji, A. and Elroy, W.D. (1959) J. Bacteriol., [77], 630—637.
32 Kawashima, N. and Asahi, T. (1961) J. Biochem. (Tokyo), [49], 52—54.
33 Kjaer, A. (1958) in Handbuch der Pflanzenphysiologie (Ruhland, W.,ed), pp. 64—88. Springer, Berlin.
33a Kredich, N.M. Becker, M.A. and Tomkins, G.M. (1969) J. Biol. Chem., [244], 2428—2439.
34 Kylin, A. (1953) Physiol. Plant., [6], 775—795.
35 Levinthal, M. (1967) Diss. Abstr., [27b], 4281.
36 Levinthal, M. and Schiff, J.A. (1968) Plant Physiol., [43], 555—562.
37 Lüttge, U. and Weigl, J. (1962) Planta, [58], 113—126.
38 Lüttge, U. and Weigl, J. (1962) Planta, [59], 15—28.
39 Lüttge, U. (1969) Aktiver Transport (Kurzstreckentransport) bei Pflanzen, Vol. VIII, in Physiologie des Protoplasmas, [7b], p. 97. Springer, Wien.
40 Mager, J. (1960) Biochim. Biophys. Acta, [41], 553—555.
41 Marx, W.M. Shoyab and Su, L.Y. (1974) Int. J. Biochem., [5], 471—477.
42 Marzluf, G.A. (1974) Biochim. Biophys. Acta, [339], 374—381.
43 Mercer, E.I. and Thomas, G. (1969) Phytochemistry, [8], 2281—2285.
44 Moore, S.J. (1963) J. Biol. Chem., [238], 235—237.
45 Murphy, M.J. and Siegel, L.M. (1973) J. Biol. Chem., [248], 6911—6919.
46 Ngo, T.T. and Shargool, P.D. (1974) Canad. J. Biochem., [52], 435—440.

47 Pardee, A.B., Prestidge, L.S., Whipple, M.B. and Dreyfus, J. (1966) J. Biol. Chem., [241], 3962–3969.
48 Roy, A.B. and Trudinger, P.A. (1970) in The Biochemistry of Inorganic Compounds of Sulphur, Cambridge University Press, Cambridge.
49 Ruhland, W. (ed) (1958) in Handbuch der Pflanzenphysiologie, Bd. IX, pp. 2–120, Springer, Berlin.
50 Shaw, W.H. and Anderson, J.W. (1972) Biochem. J., [127], 237–247.
51 Shaw, W.H. and Anderson, J.W. (1974) Biochem. J., [139], 27–35.
52 Sinha, S.K. and Cossins, E.A. (1963) Nature, [199], 119.
53 Smith, I.K. (1975) Plant Physiol., [55], 303–307.
54 Stokes, D.M. and Walker, D.A. (1971) in Photosynthesis and Photorespiration (Osmond, C.B. and Slatyer, R.O., eds), pp. 226–231, Wiley Interscience, New York.
55 Syrkin, J.K. and Gurjanowa, J.N. (1952) Ber. Akad. Wiss. USSR, [86], 107, quoted in Brodskij, A.E. (1961) Isotopenchemie, pp. 371–373, Akademie Verlag, Berlin.
56 Szczepkowski, T.W. (1958) Nature, [182], 934–935.
57 Schiff, J.A. and Hodson, R.C. (1973) Ann. Rev. Plant Physiol., [24], 381–414.
58 Schlossmann, K., Brüggemann, F. and Lynen, F. (1962) Biochem. Z., [336], 258–273.
59 Schmidt, A. (1968) Thesis University Göttingen.
60 Schmidt, A. and Trebst, A. (1969) Biochim. Biophys. Acta, [180], 529–535.
61 Schmidt, A. and Schwenn, J.D. (1971) in Proc. 2nd Int. Cong. Photosynthesis (Forti, G., Avron, M. and Melandri, A., eds), pp. 507–517, Junk, The Hague.
62 Schmidt, A. (1972) Z. Naturforsch., [27b], 183–192.
63 Schmidt, A. (1972) Arch. Mikrobiol., [84], 77–86.
64 Schmidt, A. (1973) Arch. Mikrobiol., [93], 29–52.
65 Schmidt, A., Abrams, W.R. and Schiff, J.A. (1974) Eur. J. Biochem., [47], 423–434.
66 Schwenn, J.D. (1970) Thesis University Bochum.
67 Schwenn, J.D. Lilley, R. McC. and Walker, D.A. (1973) Biochim. Biophys. Acta, [325], 586–595.
68 Schwenn, J.D. and Hennies, H.-H. (1974) Proc. 3rd Int. Cong. Photosynthesis (Avron, M., ed), pp. 629–635, Elsevier, Amsterdam.
69 Schwenn, J.D., Depka, B. and Hennies, H.H. (1976) Plant and Cell Physiol., [17], 165–176.
70 Tamura, G. (1970) Techn. Bull. Fac. Horticult., [18], Chiba University.
70a Thomas, M.D., Hendricks, R.H., Bryner, L.C. and Hill, G.R. (1944) Plant Physiol., [19], 212–226.
71 Thomas, M.D. (1958) in Handbuch der Pflanzenphysiologie (Ruhland, W., ed), Bd. IX, pp. 37–63, Springer, Berlin.
72 Thompson, J.F. (1967) Ann. Rev. Plant Physiol., [18], 59–84.
73 Torii, K. and Bandurski, R.S. (1964) Biochem. Biophys. Res. Commun., [14], 537–542.
74 Torii, K. and Bandurski, R.S. (1967) Biochim. Biophys. Acta, [136], 286–295.
75 Trebst, A. and Schmidt, A. (1969) in Progress in Photosynthesis Research (Metzner, H., ed), Vol. III, pp. 1510–1516, Lichtenstein, München.
76 Tsang, M.L.-S. and Schiff, J.A. (1973) Plant Physiol., [51], Suppl. 53.
77 Tsang, M.L.-S. and Schiff, J.A. (1975) Plant Sci. Lett., [4], 301–307.
78 Voge, H. and Libby, W. (1936) J. Am. Chem. Soc., [59], 2474.
79 Voge, H. and Libby, W. (1939) J. Am. Chem. Soc., [61], 1032–1035.
80 Weigl, J. and Ziegler, H. (1962) Planta, [58], 435–447.
81 Willenbrink, J. (1964) Z. Naturforsch., [19b], 356–357.
82 Willenbrink, J. (1967) Z. Pflanzenphysiol., Bd. 56, 427–438.
83 Wilson, L.G., Asahi, T. and Bandurski, R.S. (1961) J. Biol. Chem., [236], 1822–1830.

The Intact Chloroplast — edited by J. Barber

Chapter 10

Protein and Nucleic Acid Synthesis by Chloroplasts

R. JOHN ELLIS

Department of Biological Sciences, University of Warwick, Coventry CV4 7AL (Great Britain)

CONTENTS

Abbreviations: CCCP, *m*-chlorocarbonyl cyanide phenylhydrazone; DCMU, 3-(3,4-dichlorophenyl)-1,1-dimethylurea; GC, guanosine + cytidine; HEPES, *N*-2-hydroxyethyl-piperazine-*N*′-2-ethanesulphonic acid; SDS, sodium dodecyl sulphate; tricine, *N*-tris-(hydroxyethyl)methylglycine.

10.1. INTRODUCTION — THE CHANGING STATUS OF CHLOROPLAST AUTONOMY

Chloroplasts are important because they contain the enzymic machinery for the process of photosynthesis on which life on this planet largely depends. The extensive literature on the mechanism of photosynthesis is sufficient witness to the interest of biochemists and physiologists in this process, an interest reaching back to the beginning of the last century. An entirely separate thread of interest in chloroplasts can be discerned however. This thread originated with Strasburger [93], who observed that in some algae chloroplasts divide, and are passed to the daughter cells in cell division. This cytological evidence led to the view, first proposed by Schimper [85] and Meyer [73] in 1885, that chloroplasts do not arise de novo, but are formed by the division of pre-existing plastids. Fifteen years after this view was proposed, a most notable event took place — the rediscovery of the Mendelian laws of inheritance. The realisation by Sutton that the results of Mendel's experiments with garden peas could be explained in terms of the visible behaviour of chromosomes during meiosis, led to the nucleus being regarded as the sole carrier of the hereditary material. This splendidly simple view did not last for long. The results of experiments by Baur [3] and Correns [19] on the inheritance of plastid defects in variegated plants were difficult to explain on the basis that the genes concerned were located in the nucleus. In certain cases, the inheritance of such defects occurred only through the maternal line, i.e. it was uniparental. In other cases the inheritance was biparental, but did not obey the rules of Mendel. For a time it appeared that, far from being "nuclear", some aspects of inheritance were "unclear". It was Baur who pointed out that these results were explicable on the assumption of the genetic continuity of an extrachromosomal entity located in the plastid. Much more extensive evidence for this concept was later provided by the work of Renner [76] using the genus *Oenothera*; he suggested the term "plastome" to describe the genetic system in the plastid. Maternal inheritance is thus explicable in terms of the lack of plastids in the pollen tube of some species; the plastome of the new generation is consequently derived entirely from the mother.

These studies of variegated plants provided the first firm evidence for the existence of extrachromosomal inheritance. Since then, further evidence has accumulated with examples known in representatives of most major groups of organisms; the work of Sager on *Chlamydomonas* [79] is most notable in this regard. The concept arose that chloroplasts themselves contain genetic material controlling at least part of their development, and are thus genetically autonomous in some sense. The discovery in 1962 that chloroplasts contain both DNA and ribosomes, opened the modern era in which the *development* of chloroplasts is regarded as of equal interest to their functioning in photosynthesis.

The explosive advances in biochemical techniques and concepts that started in the 1950s have brought about an intertwining of these two threads of interest in chloroplasts. Today we realise that organelle genetic systems are a fundamental feature of the organisation of eukaryotic cells, and that therefore it is as important to ask where a chloroplast enzyme is synthesised and encoded as it is to enquire about its role in photosynthesis.

The discovery of chloroplast DNA and chloroplast ribosomes brought the idea of chloroplast autonomy into vogue to such an extent that several attempts have been made to grow isolated chloroplasts in culture. It is clear, however, from both the earlier genetic evidence and the more recent biochemical evidence, that such attempts are ill-founded. Our current dogmas maintain that for any biological system to be autonomous it must contain four components; (*a*) DNA to code for its *entire* structure; (*b*) DNA polymerase to replicate the DNA: (*c*) RNA polymerase to transcribe the DNA; (*d*) protein-synthesising machinery to translate the messenger RNAs into *all* the necessary proteins. Intermediary metabolism is not necessary in principle since a supply of small molecules could be taken up from the environment. An extensive literature establishes that chloroplasts do in fact contain these four components, the properties and functions of these are discussed in the rest of this chapter. It is equally clear that the chloroplast DNA does not code for all the chloroplast proteins nor does the protein-synthesising system make all the chloroplast proteins [12,29,31,59,79]. For example, many genes concerned with chloroplast structure and function are inherited in a Mendelian fashion, and are therefore presumed to be located in the nucleus, while there is increasing evidence that the majority of chloroplast proteins are synthesised on cytoplasmic ribosomes. It is now realised that the demonstration of cytological continuity of plastids from generation to generation is not sufficient to establish that they replicate independently of nuclear control. How, then, are we to regard the concept of organelle autonomy?

It is my contention that chloroplasts are not autonomous in any rigorous sense; the term is useful only as a quick way to describe the fact that these organelles contain *some* genes and make *some* proteins. We must regard the formation of chloroplasts as resulting from a complex interplay between the genome of the chloroplasts and the genome of the nucleus, and the fascination of this subject lies in unravelling the details of this interplay at the molecular level.

The current state of knowledge can be summarised by saying that, while some information has been accumulated about the properties of chloroplast nucleic acids and protein synthesis, only recently has any hard evidence emerged about their biological function. Developments at the genetic level in higher plants, and the establishment of isolated chloroplast systems which synthesise specific protein and RNA molecules, promise to provide increasing understanding of the precise functions of chloroplast nucleic acids and protein synthesis. The study of the synthesis of nucleic acids and proteins by *intact* isolated chloroplasts has proved especially fruitful, and will be emphasised in this review.

10.2. CHLOROPLAST DNA

10.2.1. Discovery

Both biochemical and cytological methods have been used to demonstrate the existence of DNA in chloroplasts. The first convincing evidence that chloroplasts contain DNA was published in 1962 by Ris and Plaut [77]. These authors found in both algal and higher plant chloroplasts areas of low electron density which contained fibrils sensitive to deoxyribonuclease. These fibrils were 2.5—3.0 nm in diameter, as expected for naked DNA stained with metal ions. Attempts to demonstrate chloroplast DNA by the Feulgen technique were not successfully repeated by other workers, and it is now clear that in most cases this technique is not sensitive enough to detect chloroplast DNA in situ. A much better method is electron microscopic autoradiography after exposure of cells to [^{3}H]thymidine [78]. It is important when evaluating such autoradiographic studies to consider the controls that are used. If the incorporated label is all present in DNA, it should be removed by treatment of sections with deoxyribonuclease, but not by treatment with ribonuclease. This control works well for tobacco and spinach leaves and several algae, but not for maize leaves, where deoxyribonuclease fails to remove the label from chloroplasts [66].

Most studies on chloroplast DNA since 1962 have been carried out using isolated chloroplasts. This approach suffers from the difficulty of distinguishing chloroplast DNA from DNA originating from nuclei, mitochondria, and micro-organisms which may contaminate the chloroplast pellet. The resolution of these problems is still not complete, and depends mainly on the differing properties of the DNA from the various sources. There are four characteristic properties of chloroplast DNA that have been used to identify it, namely buoyant density, ease of renaturation, lack of histones, and the absence of 5-methylcytosine; these properties are discussed below. It must be remembered that there is a possibility that some chloroplast DNA may not meet these criteria. If, for example, chloroplasts in some species contain a minor DNA component which is a nuclear transcript, contains 5-methylcytosine and renatures poorly, it is doubtful whether present techniques could identify it as chloroplastic in origin. In one species the problem of contamination by nuclei and micro-organisms can be entirely avoided; DNA has been identified in chloroplasts isolated from sterile enucleated plants of *Acetabularia* [1,36].

10.2.2. Buoyant density

It is now clear that chloroplast DNA from the algae *Euglena*, *Chlamydomonas*, and *Chlorella* has a buoyant density sufficiently different from that of nuclear DNA to allow resolution by analytical ultracentrifugation, but that in many higher plants the densities of the two DNA types are often, but not always, too close to permit this. This information has been hard-won; the story of the way in which the interpretation of band patterns in neutral caesium chloride density gradients has changed since 1963 has been told by Kirk [57] and Tewari [97]. A summary is given here since it illustrates the very real problem of establishing the identity of chloroplast DNA from higher plants by biochemical methods.

The first report of the chemical characterisation of higher plant chloroplast DNA was provided by Kirk [55]. He found that chloroplast DNA from *Vicia faba* has a guanosine + cytidine (GC) content (37.4%) which is slightly but significantly different from that of nuclear DNA (39.4%). In the same year a contrasting report was published by Chun et al. [18], who found that chloroplast preparations from *Spinacia oleracea* and *Beta vulgaris* contained two components of much higher densities than the nuclear DNA as judged by caesium chloride equilibrium density gradient centrifugation. However, the bulk of the DNA in these chloroplast preparations had a density similar to that of nuclear DNA; this band was attributed to contamination of the chloroplast pellet by nuclear fragments, and thus the chloroplast DNA was regarded as the two high-density components. There followed a spate of papers which supported the claim that, in higher plants, chloroplast DNA has an appreciably higher density than nuclear DNA. This picture changed when re-examination by more rigorous methods supported the earlier view of Kirk, and the higher density components are now attributed to contamination by mitochondria and bacteria. The present position can be summarised by saying, that in all the higher plants examined, chloroplast DNA has a base composition of 37.5 ± 1% GC and a buoyant density of 1.697 ± 0.001 $g.cm^{-3}$ [31,57,97]. Table I lists some values for the buoyant densities of chloroplast and nuclear DNA compiled from the literature. It can be seen that nuclear DNA has a smaller, larger, or similar density to chloroplast DNA from the same plant, depending on the species.

The moral to be drawn from this story is that the purity of subcellular fractions from higher plant tissue must be established by positive methods if meaningful interpretations are to be made. Microbial contamination can be reduced by surface sterilisation of fresh tissues and by the use of sterile solutions, while mitochondrial contamination can be assayed by succinic dehydrogenase or cytochrome oxidase activities. Nuclear contamination can be reduced by incubation of intact chloroplasts with deoxyribonuclease and phosphodiesterase [48]. The extent of nuclear contamination is difficult to measure, but the ease of renaturation and the absence of 5-methylcytosine and histones serve to distinguish chloroplast from nuclear DNA in all cases so far studied.

TABLE I

Buoyant densities of algal and higher-plant chloroplast and nuclear DNA

Species	Buoyant density in g.cm^{-3}		
	Chloroplast DNA	Nuclear DNA	Ref.
Vicia faba	1.696	1.696	63
Spinacia oleracea	1.697	1.694	105
Nicotiana tabacum	1.697	1.697	109
Phaseolus vulgaris	1.697	1.697	112
Triticum vulgare	1.697	1.702	106
Oenothera hookeri	1.697	1.703	48
Chlorella ellipsoidea	1.692	1.716	51
Euglena gracilis	1.685	1.707	14
Chlamydomonas reinhardi	1.695	1.723	80

10.2.3. Absence of histones and 5-methylcytosine

When released from intact isolated chloroplasts by gentle lysis, chloroplast DNA is not combined with basic proteins. This finding confirms the microscopic observations of Ris and Plaut [77] that chloroplasts contain DNA fibrils in the same form as they appear in the nucleoplasms of prokaryotic cells. It is also characteristic of chloroplast DNA from both higher plants and algae that it contains no detectable 5-methylcytosine, whereas nuclear DNA invariably does. Whitfeld and Spencer [109] regard the absence of this base as the most reliable criterion for establishing the purity of chloroplast DNA.

10.2.4. Ease of renaturation

A more useful criterion than the buoyant density for detecting chloroplast DNA from higher plants is the ease with which it will renature after heat or alkali denaturation. Nuclear DNA renatures only to a slight extent in a few hours, depending on its content of reiterated sequences. The rapid renaturation of chloroplast DNA has encouraged attempts to estimate its genome size from measurements of kinetic complexity. The kinetic complexity of a DNA sample is a measure of the size of the unique set of nucleotide sequences it contains, as judged from the rate at which the DNA renatures [15,108]. A rapid rate of renaturation implies that like sequences are present in high concentration, and therefore that the number of different unique sequences, or kinetic complexity, is small. Table II lists the kinetic complexity, in terms of molecular weight, for the chloroplast DNA from

TABLE II

Corrected kinetic and analytical complexities of chloroplast DNA

Species	Kinetic complexity (Daltons · 10^8)	Analytical complexity (Daltons per chloroplast · 10^9)	Ref.
Euglena gracilis	0.9	5.8	94
Chlamydomonas reinhardi (gamete)	0.99	4—5	107, 2
Pisum sativum	0.95	1—2	60
Lactuca sativa	0.98	2	105
Nicotiana tabacum	0.93	2	100
Oenothera hookeri, *Antirrhirum majus*, *Beta vulgaris*	0.9—1.1	1—10	48

some algal and higher plant species. Some of these values have been corrected from the published figures since the estimate of the molecular weight of the bacteriophage T4 DNA, used as standard, has been revised from $1.3 \cdot 10^8$ to $1.06 \cdot 10^8$ [20]. It is striking that the corrected values for the kinetic complexity of chloroplast DNA from the few algae and higher plants examined so far are all in the range 0.9—$1.0 \cdot 10^8$. This may be a coincidence, or it could mean that the information content of chloroplast DNA is basically similar throughout the plant kingdom, at least as regards the *amount* of information required. Whether the *type* of information required is also similar is not known; there is a need for some extensive DNA-DNA hybridization studies to establish this point. All the species so far examined belong to the Chlorophyta, with the exception of *Euglena.* It would be interesting to make such measurement of chloroplast DNA from other plant groups, especially from algae with unusually shaped chloroplasts (see Chapter 1).

Table II also lists the analytical complexities of chloroplast DNA, i.e. the amount of DNA per chloroplast. Since the kinetic complexities are always much less than the analytical complexities, there must be between 10 and 100 copies of the DNA sequences in each chloroplast. The larger the chloroplast, the more copies of DNA it contains [48]. These renaturation studies cannot rule out the possibility that microheterogeneity in nucleotide sequence exists between the copies but, if it does exist, such heterogeneity is beyond detection by current techniques.

10.2.5. Circularity

A recent discovery is that, when precautions are taken to minimise shearing, a proportion of chloroplast DNA can be isolated as circles. Circular DNA has been reported from chloroplasts of *Euglena* [38], *Pisum sativum* [60], *Spinacia oleracea* [71], *Antirrhinum majus, Oenothera hookeri*, and *Beta vulgaris* [48]. Fig. 10.1 shows a molecule of chloroplast DNA from *Spinacia oleracea.* The significance of circularity in DNA is not known, but it is a useful property since it implies that the molecule has not been degraded on isolation. The most interesting aspect of this finding is that in all the species listed the contour length of the majority of the circles is in the range 37—45 μm; this length of double-stranded DNA has a calculated molecular weight of $0.85—1.0 \cdot 10^8$ daltons, which is in the same range as the kinetic complexity of chloroplast DNA (Table II). This correspondence of length and kinetic complexity suggests that the genetic information carried by the chloroplast DNA is accommodated by the length of the circular molecule. In the case of chloroplast DNA from *Pisum sativum*, a small percentage of circles exists as dimers of contour length about 86 μm [60]; this is the largest circular molecule of DNA yet isolated.

10.2.6. Ploidy

It is clear from the data in Table II, and from the evidence for circular DNA, that each chloroplast may contain many copies of a circular genome, and thus may be genetically polyploid. Direct evidence for polyploidy has come from electron microscopic and autoradiographic studies, which show that each chloroplast can have up to 32 DNA-containing regions, the number depending on the size of the chloroplast [61]. It has been pointed out by Kirk [58] that this multiplicity increases the probability that a mutation will appear in a chloroplast in a given time, but that there is likely to be a long delay before the mutation can be expressed, since a mutation in one copy will be swamped by all the other wild-type copies. It is therefore not surprising that known non-Mendelian mutations affecting chloroplasts are small in number, and difficult to induce with mutagens. In *Chlamydomonas reinhardi* a number of mutations are known which alter the sensitivity of chloroplast ribosomes to inhibition by antibiotics such as erythromycin and carbomycin; some of these mutations show uniparental inheritance [72,79]. Those genes which are inherited in a uniparental fashion have been shown by recombinational analysis to form one linkage group. Sager has claimed that this linkage group behaves as if it were diploid in vegetative cells [79]. It is difficult to reconcile this conclusion with the evidence which suggests that there are many copies of the genome in each chloroplast. A more recent study of this paradox in *Chlamydomonas* does not support Sager's conclusion, but argues for a multiple copy model with most of the copies

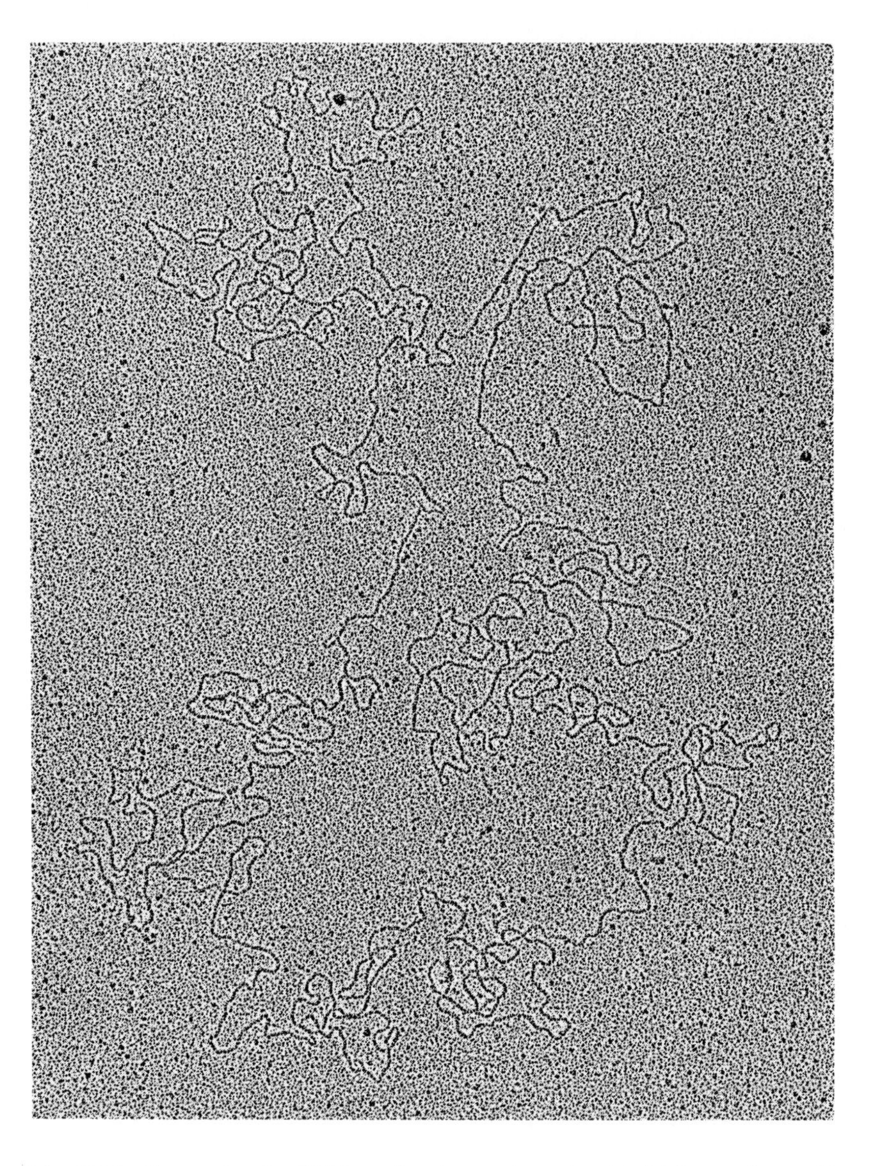

Fig. 10.1. Open circular DNA molecule, contour length 44.7 μm, from chloroplasts of *Spinacia oleracea*. Reprinted from Herrmann et al. [48] with permission.

normally being transmitted to the daughter cells by the maternal parent [37].

10.2.7. Functions

A 40 μm circle of double-stranded DNA of unique base sequence is sufficient in principle to code for about 125 proteins each of molecular weight 50 000. Since our knowledge depends ultimately on the validity of the techniques that can be used, the evidence for the functions of chloroplast DNA will be considered in terms of the four methods that have been tried so far:

(*a*) *The selective inhibition of chloroplast DNA transcription.* The antibiotic rifampicin is a potent inhibitor of the initiation of RNA synthesis in bacteria. This compound has been reported to inhibit the incorporation of labelled precursors into chloroplast ribosomal RNA, but not into cytoplasmic ribosomal RNA, in several unicellular algae, namely *Chlamydomonas* [95], *Chlorella* [13], and *Acetabularia* [35]. Rifampicin has also been reported to inhibit chloroplast RNA polymerase activity in extracts of *Chlamydomonas* [95]. These results suggest that the functional genes for chloroplast ribosomal RNA are located in chloroplast DNA and are transcribed by the chloroplast RNA polymerase.

The effect of rifampicin in higher plant systems is controversial. There have been some reports that it specifically inhibits chloroplast ribosomal RNA synthesis, but other workers could not reproduce this result [11]. There now seems general agreement that chloroplast RNA polymerase from higher plants, as normally prepared and assayed, is insensitive to rifampicin, but this may mean only that the preparations are not initiating RNA synthesis but are elongating pre-existing RNA molecules.

(*b*) *Genetic analysis of mutants.* Many mutations are known which affect chloroplast components, but the vast majority are inherited in a Mendelian fashion and are therefore presumed to be located in nuclear genes. For example, seven different genes are known which control steps in the chlorophyll biosynthetic pathway, and nuclear genes have been shown to be involved in the synthesis of phosphoribulokinase and at least five components of the photosynthetic electron transport chain [58,59,65].

In *Chlamydomonas* many mutations affecting chloroplast ribosomes and photosynthetic capacity are inherited as a single linkage group in a uniparental fashion [37,42,79]. In this alga, like gametes fuse completely, and the physical basis for uniparental inheritance has been suggested to be the destruction of the chloroplast DNA of one of the gametes by a modification-restriction system of the type known in bacteria [81,82]. At least six non-Mendelian gene loci affecting resistance of chloroplast ribosomes to antibiotics are known. A simple interpretation of these data is that up to six proteins of chloroplast ribosomes are encoded in chloroplast DNA. There is also evidence that at least one other protein of the chloroplast ribosome is encoded in a nuclear gene [72], while five other nuclear genes have been implicated in the biogenesis of chloroplast ribosomes [42].

Good evidence that a mutation inherited in a maternal or uniparental fashion actually alters the amino acid sequence of a protein has been obtained in only one case. Chan and Wildman [16] studied the inheritance of a mutation in the large subunit of Fraction I protein in *Nicotiana tabacum* at the tryptic peptide level; this mutation was inherited via the maternal line only. By contrast, mutations in the small subunit of Fraction I protein are inherited in a Mendelian fashion [53] as are mutations in the protein component of the light-harvesting complex associated with photosystem II [62]. This type of combined biochemical-genetic approach is very promising and deserves further use, especially with cultivated plants where many varieties are available. It must be emphasised that this approach should be conducted at the level of the tryptic peptide analysis of a purfied protein. It is not sufficient to show that particular proteins are absent in a mutant variety, as has been shown for some thylakoid proteins in *Antirrhinum* [46,47]; this is because absence of a protein might result from a mutation in a chloroplast gene which controls the formation of a protein encoded in the nucleus. Thus the data from *Antirrhinum* show only that chloroplast DNA is somehow involved in the synthesis of these photosystem proteins, but not that these proteins are themselves encoded in chloroplast DNA.

(*c*) *DNA-RNA hybridisation studies.* If RNA isolated from chloroplasts can be shown to hybridise to chloroplast DNA but not to nuclear DNA, this is good ground for believing that such RNA is both encoded in and transcribed from chloroplast DNA. A number of hybridisation studies have been carried out with chloroplast ribosomal RNA, which hybridises to about 4% of chloroplast DNA [31]. The most recent study is that of Thomas and Tewari [101]; they found that in a number of higher plants each circle of chloroplast DNA contains two cistrons for 16S ribosomal RNA and two cistrons for 23S ribosomal RNA. There are several reports that chloroplast ribosomal RNA will also hybridise to nuclear DNA [50,100]. The significance of these reports awaits further study, but the possibility is raised that there are two types of chloroplast ribosomal RNA, one encoded in chloroplast DNA and the other in nuclear DNA. This arrangement could be a means whereby the nucleus exerts control over events in the chloroplasts, especially if it is further postulated that chloroplast ribosomes containing nuclear RNA translate only messengers originating in the nucleus.

There is evidence that some of the transfer RNA species found in chloroplasts will also hybridise to chloroplast DNA. Tewari and Wildman [100] found that tRNA from tobacco chloroplasts would hybridise to 0.4–0.7% of chloroplast DNA. This amount of DNA would be enough to code for 20 to 30 transfer RNA molecules, each of molecular weight 25 000. This work used unfractionated tRNA however, and more studies need to be done to establish the site of encoding of individual tRNA species. Recent work in Weil's laboratory could be interpreted to indicate that a leucyl-accepting transfer RNA located in the chloroplast hybridises to nuclear DNA but not to chloroplast DNA. If this work is confirmed and extended, the possibility that RNA molecules can traverse the chloroplast envelope in a specific manner will have to be taken seriously. There are as yet no reports of the hybridisation of specific messenger RNA molecules to chloroplast DNA.

(*d*) *Identification of RNA and protein molecules synthesised by isolated chloroplasts.* This is the most direct method; if transcription coupled to translation can be obtained in isolated chloroplasts, identification of the products would simultaneously determine both the structural genes present in chloroplast DNA and the function of chloroplast ribosomes. Work in my laboratory has shown that isolated intact chloroplasts from *Pisum sativum* and *Spinacia oleracea* will synthesise discrete protein and RNA molecules, but transcription and translation are not coupled [26,31,43]. It is therefore not possible to infer that the proteins synthesised by isolated chloroplasts are encoded in chloroplast DNA. The RNA synthesised by isolated chloroplasts has been analysed by polyacrylamide gel electrophoresis (Fig. 10.2). The chief product is a species of molecular weight about $2.7 \cdot 10^6$; this has been shown by competitive hybridisation to chloroplast DNA to be a

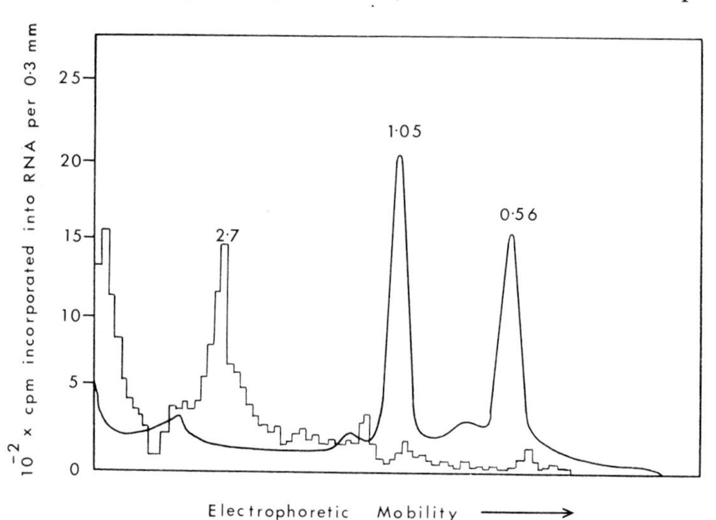

Fig. 10.2. Analysis by polyacrylamide gel electrophoresis of RNA synthesised by chloroplasts isolated from *Spinacia oleracea*. Isolated chloroplasts were incubated with [^{3}H] uridine and 15 000 lux of red light for 20 min at 20°C. Nucleic acid was extracted and run on polyacrylamide gels. The solid line represents the A_{260}, while the histogram shows the radioactivity. The figures are the molecular weights $\times$ 10^{-6} of the RNA components. Reprinted from Hartley and Ellis [43] with permission.

precursor to chloroplast ribosomal RNA. Isolated chloroplasts have not been shown convincingly to synthesise ribosomal RNA; presumably the processing system which trims the precursor molecules does not survive the trauma of isolation.

The known genes in chloroplast DNA can be summarised as follows: some chloroplast transfer RNA, chloroplast ribosomal RNA, several chloroplast ribosomal proteins, and the large subunit of Fraction I protein. These genes account for about 10% of the total potential coding capacity of a 40 μm circle of DNA. The elucidation of the function of the remaining 90% of chloroplast DNA is the most important problem in this field.

10.3. CHLOROPLAST RNA POLYMERASE

DNA-dependent RNA polymerase activity has been demonstrated in chloroplast preparations from several species of algae and higher plants [5,9,56,91,95,99]. Most workers study lysed chloroplast preparations because the nucleotide triphosphates used as substrates penetrate the chloroplast envelope at slow rates. Unlike intact chloroplasts, which use light to phosphorylate added [^{3}H]uridine (Fig. 10.2), lysed preparations do not synthesise discrete species of RNA; instead a polydisperse pattern of products ranging in size from 5S to 23S is obtained [91]. This difference between the products from intact and lysed chloroplasts may result from the dilution of controlling factors on lysis.

There are two key questions about chloroplast RNA polymerase that need to be answered: firstly, how similar is it to any nuclear polymerase, and secondly, how is its activity regulated? The enzyme is bound to the chloroplast lamellae, but a mild technique for its quantitative removal has been devised [4]. The solubilised chloroplast polymerase has been partially purified, and appears very similar to a nuclear polymerase [8,10]. Analysis of the purified chloroplast polymerase from *Zea mays* on sodium dodecyl sulphate polyacrylamide gels reveals two major protein bands with four minor bands; which, if any, of these proteins are true components of the chloroplast polymerase is not established [89]. It has been shown that the synthesis of chloroplast ribosomal RNA in dark-grown plants responds to the phytochrome system, but the nature of any mechanism which may link phytochrome to RNA polymerase is unknown. A key aim in this field should be to reconstitute a transcriptional system containing purified chloroplast DNA in circular form and purified chloroplast RNA polymerase complete with regulatory subunits. Such a system should synthesise at least the same RNA products as intact isolated chloroplasts, and allow a study of the factors affecting the initiation of specific chloroplast genes.

10.4. CHLOROPLAST DNA POLYMERASE

DNA polymerase activity has been detected in chloroplast suspensions prepared from *Spinacia oleracea* [92], *Nicotiana tabacum* [98], and *Euglena gracilis* [87]. The product renatures readily after heat denaturation, and hybridises to a much larger extent with chloroplast DNA than with nuclear DNA. It has not been established whether the chloroplast polymerase is identical with any nuclear polymerase or whether it carries out either a replicase or a repair function. The enzyme is bound to the chloroplast membranes, but in the case of *Euglena*, it has been solubilised by treatment with high concentrations of salt, and highly purified; the purified enzyme is inhibited by ethidium bromide [54]. Flechtner and Sager [34] have made the interesting observation that treatment of *Chlamydomonas* cells with ethidium bromide induces a selective and reversible inhibition of chloroplast DNA replication. Thus nuclear DNA synthesis proceeds normally while replication of chloroplast DNA is impaired; the pre-existing chloroplast DNA decreases by at least 80% in one cell generation, but this loss is reversible if the drug is removed within 12 h. This result suggests that one or a few of the chloroplast DNA copies may be protected in some way, perhaps by close attachment to chloroplast membranes.

The chloroplast DNA of *Euglena* [69] and *Chlamydomonas* [17] has been shown to replicate in a semi-conservative fashion at a different time in the growth cycle from nuclear DNA, but the factors controlling the time and rate of synthesis of chloroplast DNA are unknown.

10.5. CHLOROPLAST PROTEIN SYNTHESIS

Chloroplast preparations capable of incorporating labelled amino acids into protein have been isolated from higher plants and algae [12,29]. In such work it is vital to establish that the incorporation is due to chloroplasts and not to micro-organisms, intact leaf cells, or cytoplasmic ribosomes. The best criterion is the dependence of incorporation on an added energy source; this is ATP in the case of lysed chloroplasts, or light in the case of intact chloroplasts which can carry out photophosphorylation. Dependence on an added energy source eliminates intact leaf cells and micro-organisms as the agents of incorporation. Activity by contaminating cytoplasmic ribosomes can be ruled out by the different sensitivity of the two types of ribosome to antibiotic inhibitors. Other criteria are the sensitivity of incorporation to added ribonuclease, and to variation in the concentration of added Mg^{2+} ions; these criteria are useful when lysed chloroplasts or isolated ribosomes are being tested, but not when the incorporation is due to intact chloroplasts.

Some of the characteristics of protein synthesis by chloroplasts will now be considered.

10.5.1. Ribosomes

Lyttleton [67] was the first to show that green plant cells contain two classes of ribosome which differ in their sedimentation coefficients. Chloroplasts contain 70S ribosomes while the cytoplasm contains 80S ribosomes. In higher plants chloroplast ribosomes constitute a high proportion of the total cellular complement of leaf ribosomes, values as high as 60% having been reported.

Chloroplast ribosomes resemble those from prokaryote cells in their S value. Their RNA components are also of the same size as those found in prokaryote ribosomes, i.e. 16S and 23S. The 80S cytoplasmic ribosomes by contrast, contain 18S and 25S RNA molecules [31]. These similarities between chloroplast and prokaryote ribosomes have been much stressed in the past, but it is now clear that these similarities are of size only, and not in the primary structure of the RNA and protein components. For example, ribosomal RNA from *Escherichia coli* does not compete with chloroplast ribosomal RNA from *Euglena* for hybridisation to chloroplast DNA [86]. The protein complement of prokaryote ribosomes is also quite different from that of chloroplast ribosomes as judged by gel electrophoresis [49] and immunological tests [111]. Similarly, the proteins of chloroplast and cytoplasmic ribosomes from the same plant differ distinctly in the patterns they give on gel electrophoresis [39].

Some of the chloroplast ribosomes are bound to the internal membranes. Electron microscopy has revealed whorl-like polyribosomes attached to the granal and intergranal thylakoid membranes [33], while analysis of isolated chloroplasts shows that up to 50% of the ribosomes cannot be removed from the membranes by washing with hypotonic buffer. Tao and Jagendorf [96] have shown that some free ribosomes are lost during chloroplast isolation without the chloroplasts lysing irreversibly; when this loss is taken into account, they estimate that in chloroplasts from *Pisum sativum*, about 20% of the ribosomes are membrane-bound while the remainder are free in the stroma. There is evidence that the two classes of chloroplast ribosome synthesise different types of protein (see section 10.6).

10.5.2. Amino acid activation

Chloroplasts isolated from several algae and higher plants have been found to contain aminoacyl-transfer RNAs and the corresponding synthetases [29]. These transfer RNA and synthetase molecules can often, but not always, be distinguished from those involved in the activation of the same amino acids in the cytoplasm. Studies of the compatibility between transfer RNA and synthetases from the two cellular compartments show that the specificity of interaction ranges from none to absolute. For example, the cytoplasm of leaves of *Phaseolus vulgaris* contains a leucyl-transfer RNA synthetase that can add leucine only to the two leucyl-transfer RNA species that are common to the cytoplasm and chloroplasts, but is not able to recognise the three leucyl-transfer RNA species found only in the chloroplasts. The chloroplast enzyme, on the other hand, can aminoacylate all five leucyl-transfer RNA species. It is difficult to attach any biological significance to these variations in specificity. The simplest presumption is that in the intact cell the chloroplast synthetases aminoacylate only the transfer RNA species located in the chloroplast and these are sufficient for chloroplast protein synthesis to proceed, while the cytoplasmic synthetases aminoacylate only the transfer RNA species located in the cytoplasm. A more interesting possibility is that some of the transfer RNA species required to translate chloroplast messenger RNA are acylated in the cytoplasm; such a requirement would provide one way in which protein synthesis in the cytoplasm and chloroplast could be integrated, but there is no direct evidence to support this suggestion. Experiments with *Euglena* suggest that at least some of the chloroplast aminoacyl-transfer RNA synthetases are encoded in nuclear DNA and are synthesised on cytoplasmic ribosomes [45].

10.5.3. Initiation of chloroplast protein synthesis

Protein synthesis in bacteria is initiated by *N*-formylmethionyl transfer RNA; there is evidence that this is also true in chloroplasts, but not in the cytoplasm of eukaryotic cells. Schwarz et al. [84] found that chloroplast ribosomes from *Euglena* would translate RNA from bacteriophage f2 into viral coat protein with *N*-formylmethionine at the amino terminus. Detailed studies of initiating methionyl-transfer RNA species have since been reported for several higher plants. For example, there are two methionyl-transfer RNA species in wheat chloroplasts. One of these can be formylated by a chloroplast transformylase, whereas the other cannot, and may serve to direct methionine into internal positions in the polypeptide chains [64]. Two other methionyl-transfer RNA species are found in the cytoplasmic fraction, neither of which can be formylated. In my laboratory it has been shown that intact isolated chloroplasts from *Pisum sativum* will synthesise *N*-formylmethionylpuromycin when incubated with light, methionine, and puromycin. It is probable from this type of evidence that the initiation of protein synthesis by chloroplasts is similar to that in prokaryotes in that it uses a formylated methionyl-transfer RNA, but distinct from that in the cytoplasm which uses an unformylated methionyl-transfer RNA. The significance of this distinction is not clear; it may reflect the evolutionary origin of the compartments in eukaryotic cells, and have no contemporary functional value.

10.5.4. Energy source

In most studies of protein synthesis by isolated chloroplasts, added ATP has been used as an energy source [29]. Since isolated chloroplasts can generate ATP by photophosphorylation, it should be possible to drive protein synthesis in chloroplasts with light. Spencer [90] found that spinach chloroplasts would use light to incorporate amino acids into protein provided that ATP, inorganic phosphate and pyocyanine were added. The necessity for pyocyanine as catalyst indicates that the chloroplasts were broken, and had lost their natural catalyst, ferredoxin, by dilution. If precautions are taken to isolate chloroplasts which have their outer envelopes intact, protein synthesis is stimulated twenty-fold by light in the absence of either cofactors or catalysts of photophosphorylation (Fig. 10.3). The rates of light-driven protein synthesis by such chloroplasts are the highest yet recorded for any isolated chloroplast system. The use of intact chloroplasts for studies of protein synthesis has a major advantage over the use of broken chloroplasts in that conditions around the polysomes are more normal with respect to controlling factors. It is thus more likely that correct termination and release of polypeptide chains occurs in intact chloroplasts than in lysed preparations; such intact chloroplasts make discrete protein molecules rather than incomplete polypeptide chains (see section 10.6). A further advantage of using light as the energy source is that the broken chloroplasts, which inevitably contaminate the preparation of intact chloroplasts, cannot contribute to the incorporation in the absence of added cofactors.

It is possible that protein synthesis by chloroplasts in vivo uses ATP provided by the chloroplast itself. However, this is not the case in developing chloroplasts, since organisms such as *Chlorella*, *Chlamydomonas*, and *Pinus* do not require light to form chloroplasts. The formation of chloroplasts must therefore depend on ATP supplied by the rest of the cell. This conclusion is confirmed by the ability of intact isolated etioplasts from *Pisum sativum* to use ATP, but not light, as their source of energy for protein synthesis [88].

10.5.5. Inhibitors

It is well established that protein synthesis by isolated chloroplast ribosomes is inhibited by the same antibiotics which inhibit protein synthesis by prokaryote ribosomes [12,29]. The best studied example is chloramphenicol, which inhibits protein synthesis by isolated chloroplasts from all the species so far tested; inhibition is shown only by the D-threo isomer of chloramphenicol, and not by the other stereoisomers [23]. Three other unrelated antibiotics, spectinomycin, lincomycin, and erythromycin also inhibit protein synthesis by chloroplast ribosomes [24]. Light-driven protein

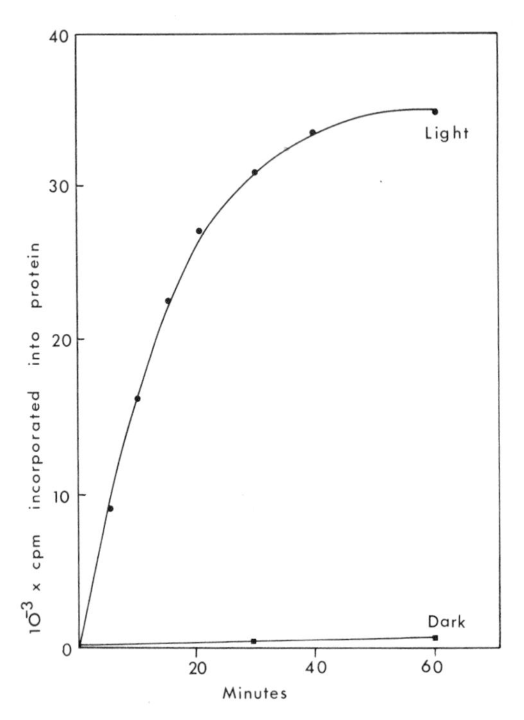

Fig. 10.3. Time course of light-driven chloroplast protein synthesis. Pea seeds *(Pisum sativum* var. Feltham First) were grown in compost for 7—10 days under a 12 h photoperiod of 2000 lux white light. The apical leaves (15 g) were homogenised for 4 sec in a Polytron homogeniser in 100 ml of a sterile ice slurry containing 0.35 M sucrose, 25 mM HEPES-NaOH, 2 mM EDTA and 2 mM sodium isoascorbate (pH 7.6). The homogenate was immediately strained through 8 layers of muslin and centrifuged at 2500 g for 1 min at 0°C. The pellet was resuspended by cotton-wool in 1—5 ml of sterile 0.2 M KCl, 66 mM tricine—KOH, 6.6 mM $MgCl_2$ (pH 8.3). Chloroplasts (300 μl) were incubated in a final volume of 500 μl with 0.5 μCi of either [^{14}C]leucine (3 μM) or [^{35}S]methionine (36 nM). Tubes were illuminated at 20°C with filtered red light at 4000 lux. Protein was extracted and counted by liquid scintillation spectrometry in toluene—0.5% PPO scintillant at 70% counting efficiency. Reprinted from Blair and Ellis [6] with permission.

synthesis by chloroplasts from *Pisum sativum* is especially sensitive to lincomycin, 50% inhibition being given by 0.2 μg/ml [30]. Protein synthesis by cytoplasmic ribosomes from plants is not inhibited by any of these antibiotics. On the other hand, cytoplasmic ribosomes are inhibited by cycloheximide, which does not affect the activity of chloroplast ribosomes. It must be emphasised that all these findings relate only to the activity of isolated sub-cellular systems, it cannot be assumed from these data alone that it is valid to use these compounds on intact cells with the expectation of inhibiting protein synthesis in one compartment but not in the other.

The protein complements of chloroplast and bacterial ribosomes are different, and thus the above similarity between them must reside in their antibiotic-binding sites, and not in the properties of the proteins as revealed by gel electrophoresis or immunological tests. There is no evidence that the mechanism of action of bacterial antibiotics on chloroplast ribosomes is similar to that on bacterial ribosomes; nor is there any understanding of the selective pressure that maintains the sensitivity of chloroplast ribosomes to these antibiotics in the face of mutation to resistance, which can, for example, be found in *Chlamydomonas.*

10.6. THE FUNCTION OF CHLOROPLAST RIBOSOMES

The high proportion of plant ribosomes which are located inside chloroplasts raises the question as to their function. Are they required in such quantities because they make a large number of different chloroplast proteins, or because they make a small number of different chloroplast proteins in large amounts? The available evidence suggests that the latter is the case [12, 29].

The problem of identifying which proteins are synthesised by chloroplast ribosomes has been tackled in two ways: (*a*) by supplying antibiotic inhibitors, especially chloramphenicol and cycloheximide, to intact cells making chloroplasts, and determining which proteins are no longer synthesised; (*b*) by studying the synthesis of specific proteins in isolated sub-cellular systems. The results and limitations of these two methods will now be discussed.

10.6.1. Evidence from in vivo inhibitor studies

It is apparent from the literature that conflicting conclusions have been drawn about the sites of synthesis of some chloroplast proteins by different workers using cycloheximide and chloramphenicol [12,29]. It cannot be overemphasised that the validity of results from experiments with any inhibitor depends absolutely on the specificity of its action in intact cells. There is evidence that both chloramphenicol and cycloheximide have effects on systems other than protein synthesis in some higher plants. The first doubts about the action of chloramphenicol were published in 1963 [22], but many reports have appeared which fail to adequately consider this problem. The data in Table III are presented to emphasise the publications that have since questioned the specificity of action of both chloramphenicol and cycloheximide in higher plant tissues. It is clear from these data that the specificity of action of these compounds must be demonstrated in the particular tissue under study, and not assumed by extrapolation from studies on other organisms. In the case of chloramphenicol a method is available for assessing its specificity. This method [22] was first suggested in 1963 and depends on the fact that the chloramphenicol molecule contains two asymmetric carbon atoms, and hence exists in four

TABLE III

Effects of chloramphenicol and cycloheximide on processes other than protein synthesis in plant tissues

Inhibitor	Species	Process inhibited	Ref.
Chloramphenicol	*Beta vulgaris* *Daucus carota* *Pisum sativum*	Sulphate uptake	22
Chloramphenicol	*Zea mays*	Oxidative phosphorylation in vitro	40,41
Chloramphenicol	*Brassica napus*	Mitochondrial electron transport in vitro	110
Chloramphenicol	*Spinacia oleracea*	Photophosphorylation and photosynthetic electron transport in vitro	104
Cycloheximide	*Beta vulagris* *Daucus carota* *Solanum tuberosum* *Pisum sativum*	Chloride uptake	32,68
Cycloheximide	*Brassica napus*	Malate oxidation in vitro	110
Cycloheximide	*Achyla bisexualis*	RNA polymerase	102,103

stereoisomeric forms; the natural antibiotic is the D-threo isomer. Systems such as ion uptake, oxidative phosphorylation and photophosphorylation are inhibited by all four stereoisomers of chloramphenicol, whereas the inhibition of protein synthesis by isolated chloroplast ribosomes is specific for the D-threo isomer [23]. This stereospecificity provides a means of establishing for any particular tissue whether chloramphenicol is inhibiting protein synthesis directly at the ribosomal level, or in addition, is affecting some other process such as the energy supply. It is strongly recommended that only if an inhibition is produced specifically by the D-threo isomer should an interpretation directly involving protein synthesis be invoked.

Another problem is that in most of the inhibitor experiments on chloroplast ribosomal function, increases in specific proteins have been measured as enzymic activities rather than as amounts of protein. Failure to observe an effect by a particular inhibitor might therefore mean that the increase in enzymic activity is due to activation of a precursor protein, rather than to de novo synthesis by either chloroplast or cytoplasmic ribosomes.

Bearing these difficulties in mind, I interpret the bulk of the published inhibitor experiments as suggesting that most of the chloroplast proteins are synthesised on cytoplasmic ribosomes [12,28,29]. In all the studies reported on several algae and higher plants, the synthesis of Fraction I protein was found to be inhibited by 70S ribosomal inhibitors. In most cases, the synthesis of the other soluble enzymes of the photosynthetic carbon dioxide reduction cycle appears to occur on cytoplasmic ribosomes; the same is true for ferredoxin and the chloroplast RNA polymerase. Besides Fraction I protein, the only other proteins that appear to be synthesised by chloroplast ribosomes are some of the chloroplast ribosomal and lamellar proteins, including the photosynthetic cytochromes. It must be emphasised, however, that these inhibitor experiments are never more than suggestive. Strictly interpreted, they never say more than that the activity of a particular group of ribosomes is required for a particular protein to accumulate in the chloroplast. This is not the same as saying that these ribosomes actually synthesise that protein, because it is possible that the apoenzyme is synthesised by one class of ribosomes but requires for its appearance in the chloroplast in an active state, additional protein(s) which are synthesised by another class of ribosomes. Exactly this situation has been demonstrated for the synthesis of cytochrome oxidase, cytochrome *b*, and the ATPase complexes of mitochondria [83].

10.6.2. Evidence from in vitro studies

This is the most direct approach to this problem, but it has been successful only recently because of the difficulty of isolating sub-cellular systems from plants which will carry out complete protein synthesis with fidelity; algal systems are still especially difficult in this respect. When we started our work at Warwick in 1970, all previous attempts to persuade isolated chloroplasts to produce discrete identifiable proteins had been unsuccessful; most papers reported that the products were polydisperse. In my laboratory we have developed a system in which intact chloroplasts from *Pisum sativum*, *Hordeum vulgare*, *Spinacia oleracea*, and *Zea mays* make discrete protein molecules, using light as the energy source [6,21,26,27,29,44,52]. This work will now be summarised. A fuller account can be found in ref. 27.

The rationale of our approach is that in order to produce identifiable proteins, we must use conditions in which correct termination and release of the polypetide chains occurs; initiation is not a requirement for the purposes of identification. It seemed to us that such conditions are likely to be met in intact chloroplasts rather than in the lysed preparations that are commonly used, although it is clear that even intact chloroplasts lose some macromolecular components during isolation [96]. We have therefore used methods that were originally developed to isolate intact chloroplasts capable of high rates of photosynthetic CO_2 fixation [75]. It is characteristic of such preparations that they will use light as the source of energy in the absence of added cofactors, such as ADP of ferredoxin. By using light as the source of energy for protein synthesis, it is therefore possible to ensure that amino acid incorporation is taking place solely in intact chloroplasts, since broken chloroplasts cannot synthesise ATP in the absence of added substrates and catalysts.

Suspensions of chloroplasts, prepared as described in Fig. 10.3, contain between 40 and 50% intact chloroplasts, as judged by phase contrast and electron microscopy, and incorporate [^{14}C]leucine, [^{14}C]phenylalanine, and [^{35}S]methionine into protein when illuminated (Fig. 10.3). The rate of incorporation falls rapidly to zero after about 20 min; this falling rate is not accompanied by lysis of the chloroplasts. A vital component of the incubation medium is the presence of K^+ ions. When chloroplasts are incubated in media containing sucrose as the osmoticum, protein synthesis is greatly reduced, while replacement by KCl by NaCl prevents all light-dependent incorporation. If chloroplasts are first lysed by resuspension in 25 mM tricine KOH, 10 mM $MgSO_4$ (pH 8.0), addition of KCl to 0.2 M does not result in any light-dependent incorporation of amino acids into protein, but incorporation can be driven by added ATP. These results suggest that KCl is acting both as an osmoticum to preserve the intactness of the chloroplasts, and as a specific cofactor for protein synthesis. Similar rates of incorporation are achieved when 0.3 M sorbitol is used as the osmoticum, but only if K^+ ions are present at an optimum concentration of about 30 mM. The specific requirement of bacterial ribosomes for K^+ or NH_4^+ ions is well known, but the importance of K^+ ions does not seem to have been appreciated by previous workers on chloroplast protein synthesis.

Some characteristics of this pea chloroplast system are listed in Table IV. Light can be partially replaced as an energy source by added ATP and an ATP-generating system, while addition of ATP as well as light gives only a slight stimulation. Inhibitors of photophosphorylation, such as m-chlorocarboxyl cyanide phenylhydrazone (CCCP) and 3-(3,4-dichlorophenyl)-1, 1-dimethylurea (DCMU), inhibit protein synthesis. as do antibiotics specific for 70S ribosomes, such as lincomycin and D-threo chloramphenicol. Added ribonuclease does not inhibit protein synthesis significantly, and I regard this as evidence that incorporation is proceeding in intact chloroplasts only. We have shown that addition of ribonuclease causes RNA to be hydrolysed to a percentage equal to the percentage of broken chloroplasts [29]. Actinomycin D at 10 μg/ml does not inhibit protein synthesis; we have found that this concentration inhibits light-driven incorporation of [^{3}H]uridine into RNA by the same chloroplast preparation by 85%. Incorporation of leucine into protein is not stimulated by the plant hormones indole-3-acetic acid or gibberellic acid, or by inorganic phosphate, cyclic AMP, $NADP^+$ or phenazine methosulphate. Addition of poly (U), which stimulates phenylalanine incorporation by intact mitochondria from *Xenopus*, does not have this effect in this chloroplast system.

Incorporation of labelled amino acids by other components of the incubation medium can be excluded. Bacterial contamination of the chloroplast preparation is minimised by using sterile glassware and media. If the prepara-

TABLE IV

Characteristics of protein synthesis by isolated pea chloroplasts

Pea chloroplasts were isolated and incubated as described in Fig. 3. Incorporation by the complete light-driven system is called 100, and represents a rate of incorporation of between 0.5 and 1.0 nmole [^{14}C]leucine per mg chlorophyll per hour. The ATP and ATP-generating system consists of 2 mM ATP, 5 mM creatine phosphate, and 100 μg/ml creatine phosphokinase. Reproduced from Blair and Ellis [6] with permission.

Energy source	Treatment	Incorporation
Light	Complete	100
None	Zero time	0.5
None	Complete	3.0
ATP + ATP-generating system	Complete	50
Light + ATP + ATP-generating system	Complete	125
Light	+ Ribonuclease (30 μg/ml)	95
Light	+ CCCP (5 μM)	6
Light	+ DCMU (1 μM)	38
Light	+ D-threo chloramphenicol (50 μg/ml)	5
Light	+ Lincomycin (5 μM)	25
Light	+ Cycloheximide (100 μg/ml)	100
Light	+ Actinomycin D (10 μg/ml)	100

tions are solubilised in 2% Triton X-100 at the end of the incubation, less than 0.1% of the radioactivity incorporated into protein is present in the pellet spun down at 10 000 g for 10 min. This indicates that incorporation is not due to bacteria, nuclei or whole leaf cells [74]. The large stimulation of incorporation by light, and the sensitivity to inhibitors of photophosphorylation argues for chloroplast, rather than mitochondrial, protein synthesis. The activity of the crude chloroplast suspensions decreases by about 50% if stored for 1 h at 0°C, so we have not attempted to purify the chloroplasts from other cellular components present in the suspension.

We concluded from all these characteristics that protein synthesis is proceeding in intact chloroplasts only, is being driven by photophosphorylation, and is probably using messenger RNA synthesised before the chloroplasts were isolated. The products of protein synthesis were analysed by electrophoresis on sodium dodecyl sulphate (SDS) polyacrylamide gels.

Isolated intact chloroplasts from *Pisum sativum* synthesise at least six discrete proteins as revealed by SDS gel electrophoresis (Fig. 10.4). Similar results have been obtained with chloroplasts from *Hordeum vulgare*, *Spinacia oleracea*, and *Zea mays*, except that in the latter case, peak B is missing. When the chloroplasts are lysed and the thylakoids are spun down, it is seen that only one of these proteins (peak B in Fig. 10.4) is soluble. This protein has been identified by tryptic peptide analysis as the large subunit of

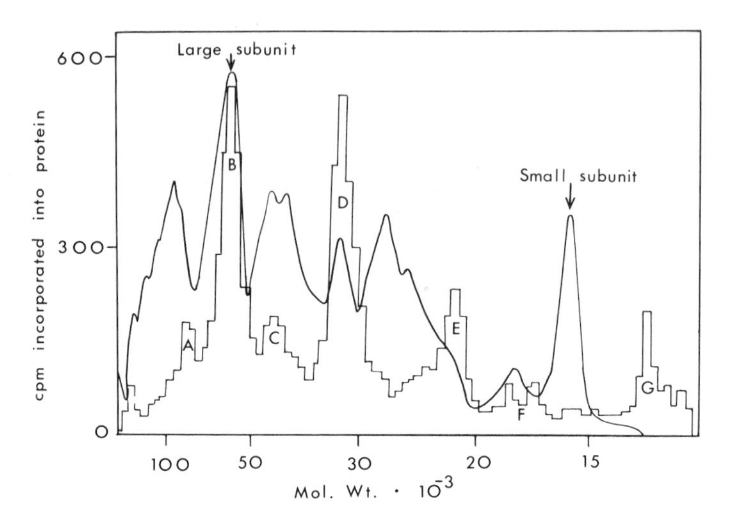

Fig. 10.4. SDS gel electrophoresis of labelled whole chloroplasts. Isolated chloroplasts were incubated with [^{35}S] methionine as described in Fig. 10.3, dissolved in SDS, and run on 15% gels (20 μg chlorophyll per gel). The smooth line represents the absorbance at 620 nm after staining with Amido Black, and the histogram the radioactivity in each 1 mm gel slice. The letters A—G mark the discrete labelled peaks. Large subunit and small subunit refer to the subunits of Fraction I protein. Reproduced from Eaglesham and Ellis [21] with permission.

Fraction I protein [6]. I regard this as the first definitive identification of a protein that is synthesised by chloroplast ribosomes. The high proportion of chloroplast ribosomes found in leaves may thus be required, not because they make a wide range of different proteins, but because Fraction I protein is one of the most abundant proteins [25]. Traces of other labelled soluble products can be detected in the gels beside the large subunit of Fraction I protein, but these are labelled so poorly as to defy analysis. The small subunit of Fraction I protein is not synthesised by isolated chloroplasts, but it has been detected as the product of protein synthesis by isolated cytoplasmic ribosomes from *Phaseolus vulgaris* leaves [38]. These results with isolated chloroplasts as regards soluble proteins thus fully confirm the results of inhibitor experiments on intact cells.

Recent work in my laboratory has shown that in *Pisum sativum* the large subunit of Fraction I protein is synthesised by the free ribosomes of the chloroplast, but not by the bound ribosomes. It is interesting that in the case of *Zea mays*, where mesophyll chloroplasts were used, the synthesis of the large subunit of Fraction I protein cannot be detected, although the synthesis of the membrane-bound proteins is easily seen. This observation might be thought to support the view that these chloroplasts lack Fraction I protein as part of their involvement in the C4 type of photosynthesis. However, it has been shown that chloroplasts can lose both their free ribosomes and Fraction I protein during isolation, without lysing irreversibly [96]; this result may indicate therefore that mesophyll chloroplasts from *Zea mays* are especially leaky in this respect, and lose all their free ribosomes on isolation.

The other proteins made by isolated pea chloroplasts are all membrane-bound, and have so far resisted identification. They do not include cytochrome *f*, the coupling factor ATPase, RNA polymerase, or photosystem I and II proteins. It is clear that they are minor components of the thylakoids. Two proteins associated with the chloroplast envelope are also labelled in isolated pea chloroplasts [52]. Inhibitor experiments with intact pea shoots have confirmed these results with isolated pea chloroplasts [28], and we can thus be confident that in this species most of the thylakoid and envelope proteins of the chloroplast are synthesised on cytoplasmic ribosomes. The possibility that a wider or different spectrum of proteins is made by chloroplast ribosomes during the development of chloroplasts from etioplasts has been examined, and found not to be the case [88].

This in vitro approach to the study of chloroplast ribosomal function gives direct and unambiguous results, and it should be extended to species lower on the evolutionary scale to see how far the pattern found in Angiosperm extends. In view of the uniformity of the size of the chloroplast genome in algae and higher plants (Table II), I am inclined to believe that the pattern seen in isolated pea chloroplasts will be universal. However, this assumes that proteins synthesised by chloroplast ribosomes are also encoded by chloroplast DNA; this may not be true if messenger RNA can cross the chloroplast envelope, but so far this is only a theoretical possibility.

Another area that may repay attention is the study of protein synthesis by plastids other than chloroplasts, especially proplastids and amyloplasts. It seems probable that these plastids synthesise a different range of proteins from chloroplasts. If these proteins are encoded in plastid DNA, they would account for some of the functions of the 90% of this genome which have not been identified.

10.7. COOPERATION BETWEEN CHLOROPLAST AND NUCLEAR GENOMES

It is clear from the available evidence that the chloroplast genome requires for its expression the cooperation of the nuclear genome. The best information about the details of this cooperation concerns the synthesis of Fraction I protein. By combining the results of Wildman's genetic studies with those of the in vitro studies of protein synthesis, a model for the synthesis of Fraction I protein can be constructed (Fig. 10.5). In this model, the large subunit is both encoded and synthesised within the chloroplast, while the small subunit is both encoded and synthesised outside the chloroplast. This model is tidy, and requires protein, but not nucleic acid, to cross the chloroplast envelope. This transport of protein must be on a large scale because it

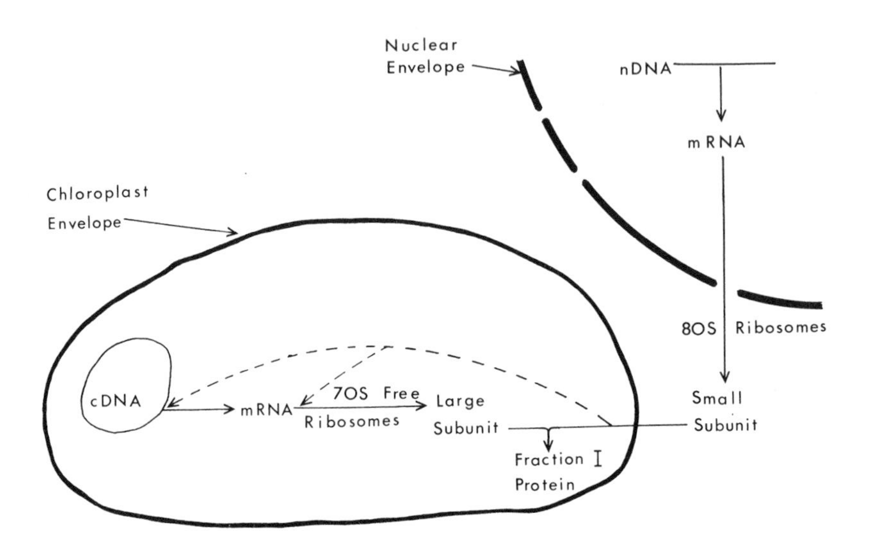

Fig. 10.5. Model for the cooperation of nuclear and chloroplast genomes in the synthesis of Fraction I protein. cDNA and nDNA stand for chloroplast and nuclear DNA, respectively. The dashed lines indicate possible sites at which small subunits may control the synthesis of large subunits. Reprinted from Ellis [28] with permission.

involves not only the small subunit of Fraction I protein but all the other proteins which, from the inhibitor evidence, are made on cytoplasmic ribosomes. The mechanism of this transport is unknown, but it must be able to distinguish between different proteins. One suggestion is that a protein exists in the outer envelope of the chloroplast which recognises a site common to all those proteins that are made on cytoplasmic ribosomes, but which are destined to function in the chloroplast.

Besides structural and enzymic proteins, it is likely that regulatory proteins also cross the chloroplast envelope. How light regulates chloroplast development is not known, but it would be reasonable to speculate in terms of proteins entering the organelle to trigger nucleic acid and protein synthesis. There is the possibility that proteins also pass out of the organelle, since recent work in *Chlamydomonas* has implicated chloroplast protein synthesis in the regulation of nuclear DNA replication [7]. I regard this question of protein transport into and out of the chloroplast as a crucial one to study if we are to understand how the chloroplast and nuclear genomes interact; almost no research has been carried out in this area. The movement of nucleic acid across the envelope remains a possibility, but no compelling evidence to suppose it occurs has been published.

Another problem that needs to be studied is the nature of the mechanism which integrates the synthesis of the two subunits of Fraction I protein in different cellular compartments. Inhibitor experiments suggest that the synthesis is tightly integrated; the large subunit does not accumulate in tissue treated with 80S ribosomal inhibitors [28]. It is suggested that a protein made on cytoplasmic ribosomes controls the synthesis of the large subunit by chloroplast ribosomes. The dashed lines in Fig. 10.5 illustrate one possibility; the small subunit is postulated to be required as a positive factor for the initiation of either transcription or translation of the large subunit messenger [28]. The messenger RNA for the large subunit has been successfully translated in a bacterial protein-synthesising system [44], so it should become possible to test this hypothesis directly in the near future.

An essential prerequisite to unravelling the molecular basis of development in plants is the ability to detect and measure specific messenger RNA molecules in different developmental situations. Experience with animal cells shows that success in isolating a specific messenger RNA depends on using a tissue which is synthesising one or a few proteins in much greater quantities than any others. Fraction I protein is such a protein in plant tissues, and the fact that its synthesis requires the cooperation of both the nuclear and chloroplast genomes means that studies on the synthesis of this protein are of relevance not only to the problem of how chloroplasts develop, but also to the more general problem of how plants develop.

REFERENCES

1 Baltus, E. and Brachet, J. (1963) Biochim. Biophys. Acta, [76], 490—492.
2 Bastia, D., Chiang, K., Swift, H. and Siersma, P. (1971) Proc. Natl. Acad. Sci. USA, [68], 1157—1161.
3 Baur, E. (1909) Z. Vererbungslehre, [1], 330—451.
4 Bennett, J. and Ellis, R.J. (1973) Biochem. Soc. Trans., [1], 892—894.
5 Berger, S. (1967) Protoplasma, [64], 13—25.
6 Blair, G.E. and Ellis, R.J. (1973) Biochim. Biophys. Acta, [319], 223—234.
7 Blamire, J., Flechtner, V.R. and Sager, R. (1974) Proc. Natl. Acad. Sci. USA, [71], 2867—2871.
8 Bogorad, L., Mets, L.J., Mullinix, K.P., Smith, H.J. and Strain, G.C. (1973) in Nitrogen Metabolism in Plants (Goodwin, T.W. and Smellie, R.M.S., eds), Vol. 38, pp. 17—42, Biochem. Soc. Symp., Biochem. Soc., London.
9 Bottomley, W. (1970) Plant Physiol., [45], 608—611.
10 Bottomley, W., Smith, H.J. and Bogorad, L. (1971) Proc. Natl. Acad. Sci. USA, [68], 2412—2416.
11 Bottomley, W., Spencer, D., Wheeler, A.M. and Whitfield, P.R. (1971) Arch. Biochem. Biophys., [143], 269—275.
12 Boulter, D., Ellis, R.J. and Yarwood, A. (1972) Biol. Rev., [47], 113—175.
13 Brandle, G. and Zetsche, K. (1971) Planta, [99], 46—55.
14 Brawerman, G. and Eisenstadt, J. (1964) Biochim. Biophys. Acta, [91], 477—485.
15 Britten, R.J. and Kohne, D.E. (1968) Science, [161], 529—540.
16 Chan, P. and Wildman, S.G. (1972) Biochim. Biophys. Acta, [277], 677—680.
17 Chiang, K.S. and Sueoka, N. (1967) Proc. Natl. Acad. Sci. USA, [57], 1506—1513.
18 Chun, E.H.L., Vaughan, M.H. and Rich, A. (1963) J. Mol. Biol., [7], 130—141.
19 Correns, C. (1909) Z. Vererbungslehre, [1], 291—329.

20 Dubin, S.B., Benedek, G.B., Bancroft, G.C. and Friefelder, D. (1970) J. Mol. Biol., [54], 547—556.
21 Eaglesham, A.R.J. and Ellis, R.J. (1974) Biochim. Biophys. Acta, [335], 396—407.
22 Ellis, R.J. (1963) Nature, [200], 596—597.
23 Ellis, R.J. (1969) Science, [163], 477—478.
24 Ellis, R.J. (1970) Planta, [91], 329—335.
25 Ellis, R.J. (1973) Comment. Plant Sci., [4], 29—38.
26 Ellis, R.J. (1974) Biochem. Soc. Trans., [2], 179—182.
27 Ellis, R.J. (1975) in Membrane Biogenesis: Mitochondria, Chloroplasts and Bacteria (Tzagoloff, A., ed), pp. 247—278, Plenum, New York.
28 Ellis, R.J. (1975) Phytochemistry, [14], 89—93.
29 Ellis, R.J., Blair, G.E. and Hartley, M.R. (1973) Biochem. Soc. Symp. [38], 137—162.
30 Ellis, R.J. and Hartley, M.R. (1971) Nature, [233], 193—196.
31 Ellis, R.J. and Hartley, M.R. (1974) in Biochemistry of Nucleic Acids (Burton, K., ed), MTP International Review of Science, Biochemistry Series One, Vol. 6, pp. 323—348. Butterworth, London.
32 Ellis, R.J. and MacDonald, J.R. (1970) Plant Physiol., [46], 227—232.
33 Falk, H. (1969) J. Cell. Biol., [42], 582—587.
34 Flechtner, V.R. and Sager, R. (1973) Nature New Biol., [241], 277—279.
35 Galling, G. (1971) Planta, [98], 50—62.
36 Gibor, A. and Izawa, M. (1963) Proc. Natl. Acad. Sci. USA [50], 1164—1169.
37 Gillham, N.W., Boynton, J.E. and Lee, R.W. (1974) Genetics, [78], 439—457.
38 Gray, J.C. and Kekwick, R.G.O. (1974) Eur. J. Biochem., [44], 491—500.
39 Gualerzi, C. and Cammarano, P. (1970) Biochim. Biophys. Acta, [199], 203—213.
40 Hanson, J.B. and Hodges, R.K. (1963) Nature, [200], 1009.
41 Hanson, J.B. and Krueger, W.A. (1966) Nature, [211], 1322.
42 Harris, E.H., Boynton, J.E. and Gillham, N.W. (1974) J. Cell. Biol., [63], 160—179.
43 Hartley, M.R. and Ellis, R.J. (1973) Biochem. J., [134], 249—262.
44 Hartley, M.R., Wheeler, A. and Ellis, R.J. (1975) J. Mol. Biol., [91], 67—77.
45 Hecker, L.I., Egan, J., Reynolds, R.J., Nix, C.E., Schiff, J.A. and Barnett, W.E. (1974) Proc. Natl. Acad. Sci. USA, [71], 1910—1914.
46 Hermann, F. (1971) FEBS Lett., [19], 267—269.
47 Hermann, F. (1972) Exptl. Cell. Res., [70], 452—453.
48 Herrmann, R.G., Bohnert, H.J., Kowallik, K.V. and Schmitt, J.M.(1975) Biochim. Biophys. Acta, [378], 305—317.
49 Hoober, J.K. and Blobel, G. (1969) J. Mol. Biol., [41], 121—138.
50 Ingle, J., Possingham, J.V., Wells, R., Leaver, C.J. and Loening, V.E. (1969) in Control of Organelle Development, Symposium 24 of the Society for Experimental Biology (Miller, P.L., ed), pp. 303—325, Cambridge University Press, Cambridge.
51 Iwamura, T. and Kuwashima, S. (1969) Biochim. Biophys. Acta, [174], 330—339.
52 Joy, K.W. and Ellis, R.J. (1975) Biochim. Biophys. Acta, [378], 143—151.
53 Kawashima, N. and Wildman, S.G. (1972) Biochim. Biophys. Acta, [262], 42—49.
54 Keller, S.J., Bredenbach, S.A. and Meyer, R.R. (1973) Biochem. Biophys. Res. Commun., [50], 620—628.
55 Kirk, J.T.O., (1963) Biochim. Biophys. Acta, [76], 417—424.
56 Kirk, J.T.O. (1964) Biochem. Biophys. Res. Commun., [16], 233—238.
57 Kirk, J.T.O. (1971) in Autonomy and Biogenesis of Mitochondria and Chloroplasts, (Boardman, N.K., Linnane, A.W. and Smillie, R.M., eds), North-Holland, Amsterdam.
58 Kirk, J.T.O. (1972) Sub-Cell. Biochem., [1], 333—361.
59 Kirk, J.T.O. and Tilney-Bassett, R.A.E. (1967) The Plastids: Their Chemistry, Structure, Growth and Inheritance, Freeman, London.

60 Kolodner, K.K. and Tewari, K.K. (1972) J. Biol. Chem., [247], 6355—6364.
61 Kowallik, K.V. and Herrmann, R.G. (1972) J. Cell. Sci., [11], 357—377.
62 Kung, S.D., Thornber, J.P. and Wildman, S.G. (1972) FEBS Lett., [24], 185—188.
63 Kung, S.D. and Williams, J.P. (1969) Biochim. Biophys. Acta, [195], 434—445.
64 Leis, J.P. and Keller, E.B. (1971) Biochemistry, [10], 889—894.
65 Levine, R.P. and Goodenough, V.W. (1970) Ann. Rev. Genet., [4], 397—408.
66 Lima-de-Faria, A. and Moses, M.J. (1965) Hereditas, [52], 367—378.
67 Lyttleton, J.W. (1962) Exp. Cell. Res., [26], 312—317.
68 MacDonald, I.R. and Ellis, R.J. (1969) Nature, [222], 791—792.
69 Manning, J.E. and Richards, O.C. (1972) Biochemistry, [11], 2036—2043.
70 Manning, J.E., Wolstenholme, D.R., Ryan, R.S., Hunter, J.A. and Richards, O.C. (1971) Proc. Natl. Acad. Sci. USA, [68], 1169—1173.
71 Manning, J.E., Wolstenholme, D.R. and Richards, O.C. (1972) J. Cell. Biol., [53], 594—601.
72 Mets, L. and Bogorad, L. (1972) Proc. Natl. Acad. Sci. USA, [69], 3779—3783.
73 Meyer, A. (1883) Bot. Ztg., [41], 489—498.
74 Parenti, F. and Margulies, M.M. (1967) Plant Physiol., [42], 1179—1186.
75 Ramirez, J.M., Del Campo, F.F. and Arnon, D.I. (1968) Proc. Natl. Acad. Sci. USA, [59], 606—611.
76 Renner, O. (1929) in Handbuch der Vererbungswissenschaften (Baur, E., and Hartmann, M., eds), Vol. II A, pp. 1—53, Gebt. Borntraeger, Berlin.
77 Ris, H. and Plaut, W. (1962) J. Cell. Biol., [13], 383—391.
78 Rose, R.J., Cran, D.G. and Possingham, J.V. (1974) Nature, [251], 641—642.
79 Sager, R. (1972) Cytoplasmic Genes and Organelles, Academic Press, New York.
80 Sager, R. and Ishida, M.R. (1963) Proc. Natl. Acad. Sci. USA, [50], 725—730.
81 Sager, R. and Lane, D. (1972) Proc. Natl. Acad. Sci. USA, [69], 2410—2413.
82 Sager, R. and Ramanis Z. (1974) Proc. Natl. Acad. Sci. USA, [71], 4698—4702.
83 Schatz, G. and Mason, T.L. (1974) Ann. Rev. Biochem., [43], 51—87.
84 Schwarz, J.H., Meyer, R., Eisenstadt, J.M. and Brawerman, G. (1967) J. Mol. Biol., [25], 571—574.
85 Schimper, A.F.W. (1885) Jb. Wiss. Bot., [16], 1—247.
86 Scott, N.S., Munns, R., Graham, D. and Smillie, R.M. (1971) in Autonomy and Biogenesis of Mitochondria and Chloroplasts (Boardman, N.K., Linnane, A.W. and Smillie, R.M., eds), North-Holland, Amsterdam.
87 Scott, N.S., Shah, V.C. and Smillie, R.M. (1968) J. Cell. Biol., [38], 151—157.
88 Siddell, S.G. and Ellis, R.J. (1975) Biochem. J., [146], 675—685.
89 Smith, H.J. and Bogorad, L. (1974) Proc. Natl. Acad. Sci. USA, [71], 4839—4842.
90 Spencer, D. (1965) Arch. Biochem. Biophys., [111], 381—390.
91 Spencer, D. and Whitfield, P.R. (1967) Arch. Biochem. Biophys., [121], 336—345.
92 Spencer, D. and Whitfield, P.R. (1969) Arch. Biochem. Biophys., [132], 477—488.
93 Strasburger, E. (1882) Arch. F. Mikrosk. Anat. Entwmech., [21], 476—590.
94 Stutz, E. (1970) FEBS Lett., [8], 25—28.
95 Surzycki, S.J. (1969) Proc. Natl. Acad. Sci. USA, [63], 1327—1334.
96 Tao, K. and Jagendorf, A.T. (1973) Biochim. Biophys. Acta, [324], 518—532.
97 Tewari, K.K. (1971) Ann. Rev. Plant Physiol., [22], 141—168.
98 Tewari, K.K. and Wildman, S.G. (1967) Proc. Natl. Acad. Sci. USA, [58], 689—696.
99 Tewari, K.K. and Wildman, S.G. (1969) Biochim. Biophys. Acta, [186], 358—372.
100 Tewari, K.K. and Wildman S.G. (1970) in Control of Organelle Development, Symposium 24 of the Society for Experimental Biology (Miller, P.L., ed), pp. 147—179. Cambridge University Press, Cambridge.
101 Thomas, J.R. and Tewari, K.K. (1974) Proc. Natl. Acad. Sci. USA, [71], 3147—3151.
102 Timberlake, W.E. and Griffin, D.H. (1974) Biochim. Biophys. Acta, [349], 39—46.

103 Timberlake, W.E., McDowell, L. and Griffin, D.H. (1972) Biochem. Biophys. Res. Commun., [46], 942—947.
104 Wara-Aswapati, O. and Bradbeer, J.W. (1974) Plant Physiol., [53], 691—693.
105 Wells, R. and Birnstiel, M. (1969) Biochem. J., [112], 777—786.
106 Wells, R. and Ingle, J. (1970) Plant Physiol., [46], 178—179.
107 Wells, R. and Sager, R. (1971) J. Mol. Biol., [58], 611—622.
108 Wetmur, J.G. and Davidson, J. (1968) J. Mol. Biol., [31], 349—370.
109 Whitfeld, P.R. and Spencer, D. (1968) Biochim. Biophys. Acta, [157], 333—343.
110 Wilson, S.J. and Moore, A.L. (1973) Biochim. Biophys. Acta, [292], 603—610.
111 Wittman, H.G. (1970) in Organisation and Control in Prokaryotic and Eukaryotic Cells, Symposium 20 of the Society of General Microbiology (Charles, H.P. and Knight, B.C.J.G., eds), pp. 55—76, Cambridge University Press, Cambridge.
112 Wolstenhome, D.R. and Gross, N.J. (1968) Proc. Natl. Acad. Sci. USA, [61], 245—252.

The Intact Chloroplast — edited by J. Barber

Chapter 11

The Cooperative Function of Chloroplasts in the Biosynthesis of Small Molecules

RACHEL M. LEECH and D.J. MURPHY

Department of Biology, University of York, York (Great Britain)

CONTENTS

Abbreviations: ACP, acyl carrier protein; C12:3, *cis*-3,6,9-dodecatrienoic acid; C16:0, palmitic acid (hexadecenoic); C16:1, Δ-3-trans-hexadecenoic acid; C16:3, *cis*-7,10,13-hexadecatrienoic acid; C18:1, oleic acid (*cis*-9-octadecenoic); C18:2, linoleic acid (*cis*-9, 12-octadecadienoic); C18:3, linolenic acid (*cis*-9,12,15-octadecatrienoic acid), chl, chlorophyll; CoA, Coenzyme A; DCMU, 3-(3,4-dichlorophenyl)-1,1-dimethylurea; DG, diglyceride; DGD, digalactosyl diglyceride; DHAP, dihydroxyacetone phosphate; FMN, flavin mononucleotide; LPA, lysophosphatidic acid; MGD, monogalactosyl diglyceride; PA, phosphatidic acid; PAPS, phosphoadenine phosphosulphate; PC, phosphatidyl choline; PE, phosphatidyl ethanolamine; PEP, phosphoenol pyruvate; PG, phosphatidyl glycerol; SL, sulphoquinovosyl diglyceride; UDP-gal, uridine diphosphate galactose.

11.1. INTRODUCTION

Photosynthesising isolated chloroplasts incorporate $^{14}CO_2$ into a wide range of sugar phosphates at similar rates and in similar proportions to photosynthesising leaf cells ([15,23,92,192,193,195] and Chapter 7 of this volume). More recently, the terminal steps in a variety of macromolecular syntheses have also been demonstrated to occur in isolated chloroplasts. The incorporation of amino acids into proteins ([21,44] and Chapter 10 of this volume), of nucleotide triphosphates into nucleic acids ([63] and Chapter 10 of this volume) and of UDP galactose (UDP-gal) into galactolipids [39,68, 138] have all been shown unequivocally to occur in isolated plastids. In contrast, less is known about the intermediate syntheses involved in the metabolic conversion of the photosynthetic carbon chains to the precursors required for macromolecular syntheses and the location of these processes in the leaf cell has, until recently, been little investigated. The availability of suspensions of chloroplasts capable of high rates of CO_2 photoreduction has stimulated endeavours to describe the location of the enzymes and metabolic controls involved in the integration of the pathways involved in the conversion of photosynthetically produced sugar phosphates to a variety of macromolecules.

Recently several groups have specifically studied two particular aspects of cellular metabolism in photosynthesising tissue:

(1) The reduction of inorganic nitrogen to α-amino compounds.

(2) The biosynthesis of precursors for lipid biosynthesis.

In this chapter the results of these experiments are reviewed but in general, consideration is restricted to studies of organelles from higher plant leaves showing C_3 (Calvin)-photosynthesis. For the biosynthesis of all the precursors about which information is available, collaborative integration of chloroplast and cellular function most certainly occurs and it is now abundantly clear that the mature chloroplast and the other components of the green leaf cell interact and collaborate in a myriad of enzymic syntheses. The details of the collaborative function of the chloroplast in the synthesis of two types of small molecules, amino acids and membrane lipids, is considered in this chapter.

11.2. THE REDUCTION OF INORGANIC NITROGEN AND ITS ASSIMILATION IN GREEN CELLS

In illuminated algal cells [12–14,171,173] and in leaves [9,20,24,112, 131,155] fed with $^{14}CO_2$ the photosynthetically reduced carbon moves rapidly into amino acids. The now classical method of experimentation employed, depends on the observation of the appearance of radiocarbon in individual compounds as a function of time of exposure of the plant to $^{14}CO_2$ during steady-state photosynthesis. In these condition the synthesis of amino acids in algal cells can account for as much as 30% of all carbon fixed [171]. Alanine, aspartate, serine, glycine and later glutamate [13,14, 171,173] are among the products to become labelled with ^{14}C during the first few minutes of photosynthesis: other amino acids become labelled considerably later. The net increase of ^{14}C in alanine, serine and aspartic acid reaches maximum rates by 5 min after the introduction of $^{14}CO_2$ i.e. at the same time as intermediates of the carbon reduction cycle become saturated with ^{14}C. The rapidity of labelling of the amino acids and the dependence of their continuing synthesis on light, led to the early suggestion that amino acid biosynthesis takes place in the chloroplast itself. This suggestion can only be directly substantiated by detailed investigation of the precursor pools and synthetic capacities of isolated chloroplasts themselves. Although isolated chloroplasts can now be prepared which are capable of rates of $^{14}CO_2$ fixation commensurate with the rates for intact leaf tissue ([92] and Chapter 7 of this volume), the proportion of labelling found in amino acids is small (<10%). Virtually all the experiments using isolated chloroplasts have been performed using chloroplasts from the leaves of higher plants. Enzyme distribution and labelling studies with isolated chloroplasts have been addressed to the solution of several major questions raised by experiments with whole cells:

(i) The identification and localization of the reactions leading to the reduction of nitrate to nitrite and ammonia.

(ii) The identification and localization of the primary amination reaction(s).

(iii) The location of the substrate pools needed for continuing synthesis of amino acids.

(iv) The identification and localization of secondary reactions leading to subsequent amino acid and amide formation.

(v) The determination of the role(s) of light in amino acid biosynthesis.

(vi) The determination of the quantitative contribution of chloroplast biosynthesis to in vivo amino-nitrogen incorporation in the plant.

11.2.1. The identification and localization of the enzyme(s) responsible for nitrate and nitrite reduction

The synthesis of amino acids from inorganic precursors clearly depends not only on the availability of the appropriate carbon skeletons but also on a supply of nitrogen in a suitable form. Most of the nitrogen absorbed in a plant growing under field conditions is in the form of nitrate which must be reduced to ammonia or ammonium ion prior to incorporation into amino acids. Until recently it was assumed that in leaves, as in other parts of the plant, the assimilation of nitrate to amino-nitrogen occurred in a step-wise manner involving only the enzymes nitrate reductase, nitrite reductase and glutamate dehydrogenase functioning in a coordinated enzymic sequence.

However, the special characteristic of nitrate reduction in green tissue, in particular in leaves, is the light dependence of nitrate reduction and a comprehensive review of this subject is available [16].

The intracellular location of the enzymes responsible for nitrate reduction in the leaf has however been the subject of considerable debate [51], but several recent papers provide more definitive evidence on enzyme localization from experiments in which improved cell fractionation procedures have been employed [121,129,151].

11.2.1.1. Nitrate reductase

The first step in the assimilatory reduction of nitrate to ammonia is the reduction of nitrate to nitrite catalysed by nitrate reductase, a flavo-molybdo protein [27,146]. The subcellular distribution of nitrate reductase in photosynthesising cells has been the subject of considerable recent controversy [16,51]. However, as originally suggested by Beevers and Hageman [16] it now seems most likely that nitrate reductase is associated with the membranes of the chloroplast envelope. In a series of carefully designed experiments, Rathnam and Das [151] have provided the first direct demonstration of the association of nitrate reductase activity with isolated envelope membranes of chloroplasts. They showed that over 80% of the nitrate reductase activity of intact purified chloroplast from millet *(Eleusine caracana)* is associated with envelope membrane fraction. Since the chloroplast nitrate reductase constituted over 50% of the activity of the cell homogenate, the chloroplast envelope would appear to be a major site of nitrate reduction in these leaf cells. The rate of nitrate reduction recorded, 8 μmole nitrate reduced. mg chl^{-1}. h^{-1}, compares favourably with the rates of nitrite reduction and amino acid formation obtained with photosynthesising chloroplasts (see later). In the light of the findings of Rathnam and Das [151], the contradictions of the earlier conflicting reports on the intracellular location of nitrate reductase in leaf cells can now be resolved. Grant et al. [56] had previously concluded that 25% of the total leaf nitrate reductase activity in spinach and sunflower leaves is associated with the chloroplast and Coupe et al. [31] using non-aqueous density gradients found nitrate reductase activity in isolated barley chloroplasts. However, other careful studies [38,154,179] in which partial purification of the chloroplasts by washing or density gradient centrifugation was carried out before assay of the enzyme activity led to the opposite conclusion, i.e. that nitrate reductase is absent from chloroplasts. Differential release of nitrate reductase from the envelope during plastid isolation could clearly account for the contradictions in the various experimental findings. The results of Swader and Stocking [179] in particular show evidence of loss of activity from *Wolffia* chloroplasts after partial purification of chloroplasts by washing. This contrasted with chloroplasts which still showed nitrate reductase activity after isolation by a rapid method similar to that used by Jensen and Bassham [92]. The latter chloroplast suspensions were able to utilize nitrate as a terminal electron acceptor.

For several years it has been known that nitrate can act as a terminal electron acceptor in an artificial light-driven cell-free system containing grana, partially purified nitrate reductase. FMN and NADPH [35,118]. Such systems have been useful in analysing the cofactor and physical requirements for the nitrate reductase activity but on their own do not prove that such reactions occur in vivo and could be misleading in considerations of intracellular enzyme location. The current challenge is now to find methods of preparing chloroplasts capable of reducing both nitrate to nitrite and nitrite to amino-nitrogen at rates commensurate with those observed in the intact leaf. It would be particularly interesting to know with which membrane of the chloroplast envelope the enzyme is associated.

Losada's group has investigated the properties of the nitrate reductase from leaves and partially characterised the enzyme from unicellular green algae. The nitrate reductase from *Chlorella* is a flavo-molybdo protein, NADH-dependent and is an enzyme complex of high molecular weight [130,188]. The isolated enzyme from *Chlamydomonas* can be reversibly inactivated by ammonia [116,117]. Studies with the isolated enzyme suggest that two different enzymic activities operate sequentially during the transfer of electrons from NADH to nitrate: an NAD-dependent NADH-diaphorase and the molybdo protein nitrate reductase proper which can also use exogenous flavin nucleotides as electron donors [152,188]. The two enzyme activities can also be selectively inhibited [189]. The nitrate reductase from *Chlorella vulgaris* has been studied by Vennesland and her co-workers [93,172,190,191]. As extracted, the enzyme is largely inactive but becomes activated on adding nitrate and phosphate buffer at low pH.

11.2.1.2. Nitrite reductase

The association of nitrite reduction in leaves with the electron transport reactions of photosynthesis is indicated by its light-dependence, its inhibition by 3-(3,4-dichlorophenyl)-1,1-dimethylurea (DCMU), its insensitivity to uncouplers and its independence of CO_2 fixation [125]. There is now ample evidence that the light-dependent reduction of nitrite to amino-nitrogen occurs in isolated envelope-bound chloroplasts [32,56,121,126,154, 179] and the physiological electron donor appears to be reduced ferredoxin [76,94,125,142]. The presence of nitrite reductase in chloroplasts has been confirmed by careful enzyme location studies using both non-aqueously [154] and aqueously [32,57,121,126,129,170] isolated chloroplasts, by the stoichiometric conversion of nitrite to amino-nitrogen by isolated chloroplasts [121] and by the demonstration of the utilization of nitrite during non-cyclic photo-induced electron transport accompanied by the evolution of O_2 in the expected stoichiometric ratio of 1 mole NO_2 reduced per 1.5 mole O_2 evolved [179]. Osmotically ruptured chloroplasts are incapable of reducing nitrite [129,179] presumably because of the rapid loss of soluble ferredoxin on breakage as well as the loss of nitrite reductase itself. Spinach leaf nitrite reductase has recently been isolated, purified to electrophoretic homogeneity and characterised [27]. It contains 2 Fe atoms per molecule but no flavin and has a molecular weight of 63 000 daltons. Cell-free systems containing grana require the addition of nitrate reductase, FMN and NADPH before nitrite reduction occurs [35,118]. In this system a variety of other electron donors, for example H_2 + hydrogenase or NADPH-ferredoxin + NADP reductase, can substitute ferredoxin.

11.2.2. The identification and localization of the enzymes responsible for the primary amination reactions

After NO_3^- has been reduced to NH_4^+, the ammonium ion can be incorporated into organic molecules. The enzymes which may be involved in such amination reactions will now be considered.

11.2.2.1. Glutamate dehydrogenase

Identification of the primary amination reaction(s) involved in amino acid biosynthesis during photosynthesis is only possible if the products of fixation can be isolated and analyzed after very short periods (less than 1 min) of exposure to tracers. For in vivo studies steady-state photosynthesis and rapid sampling can be much more rigorously controlled using populations of unicellular algae than is possible with leaf tissue, so much of the physiological information available has been derived from algal systems. In a series of elegant kinetic experiments, Bassham's group [13,14,171] have followed the time course of labelling of amino acids in photosynthesising *Chlorella* cells fed with either ^{15}N ammonium ions or $^{14}CO_2$ under conditions where the specific activity of the tracer was maintained at a constant level. The rate of labelling of the amino group of glutamic acid (1.5 μmoles/min/ml algal cells) accounted for most of the NH_4^+ uptake in these experiments [14]. (A small pool of actively turning-over glutamine was also identified accounting for 4% of the $^{15}NH_4$ incorporated). The initially labelled pool of glutamate became saturated after 15 min and a second pool, previously identified in ^{14}C tracer experiments [13,171], became saturated more slowly. The total rate of incorporation into glutamate increased with time: between 40 and 60 sec after fixation began, the rate of synthesis was 0.2 μmoles/min/ml algal cells but, after 5 min of photosynthesis the rate increased to 0.7 μmoles/min/ml algae, accounting for 4.5% of all the C fixed. The rate of labelling of all other amino acids with ^{15}N was considerably slower than for glutamate and in particular the labelling pattern for alanine was characteristic for a secondary formed product, i.e. it was not consistent with primary reductive amination of pyruvate. As a result of their experiments Bassham and Kirk [14] suggested that a single primary amination reaction occurs during the biosynthesis of amino acids during photosynthesis and leads to the formation of glutamate. Aspartate, alanine and other amino acids were regarded as secondary products formed from glutamate, presumably by amino transfer reactions. In experiments using $^{14}CO_2$, two pools of glutamate of similar size were identified in the *Chlorella* cell [171]. Concurrent $^{15}NH_4$ and $^{14}CO_2$ incorporation appears to occur in both pools but the less actively turning over pool only begins to accumulate ^{14}C label after 30 min of photosynthesis [171]. Partial degradation of the total glutamate pool in the cells revealed a very interesting fact, i.e. that the glutamate was not uniformly labelled with ^{14}C: more than 20% of the label was found in the C-1 of the molecule. The suggestion was therefore made that there are two sites of glutamate synthesis in the cell, one probably in the chloroplast and the second elsewhere in the cell but utilizing photosynthetically labelled intermediate(s). The several recent demonstrations of the very rapid transfer of photosynthetically formed triosephosphate from the chloroplast at rates exceeding 500 μmoles.mg chl^{-1}. h^{-1} (see ref. 55 for a review and also Chapters 5 and 6 of this volume) show that a supply of carbon skeletons for extra-chloroplastic amino acid synthesis is readily available in the photosynthesising cell.

With attached leaves it is more difficult to carry out short-term exposure experiments over the initial fixation period when the primary amination reactions take place. In tomato leaves 40% of the ^{14}C label is in amino acids after 3 min exposure [112]. In our laboratory recently, using an apparatus constructed to allow exposure of leaves to $^{14}CO_2$ at constant specific activity for periods of less than 1 min in spinach and maize, 25% of the ^{14}C is found in amino acids in 2 min [28]. Reductive amination to glutamate has been the most frequently investigated synthesis following Bassham and Jenson's suggestion [12] that it is the major route of amino incorporation into carbon chains in photosynthesising cells. [5-^{14}C]α-Oxoglutarate is converted to glutamate by isolated intact chloroplasts at rates of 1 to 2 μmoles.mg $chl^{-1}.h^{-1}$ in a light-dependent reaction [52,186,187]: glutamate dehydrogenase (EC 1.4.1.2) might be expected to be the enzyme responsible for catalysing this reaction in chloroplasts. However Ritenour et al. [154], after a careful examination of subcellular fractions from foxtail and maize leaf tissue, came to the conclusion that the NAD(H)-dependent glutamate dehydrogenase activity in leaves is entirely extrachloroplastic. Only soluble dehydrogenases are detectable by the conventional spectrophotometric assay procedure used by these workers and the application of an alternative assay [109] later revealed a glutamate dehydrogenase bound to the photosynthetic membranes of the chloroplast and able to utilize NADP(H). While it is still bound to lamellar membranes, the recorded activities of this enzyme are low, e.g. 2.0 μmoles oxoglutarate aminated.mg $chl^{-1} \cdot h^{-1}$ [109] and 13 μmoles oxoglutarate aminated·mg $chl^{-1} \cdot h^{-1}$ [121] but sufficient to account for the rates of assimilation of α-oxoglutarate by isolated chloroplasts. After its solubilization from the photosynthetic membranes by 0.1% Triton X-100

[107] the properties of the enzyme have been examined in more detail. The lettuce chloroplast enzyme, after solubilization, utilized NAD(H) and NADP(H) without preference and its K_m-ammonia was 5—6 · 10^{-3} M, i.e. half the value for leaf mitochondrial glutamate dehydrogenase (NADH) [107].

A single amination reaction catalysed by a chloroplast NADP(H)-dependent glutamate dehydrogenase could certainly account quantitatively for the light-dependent glutamate synthesis in vivo [13,14,171,173] and in vitro [52,92]. On this interpretation photosynthetic electron transport reactions would supply the reducing equivalents required in reductive amination which would be light-dependent. Quantitatively the higher rates of activity recorded for the membrane bound enzyme are just sufficient to account for amino-incorporation in photosynthesising algae or leaves.

11.2.2.2. Glutamate, aspartate and alanine synthesis

Following the initial glutamate synthesis, a series of rapid amino-transferase reactions could account for the sequential labelling of first alanine and aspartate and later glutamate during in vivo photosynthesis. As illustrated in Fig. 11.1, the small endogenous pool (6—7 nmoles · mg chl^{-1}) of α-oxoglutarate would be unlabelled and so the glutamate synthesised would also be unlabelled [51]. Aspartate and alanine formation has been frequently demonstrated to occur when oxaloacetate and pyruvate respectively are supplied to isolated chloroplasts [102,156]. These syntheses appear to depend on the activity of two amino transferases glutamate-oxaloacetate aminotransferase (EC 2.6.1.1) and glutamate-pyruvate amino-transferase (EC 2.6.1.2), both located in chloroplasts [102,159,202] and the only plastid aminotransferases transferring amino-groups from glutamate. If the keto-acid substrates for the aminotransferase reactions are themselves rapidly labelled during photoassimilation, then the amino-acids derived from them, for example aspartate and alanine, will also be labelled. Several lines of evidence suggest this rapid labelling of "respiratory" intermediates takes place largely outside the chloroplast [72]. Phosphorylated intermediates, particularly triosephosphates, leave the chloroplast and are rapidly metabolised in the cytoplasm and mitochondria to dicarboxylic acids [78,79,174,194]. Further, a dicarboxylate translocase has been shown to mediate the rapid transfer of oxaloacetate back into the chloroplast where it considerably stimulates photosynthetic oxygen evolution ([74] and Chapters 5 and 6 of this volume). That the continuing synthesis of amino-acids within the chloroplasts depends on a supply of carboxylic acids is demonstrated by the addition of an external oxaloacetate-generating system, including partially purified PEP-carboxylase, to photosynthesising chloroplasts when the ratio of amino acids:sugar phosphates which become labelled increases from 2:1 to 31:1 [102].

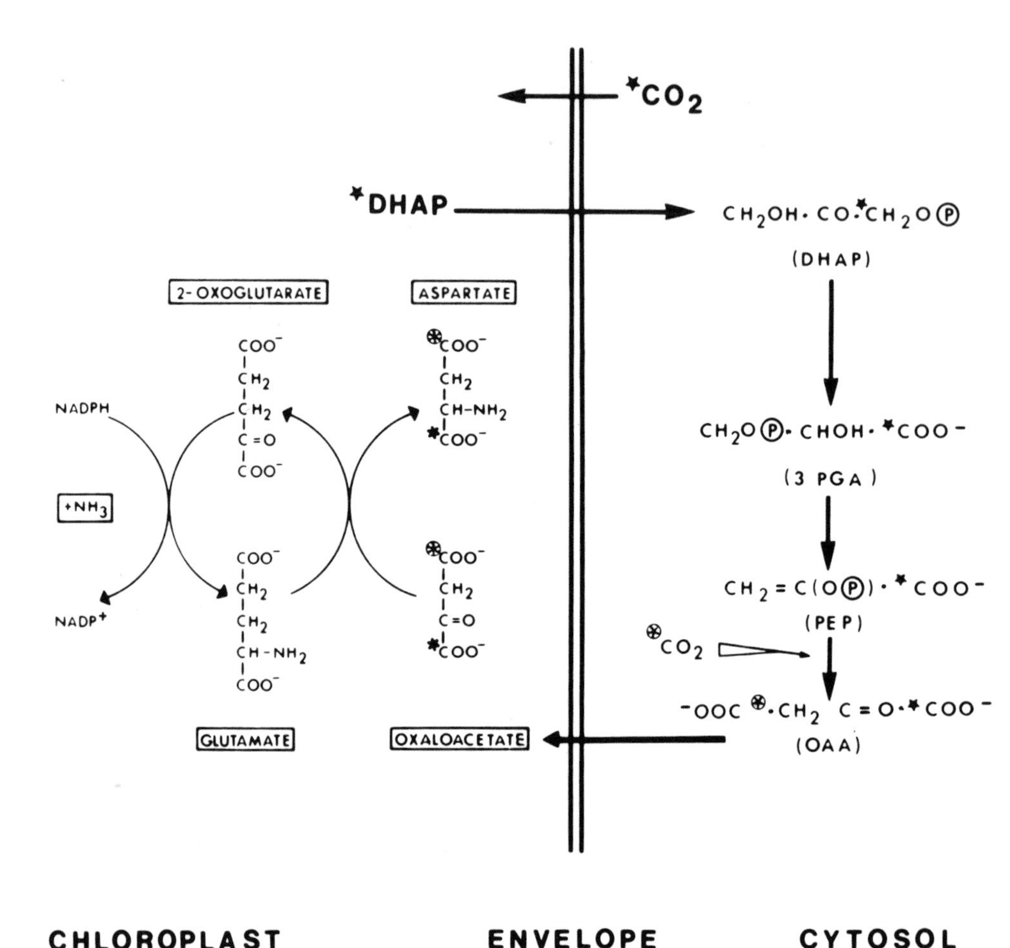

Fig. 11.1. Diagram illustrating the theoretical labelling pattern for amino-acids synthesised in the chloroplast utilizing oxaloacetate precursors from the cytosol. Photosynthetically produced triosephosphate leaves the chloroplast and is converted in the cytosol to phosphoenol pyruvate (PEP) which is carboxylated to yield oxaloacetate. The oxaloacetate re-enters the chloroplast and amino-transferase catalyses the transfer of amino groups from glutamate to yield aspartate labelled in the C1 and C4 atoms.

11.2.2.3. Glutamine synthetase and glutamate synthase

Until recently glutamate dehydrogenase was the only postulated catalyst for primary amination in the chloroplast. A few reports of the presence of small levels of aspartate dehydrogenase [159] in plastids have not been substantiated. Several features of the chloroplast glutamate dehydrogenase however appear difficult to reconcile with its proposed role in catalysing the primary amination reaction. The high K_m for ammonia of the solubilized chloroplast glutamate dehydrogenase [107], its rather low activity in chloroplast suspensions and particularly the repeated inability of many workers to obtain the expected increase of glutamate synthesis at high ammonium concentrations [52,73,102,159], stimulated the search for alternative catalysts of primary amination processes in the chloroplast. While the characteristics of glutamate dehydrogenase do not rule out the possibility that this enzyme functions in the chloroplast, alternative pathways may also be involved. An alternative mechanism for NH_4^+ incorporation into organic molecules is by amidation. In early experiments with whole *Chlorella* cells [14], a small amount of labelling from $^{15}NH_4Cl$ appeared rapidly in the amide group of glutamine: the kinetics indicated the presence of a small actively turning-over pool of glutamine which quickly saturated. Although in earlier published experiments only small amounts of glutamine were synthesised when [5-^{14}C]α-oxoglutarate was supplied as a carbon source to isolated chloroplasts [52], in later (unpublished) experiments larger syntheses were found. [^{14}C]Glutamate is also readily incorporated into glutamine by isolated spinach chloroplasts [129]; indeed the ^{14}C of the added glutamate appears only in glutamine. Mitchell and Stocking [129] have recently published a careful investigation of glutamine synthesis in isolated pea chloroplasts. The synthesis is light-dependent and is stimulated by exogenously supplied NH_4Cl (10^{-4} M to 10^{-2} M). Chloroplasts are also able to assimilate NO_2^- (10^{-3} M) via ammonia to glutamine at a rate of approx. 4 μmoles·mg chl^{-1} · h^{-1}, i.e. at half the rate for the direct assimilation of NH_4^+ into glutamine in the same chloroplasts [129]. The stoichiometric requirement for ATP in the synthesis of glutamine suggested a direct coupling with photophosphorylation and a careful examination of the spectral properties of the effective light and of photosynthetic inhibitors on glutamine synthesis, clearly demonstrated that in some circumstances the ATP may be generated by cyclic electron flow in the chloroplast. The lack of inhibition of glutamine synthesis by DCMU (02. μM) is surprising and not so far explained.

The demonstration of glutamine synthesis in isolated chloroplasts and the discovery that a major route for accumulation of ammonia into the α-amino group in some bacteria involves glutamine as an intermediate (see ref. 22 for a review), stimulated the search for a similar system in green leaf tissue. The so-called "GOGAT" system in bacteria operates via the linked enzymes L-glutamate-ammonia ligase (glutamine synthetase) (ADP, EC 6.3.1.2) and glutamine (amide) 1,2-oxoglutarate aminotransferase (glutamate synthase) (EC 2.6.1.53).

Two independent reports of glutamine synthetase activity in spinach and broad-bean chloroplasts were published in 1973 [71,139]. The enzyme catalyses the ATP-dependent addition of ammonia to glutamate to produce the amide. Using NADP-triosephosphate dehydrogenase and alkaline fructose diphosphatase as chloroplast marker enzymes, O'Neal and Joy [139] showed that between 32% and 66% of the glutamine synthetase activity of leaf homogenates was associated with the chloroplast fraction. The affinity of this glutamine synthetase for ammonia is about 100-fold more [141] than the comparable affinity of isolated chloroplast glutamate dehydrogenase [107], suggesting it may play a considerable role in the assimilation of inorganic nitrogen in plastids. Although rather unstable, glutamine synthetase from pea shoots has been purified to electrophoretic homogeneity by O'Neal and Joy [140]. It appears to have a molecular weight of about 360 000 and in this respect resembles the enzyme from pea seeds and mammalian sources. The demonstration of glutamine synthesis in isolated chloroplasts and the isolation and characterisation of glutamine synthetase stimulated the search for the enzyme glutamate synthase in chloroplasts. The presence of this enzyme has now been reported for chloroplasts [106], blue-green algae [105], and *Chlorella* [104]. The most interesting feature of the spinach chloroplast glutamate synthase is its ability to use reduced ferredoxin as an electron donor [106] and the operation of a light-dependent GOGAT cycle in chloroplasts would certainly be consistent with the rapid turnover of a small pool of glutamine noted by several workers [14,95,96] and the light-stimulated and PGA-depressed utilisation of α-oxoglutarate for glutamate biosynthesis in isolated chloroplasts [52]. However, at present the highest published rates of catalysis for NADP-glutamate dehydrogenase [102,107,121] and for ferredoxin-dependent glutamate synthase [102] are of the same order of magnitude and both are 1000-fold less than the rate of glutamine synthesis when exogenous glutamate is supplied. Further work, in particular in characterising the mode of operation of the enzymes within the thylakoids (glutamate dehydrogenase) and stroma (glutamate synthase) is needed, before the contribution of the alternative systems to ammonia assimilation in vivo can be quantitatively assessed. Glutamine synthetase and glutamate synthase have also been recently identified in roots [128] and in plant tissue cultures [42] so such enzymes may be of general significance in the assimilation of ammonium ions in plants and not merely of significance in photosynthetic cells. It is possible that different modes of glutamate synthesis may predominate in different cellular or environmental conditions or perhaps under different conditions of nitrogen stress.

11.2.3. The carbon substrate for further amino acid biosynthesis

11.2.3.1. Glycine and serine biosynthesis

Glycine and serine are the only amino acids in addition to alanine, aspartate and glutamate, which become rapidly labelled both when $^{14}CO_2$ is fed to photosynthesising cells [13,14,95,96,171,173] and also when $^{14}CO_2$ is fed to isolated chloroplasts [50,102]. The mechanism and subcellular localization of glycine and serine synthesis in photosynthesising tissues have been intensively studied, particularly in relation to photorespiration. In leaves, serine and glycine both become extremely rapidly labelled from $^{14}CO_2$, the label appearing in these amino acids before it appears in glutamate. The labelling pattern within the glycine and serine molecules is consistent with their derivation from glycollate, which is synthesised and accumulates in isolated chloroplasts [145], but there is no unanimous agreement about whether chloroplasts alone can synthesise serine and glycine. Evidence from careful aqueous cell-fraction studies points to the synthesis of these two amino acids being a collaborative process, involving mitochondrial and peroxisomal enzymes in addition to the chloroplast. Tolbert [180] has excellently reviewed the experimental findings supporting this scheme and little can be usefully added to his account. According to Tolbert's proposal, glycollate leaves the chloroplast and is oxidised to glyoxylate in the peroxisomes, glyoxylate is aminated by peroxisomal aminotransferase to glycine and glycine converted to serine in the mitochondria. However, there is also some evidence [155,161] from studies using non-aqueously isolated chloroplast suspensions which suggests the chloroplast itself may have a limited capacity for glycine and serine biosynthesis. The distribution of $[^{14}C]$glycine in fractions isolated from leaves previously fed with $^{14}CO_2$ did not parallel the distribution of the perioxisomal enzyme, glycollate oxidase, and even in extracts of chloroplasts deprived of catalase, glycine and serine can still be synthesised [161].

11.2.3.2. Other amino-acid syntheses

Many additional amino-transferases are present in chloroplasts which can transfer amino-groups from either alanine or aspartate to a variety of keto acid acceptors [102,159]. The activities of these enzymes could account for the synthesis of all common protein amino acids, apart from leucine, provided the keto acids are available. Many alternative metabolic pathways are well-documented for the biosynthesis of amino acids, particularly aromatic amino acids, but as yet virtually nothing is known of the subcellular location of the key enzymes of amino acid biosynthesis in plants. Recently [127] clear evidence from density gradients centrifugation studies has shown that some of the activity of acetolactate synthase is present in plastids. Further investigations using similar techniques are urgently needed.

11.2.3.3. The regulation of amino acid biosynthesis in chloroplasts

There is no evidence that repression or induction mechanisms play a part in regulating amino acid biosynthesis under normal conditions in higher plants. The evidence available would suggest that the enzymes within a membrane-bound organelle, such as a plastid, are sensitive to the end-product amino acid concentration within the organelle. Transport processes of amino acids into the plastids will certainly affect their concentration and are dealt with in Chapter 6 of this volume.

11.2.4. The carbon substrates for further amino acid biosynthesis

While considerable effort has been devoted to the discovery of the form in which inorganic nitrogen enters the chloroplast and the mechanisms of its reduction and incorporation into organic molecules, less attention has been given to the origin and nature of the carbon skeletons which are essential for the continuing synthesis and accumulation of small molecules containing α-amino nitrogen, in particular the amino acids. For glutamate and glutamine biosynthesis a constant supply of α-oxoglutarate is required which may be indirectly derived from photosynthetically assimilated intermediates and exported from the chloroplasts and further metabolised in the cytoplasm. The endogenous pools of α-oxoglutarate in the chloroplast are 7.0 nmoles·mg chl^{-1} in the light and approx. 14 nmoles·mg chl^{-1} in the dark.

Further synthesis of additional amino acids or amides would require a continuing supply of their corresponding keto acids as already discussed for aspartate synthesis. In summary, the evidence suggests that nitrate is reduced to nitrite at the chloroplast envelope membrane and the nitrite is further reduced to ammonium within the chloroplast. Chloroplasts have enzymes which catalyse the reductive amination of α-oxoglutarate to glutamate and the amidation of glutamate to glutamine. They are also able to synthesise two molecules of glutamate from a molecule of glutamine and one of α-oxoglutarate. The relative significance and control of the alternative

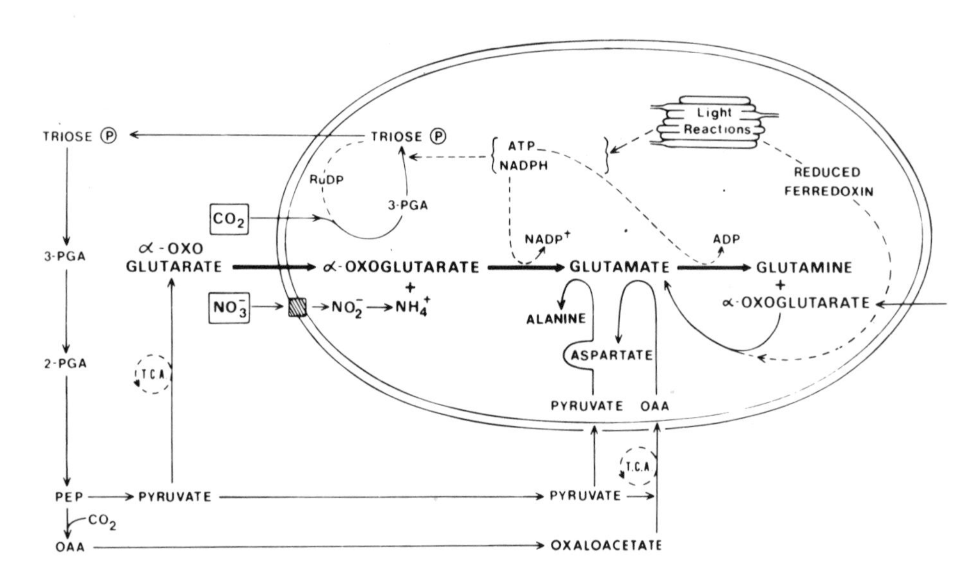

Fig. 11.2. Diagram illustrating possible routes associated with the chloroplast for the assimilation of nitrate and the synthesis of glutamate and glutamine. For details see the text.

syntheses is currently being actively investigated. The initial formation of glutamate and the subsequent formation of other amino acids catalysed by amino-transferases depends on a continuing supply of the appropriate carbon skeletons mainly from the cytoplasm.

These series of reactions are shown diagramatically in Fig. 11.2 and illustrate several aspects of the collaborative function of the chloroplast in the biosynthesis of some amino-acids. The lack of permeability of the chloroplast envelope to many exogenously supplied amino acids is discussed in Chapter 6 of this volume.

11.3. THE BIOSYNTHESIS OF THE ACYL LIPIDS OF CHLOROPLASTS

11.3.1. The lipids of chloroplasts

11.3.1.1. Composition

Lipids make up 10% of the dry weight of the green leaves of higher plants [100] where they are concentrated in the membraneous organelles, in particular the plastids. 35% of chloroplast dry weight and 50% of the photosynthetic thylakoid membrane is lipidic [144]. The lipid compositions of chloroplast thylakoid and envelope membranes, "mitochondrial" and "microsomal" fractions from *Vicia faba* leaves are compared in Table I

TABLE I

The lipid composition of subcellular fractions isolated from leaves of *Vicia faba* L. (From Mackender and Leech [120])

The values have been corrected for the contribution of the lipids of a small proportion of lamellae fragments to the envelope mitochondrial and microsomal fractions.

Subcellular fraction	Lipid				
	MGD	DGD	PC	PG	PE
	Moles per 1000 moles of lipid analysed				
Chloroplast envelope	291	324	296	89	0
Chloroplast lamellae	654	262	28	55	0
"Mitochondrial" (20 000 *g*)	102	128	435	78	257
"Microsomal" (100 000 *g*)	82	204	476	96	142

[120]. Although the significance of the differences in lipid composition is not yet understood, almost certainly they reflect differences in function of the different membranes. The major chloroplast lipids are the glycolipids (Fig. 11.3), monogalactosyl diglyceride (MGD), digalactosyl diglyceride (DGD) (which account for 30% and 15% respectively of the total lipids) and sulphoquinovosyl diglyceride (SL) (5%). Phospholipids account for 10% of the chloroplast lipids and the remainder (30%) are neutral lipids (1.2%)

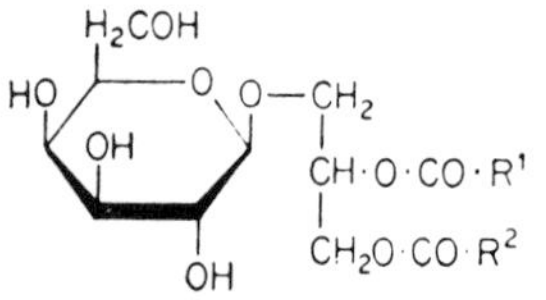

Monogalactosyl diglyceride [β-D-galactosyl-(1→1′)-2′, 3′-diacyl-D-glycerol]

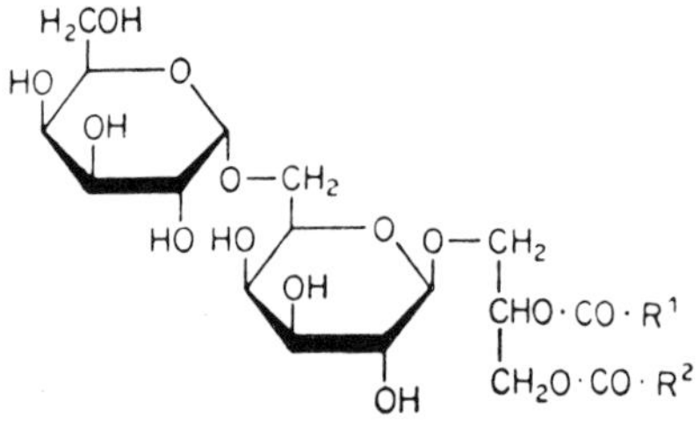

Digalactosyl diglyceride [α-D-galactosyl-(1-6)-β-D-galactosyl-(1-1′)-2′, 3′-diâcyl-D-glycerol]

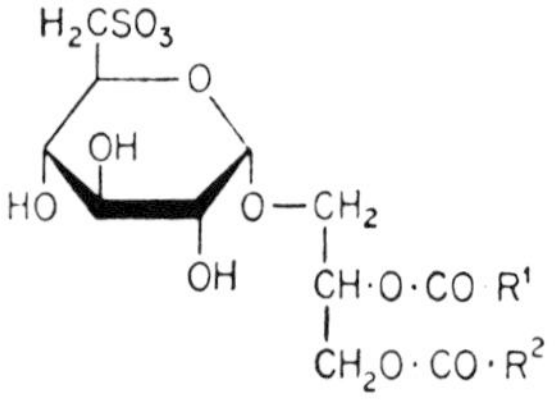

Sulphoquinovosyl diglyceride [6-Sulpho-α-D-quinovosyl-(1-1′)-2′, 3′-diacyl-D-glycerol]

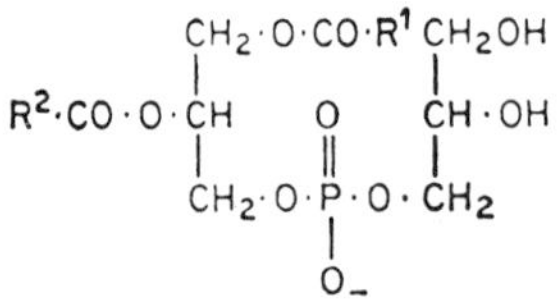

Phosphatidyl glycerol

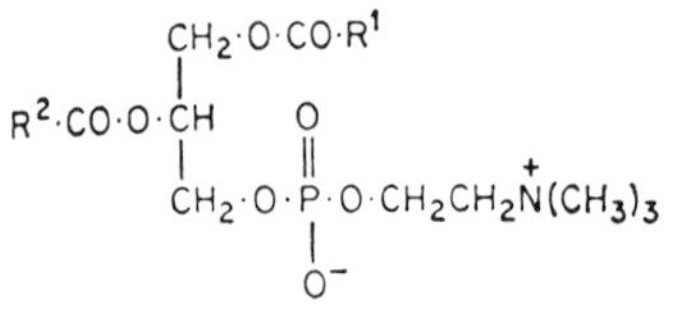

Phosphatidyl choline (Lecithin)

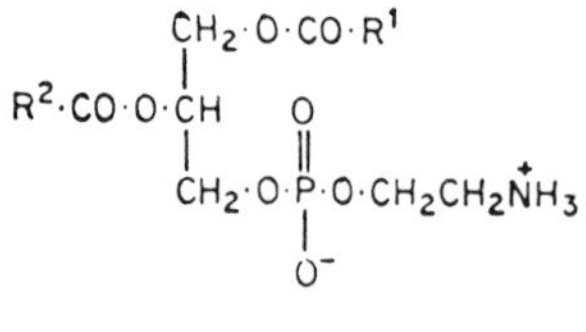

Phosphatidyl ethanolamine

Fig. 11.3. The structures of the major chloroplast lipids.

pigments (26%) and sterol glycosides (2%). The major fatty acids of the chloroplast lipids are characteristically polyunsaturated and most are octadecenoic acids (C18:1, C18:2, C18:3). The most unsaturated acid, α-linolenic acid (C18:3) can account for over 90% of chloroplast fatty acids [81]. Of particular interest is trans-3-hexadecenoic acid (C16:1) which is found specifically in photosynthetic tissue [88] and then only acylated to phosphatidyl glycerol [4,67]. The fatty acid composition of the lipids of maize chloroplasts is given in Table II. While much of the analytical work so far has been concentrated on the characterisation of the membrane lipids, the plastoglobuli and the chloroplast stroma have received less attention. Plastoglobuli are present in plastids of all types [11] but are more common in developing or senescing chloroplasts [114] where they constitute a store of lipid material, sometimes including pigments [115], which may account for as much as 10% of the total chloroplast lipid. Stroma lipids make up a high proportion of the plastid lipid complement, up to 25% in some species [134,148,149] and as this fraction may well contain lipoprotein complexes in transit between different chloroplast membrane systems [1] it certainly merits further attention.

The acyl lipids of chloroplasts (Fig. 11.3) have a three-carbon glycerol "backbone", two fatty acid side-chains attached to this backbone and a polar head group containing either galactosyl, sulphoquinovosyl or phosphate ester moieties. Glycerol-3-phosphate, fatty acids and the polar

TABLE II

Fatty acid composition of the individual lipids isolated from chloroplasts of maize *(Zea mays)*

The values are the means of the results of the analyses of two different chloroplast preparations. Each analysis was performed in duplicate (Rumsby and Leech, unpublished data).

	Fatty acids present (moles %)							
	14:0	14:1	16:0	16:1	18:0	18:1	18:2	18:3
Monogalactosyl diglyceride	0.3	0	2.0	0.4	0.4	2.2	5.6	89.1
Digalactosyl diglyceride	0.3	0.1	8.4	0	1.1	8.8	2.5	78.8
Sulpholipid	6.0	0	42.3	0	2.3	3.6	9.3	36.5
Phosphatidyl choline	12.6	0	33.3	2.4	10.0	8.1	17.7	16.2
Phosphatidyl ethanolamine	15.3	2.4	17.7	4.4	3.9	11.0	5.6	39.7
Phosphatidyl glycerol	4.3	1.4	38.7	7.6	2.7	6.5	10.2	28.6

group moieties are also metabolic intermediates and therefore there are likely to be several independent pools of these components in the cell turning-over at different rates. In addition, components of lipid molecules turnover *within* the molecule at rates which vary with the age of the tissue and environmental conditions [182]. The turnover rates may also differ in the same lipid in different intracellular locations [2]. These situations complicate both the design and interpretation of experiments aimed at determining the mode of synthesis of lipids in photosynthesising cells.

11.3.1.2. The organization of lipids in chloroplast membranes

(i) Chloroplast thylakoids. The preponderance of glycolipids containing a high proportion of polyunsaturated fatty acids in chloroplast thylakoids has raised questions about the significance of these characteristic lipids in photosynthetic function [175,198]. In particular the high proportion of α-linolenic acid appeared in early analyses to be characteristic of many photosynthetic membranes [137], notably of most organisms evolving O_2 during photosynthesis [45,53]. One notable exception is the blue-green alga *Anacystis nidulans* which lacks polyunsaturated fatty acids (see ref. 137 for a review). However, several lines of evidence suggest that glycolipids are unlikely to be involved directly in the light reactions or electron transport reactions of the thylakoids [6,19]. For example, large amounts of galactolipid can be removed from photosynthetic membranes by treatment with lipase without marked effect on electron transport [6] and etioplasts and other organelles without photosynthetic function also have a high content of galactolipids containing polyunsaturated fatty acids [120]. Rather it would seem that the thylakoid membrane lipids possess those particular physicochemical characteristics which can provide a suitable environment for the assembly and function of the pigment/protein complexes of the photosynthetic photosystems. The fluid mosaic model of membrane structure was first proposed by Singer in 1971 [166] for animal membranes [169] at a time when a growing body of experimental evidence from chloroplast membranes also seemed at variance with the older "protein sandwich" type of model (for reviews, see refs. 167,168). Kirk [101] proposed a model for chloroplast thylakoid structure in which "extrinsic" proteins (such as the chloroplast coupling factor) and "intrinsic" proteins, such as those associated with the photosystems, are embedded to varying degrees in a fluid lipid bilayer. In this earlier model no proposals were made about specific associations between the chlorophyll pigments and the membrane proteins. Recently Anderson [5] has proposed a new model for the chloroplast thylakoid membrane in which she extends the fluid mosaic membrane model to the special case of thylakoids where only limited mobility of the pigment molecules would be consistent with photofunction. Anderson's model (reproduced in a modified form in Fig. 11.4 proposes

that the thylakoid is composed of a lipid bilayer consisting of the galactolipids which stabilise the membrane [19], particularly by virtue of their polyunsaturated fatty acid chains, which are well known to contribute to the stability of animal membranes [59]. The intrinsic proteins in Anderson's model are visualised as having associated with them a monomolecular layer of fixed or "boundary" lipid which moves around with the proteins [26]. Anderson suggests that the chlorophylls may be boundary lipids. Because of their relatively saturated, and hence less mobile fatty acid components, phospholipids, in particular phosphatidyl glycerol, may also be boundary lipids in chloroplast thylakoids. The Anderson model certainly provides inspiration for further experimentation. Two systems which may prove particularly useful in providing information about the relationship of specific lipid composition to photosynthetic function are tissues in which the assembly of the photosynthetic membranes is occurring as in the naturally developing leaves of some monocotyledons [110,111], and also artificially made lipid bilayers in which pigment and proteins can be implanted and their relationships determined.

(ii) Chloroplast envelope membranes. The chloroplast thylakoids and chloroplast envelope membranes have very different compositions (See Table I). The notable distinction of the chloroplast envelope membranes is their high content of phospholipids, particularly phosphatidyl choline (PC). The

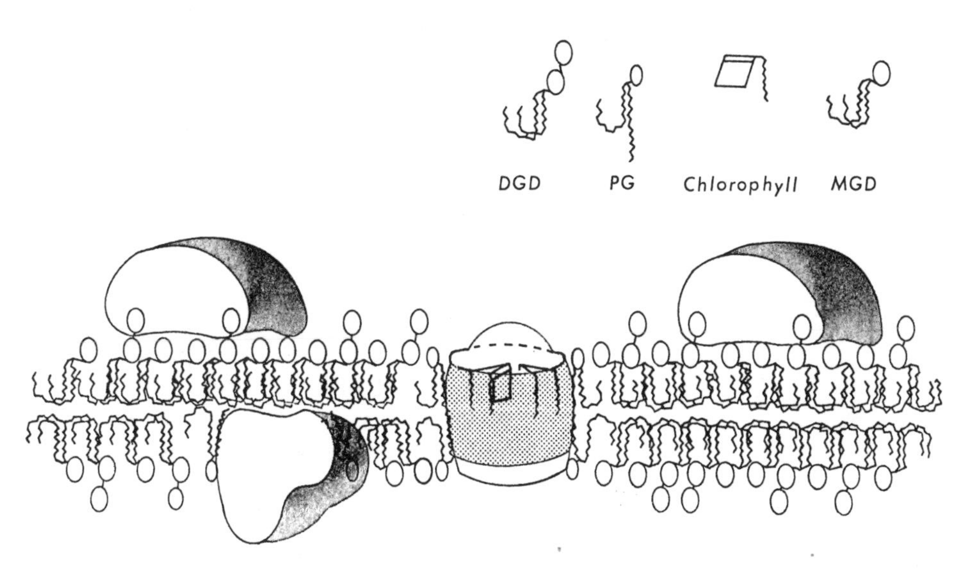

Fig. 11.4. Model showing a possible arrangement of lipids and protein components in the chloroplast thylakoid membrane. The diagram is based on the ideas of Singer [166], Anderson [5] and our own.

relationship between the lipid composition and function of the chloroplast envelope membranes remains to be investigated.

11.3.2. Lipid biosynthesis during photosynthesis

Dihydroxyacetone phosphate (DHAP), an important precursor in glycerol-phosphate biosynthesis is exported rapidly from photosynthesising isolated chloroplasts at rates exceeding 500 μmoles. mg chl^{-1} . h^{-1} [72,78]. This rapid export, which begins within seconds of the onset of photosynthetic carbon reduction, complicates the interpretation of experiments in which light-dependent synthesis of secondary photosynthetic products is followed in whole cells. Few detailed analyses of the synthesis of lipids in photo-synthesising systems have been carried out. In early experiments with *Chlorella* cells [48], $^{14}CO_2$ was rapidly incorporated into galactolipids after 15—60 sec, whereas incorporation into phospholipids occurred between 5 min and 60 min after the onset of photosynthesis. 50% of the radioactivity in the lipid fraction was found in the galactose moiety of the galactolipids after 5 min [17]. In contrast, in detached pumpkin leaves pulse-labelled for 5 min with $^{14}CO_2$, PC was the first lipid to become labelled and the label was concentrated in the fatty acid moieties [157]. These experiments with leaves were much more long-term than the *Chlorella* studies and labelled PC first appeared 1 h after the start of the cold chase. In contrast, exposure of whole young maize leaves of 7-day-old plants to $^{14}CO_2$ in the light results in the incorporation of $^{14}CO_2$ into the leaf PC after 20 sec [134]. After 20 min, 1.5% of the newly incorporated carbon is found in the lipid fraction and the proportion rises to 5% after 1 h [134,157].

11.3.2.1. The synthesis of chloroplast lipids

While the high lipid content of chloroplasts does not of itself prove that these organelles have an independent system for the synthesis of lipids and their fatty acids, there is direct evidence that many such syntheses do occur in plastids.

11.3.2.2. The biosynthesis of lipid precursors

(i) Synthesis of sn-glycerol-3-phosphate. "Microsomal" fractions, and to a lesser extent chloroplast fractions, from leaves will incorporate [^{32}P]-glycerol-phosphate into phosphatidic acid and phosphatidyl glycerol (PG) [160]. The incorporation of [^{14}C]glycerol-phosphate via lysophosphatidic acid (LPA), phosphatidic acid (PA) and diglyceride (DG) into MGD has also been demonstrated to occur in plastid preparations [40] (Fig. 11.5). Both these findings suggest that glycerol-3-phosphate can be an intermediate in lipid biosynthesis in plastids. An alternative pathway [3], which occurs in liver microsome fractions and has not so far been looked for in plants, involves the direct acylation of DHAP to acyl-DHAP, then reduced to lysophosphatidic acid.

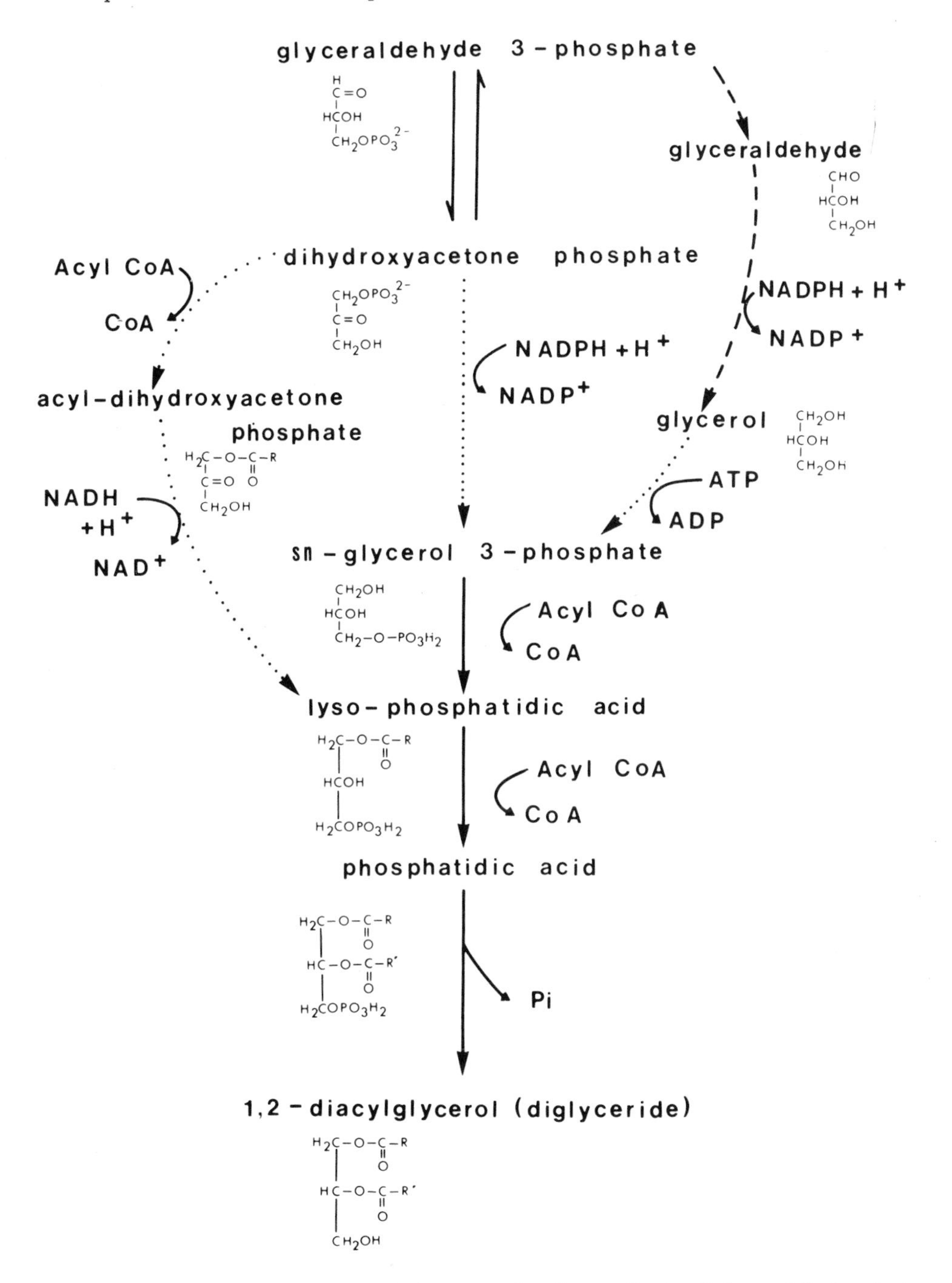

Fig. 11.5. Diagram illustrating possible routes for the synthesis of glycerol-3-phosphate and diglyceride. ——→, demonstrated to occur in chloroplasts; — — →, demonstrated to occur in leaves; · · · · →, demonstrated only in animal systems. See text for details.

The location of the enzymes required for glycerol-3-phosphate synthesis in leaf cells is clearly relevant in discussions of lipid biosynthesis. Two alternative routes for the synthesis of glycerol-3-phosphate have been described in animal systems (see Fig. 11.5). The major pathway is from DHAP in a reaction catalysed by glycerol phosphate dehydrogenase (EC 1.1.1.8). This enzyme, and also an NADP-linked enzyme have been sought, but not found in isolated chloroplasts [113] or leaves [103]. A second possible route for glycerol-3-phosphate synthesis would be from glyceraldehyde-3-phosphate via glyceraldehyde and glycerol. Information about the occurrence of the enzymes of this pathway in fractions from leaves is presently incomplete. In sugar cane leaves activity of the first enzyme of the pathway, glyceraldehyde-3-phosphatase (EC 2.7.1.28), has been shown to be associated with the plastid fraction [150]. In *Vicia faba* leaves the second enzyme, glyceraldehyde reductase (EC 1.1.1.21) is associated with the 40 000 *g* supernatant but not the plastid fraction [103] and the enzyme is NADP-linked. There is no evidence that glycerol itself can be incorporated into lipids; indeed leaf microsomal fractions have been shown to incorporate glycerol-phosphate but not glycerol into lipids [29]. However glycerol kinase, the enzyme converting glycerol to glycerol-3-phosphate, has not been demonstrated to be present in any leaf fractions. Re-examination of the activities of the so-called "microsomal" and chloroplast envelope fractions may prove rewarding, particularly since the spectrophotometric assays previously employed would appear not to be sufficiently sensitive [103] to detect enzyme activities responsible for the very low published rates of lipid synthesis (1—100 pmoles. mg protein^{-1} min^{-1} observed in isolated chloroplasts.

(ii) Synthesis of UDP-galactose. The galactosyl donor in the biosynthesis of the galactolipids is UDP-galactose [135]. Two alternative pathways of UDP-gal biosynthesis have recently been re-examined in leaves (see Fig. 11.6). UDPG pyrophosphorylase (EC 2.7.7.9) and UDP-gal-4-epimerase (EC 5.1.3.2), two enzymes of the pathway from glucose-1-phosphate, were found in *Vicia faba* leaves [103]. Since these enzymes followed the distribution of pyruvic kinase, a cytoplasmic enzyme, during differential centrifugation, it seems likely that both are present in the cytosol rather than in the plastids. Isolated chloroplasts can readily incorporate ^{14}C into glucose-1-phosphate [49] so the synthesis of UDP-gal appears to be another example of a collaborative cellular process in which photosynthetically assimilated intermediates are further interconverted in the cytosol.

The enzymes of the alternative pathway of UDP-gal synthesis, which involves the prior phosphorylation of galactose (Fig. 11.6), could not be demonstrated in *Vicia faba* leaf homogenates [103], although previous tracer experiments have indicated that they are present in leaves [55,82].

(iii) Synthesis of sulphoquinovose. Sulphoquinovosyl diacylglycerol,

usually known as the sulpholipid, is uniquely associated with the chloroplast in leaf cells [3] but the biosynthesis of its precursor 6-sulphoquinovose, and the incorporation of this precursor into diglyceride has received little attention. Neither the location nor the pathway of sulpholipid biosynthesis in higher plants has been characterised. The characteristics of incorporation of $^{35}SO_4^{2-}$ into the sulpholipid of germinating alfalfa [64] support the operation of one possible pathway of sulpholipid biosynthesis from phosphoadenosine phosphosulphate (PAPS) and 2-phospho-3-sulphoacetate as demonstrated in *Euglena gracilis* [33,34,163] and the diatom *Navicula pelliculosa* [25].

(iv) Synthesis of CDP-choline. PC and PG are the phospholipids present in the highest concentration in chloroplast envelope and thylakoid membranes respectively (see Table I). Glycerol-3-phosphate is the precursor of the polar moiety of PG, but choline formation is required for the continuing synthesis of CDP-choline required in PC biosynthesis. There are no data on choline biosynthesis in leaves. Serine and glycine are important precursors of choline and since recent careful experiments with isolated chloroplasts [50] showed insignificant amounts of these amino acids were formed during $^{14}CO_2$ photosynthesis (see section 11.2.3.1), there is no evidence that even these early precursors of choline are synthesised in the chloroplast.

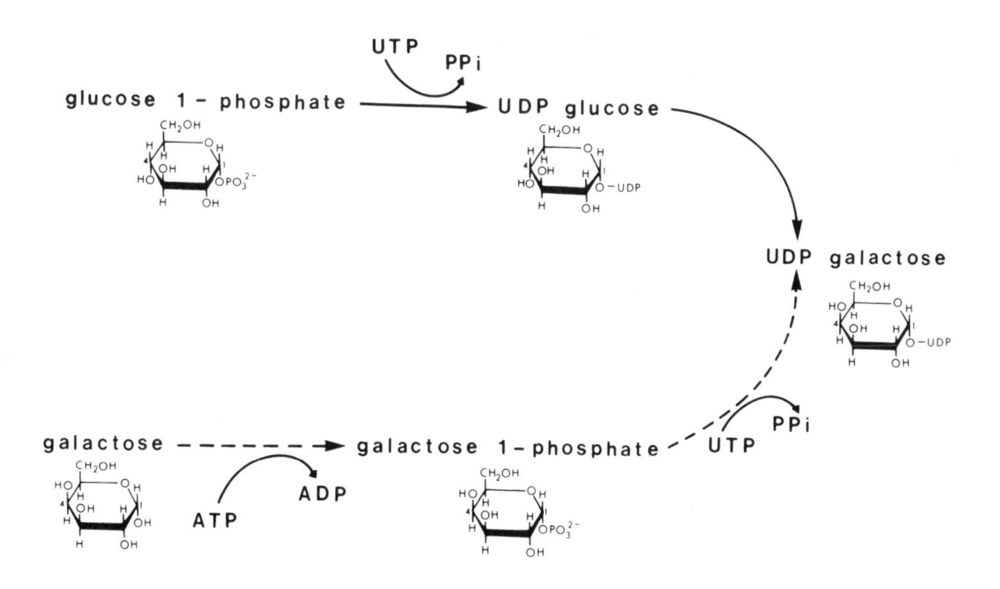

Fig. 11.6. Diagram illustrating two possible routes for the synthesis of UDP-galactose. ⟶, demonstrated to occur in chloroplast; — — →, demonstrated to occur in leaves.

(v) Fatty acid biosynthesis. Fatty acid synthesis takes place by the step-wise addition of 2-carbon units to a growing acyl chain. Acetyl-CoA is the precursor and the 2-carbon units are added on the reaction with the 3-carbon compound malonyl-CoA, which is decarboxylated in the process. The CO_2 formed is used to regenerate malonyl CoA. In *E.coli* the fatty acid synthetase is a multienzyme complex containing seven enzyme sub-units, including the acyl carrier protein (ACP) which is the final acceptor of the newly completed acyl chain. Largely through the studies of Stumpf and his co-workers (see refs. 175,177,178) on cell free preparations, considerable information on the fatty acid synthesis in higher plants is available. Syntheses of fatty acids from acetate, malonyl-CoA and acetyl-CoA seem to operate in a manner analogous with the systems in *E.coli*, although few of the enzymes from higher plants have been completely characterised [8,175, 196,197]. The work earlier to 1971 has been reviewed [51].

The label from photosynthetically assimilated $^{14}CO_2$ does not appear in the fatty acids of leaf lipids for several hours [157] and to-date $^{14}CO_2$ incorporation into lipid fatty acids in isolated chloroplasts has not been demonstrated. Control rates of fatty acid synthesis, in the absence of an exogenous carbon source, are negligible. Since acetyl-CoA only crosses the chloroplast envelope at very low rates [77], [1—^{14}C]acetate is normally used as the carbon source in studies of fatty acid biosynthesis in isolated chloroplasts. Since isolated chloroplasts are unable to incorporate $^{14}CO_2$ into fatty acids under similar conditions to those in which they assimilate [^{14}C]acetate [47], it has been suggested that chloroplasts may lack the enzymes required for the conversion of 3-phosphoglycerate to acetyl-CoA [162]. However, it has recently been found that isolated spinach chloroplasts will incorporate label from [2-^{14}C]pyruvate into fatty acids [201] (but not from [1-^{14}C]pyruvate as the 1-carbon is lost in the decarboxylation reaction from pyruvate → acetyl CoA). Since isolated chloroplasts can synthesise pyruvate from PGA (via PEP) this raises the possibility that they can synthesise all the acetyl-CoA needed for fatty acid biosynthesis without recourse to cytoplasmic enzymes. The light-stimulated incorporation of [^{14}C]acetate into fatty acids by isolated intact chloroplasts has been demonstrated frequently and with numerous species of chloroplast [8,70,97,176]. The synthesis of C16:1 has also recently been demonstrated in chloroplasts [65]. However, several features of these syntheses are distinctive. The rates of fatty acid synthesis are low compared with in vivo rates and the fatty acids synthesised in vitro are more saturated and are produced in proportions which differ greatly from the endogenous fatty acid ratios of the chloroplast lipids [181,185]. The main products of chloroplast fatty acid biosynthesis in vitro are palmitate and oleate which together account for approx. 60—90% of the fatty acids synthesised. In contrast the major fatty acid of chloroplasts, α-linolenic acid (C18:3), is synthesised by isolated chloroplasts in only pmole quantities per mg chl·h^{-1}. Indeed, despite numerous efforts over many years, no chloroplast suspension has been obtained which is capable of sustained α-linolenate biosynthesis from any precursor. Recently considerably higher rates of fatty acid biosynthesis have been demonstrated using plastid suspensions from avocado pear *(Persea americana)* mesocarp tissue [196,197]. Although the mesocarp tissue is not a typical photosynthetic tissue and its lipids do not contain α-linolenate, its plastids show a higher degree of structural organisation and chlorophyll content than those in the non-photosynthetic tissue of the same fruit. Although the mesocarp plastids are certainly not typical chloroplasts, the advantages of using such a soft tissue is the ease with which the plastids can be isolated from it, presumably with relatively little damage. Such plastids are able to synthesise all their native fatty acids without the addition of other cellular components, and quantitatively appear to constitute the principal site of fatty acid synthesis in the mesocarp. The range of fatty acids formed depends on the nature of the substrate provided. This fatty acid synthetic capability shown by mesocarp plastids raises the question of whether leaf chloroplasts have a similar potential for fatty acid biosynthesis which is destroyed during plastid isolation. At the present time only the limited capacity of leaf chloroplasts for fatty acid biosynthesis already demonstrated can be assumed. It should be remembered that the biosynthesis of a C18 fatty acid from acetyl-CoA requires 17 ATP molecules and 18 molecules of NADPH and, under conditions where other rapid syntheses are also occurring, in particular of sugar phosphates, sufficient ATP-reducing equivalents may not be available to support both photoreduction of CO_2 and fatty acid biosynthesis simultaneously. Another possible explanation for the lack of detection of fatty acid biosynthesis in vitro is the low turnover rates of some fatty acids in membrane lipids, in particular C18:3 [2,182].

(vi) Fatty acid desaturation and elongation. Chloroplasts contain a de novo fatty acid synthetase which synthesises palmitate (C16:0) and is immunologically distinct from the synthetase of the cytosol [66]. Since the synthetase activity can be readily removed from osmotically ruptured chloroplasts, it would appear to be a stroma enzyme [86,98]. Recently Weaire and Kekwick have isolated a second synthetase from avocado mesocarp plastids which appears to be membrane-bound and, in contrast to the soluble enzyme, avidin-insensitive [197]; Further elongation and desaturation of palmitate requires separate elongases and desaturases: the location and mode of operation of these enzymes is still a matter of considerable speculation. In plants the synthesis of unsaturated fatty acids requires molecular O_2 in contrast to their synthesis in *E.coli* which is anaerobic [45]. Leaf tissue will readily incorporate [^{14}C]acetate into polyunsaturated fatty acids including α-linolenate (60—80% of the chloroplast lipid fatty acids), but in isolated chloroplasts further elongation and desaturation to yield linoleate (C18:2) and linolenate (C18:3) occurs only at very low rates. The major area of conflict concerns the in vivo pathway for desaturation and

elongation. Since oleate is one of the two major products of chloroplast fatty acid synthesis from [^{14}C]acetate, it is clear that chloroplasts do contain the palmityl elongase [8,53,70,97]. The enzyme from spinach chloroplast stroma is ACP-dependent and cerulenin-insensitive, unlike the de novo fatty acid synthetase which is very sensitive to cerulenin [87,90,91]. The original proposal for the mechanism of synthesis of the polyunsaturated octadecenoic acids [62] suggested a progressive desaturation from oleate C18:1 → linoleate (C18:2) → linolenate (C18:3) [62]. The first stage in this sequence, oleate desaturase, has been demonstrated in isolated plastids by several workers [70,97]; the second step to linolenate has been far more intractable. Results of experiments in which [^{14}C]acetate, [^{14}C]oleate and [^{14}C]linoleate were fed separately to either pea leaves or pea chloroplasts gave some insight into the complexity of the problem [185]. Reductive ozonolysis of the linolenate formed following acetate or linolenate incorporation, showed a labelling pattern consistent with the progressive desaturation proposal [62]. In these experiments no reductive ozonolysis of the products of [^{14}C]oleate incorporation was carried out. However, Kannagara et al. have published results from similar experiments using spinach leaf discs where the products of [^{14}C]oleate incorporation were subjected to reductive ozonolysis: in these analyses the labelling pattern within the C18:3 molecules was totally inconsistent with its derivation from oleate (C18:1). It was suggested rather than linolenate biosynthesis had occurred by a very different mechanism in which C_2 units (in this case presumably derived from oleate) were added to an endogenous acceptor, presumed to be a C16 component. Since some synthesis of C18:3 can occur under anaerobic conditions at the same time as oleate formation is partially inhibited by cyanide [85], Kannagara et al. have suggested that C16:3 [99] may be the direct precursor of C18:3, i.e. that desaturation occurs prior to elongation. To-date no other laboratory has published confirmatory data supporting this suggestion: indeed a demonstration of linolenate biosynthesis under conditions where oleate biosynthesis is *completely* inhibited would give rather more convincing support for the suggestion concerning the alternative pathway. Results of experiments in which microdroplets containing [^{14}C] 18:2 and [^{14}C] 18:1 were sprayed on a variety of plant species have failed to discriminate convincingly between the two alternatives [30]. The incorporation of ^{14}C label from 18:1 or 18:2 into 18:3 is not a definitive proof that these fatty acids are acting as precursors of 18:3 unless reductive ozonolysis is carried out on the products in order to eliminate the possibility that label is being transferred from breakdown products (e.g. C_2 units) of the presumed precursors. Indeed there is a real possibility that, as in *Penicillium* [153], the two alternative pathways for C18:3 biosynthesis, one from C16:3 and one from C18:2 could occur together in leaves. The relative importance of the pathways may differ in different species and at different developmental stages in the plant life cycle, but it is clear that more definitive experiments are urgently needed using a variety of plant species and ages to provide the necessary information.

(vii) The acyl carrier role of phosphatidyl choline. The probable importance of phosphatidyl choline in chloroplast lipid metabolism is belied by the small proportions in which this lipid is found in whole chloroplasts [174]. PC accounts for less than 1% of the lamellae lipids but it is a major constituent of the chloroplast envelope membranes (Table I). A metabolic role for PC in photosynthetic tissue was originally proposed following experiments with *Chlorella* [61,136] in which [^{14}C]oleate fed to the alga became rapidly incorporated into phosphatidyl choline, mainly at the 2-glycerol position. In addition, subcellular fractions enriched with chlorophyll containing membranes converted oleoyl-PC to linoleoyl-PC suggesting desaturation occurred *after* incorporation of the acyl chain into the lipid. Further experiments [60] showed that dioleyl-PC is the phosphatidyl choline species which most readily undergoes desaturation. [^{14}C]Oleate is incorporated first into the 2-position, then into the 1-position and also progressively desaturated to linoleate and linolenate. In higher plant leaves also, phosphatidyl choline is often the first acyl lipid to become labelled with ^{14}C in long-term feeding experiments with 10-min pulse label followed by several hours cold chase [157]. The labelled fatty acids appear subsequently in the galactolipids. Experiments with *Euglena* [146,147] apple parenchyma [124], isolated pea chloroplasts [75,143], spinach leaves [38] and maize leaves [134] suggest that this sequential labelling pattern from $^{14}CO_2$ is common to several photosynthetic tissues. The demonstration that the fatty acids in PC from spinach chloroplasts are not distributed randomly within the molecule but that the unsaturated acyl chains are concentrated at the 2-glycerol position [36,37], supports the view that this is the primary site of desaturation in the molecule. The glycerol and phosphate moieties of the PC molecule in leaves turn over relatively slowly as shown by the low rates of [^{32}P]glycerol phosphate incorporation into PC in spinach leaf homogenates [160] and ^{14}C from [^{14}C]glycerophosphate into PC in isolated spinach or maize chloroplasts [40]. The contrast between the initial rapid labelling and turnover of the fatty acids of the PC molecule and the delayed labelling in the fatty acids of the galactolipids, has led to the suggestion that the transfer of saturated fatty acids occurs from PC to other lipids in leaves. Fatty acid transfer via PC is known to take place in several animal systems [54,108]. Acyl transferase activity has yet to be demonstrated in chloroplasts and, indeed, plastids may not be the only or even the major site of such activity in leaves.

11.3.2.3. The assembly of the lipid molecule

(i) Biosynthesis of diglycerides. Evidence for the synthesis of glycerol-3-phosphate, the immediate precursor of the diglycerides in chloroplasts has

already been reviewed. Isolated spinach and maize chloroplasts are able to incorporate [^{14}C]glycerol-3-phosphate into diglyceride [40,123], implying they possess a fatty acyl-CoA esterase, which converts glycerol-3-phosphate to phosphatidic acid, and a phosphatidate phosphatase which converts phosphatidic acid to diglyceride (Fig. 11.5). This pathway has been demonstrated by several groups of workers using other plant systems [18,82, 170]. In leaf homogenates from *Vicia faba*, phosphatidate phosphatase activity is associated both with the 15 000 *g* pellet and also with the 40 000 *g* supernatant [103]. A mitochondrial origin for the membrane-bound activity is unlikely because the activity does not sediment with mitochondrial marker enzymes and the possibility that the activity is associated with chloroplast envelope membranes (which will sediment at 50 000 *g*) cannot be ruled out.

(ii) The incorporation of UDP-gal into galactolipids. Chloroplasts can readily synthesise galactolipids when supplied with the appropriate precursors. Mudd and his coworkers [133,138] have examined galactolipid synthesis and greatly extended earlier work [135]. They demonstrated that [^{14}C]galactose, derived from UDP-gal, was incorporated into MGD and DGD by isolated chloroplasts. The chloroplasts presumably contained acceptor molecules, and the evidence suggested that the endogenous MGD was being further galactosylated to DGD, i.e. MGD, and not diglyceride, was the acceptor for the second galactosylation. Recently the existence of two separate galactosylation enzymes in chloroplasts has been confirmed [200]. Using acetone powders derived from beans, it has been shown that unsaturated diglycerides, containing diolein rather than dipalmitin, preferentially act as acceptors for galactosyl transfer, and that the efficiency of the acceptor molecule increases with increasing levels of diglyceride unsaturation [133]. This clear preference for highly unsaturated diglycerides is of great interest in view of the preponderance of polyunsaturated α-linolenate (over 95% of total fatty acids) in naturally occurring chloroplast MGD. In contrast Eccleshall and Hawke [43] found a complete lack of specificity in the fatty acid composition of diglyceride acceptors in a similar system derived from spinach chloroplasts. The reason for the different finding is unclear but it is possible that the use of detergent in the preparation of the acetone powder may have in some way interfered with the specificity of the syntheses [43]. The results of numerous attempts to locate the site(s) of UDP-gal incorporation into galactolipids in green cells are of considerable interest since they clearly illustrate the problems involved in such work. Recently the isolation of chloroplast envelope fractions [41,119] has enabled some confusions to be resolved.

Early reports had suggested that the chloroplast was the main site of galactolipid biosynthesis in leaves, but DGD synthetase was demonstrated to be also associated with the 100 000 *g* supernatant [133] and monogalactosyl synthetase with both 20 000 *g* and 100 000 *g* pellets. A recent report [83] gives evidence that the highest specific activity for UDP-gal incorporation is found in the 40 000 *g* pellet (from leaves). The confusion associated with these earlier reports was resolved when it was unequivocally demonstrated that an isolated chloroplast envelope membrane fraction could incorporate UDP-gal into galactolipid with a specific activity twenty times greater than the activity of a suspension of intact chloroplasts [39]. Confirmation of this result has subsequently come from a number of sources [84,199]. The ease with which the chloroplast envelope ruptures in all aqueous isolation procedures leads to its separation from the rest of the chloroplast and its fragmentation. Depending on the size and physical characteristics of the chloroplast envelope membrane fragments, they will sediment with organelle and membrane fractions between 4000 and 100 000 *g*. Another variable will be the ease with which different enzyme activities can be removed from the membrane. DGD synthetase, for example, appears to be more easily removed in aqueous media than MGD synthetase.

(iii) The biosynthesis of phospholipids. (a) Phosphatidyl choline. CDP-choline incorporation in animal systems proceeds equally rapidly whether saturated or unsaturated diglycerides are supplied [132] and the major activity in leaf homogenates appears to be associated with the "100 000 *g*" pellet fraction [37], although some activity is also associated with the plastid fraction. Transfer of PC across the chloroplast envelope presents little difficulty but in view of the recent demonstration that galactolipid biosynthesis is associated with the envelope membranes themselves [39], the possibility of PC synthesis in envelope fractions should certainly be re-investigated.

(b) Phosphatidyl glycerol. In early experiments Sastry and Kates [160] showed detectable amounts of PG were formed in spinach leaf homogenates supplied with glycerophosphate, CoA, ATP and Mg^{2+}. About 10% of the activity was associated with the chloroplast fraction which also synthesised proportionately more PG. An enzyme which catalyses the incorporation of [1,3—^{14}C]glycerol-3-phosphate into CDP-diglyceride has been found in spinach leaves [122]. Although most of the activity was associated with the 40 000 *g* pellet, the presence of some catalytic capacity in the 1200 *g* and 15 000 *g* pellet and in the 90 000 *g* supernatant again emphasises the necessity of re-examining chloroplast envelope fractions critically for their capacity for phospholipid biosynthesis. CDP-diglyceride pyrophosphorylase has been found in spinach leaf mitochondria and the 40 000 *g* pellet [10]. The negative finding for chloroplasts should be treated with caution since it is possible that the relative impermeability of chloroplasts to CTP [76] may have interfered with the assay for this enzyme in chloroplasts. It is unfortunate that so little information is available on the synthesis of this important chloroplast lipid.

11.3.3. Chloroplast lipid biosynthesis as a collaborative cellular function

A summary of the information available is collected together diagramatically in Fig. 11.7. The evidence suggests that diglyceride(s) are probably synthesised in the chloroplast, but UDP-gal is almost certainly synthesised in the cytoplasm and transported into the chloroplast. The origin (and nature) of the carbon precursor(s) for fatty acid biosynthesis is unresolved but de nova fatty acid synthesis and probably all the desaturation and elongation steps can occur in the chloroplasts when acetate is supplied exogeneously. Galactolipid biosynthesis occurs in the chloroplast envelope. Further experimentation should reveal whether the envelope is also the site of acyl transfer via PC and the terminal stages in diglyceride synthesis, as seems possible from the present data. Present indications are that the phospholipids are also probably synthesised outside the chloroplast and transported in via acyl carrier proteins. Clearly lipid biosynthesis is a coordinated process, tightly regulated in vivo. Some of the modes of control are already indicated. For example, since the MGD:DGD ratio changes from 1:1 to 2:1 during development [110,111], it seems likely that the second galactosylation step is controlled independently of the first; indeed when the supply of galactose is restricted [18,170] the MGD:DGD ratio rises. The supply of precursors from the cytosol would be another control point and the apparent specificity of galactosylation for unsaturated diglycerides suggests the activity of the cellular desaturase systems may also control galactolipid synthesis. The rapid turnover of the fatty acids within leaf lipids, particularly C18:2 suggests that regulation of the synthesis of these components occurs after incorporation into the lipid molecule. C18:3 turns over far more slowly and therefore is less susceptible to changes in pool size.

This article has been restricted to a consideration of lipid biosynthesis in chloroplasts which are photosynthesising, since most of the available information has been obtained from mature plastids. However, the most active phase of increase in plastid lipids occurs during development. Both in leaves developing in a normal diurnal light regime [110,111] and in etiolated leaves exposed to the light [58,158,183,184], dramatic increases, in particular in the concentration of galactolipids and phosphatidyl glycerol, occur. Simultaneously the proportion of polyunsaturated fatty acids, particularly α-linolenic acid, increases by about 60%. The pattern of products of acetate incorporation differ greatly for both leaf sections of different age and also for the suspensions of plastids isolated from them [7,69]. There are also clear indications that the release of free fatty acids contributes to senescence [80,164,165]. Developmental systems should prove to be very valuable experimental ones in the future. It should also be borne in mind that large diurnal changes in leaf lipids may occur as shown recently in bracken [89] and certainly the species differences in fatty acid ratios can be very great [100]. The studies on chloroplast desaturase

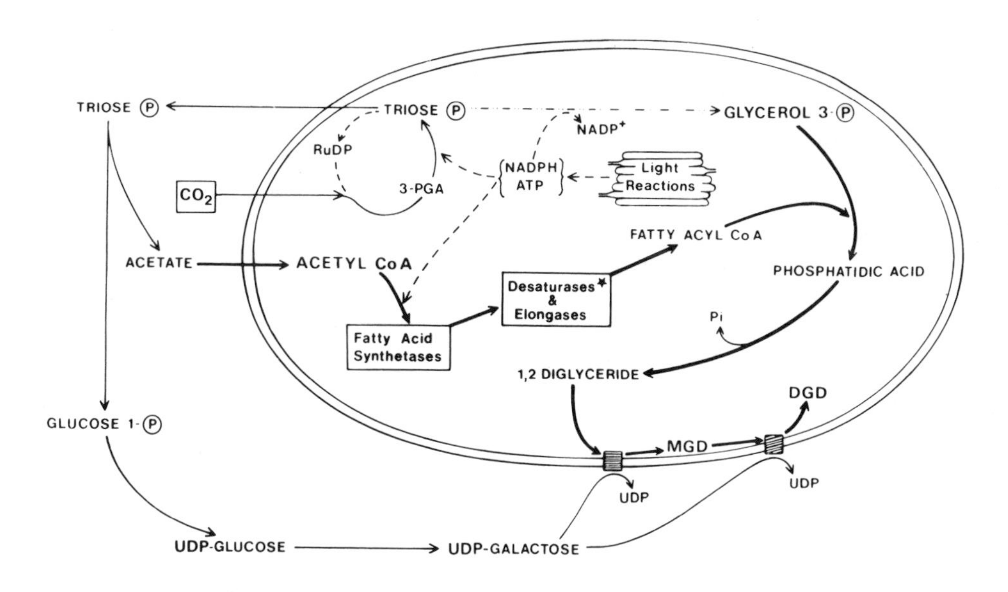

Fig. 11.7. Diagram illustrating possible routes of galactolipid biosynthesis associated with the chloroplast. ——→, demonstrated to occur in chloroplasts; — — — →, demonstrated to occur in leaves; — — · · — →, demonstrated only in animal systems.

systems, for example from pea and spinach [85,185], have given consistently different results and the possibility that these differences may be genuine specific ones should be investigated by looking at other systems.

11.4. CONCLUSIONS

The biosynthesis of small molecules such as amino acids and lipid precursors in green cells illustrates very clearly the collaborative function of chloroplasts as cellular organelles. Although chloroplasts are by far the largest and most numerous organelle of leaf cells, and are uniquely capable of mediating the photoassimilation of carbon chains, they clearly cooperate with other cellular components in biosynthesis in an interdependent manner. The details of the nature and the controls involved in the collaborative metabolic syntheses will require a more sophisticated approach to intracellular localisation studies than has been employed so far. It will be essential to know something of the permeability changes brought about by isolation and in particular to consider the chloroplast envelope membranes not merely as selectively permeable membranes, but also as a site for biochemical synthesis. The obvious structural heterogeneity of the plastids in isolated suspensions commonly used, also needs to be examined in relation to heterogeneity of function. Synthetic differences in chloroplasts from C4 plants and plastids of different ages and from different species, for example, should be valuable to compare. Finally we may look to experiments in which isolated purified organelles are recombined in isolation and their collaborative functions in biosynthesis examined under controlled conditions to provide new insights into the cellular biology of the green leaf cell.

11.5. ACKNOWLEDGEMENT

We are very grateful to Dr. Curtis V. Givan for reading the manuscript and for his many helpful criticisms.

REFERENCES

1 Ben Abdelkader, A. and Mazliak, P. (1970) Eur. J. Biochem., [15], 250—262.
2 Ben Abdelkader, A. and Mazliak, P. (1971) Physiol. Veg., [8], 1121.
3 Agranoff, B.W. and Hajra, A.K. (1968) J. Biol. Chem., [243], 1617—1622.
4 Allen, C.F., Good, P., Davis, H.F. and Fowler, S.D. (1964) Biochim. Biophys. Acta, [15], 424—430.
5 Anderson, J.M. (1975) Nature, [253], 536—537.
6 Anderson, J.M., McCarty, R.E. and Zimmer, E.A. (1974) Plant Physiol., [53], 699—704.
7 Appelqvist, L.A., Boyton, J.E., Stumpf, P.K. and von Wettstein, D. (1968), J. Lipid Res., [9], 425—436.
8 Appelqvist, L.A., Stumpf, P.K. and von Wettstein, D. (9167) Plant Physiol., [43], 163—187.
9 Atkins, C.A. and Canvin, D.T. (1971) Can. J. Bot., [49], 1225—1234.
10 Bahl, J., Guillot-Salomon, T. and Douce, R. (1970) Physiol. Veg., [8], 55—74.
11 Bailey, J.L., Thornber, J.P. and Whyborne, A.G. (1967) in Biochemistry of Chloroplasts (Goodwin, T.W., ed), Vol. 1, pp. 243—256, Academic Press, London.
12 Bassham, J.A. and Jensen, R.G. (1967) in Harvesting the Sun (San Pietro, A., Greer, and Arny, eds), pp. 79—110, Academic Press, New York.
13 Bassham, J.A. and Kirk, M. (1960) Biochim. Biophys. Acta, [43], 447—464.
14 Bassham, J.A. and Kirk, M. (1964) Biochim. Biophys. Acta, [90], 553—562.
15 Bassham, J.A. and Kirk, M. (1968) in Comparative Biochemistry and Biophysics of Photosynthesis (Shibata, K., Takamiya, A., Jagendorf, A.T. and Fuller, R.C., eds), pp. 365—378, University of Tokyo Press, Tokyo.
16 Beevers, L. and Hageman, R.H. (1969) A. Rev. Plant Physiol., [20], 495—522.
17 Benson, A.A., Wiser, R., Ferrari, R.A. and Miller, J.A. (1958) J. Am. Chem. Soc., [80], 4740.
18 Bishop, D.G. and Smillie, R.M. (1970) Arch. Biochem. Biophys., [137], 179—189.
19 Bishop, D.G., Anderson, K.S. and Smillie, R.M. (1970) Biochim. Biophys. Acta, [231], 412—414.

20 Bishop, P.M. and Whittingham, C.P. (1968) Photosynthetica, [2](1), 31—38.
21 Blair, G.E. and Ellis, R.J. (1973) Biochim. Biophys. Acta, [319], 223—234.
22 Brown, C.M., MacDonald-Brown, D.S. and Meers, J.L. (1974) Advan. Microbiol. Physiol., [11], 1—52.
23 Bucke, C., Walker, D.A. and Baldry, C.W. (1966) Biochem. J., [101], 636—641.
24 Burma, D.P. and Mortimer, D.C. (1957) Can. J. Biochem. Physiol., [35], 835—843.
25 Busby, W.F. (1966) Biochim. Biophys. Acta, [121], 160—161.
26 Capaldi, R.A. (1974) Scient. Am., [230], 26—33.
27 Cardeñas, J., Barea, J.L., Rivas, J. and Moreno, C.G., (1972) FEBS Lett., [23], 131—135.
28 Chapman, D.J. and Leech, R.M. (1975) Plant Physiol., (in press).
29 Cheniae, G.M. (1965) Plant Physiol., [40], 235—243.
30 Cherif, A., Dubaq, J.P., Mache, R., Oursel, A. and Trémolières, A. (1975) Phytochemistry, [14], 703—706.
31 Coupe, M., Champigny, M.L. and Moyse, A. (1967) Physiol. Veg., [5], 271—291.
32 Dalling, M.J., Tolbert, N.E. and Hageman, R.H. (1972) Biochim. Biophys. Acta, [238], 505—512.
33 Davies, W.H., Mercer, E.I. and Goodwin, T.W. (1965) Phytochemistry, [4], 741—749.
34 Davies, W.H., Mercer, E.I. and Goodwin, T.W. (1966) Biochem. J., [98], 369—374.
35 Del Campo, F.A., Paneque, A., Ramirez, J.M. and Losada, M. (1963) Biochim. Biophys. Acta, [66], 450—452.
36 Devor, K.A. and Mudd, J.B. (1971) J. Lipid Res., [12], 396—402.
37 Devor, K.A. and Mudd, J.B. (1971) J. Lipid Res., [12], 403—411.
38 Devor, K.A. and Mudd, J.B. (1971) J. Lipid Res., [12], 412—419.
39 Douce, R. (1974) Science, [183], 852—853.
40 Douce, R. and Guillot-Salomon, T. (1970) FEBS Lett., [11], 121—134.
41 Douce, R., Holtz, R.B. and Benson, A.A. (1973) J. Biol. Chem., [248], 7215—7222.
42 Dougall, D.K. (1974) Biochem. Biophys. Res. Commun., [58], 638—646.
43 Eccleshall, T.R. and Hawke, J.C. (1971) Phytochemistry, [10], 3035—3045.
44 Ellis, R.J., Blair, G.E. and Hartley, M.R. (1973) Biochem. Soc. Symp., [38], 137—162.
45 Erwin, J. and Bloch, K. (1964) Science, [143], 1006—1012.
46 Evans, H.J. and Nason, A. (1953) Plant Physiol., [28], 233—254.
47 Everson, R.G. and Gibbs, M. (1967) Plant Physiol., [42], 1153—1154.
48 Ferrari, R.A. and Benson, A.A. (1961) Arch. Biochem. Biophys., [93], 185—192.
49 Gibbs, M. (1971) in Structure and Function of Chloroplasts (Gibbs, M., ed), pp. 170—175, Springer, Berlin.
50 Gimmler, H., Schäfer, G., Kraminer, H. and Heber, U. (1974) Planta, [120], 47—61.
51 Givan, C.V. and Leech, R.M. (1971) Biol. Rev., [46], 409—428.
52 Givan, C.V., Givan, A. and Leech, R.M. (1970) Plant Physiol., [45], 624—630.
53 Givan, C.V. and Stumpf, P.K. (1971) Plant Physiol., [47], 510—515.
54 Glomset, J.A. (1968) J. Lipid Res., [9], 155—167.
55 Goring, H. and Reckin, E. (1968) Flora, [159], 82—103.
56 Grant, B.R., Atkins, C.A. and Canvin, D.T. (1970) Planta, [94], 60—72.
57 Grant, B.R. and Canvin, D.T. (1970) Planta, [94], 60—72.
58 Gray, I.K., Rumsby, M.G. and Hawke, J.C. (1967) Phytochemistry, [6], 107—113.
59 Guanierni, M. and Johnson, R.M. (1970) Adv. Lipid Res., [8], 115—174.
60 Gurr, M.I. and Brawn, P. (1970) Eur. J. Biochem., [17], 19—22.
61 Gurr, M.I., Robinson, M.P. and James, A.T. (1969) Eur. J. Biochem., [9], 70—78.
62 Harris, R.V. and James, A.T. (1965) Biochim. Biophys. Acta, [106], 456—473.
63 Hartley, M.R. and Ellis, R.J. (1973) Biochem. J., [134], 249—262.
64 Harwood, J.L. (1975) Biochim. Biophys. Acta, in press.
65 Harwood, J.L. and James, A.T. (1975) Eur. J. Biochem., [50], 325—334.

66 Harwood, J.L. and Stumpf, P.K. (1972) Lipids [7], 8—19.
67 Haverkate, F. and Van Deenen, L.L.M. (1964) Biochim. Biophys. Acta, [84], 106—108.
69 Hawke, J.C., Leese, B.M. and Leech, R.M. (1975) Phytochemistry, in press.
69 Hawke, J.C., Rumsby, M.G. and Leech, R.M. (1974) Plant Physiol., [53], 555—561.
70 Hawke, J.C., Rumsby, M.G. and Leech, R.M. (1974) Phytochemistry, [13], 403—413.
71 Haystead, A. (1973) Planta, [111], 271—274.
72 Heber, U. (1974) A. Rev. Plant Physiol., [25], 393—421.
73 Heber, U., Hallier, U.W. and Hudson, M.A. (1967) Z. Naturforsch., [22], 1200—1215.
74 Heber, U. and Krause, G. (1972) in Photosynthesis and Photorespiration (Hatch, M.D., Osmond, C.B. and Slatyer, R.O., eds), pp. 218—225, Wiley, New York.
75 Heise, K.P. and Jacobi, G. (1973) Planta, [111], 137—148.
76 Heldt, H.W. (1969) FEBS Lett., [5], 11—14.
77 Heldt, H.W. and Rapley, L. (1970) FEBS Lett., [7], 139—142.
78 Heldt, H.W. and Rapley, L. (1970) FEBS Lett., [10], 143—148.
79 Heldt, H.W., Sauer, F. and Rapley, L. (1971) in Proc. 2nd Int. Cong. Photosynthesis (Forti, G., Avron, M. and Melandri, A., eds), pp. 1345—1355, Junk, The Hague.
80 Hitchcock, C. and James, A.T. (1964) J. Lipid Res., [5], 593—599.
81 Hitchcock, C. and Nichols, B.W. (1971) in Plant Lipid Biochemistry, p. 94, Academic Press, London.
82 Hoffman, F., Kull, U. and Jeremias, K. (1971) Z. Pflanzenphysiol., [64], 223—231.
83 van Hummel, H.C. (1974) Z. Pflanzenphysiol., [71], 228—241.
84 van Hummel, H.C., Hulsebos, T.J.M. and Wintermans, J.F.G.M. (1975) Biochim. Biophys. Acta, [380], 219—226.
85 Jacobson, B.S., Kannangara, C.G. and Stumpf, P.K. (1973) Biochem. Biophys. Res. Commun., [51], 487—493.
86 Jacobson, B.S., Kannangara, C.G. and Stumpf, P.K. (1973) Biochem. Biophys. Res. Commun., [52], 1190—1198.
87 Jacobson, B.S., Jaworski, J.G. and Stumpf, P.K. (1974) Plant Physiol., [54], 484—486.
88 James, A.T. and Nichols, B.W. (1966) Nature, [210], 372—375.
89 Jarvis, M.C. and Duncan, H.J. (1975) Phytochemistry, [14], 77—78.
90 Jaworksi, J.G., Goldschmidt, E.E. and Stumpf, P.K. (1974) Arch. Biochem. Biophys., [163], 769—776.
91 Jaworski, J.G. and Stumpf, P.K. (1974) Arch. Biochem. Biophys., [162], 166—173.
92 Jensen, R.G. and Bassham, J.A. (1966) Proc. Natl. Acad. Sci. USA, [56], 1095—1101.
93 Jetschmann, K., Solomonson, L.P. and Vennesland, B. (1972) Biochim. Biophys. Acta, [275], 276—278.
94 Joy, K.W. and Hageman, R.H. (1966) Biochem. J., [100], 263—273.
95 Kanazawa, T., Kanazawa, K., Kirk, M.R. and Bassham, J.A. (1972) Biochim. Biophys. Acta, [256], 656—669.
96 Kanazawa, T., Kirk, M.R. and Bassham, J.A. (1970) Biochim. Biophys. Acta, [205], 401—408.
97 Kannangara, C.G. and Stumpf, P.K. (1972) Arch. Biochem. Biophys., [148], 414—424.
98 Kannangara, C.G., Jacobson, B.S. and Stumpf, P.K. (1973) Plant Physiol., [52], 156—161.
99 Kannangara, C.G. Jacobson, B.S. and Stumpf, P.K. (1973) Biochem. Biophys. Res. Commun., [52], 648—655.
100 Kates, M. (1970) Adv. Lipid Res., [8], 225—265.
101 Kirk, J.T.O. (1971) Adv. Rev. Biochem., [40], 161—196.

102 Kirk, P.R. and Leech, R.M. (1972) Plant Physiol., [50], 228—234.
103 Königs, B. and Heine, E. (1974) Planta, [118], 159—161.
104 Lea, P.J. and Miflin, B.J. (1975) In press.
105 Lea, P.J. and Miflin, B.J. (1975) In press.
106 Lea, P.J. and Miflin, B.J. (1974) Nature, [251], 614—616.
107 Lea, P.J. and Thurman, D.A. (1972) J. Exp. Bot., [23], 440—449.
108 LeBreton, E. and Pantaléon, J. (1947) C.R. Acad. Sci. Paris, [137], 609—612.
109 Leech, R.M. and Kirk, P.R. (1968) Biochem. Biophys. Res. Commun., [32], 685—690.
110 Leech, R.M. Rumsby, M.G. and Thomson, W.W. (1973) Plant Physiol., [52], 240—245.
111 Leech, R.M., Rumsby, M.G., Thomson, W.W., Crosby, W. and Wood, P. (1972) in Proc. 2nd Int. Cong. Photosynthesis (Forti, G., Avron, M. and Melandri, A., eds), pp. 2479—2487, Junk, The Hague.
112 Lee, R.B. and Whittingham, C.P. (1974) J. Exp. Bot., [25], 277—287.
113 Leese, B.M. (1972) D. Phil. Thesis, University of York, U.K.
114 Lichtenthaler, H.K. (1966) Ber. Deut. Bot. Ges., [79], 82—88.
115 Lichtenthaler, H.K. (1967) Z. Pflanzenphysiol., [56], 273—281.
116 Losada, M. (1975) in Metabolic Interconversion of Enzymes 1973; 3rd Int. Symp.
117 Losada, M., Herrera, J., Maldonado, J.M. and Paneque, A. (1973) Plant Sci. Lett., [1], 31—37.
118 Losada, M., Ramirez, J.M., Paneque, A. and Del Campo, F.F. (1965) Biochim. Biophys. Acta, [109], 86—96.
119 Mackender, R.O. and Leech, R.M. (1970) Nature, [228], 1347—1349.
120 Mackender, R.O. and Leech, R.M. (1974) Plant Physiol., [53], 496—502.
121 Magalhães, A.C., Neyra, C.A. and Hageman, R.H. (1974) Plant Physiol. [53], 411—415.
122 Marshall, M.O. and Kates, M. (1972) Biochim. Biophys. Acta, [260], 558—570.
123 Mazliak, P. (1967) Phytochemistry, [6], 687—702.
124 Mazliak, P. (1967) Phytochemistry, [6], 941—956.
125 Miflin, B.J. (1972) Planta, [105], 225—233.
126 Miflin, B.J. (1974) Planta, [116], 187—196.
127 Miflin, B. and Beevers, H. (1974) Plant Physiol., [53], 870—874.
128 Miflin, B. J. and Lea, P.J. (1975) in press.
129 Mitchell, C.A. and Stocking, C.R. (1975) Plant Physiol., [55], 59—63.
130 Moreno, C.G., Aparicio, P.J., Palacian, E. and Losada, M. (1972) FEBS Lett., [26], 11—14.
131 Mortimer, D.C. (1959) Can. J. Bot., [37], 1191—1201.
132 Mudd, J.B., van Golde, L.M.G. and van Deenen, L.L.M. (1969) Biochim. Biophys. Acta, [176], 547—556.
133 Mudd, J.B., van Vliet, H.H.D.M. and van Deenen, L.L.M. (1969) J.Lipid Res., [10], 623—630.
134 Murphy, D.J. and Leech, R.M. (1975) unpublished.
135 Neufeld, E.F. and Hall, C.W. (1964) Biochem. Biophys. Res. Commun., [14], 503—508.
136 Nichols, B.W., James, A.T. and Breuer, J. (1967) Biochem. J., [104], 486—496.
137 Nichols, B.W., Stubbs, J.M. and James, A.T. (1967) in Biochemistry of Chloroplasts (Goodwin, T.W., ed), Vol. II, pp. 677—691, Academic Press, London.
138 Ongun, A. and Mudd, J.B. (1968) J. Biol. Chem., [243], 1558—1566.
139 O'Neal, D. and Joy, K.W. (1973) Nature New Biol., [246], 61—62.
140 O'Neal, D. and Joy, K.W. (1973) Arch. Biochem. Biophys., [159], 113—122.
141 O'Neal, D. and Joy, K.W. (1974) Plant Physiol., [54], 773—779.

142 Paneque, A., Ramirez, J.M., Del Campo, F.F. and Losada, M. (1964) J. Biol. Chem., [239], 1737–1741.
143 Panter, R.A. and Boardman, N.K. (1973) J. Lipid Res., [14], 664–671.
144 Park, R.B. and Pon, N.G. (1963) J. Mol. Biol., [6], 105–114.
145 Plaut, Z. and Gibbs, M. (1970) Plant Physiol., [46], 488–490.
146 Pohl, P. (1973) Z. Naturforsch., [28], 264–269.
147 Pohl, P. (1973) Z. Naturforsch., [28], 270–284.
148 Poincelot, R.D. (1971) Biochim. Biophys. Acta, [239], 57–60.
149 Poincelot, R.P. (1973) Arch. Biochem. Biophys., [159], 134–142.
150 Randall, D.D. and Tolbert, N.E. (1971) J. Biol. Chem., [246], 5510–5517.
151 Rathnam, C.K.M. and Das, V.S.R. (1974) Can. J. Bot., [52], 2599–2605.
152 Relimpio, A. Mª., Aparicio, P.J., Paneque, A. and Losada, M. (1971), FEBS Lett., [17], 226–230.
153 Richards, R.L. and Quackenbush, F.W. (1974) Arch. Biochem. Biophys., [165], 780–786.
154 Ritenour, G.L., Joy, K.W., Bunning, J. and Hageman, R.H. (1967) Plant Physiol., [42], 233–237.
155 Roberts, G.R., Keys, A.J. and Whittingham, C.P. (1970) J. Exp. Bot., [21], 683–692.
156 Rosenberg, L.L., Capindale, J.B. and Watley, F.R. (1958) Nature, [181], 632–633.
157 Roughan, P.G. (1970) Biochem. J., [117], 1–8.
158 Roughan, P.G. and Robinson, N.K. (1972) Plant Physiol., [50], 31–34.
159 Santarius, K.A. and Stocking, C.R. (1969) Z. Naturforsch., [24b], 1170–1179.
160 Sastry, P.S. and Kates, M. (1966) Can. J. Biochem., [44], 459–467.
161 Shah, S.P.J. and Cossins, E.A. (1970) Phytochemistry, [9], 1545–1551.
162 Sherratt, D. and Givan, C.V. (1973) Planta, [113], 47–53.
163 Shibuya, I., Yang, T. and Benson, A.A. (1963) in Studies on Microalgae and Photosynthetic Bacteria (Ashida, J., ed), p. 627, University of Tokyo Press, Tokyo.
164 Siegenthaler, P.A. (1972) Biochim. Biophys. Acta, [275], 182–191.
165 Siegenthaler, P.A. (1974) FEBS Lett., [39], 337–340.
166 Singer, S.J. (1971) in Structure and Function of Biological Membranes (Rothfield, L.I., ed), pp. 145–222, Academic Press, London.
167 Singer, S.J. (1974) Adv. Immunol., [19], 1–66.
168 Singer, S.J. (1974) A. Rev. Biochem., [43], 805–833.
169 Singer, S.J. and Nicholson, G.L. (1972) Science, [175], 720–731.
170 Smillie, R.M., Bishop, D.G., Gibbons, G.C., Graham, D., Grieve, A.M., Raison, J.K. and Reger, B.J. (1971) in Autonomy and Biogenesis of Mitochondria and Chloroplasts (Boardman, N.K., Linnane, A.W. and Smillie, R.M., eds), pp. 422–433, North Holland, Amsterdam.
171 Smith, D.C., Bassham, J.A. and Kirk, M. (1961) Biochim. Biophys. Acta, [48], 299–313.
172 Solomonson, L.P. and Vennesland, B. (1972) Biochim. Biophys. Acta, [267], 544–557.
173 Stepka, W., Benson, A.A. and Calvin, M. (1948) Science, [108], 304–305.
174 Stocking, C.R. and Larson, S. (1969) Biochem. Biophys. Res. Commun., [37], 278–282.
175 Stumpf, P.K., Brooks, J., Galliard, T., Hawke, J.C. and Simoni, R. (1967) in Biochemistry of Chloroplasts (Goodwin, T.W., ed), Vol. II, pp. 213–239. Academic Press, London.
176 Stumpf, P.K. and Boardman, N.K. (1970) J. Biol. Chem., [245], 2579–2587.
177 Stumpf, P.K. and James, A.T. (1962) Biochim. Biophys. Acta, [57], 400–402.
178 Stumpf, P.K. and James, A.T. (1963) Biochim. Biophys. Acta, [70], 20–32.

179 Swader, J.A. and Stocking, C.R. (1971) Plant Physiol., [47], 189—191.
180 Tolbert, N.E. (1971) Ann. Rev. Plant Physiol., [22], 45—74.
181 Trémolières, A. (1970) Ann. Biol., [9], 113—156.
182 Trémolières, A. (1970) C. R. Acad. Sci. Paris, [272], 2777—2780.
183 Trémolières, A. and Lepage, P. (1971) Plant Physiol., [47], 329—334.
184 Trémolières, A. and Mazliak, P. (1970) Physiol. Veg., [8], 135—150.
185 Trémolières, A. and Mazliak, P. (1974) Plant Sci. Lett., [2], 193—201.
186 Tsenova, E.N. and Vaklinova, S. (1970) C.R. Acad. Bulg. Sci., 727—730.
187 Tsenova, E.N. and Vaklinova, S. (1971) in 2nd Int. Cong. Photosynthesis (Forti, G., Avron, M. and Melandri, A., eds), pp. 2199—2204, Junk, The Hague.
188 Vega, J., Herrera, J., Aparicio, P.J., Paneque, A. and Losada, M. (1971) Plant Physiol., [48], 294—299.
189 Vega, J.M., Herrera, J., Relimpio, A.M. and Aparicio, P.J. (1972) Physiol. Veg., [10], 637—652.
190 Vennesland, B. and Jetschmann, C. (1972) Biochim. Biophys. Acta, [227], 554—564.
191 Vennesland, B. and Solomonson, L.P. (1972) Plant Physiol., [49], 1029—1031.
192 Walker, D.A. (1964) Biochem. J., [92], 22C—23C.
193 Walker, D.A. (1970) Nature, [226], 1204—1208.
194 Walker, D.A. (1974) Med. Tech. Publ. Int. Rev. Sci. Biochem., Ser. I, Vol. II, pp. 1—49.
195 Walker, D.A. and Crofts, A.R. (1970) Ann. Rev. Biochem., [39], 389—428.
196 Weaire, P.J. and Keckwick, R.G.O. (1975) Biochem. J., [146], 425—437.
197 Weaire, P.J. and Keckwick, R.G.O. (1975) Biochem. J., [146], 439—445.
198 Weier, T.E. and Benson, A.A. (1967) in Biochemistry of Chloroplasts (Goodwin, T.W., ed), Vol. I, pp. 91—115, Academic Press, London.
199 Williams, J.P. (1975) Personal communications.
200 Williams, J.P., Kahn, M. and Leung, S. (1975) J. Lipid Res., [16], 61—66.
201 Yamada, M. and Nakamura, Y. (1975) Plant and Cell Physiol., [16], 151—162.
202 Yamazaki, R.K. and Tolbert, N.E. (1970) J. Biol. Chem., [245], 5137—5144.

The Intact Chloroplast — edited by J. Barber

Chapter 12

Division of Labour Between Chloroplast and Cytoplasm

J.A. RAVEN

Department of Biological Sciences, University of Dundee, Dundee DD1 4HN (Great Britain)

CONTENTS

Abbreviations: CCCP, carbonyl cyanide *m*-chlorophenyl hydrazone; cyt, cytochrome; DCMU, 3′-(3, 4-dichlorophenyl)-1′,1′-dimethylurea; DSPD, disalicylidenepropanediamine; Fd, ferredoxin; FP, flavoprotein; PCy, plastocyanin; 3-PGA, 3-phosphoglycerate; PQ, plastoquinone; PSCRC, photosynthetic carbon reduction cycle; TCAC, tricarboxylic acid cycle; UQ, ubiquinone.

12.1. INTRODUCTION

This chapter sets out to explore the energetic role of the chloroplast in the green cell. The major function of the chloroplast is to convert solar energy and carbon dioxide into the chemical energy of reduced carbon compounds, which provide the carbon and energy source for most of the life on earth. Thus a simple view is that the chloroplast is a "sugar factory", producing carbohydrate which the phototrophic cell (or organism) used as a heterotrophic cell uses exogenous sugar.

The heterotrophic cell uses sugar as a source of not only the carbon-containing monomers for cell synthesis, but as the source of the ATP and reductant used in cell synthesis and maintenance. The illuminated photosynthetic cell has an alternative method of supplying ATP and reductant to these biosynthetic and maintenance processes. The light reactions of photosynthesis produce ATP and reductant; much of this is used in CO_2 fixation, but it is possible that some can be used directly for other energy-reducing reactions. This provides a short-circuit of the use of these co-factors to fix CO_2 into storage carbohydrates, with subsequent regeneration of the ATP and reductant by the respiratory metabolism of the stored carbohydrate. These alternatives are presented in Fig. 12.1, and are discussed by Hoch and Owens [88], Hoch et al. [89], Kok [119], Raven [179,180,183] and by Ried [196].

I wish to discuss the evidence that ATP and reductant generated in the light reactions of photosynthesis can be used in vivo for processes other than CO_2 fixation, that such use is quantitatively important under natural conditions and is significant in the life of the green cell. Much of the discussion will involve evidence obtained with unicellular algae. This is partly because much of the relevant work has been carried out with these organisms, coupled to the fact that unlike multicellular plants there are no complications due to non-green cells. The special problems associated with multicellular photolithotrophs are discussed in section 12.5.3.

12.2. CLASSIFICATION AND QUANTITATION OF ENERGY-REQUIRING REACTIONS

12.2.1. Growth, maintenance and luxury accumulation

These three energy-requiring processes will be considered in turn.

(a) Energy for growth. Use of energy for growth implies a stoichiometric relationship between energy consumption and growth. Examples are the use of ATP for the biochemical and biophysical (transport and mechanical work) processes involved in growth, and the use of reductant for reductive biosynthesis.

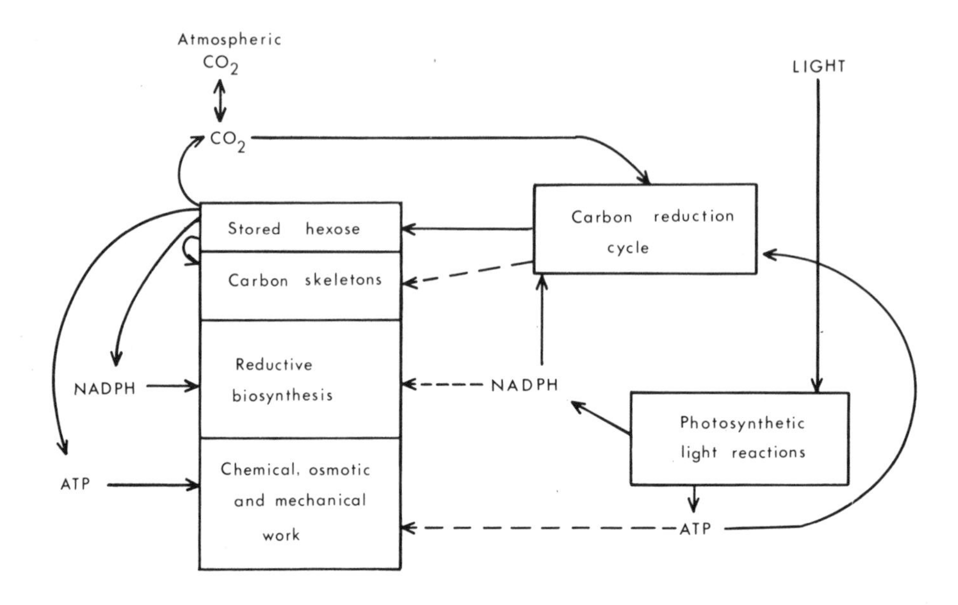

Fig. 12.1. Possible relations between photosynthesis, respiration and energy-requiring processes (growth, maintenance and luxury accumulation) in the green plant cell.

(b) Energy for maintenance. Energy for maintenance is consumed in maintaining the organism in a viable condition (e.g. turnover of macromolecules, and maintenance of solute content in a "pump and leak" situation. Energy consumption in maintenance is proportional to biomass; this contrasts with the situation for growth-associated energy consumption (see refs. 12,59,165, 164).

(c) Energy for luxury accumulation A process which does not fit readily into categories *(a)* or *(b)* is the process of luxury accumulation [186]. Here energy is used to accumulate some cell constituent to a higher level than is found in normal cell growth. This process can be either superimposed on growth (e.g. polysaccharide accumulation when photosynthetic carbon fixation at high irradiance outstrips the rate at which growth can consume it [37,142], or occurs in the absence of growth (e.g. phosphate accumulation as polyphosphate in algae when light energy and phosphate are available but CO_2 is deficient [239].

My aim in this Chapter is to see how far the direct use of photo-produced ATP and reductant is involved in supplying energy to these three categories of processes. In terms of deciding whether a given process in vivo can be powered in this way it is important to remember that growth-associated processes must show characteristics of overall growth (sections 12.3 and

12.4). Thus ATP-requiring processes during phototrophic growth will show characteristics of non-cyclic photophosphorylation (section 12.5.3), because overall growth shows this characteristic. The conditions imposed to isolate cyclic photophosphorylation (section 12.4.4) necessarily dissociate metabolism from growth, i.e. the energy-requiring processes are being investigated under maintenance or luxury accumulation conditions. The difficulties of bringing about this dissociation from growth for energy-supplying processes involving non-cyclic electron flow are discussed in section 12.3.3.2.

12.2.2. Quantitation of energy requirements for growth

12.2.2.1. Reductant requirements

Penning de Vries et al. [165] tabulate the reductant requirements for cell synthesis, based on cell composition and known biosynthetic pathways. I shall adopt a more empirical stratagem to compute the *minimum* reductant requirement for growth based on a comparison of the substrates and products of growth.

For the heterotrophic growth of *Chlorella* on glucose, with ammonium as N source, the reduction level of the cell is greater than that of the substrates. For every mole of C assimilated into cell material, this greater reduction level amounts quantitatively to 0.4 [H] above that supplied in the assimilated C and N. It is supplied by the oxidation of 0.1 mole of substrate (glucose) C to CO_2. This conclusion from cell composition studies is confirmed by the finding that for every mole C assimilated into cell materials, 0.1 mole CO_2 is evolved in excess of O_2 taken up [147,200]. When nitrate is the N source for heterotrophic growth, this ratio of reductant consumed per mole C incorporated increases to 1.6 [H] (data of Samejima and Myers, [200] i.e. an excess of CO_2 produced over O_2 consumed of 0.4 per mole C assimilated.

For photolithotrophic growth producing cells of similar composition, the overall reductant requirement will be increased by the reductant requirements for CO_2 reduction to the carbohydrate level in each case, i.e. 4 [H] per C assimilated. Thus the total reductant requirement for phototrophic growth with ammonium as N source is 4.4 [H] per C assimilated into cell material, and 5.6 [H] per C assimilated when nitrate is N source. This reductant must all ultimately come from photoreactions, using water as electron donor. The basic 4 [H] per C assimilated must be used directly as photoreductant; the residual 0.4—1.6 [H] per C assimilated can either be supplied as photoreductant directly, or by reoxidation of photosynthate produced by further use of photoreductant in CO_2 fixation (see Fig. 12.1, and Table I).

12.2.2.2. ATP requirements

ATP requirements for cell growth cannot be determined from measurements of growth substrate and product stoichiometries, as was done for reductant in section 12.2.2.1. Some biochemical insight is needed. One procedure is to attempt to compute ATP requirement for growth from cell composition, and the known in vitro ATP requirements for the biochemical (synthesis of monomers and polymers) and biophysical (active transport, cell and nuclear division) processes of growth. Such analyses have been attempted by e.g. Forrest and Walker [59] and Penning de Vries et al. [165]. While there are still some gaps in our knowledge of the biochemical ATP requirements, a major difficulty is the quantitation of the energy requirement for transport (see refs. 186, 187). Generally transport at cell membranes other than the plasmalemma and tonoplast is ignored; even at these two membranes the energetic stoichiometry of transport is poorly understood [186, 187]. It is therefore likely that the ATP requirements are underestimated.

An alternative method [59] is to attempt to use the rate of ATP-*producing* reactions during growth to estimate the ATP requirements for growth. This will tend to *over*-estimate the ATP requirement to the extent that "idling respiration" occurs. Beevers [12] defines "idling respiration" as respiration not coupled to energy-requiring processes, with which may be included respiration coupled to wasteful ATP hydrolysis or NADPH oxidation. From estimates of the O_2 uptake associated with aerobic growth, and measured or assumed ATP/2e ratios in oxidative phosphorylation, minimal ATP requirements for growth on glucose of some 2 ATP per mole C assimilated may be derived [59] (Table I).

For photolithotrophic growth in plants using the unadorned photosynthetic carbon reduction cycle (see ref. 184 for evidence that this is the pathway used in unicellular algae), 3 ATP are needed per C assimilated (see Table I). This is directly photoproduced ATP, since isolated chloroplasts, in which there is no other means of making ATP, can fix CO_2 at high rates with high quantum yields ([80] and Chapter 7 of this volume). The additional 2 ATP per C assimilated which is required for growth in photolithotrophs can either be supplied by direct use of photoproduced ATP, or by oxidative phosphorylation at the expense of photosynthate produced by further investment of photoproduced ATP and reductant in CO_2 fixation (Fig. 12.1).

ATP is also consumed in maintenance processes (section 12.2.1). The difficulties associated with estimating ATP requirements for maintenance are discussed by Forrest and Walker [59] and by Penning de Vries [163,164]. Estimates from the observed rate of oxygen uptake under non-growing conditions, and an assumed ATP/2e ratio of 3, and from estimates of the rate of ATP-requiring processes, suggest that 0.04—0.2 ATP per mole C biomass per day are required for maintenance in plants [141,163,164].

12.2.2.3. The minimal respiratory gas exchange required for growth

Even if all the ATP and reductant for growth are supplied directly as photoproducts in a photolithotrophic cell cultured under continuous illumination, some O_2 uptake and CO_2 production must still occur.

TABLE I

Reductant and ATP requirements for growth with various energy, carbon and nitrogen sources, summarising evidence discussed in Section 12.2

This ignores maintenance ATP requirements; for a cell with a doubling time of 1 day, this adds another 0.1 ATP per mole C assimilated to the ATP requirement.

Energy, carbon, nitrogen sources	Equivalents of reductant used during assimilation of 1 mole of C substrate	Moles ATP needed per mole C assimilated	Comments
Light CO_2 ammonium	4.4	5	Assumes 2 NADPH and 3 ATP are needed to fix 1 CO_2 into carbohydrate, i.e. no extra energy needed for the glycolate or C4 pathways; approximately true of algae grown in 5% CO_2 [74,134,222,223]
Light CO_2 nitrate	5.6	5	
Glucose glucose ammonium	0.4	2	With ATP/2e ratio of 3 in oxidative phosphorylation, involves oxidation of 0.43 moles C from glucose per mole C assimilated
Glucose glucose nitrate	1.6	2	With ATP/2e ratio of 3 in oxidative phosphorylation, involves oxidation of 0.73 moles C from glucose per Mole C assimilated

Oxygenase [76,77] and decarboxylase [182,233] reactions are essential for certain eukaryote carbon skeleton biosyntheses, quite apart from their involvement in reductant or ATP generation. In quantitative terms, the minimal CO_2 production for these biosyntheses are, for *Chlorella* cells of the normal composition given by Tamiya [219] some 0.1 CO_2 produced and 0.01 O_2 taken up per net C assimilated [77,182]. Thus 0.4 [H] are generated per C assimilated in these decarboxylating carbon skeleton interconversions. These are largely produced as NADH in the mitochondrion, since much of the CO_2 is produced by the decarboxylation of pyruvate and operation of the TCAC [182,233]. This probably means that it is used in oxidative phosphorylation rather than in reductive biosynthesis (see refs. 7,192).

12.2.2.4. Intracellular location of energy-requiring reactions

The light reactions of photosynthesis generate ATP and reductant in the chloroplast; oxidative phosphorylation produces ATP in the mitochondrion which is readily available to the cytoplasm, and the oxidative pentose phosphate pathway produces reductant in both the cytoplasm and the chloroplast (see sections 12.3.2, 12.3.3.1, 12.4.3, 12.4.4, 12.4.5). In order to find out if shuttles ([81] and Chapters 5 and 6) are required to move reductant and "high-energy phosphate" across the chloroplast outer membranes in vivo, it is necessary to know the intracellular location of the ATP-requiring and reductant-requiring processes.

The reductant used in CO_2 fixation and nitrite reduction is used in the chloroplast, while nitrate reduction occurs in the cytoplasm, sulphate reduction, lipid synthesis and reductive amination probably occur in both compartments ([14,64,124,135,184,204] and Chapters 9, 11 of this volume). Thus the large additional reductant requirement involved in growth with nitrate as N source is largely (75%) intrachloroplastic, while other reductive biosyntheses are more evenly distributed between chloroplast and cytoplasm.

As regards ATP, it appears that the requirement for ATP for processes other than CO_2 fixation is largely extrachloroplastic. From the distribution of macromolecules between cytoplasm and chloroplast, and knowledge of the intracellular site of their synthesis, it may be concluded that only 25—33% of the ATP used in biochemical synthesis is used in the chloroplast [26,78,84,210]. It is likely that most of the ATP used in biophysical reactions is used outside the chloroplast [73,186,187]. Thus it is likely that not more than 25% of the ATP used for processes other than CO_2 fixation (i.e. some 0.5 ATP per C assimilated) is used in the chloroplast.

12.2.2.5. Conclusions

Table I summarises the quantitative conclusions arrived at in sections 12.2.2.1—12.2.2.4 as regards quantitative aspects of reductant and ATP requirements for growth and maintenance of *Chlorella* cells under various nutritional conditions. It must be emphasised that these are minimal estimates. The data reviewed by Tamiya [219] and Cook [38] show that most energy-requiring growth processes in photolithotrophic growth occur in the light phase of a synchronous light-dark cycle, i.e. when the direct use of photoproducts is possible.

12.3. REDUCTANT SUPPLY

12.3.1. General considerations

As was pointed out in section 12.2.2.1, all the reductant used in photolithotrophic growth comes from water via non-cyclic electron flow. This section sets out to determine how much of the reductant used in reductive biosynthesis (other than CO_2 fixation in photolithotrophic cells) comes from direct use of photoreduced reductant, and how much by reductant regeneration from fixed carbon. Two general approaches are open. One is to investigate the capacity of the "dark regeneration" pathways in the green cell, and to determine how active they are under growth conditions (section 12.3.2). A low activity provides evidence (albeit negative) for direct use of photoreductant. The other and more direct approach is to look for evidence of the use of photoproduced reductant in biosynthesis (section 12.3.3). However, this is complicated by the fact (section 12.2.1) that any reductive biosynthesis which is stoichiometric with photolithotrophic growth must, directly or indirectly, have the characteristics of non-cyclic electron flow.

12.3.2. The contribution of "dark" reductant generation processes in illuminated green cells

Reductant generation for biosynthesis in non-green plant tissues appears to involve the oxidative pentose phosphate pathway and (especially for lipid synthesis) the triose phosphate dehydrogenase reaction of glycolysis [7,182, 201,235].

In the illuminated green cell the operation of the triosephosphate dehydrogenase reaction in the cytoplasm is involved in shuttling reductant and ATP out of chloroplasts ([81]; Chapters 5 and 6). Thus any investigation of the use of this reaction in generating reductant from stored carbohydrate must distinguish this activity from its role in the direct use of photoproduced co-factors in the cytoplasm. Experiments in which high concentrations of $^{14}CO_2$ are fed to *Chlorella* show that there is very little dilution of specific activity in the C_3 compounds involved in this reaction from unlabelled stored carbohydrate [105,162]. However, the dilution due to input of cold carbon at a rate sufficient to generate a significant amount of NADH might not be detectable in terms of the tracer carbon throughput, so these experiments cannot completely eliminate a carbon flux from reserves through the triosephosphate dehydrogenase reaction in illuminated algae. Further, Fock, et al. [58] found that 3-phosphoglycerate (3-PGA) in *Chlorella* fed lower CO_2 concentrations in 80% O_2 never reached the specific activity of the supplied $^{14}CO_2$. While this is consistent with a flux of reduced carbon from reserves (unlabelled) into 3-PGA, it is also consistent with refixation of unlabelled CO_2 generated within the cell by reactions not involving triosephosphate dehydrogenase. The high respiratory rate of the cells used in these experiments is another complicating factor.

A more favourable case for analysis is the oxidative pentose phosphate pathway in which there is a unique intermediate i.e. 6-phosphogluconate. The two enzymes specific to this pathway, i.e. glucose-6-phosphate dehydrogenase and 6-phosphogluconate dehydrogenase are present in both the plastids and the cytoplasm of cells from the chlorophyte line of evolution, but are absent from the chloroplasts of Euglenoids [4,75,137, 182,205,210].

Evidence consistent with a lower carbon flux through this pathway in the light than in the dark in photolithotrophic algae has been discussed by Raven [182]. While this evidence is not conclusive, it is likely that neither recent nor reserve photosynthate generates much reductant by this pathway in illuminated green cells. This is consistent with the control properties of glucose-6-phosphate dehydrogenase from green tissues which suggest that it is less active under photosynthetic conditions in vivo [4,102].

As regards the maximum catalytic capacity of the oxidative pentose phosphate pathway in photolithotrophic cells, it has been shown that the activity of the two dehydrogenases of the pathway in cell extracts is lower in phototrophically than in heterotrophically grown algal cells, despite the generally higher growth rate (and hence demand for reductant) under photoconditions ([47,50,96,155]; Table V). Thus the capacity of the oxidative pentose pathway is much lower relative to reductant requirement under photolithotrophic growth conditions.

The available evidence suggests that dark reductant generation processes are relatively inactive in the illuminated green cell, thus supporting the view that direct use of photoproduced reductant is important under these conditions. Unfortunately, no treatment is known which completely and selectively inhibits the dark reductant synthesis pathways; this complicates the analysis of reductant supply in illuminated green cells compared with ATP supply, where selective inhibition of dark ATP synthesis is possible (see sections 12.4.2, 12.4.3, 12.4.4 and 12.4.5).

12.3.3. The direct use of photoproduced reductant for reductive biosynthesis

12.3.3.1. Properties of non-cyclic electron flow

Non-cyclic electron flow with ferredoxin or $NADP^+$ as the redox product interacting with reductive biosynthesis in vivo, is commonly regarded as involving two light reactions in series (Fig. 12.2) [72,183,226]. It only occurs in photosynthetic (green) cells, is light-dependent, has an action spectrum showing a "red drop", a minimum quantum requirement of two quanta absorbed per electron transferred, and shows the Emerson enhancement effect [143,183].

The reaction sequence is located in the chloroplast thylakoids, with the reductant being directly available to reductive biosyntheses in the stroma, and indirectly (via shuttles) to the cytoplasm ([183] and Chapters 5,6,7).

As estimated from the maximum rate of O_2 evolution or CO_2 fixation at light and CO_2 saturation and at the temperature optimum for photosynthesis, non-cyclic electron flow in vivo can occur in genetically sun-adapted angiosperms and microalgae at up to 1000 μmoles $O_2 \cdot$ mg chl^{-1} .h^{-1} (see Table 5 of ref. 191 and also ref. 51). For chlorococcalean microalgae, rates up to 500 μmoles.mg chl^{-1} .h^{-1} have been found in *Ankistrodesmus* [227] (Table II).

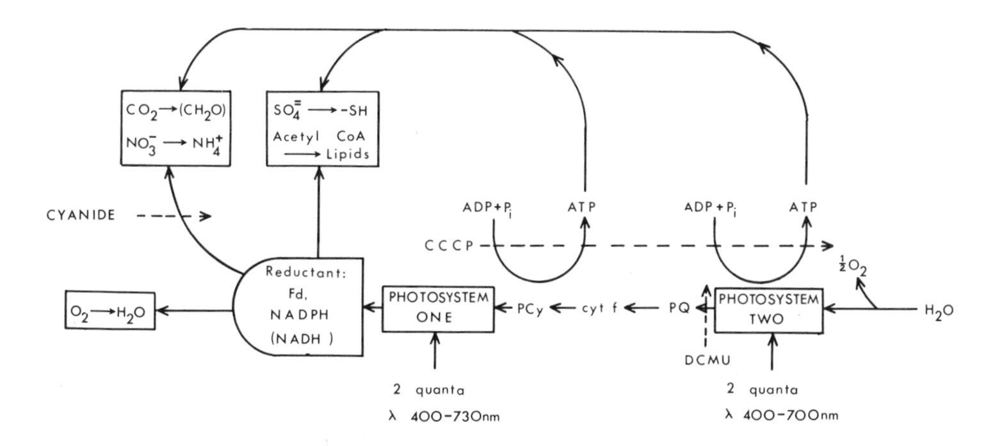

Fig. 12.2. Scheme of non-cyclic electron flow and coupled phosphorylation in plant chloroplasts; for further details see refs. 184, 226, and Chapter 4. Dotted lines indicate sites of inhibitor action.

NADPH is widely assumed to be the major product of non-cyclic electron transport in vivo, and to be the reductant used in CO_2 fixation ([183,184] but see ref. 16). Attempts to measure rates of net $NADP^+$ reduction in dark-light transients have yielded rates much lower than those of concurrent non-cyclic electron flow; these results have been rationalised in terms of concurrent NADPH oxidation in PGA reduction [83,153]. The $NADPH/NADP^+$ ratio in either the chloroplast or cytoplasm during steady-state light conditions is not consistently higher than in the dark (e.g. refs. 70,83,153 and Chapter 5), although the ratio for free rather than total nucleotide could be higher [83]. The measured turnover of $NADP^+$ being commensurate with the rate of non-cyclic electron transport is capable of being rationalised in terms of $NADP^+$ as a major intermediate in the pathway. However, it may not be possible to explain the *magnitude* of light stimulation of a process using photoreductant in terms of the measured in vivo increase in NADPH level in the light, and the in vitro relation between reaction rate of the enzyme and NADPH concentration. The analogous problem with ATP is discussed in section 12.4.4.

12.3.3.2. The nature of light effects on reductive biosyntheses in green plant tissues

Light generally stimulates the rate of reductive biosynthesis in green plant tissues [176]. Examples are the reduction of nitrate and nitrite (e.g. refs. 32,140 and Chapter 11), sulphate (e.g. refs. 121,238 and Chapter 9), and the reductive conversion of acetate to fatty acids ([69,240] and Chapter 11;

note that the conversion of carbohydrate to lipid does not involve a net input of reductant unless the CO_2 produced is refixed in photosynthesis [193]). Before these can be accepted as examples of the direct use of photo-reductant, a number of criteria must be satisfied.

One is that the predominant light effect must not be on the entry of the substrate in the cell. This is a particular difficulty with nitrate and nitrite where entry and reduction are closely coupled in many organisms (e.g. refs. 186,229). However, in some cases, e.g. the cases cited above on acetate and sulphate metabolism, and nitrite reduction in *Ditylum brightwelli* (see ref. 186), influx and reduction can be distinguished.

A second difficulty involves dissociation of the light effect from that on photolithotrophic growth. In the case of nitrate and nitrite reduction, the light stimulation is considerably inhibited by the absence of CO_2 (see ref. 140). However, in a number of cases a substantial light-stimulated reduction occurs in the absence of CO_2, particularly with nitrite (e.g. refs. 25,32, 111). Kylin [121] showed that light can stimulate sulphate reduction in *Scenedesmus* in the absence of CO_2, while Goulding and Merrett [69] found a similar effect for conversion of acetate to lipid in *Chlorella;* however, in this latter case the cells showed a light-stimulated growth on acetate in the absence of CO_2, so a coupling to growth cannot be eliminated.

The third criterion is that the observed light stimulated reduction, independent of light effects on uptake and growth, should have the characteristics of non-cyclic electron flow in terms of action spectrum, sensitivity to DCMU and stoichiometry with O_2 evolution, and should be attributable to reductant supply rather than ATP supply if the synthesis needs both photoproducts (Fig. 12.2).

Many workers have applied these criteria to nitrate and nitrite reduction, with varying results, in terms of agreement with expectations for coupling to non-cyclic electron flow. Thus in a number of cases the action spectrum shows a stimulation in blue light not attributable to chlorophyll (see ref. 175); the O_2 evolution is often less than that expected for stoichiometry with the nitrate or nitrite reduced [140], and the inhibition by DCMU is often less than that of concurrent O_2 evolution [136,140]. So while many of the data are consistent with an involvement of non-cyclic electron flow in nitrate and nitrite reduction (e.g. refs. 151,228), the situation is clearly complex.

Light-dependent conversion of acetate to lipid in intact *Chlorella* is inhibited by DCMU [69], although no comparative studies on the effect of DCMU on photosynthesis are reported. Light-stimulated acetate incorporation into fatty acids by isolated chloroplasts shows features (relative lack of sensitivity to removal of short wavelength light or the presence of DCMU) which suggest that light stimulation is not explicable entirely in terms of photoreductant generation by non-cyclic electron flow [109,216]. However, the very low rates found in these experiments [55] mean that a very low

TABLE II

Quantum requirement and maximum in vivo capacity of reductant and ATP generating processes, and of C assimilation in photolithotrophic growth of *Chlorella, Ankistrodesmus* and *Scenedesmus* (excluding the high temperature strains; see Table IV)

Rows 1—5 from data discussed in sections 12.3 and 12.4.

Process	Quantum requirement: quanta absorbed per electron transferred or per molecule ATP made or C assimilated	Maximum in vivo capacity, μequiv. transferred, μmole ATP made or μmole C assimilated. mg $chl^{-1}.h^{-1}$	Comments
Fermentative ATP synthesis	26(6)	20	High quantum requirement refers to ATP made from glucose breakdown, assuming 8 quanta to fix 1 C. Lower includes excess of ATP made in non-cyclic photophosphorylation over that used in CO_2 fixation
Oxidative phosphorylation	1.33 (1.15)	200	
Cyclic photophosphorylation	1.0	200	Quantum requirement per quantum absorbed by System I assumes optimal spillover of excitation energy from pigment System II to pigment System I during simultaneous non-cyclic and cyclic photophosphorylation [143,181]
Non-cyclic electron flow	2.0	2000	See Myers [142] for discussion of the excess capacity of non-cyclic electron flow and photophosphorylation found in short-term photosynthesis experiments, compared to that used in growth (Rows 6 and 7)
Non-cyclic photophosphorylation	2.0	2000	
C assimilation in photolithotrophic growth, ammonium as N source	10	300	Quantum requirement from Kok [117], van Oorschot [231]. See Syrett [217] for equality of light-saturated growth rates on the two N sources. C assimilation rate computed for a doubling time of 5 h (see Table IV) [50,91,129] assuming the ratio of C to chlorophyll in cell dry matter is 25 (50%C, 2% chlorophyll)
C assimilation in photolithotrophic growth, nitrate as N source	12	300	

photosystem II activity could supply enough reductant for lipid synthesis (cf. ref. 228).

Sulphate reduction in *Navicula* as indicated by incorporation of exogenous sulphate into sulpholipid, was light-stimulated by a DCMU-sensitive process even in the absence of CO_2 [156]. This is consistent with non-cyclic electron flow involvement in reduction of "active sulphate" ([204] and as discussed in Chapter 9) to form a sulphonate.

12.3.4. Conclusions on reductant supply in the illuminated green cell

The capacity of the oxidative pentose phosphate pathway is lower in photolithotrophic than heterotrophic algae, and the extent of reductant generation by this pathway and by glycolytic carbon supply to triosephosphate dehydrogenase is lower in illuminated than darkened autotrophic cells. This is despite the faster rate of reductive biosynthesis in the light [38,219]. This shortfall in reductant generation can probably be made up by direct use of photoreductant in many cases, although the situation is frequently more complex than a straightforward involvement of non-cyclic electron flow. Some quantitative conclusions are presented in Table II.

12.4. ATP SUPPLY

12.4.1. General considerations

As was pointed out in section 12.2.2.3, quantitation of ATP turnover in vivo is more difficult than that of net reductant requirement. If in vitro values of ATP/2e values for oxidative and photosynthetic electron flow are used, ATP synthesis rate can be estimated from respiratory O_2 uptake and photosynthetic O_2 evolution, (see refs. 183,227). For illuminated cells, tracer O_2 measurements are needed if the possible occurrence of O_2-linked non-cyclic photophosphorylation (Section 12.4.5) is not to be overlooked; flash spectrophotometric techniques are required in order to quantitate cyclic photosynthetic electron transport, and also measure O_2-linked non-cyclic electron transport [22].

More direct estimation of ATP turnover in vivo under various conditions involves either measurements of changes in ATP or P_i levels during such transitions as dark to light, dark-N_2 to dark-air, or dark-N_2 to light-N_2 [209]. This method suffers from the same drawback as the analagous method for measuring NADPH turnover, i.e. occurrence of ATP utilizing or P_i regeneration reactions during the change in conditions; this leads to an underestimate of turnover.

Measurement of incorporation of exogenous $^{32}P_i$ into organic compounds in the steady-state has also been used to estimate ATP turnover [183,209].

Quantitative estimates with this method are complicated by a rate-limitation by the rate of uptake of $^{32}P_i$ into the cells, and the possible occurrence in vivo of phosphate pools of different specific activities [183,209]. Similar complications occur when the method is applied to isolated, intact chloroplasts [104]. This method is, nevertheless, very useful as a qualitative or semi-quantitative indicator of different types of phosphorylation [103,104, 183,209].

Finally, minimal estimates of the capacity of a given type of phosphorylation may be obtained by the use of some ATP-using reaction of known stoichiometry under conditions which favour that type of phosphorylation. A widely used reaction is glucose assimilation which uses 3 ATP (or 2 ATP and 1 ATP precursor [120]) per glucose assimilated into starch [220].

The various kinds of phosphorylation which occur in green plant cells will be considered in turn; methods by which the process can be isolated characterised and quantified will be considered, along with estimates of quantum yield and its contribution to the overall ATP balance of the growing cell.

12.4.2. Substrate level phosphorylation in glycolysis fermentation

The two ATP-generating reactions of glycolysis (PGA kinase and pyruvate kinase) may be isolated from other ATP-generating reactions by maintaining green cells under dark-anaerobic conditions. The complete glycolytic sequence occurs in the cytoplasm, while phosphofructokinase and the enzymes for conversion of PGA to pyruvate appear to be absent from plastids (see refs. 168,182,210). An important point about glycolytic enzymes in plastid-containing cells is that there are often different isozymes in the two compartments for those enzymes which are found in both chloroplast and cytoplasm [3,132]. These enzymes are involved in photosynthesis in the chloroplast, and (in the case of the enzymes interconverting PGA and triosephosphates) in the shuttle transferring high-energy phosphate and reductant across the chloroplast membrane as well as in cytoplasmic glycolysis.

As regards the function of fermentation per se in the green cell, no instance of dark growth supported by fermentative ATP synthesis has been reported [38,45,62,112], although viability can be maintained during prolonged dark anaerobiosis in many algae.

The capacity for fermentation in green cells is low; in the presence of exogenous glucose, *Chlorella* and *Scenedesmus* can ferment at 5—15 μmoles ATP. mg $chl^{-1}.h^{-1}$ [15,62,87,218] (see Table II). As might be expected from this low rate, the rates of ATP-requiring reactions under dark-anaerobic conditions are low (e.g. refs. 186, 187). These low rates may be related to the toxicity of the fermentation products such as ethanol and organic acids [44,46,112,194]. This might account for the inability of fermentation to support growth [62], although O_2 requirements for biosynthesis may also be a factor [152] (see section 12.2.2.3).

The role of fermentation in ATP generation in the photolithotrophic growth is negligible; the initially low capacity is further lowered by the occurrence of the Pasteur effect (inhibition of carbohydrate breakdown by oxidative phosphorylation) and an analogous inhibition by photophosphorylation in green tissues [27,87,189]. These regulatory reactions restrict the rate of pyruvate synthesis to the rate at which it is removed in biosynthesis, with no excess for ethanol or lactate synthesis.

12.4.3. Oxidative phosphorylation

This is a mitochondrial process; mitochondria isolated from green tissues have similar properties to those found in other plant tissues [97,128] (Fig. 12.3). Chloroplasts rigorously purified from mitochondria lack oxidative phosphorylation [126]. These experiments involved "stripped" chloroplasts; subsequent analysis of purified intact chloroplasts (e.g. refs. 137,224) and chloroplast outer envelopes [49] confirms the absence from chloroplasts of respiratory cytochromes and of the mitochondrial coupling factor.

Oxidative phosphorylation can be separated from photosynthetic phosphorylations (sections 12.4.4 and 12.4.5) by conducting experiments aerobically in the dark. The occurrence of the Pasteur effect (section 12.4.2) restricts the substrate level phosphorylations of glycolysis to supplying less than 10% of the total aerobic ATP synthesis.

The use of the methods outlined in section 12.2.1 to estimate the capacity for oxidative phosphorylation in phototrophically grown *Chlorella* yields values of up to 200 μmoles ATP. mg chl^{-1} .h^{-1} (e.g. refs. 61,106,107,128, 141,202,220; Table II). The rate in the presence of glucose was used as being more representative of the capacity for oxidative phosphorylation than the

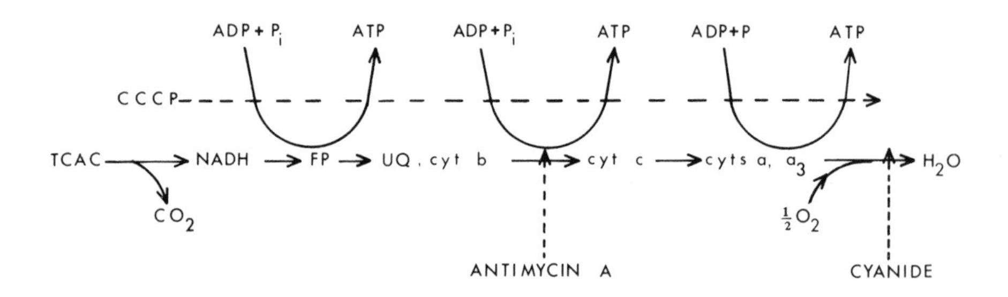

Fig. 12.3. Scheme of oxidative electron flow and coupled phosphorylation in plant mitochondria; for further details see refs. 97, 128. Dotted lines indicate sites of inhibitor action.

endogenous rate; the glucose-stimulated rate is similar to the uncoupler-stimulated rate [197] (cf. ref. 13). The quantum requirement of ATP synthesis by oxidative phosphorylation can be computed as 1.33 quanta per ATP, assuming 8 quanta to reduce 1 CO_2 to the level of hexose [118] and 36 ATP per hexose oxidised (i.e. glycolytically generated NADH oxidised by mitochondria with an ATP/2e of 2). If the possible excess of 1 ATP per CO_2 fixed (section 12.4.5) is taken into account, this is reduced to 1.15 quanta per mole (Table II). In the shade alga *Hydrodictyon africanum* the capacity for oxidative phosphorylation on a chlorophyll basis is much lower; about 20 μmoles ATP. mg chl^{-1} .h^{-1} [189,191].

Regarding the possible role of oxidative phosphorylation in ATP supply to the illuminated green cell: attempts may be made to apply criteria similar to those used in considering the "dark" reductant generation pathways. There is abundant evidence for the operation in the illuminated green cell of the TCAC; this pathway is essential for the synthesis of certain essential carbon skeletons for growth [33—35,182,233] (see section 12.2.2.4). In order to fulfil this requirement, recycling of carbon is not required; indeed certain obligate autotrophs have an incomplete TCAC which presumably functions largely as a carbon skeleton synthesis pathway (see ref. 182). However, there is evidence for recycling of the TCAC in illuminated *Chlorella* cells [171]. This suggests (see section 12.2.2.4, and ref. 182) that the rate of NADH synthesis in mitochondria in the light is at least as high as that implied in the requirements for C skeletons produced by the TCAC, and indeed of the rate in the dark.

However, evidence from O_2 exchange studies suggests that the rate of O_2 uptake mediated by cytochrome oxidase in green cells is not greater than that in the dark, and is frequently lower [98,182,198]. This is often attributed to inhibition of mitochondrial ATP generation by photophosphorylation, although the precise mechanism is difficult to formulate. While these experiments may not pertain to growth conditions, they do suggest that the rate of ATP synthesis by oxidative phosphorylation in the light is unlikely to be greater than that found in the immediately preceding or succeeding dark period, and is definitely not as great as the capacity for oxidative phosphorylation discussed above.

A relatively smaller contribution of oxidative phosphorylation to photolithotrophic than to heterotrophic growth is supported by a number of estimates of oxidative phosphorylation capacity in photolithotrophic compared with heterotrophic algal cells. Thus certain strains of *Chlorella* have a higher capacity for oxidative phosphorylation when grown heterotrophically (measured as O_2 uptake in the presence of glucose per cell or per mg dry weight: cf. ref.57) than when grown photolithotrophically, despite a faster growth rate under photolithotrophic conditions [53,170, 212]. Similarly, succinic dehydrogenase activity in *Chlamydomonas reinhardii* y-1 is higher in heterotrophic than in photolithotrophic cells

[154]. In each case a smaller capacity for oxidative phosphorylation under photolithotrophic conditions is correlated with a faster growth rate and hence ATP requirement (see Table V). This lower capacity for oxidative phosphorylation in photolithotrophic compared with heterotrophic cells is supported by quantitative electron microscopy, which shows that mitochondria comprise almost 20% of the cell volume in the apochlorotic alga *Polytomella* [29] but only about 5% in phototrophic *Chlamydomonas* and *Chlorella* [9,206].

While many algae and probably all higher plants can grow heterotrophically with oxidative phosphorylation as their ATP source, there are a number of obligately phototrophic algae which cannot grow heterotrophically [50]. The role of a low capacity for oxidative phosphorylation in causing obligate phototrophy is still a matter of dispute [50,110]. While some evidence is consistent with a decreased capacity for oxidative ATP synthesis in a number of obligately photolithotrophic eukaryotic algae (e.g. refs. 1,19,20,39,40,127,149), in no case is the evidence conclusive that this is the cause of the obligate photolithotrophy. A well established case of severe impairment of oxidative phosphorylation in an obligate phototroph would be very useful in investigating the role of various ATP synthesising mechanisms in illuminated green cells.

12.4.4. Cyclic photophosphorylation

This is characterised in vivo by the occurrence of the processes discussed in section 12.4.1 under conditions which exclude other mechanisms at ATP synthesis. Cyclic photophosphorylation is coupled to cyclic electron flow which requires photoreaction I of photosynthesis alone, and does not involve the net input of electrons from a donor or their removal by an acceptor in the steady state [183] (Fig. 12.4; Chapters 4 and 5). The in vivo demonstration of cyclic photophosphorylation requires the elimination of other ATP syntheses coupled to electron transport, i.e. oxidative phosphorylation (section 12.4.3) and non-cyclic photophosphorylation (section 12.4.5). These can both be eliminated by removing electron acceptors (O_2, CO_2, NO_3^-) and electron donation by photosystem II. This can be achieved by the removal of short-wavelength light, the addition of DCMU (but see ref. 103), or the use of mutants lacking photosystem II activity. The necessity to combine the removal of electron acceptors and photosystem II activity in order to eliminate the other redox-chain-coupled ATP syntheses is pointed out by Ullrich [227], Raven [182], Ullrich-Eberius [229] and Johansen and Luttge [101]. This treatment also eliminates possible cyclic photophosphorylations involving photosystem II alone, or both photosystems [225]. As was pointed out in section 12.4.2, cyclic photophosphorylation substantially inhibits fermentation, the only other source of ATP under these conditions.

Cyclic photophosphorylation isolated in this way, and assayed by the

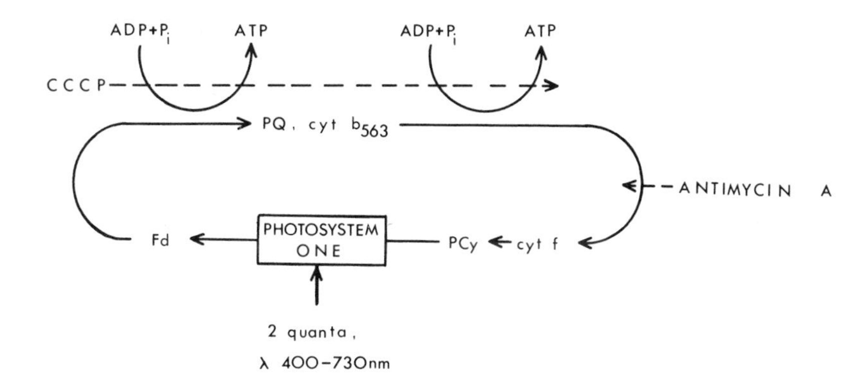

Fig. 12.4. Scheme of cyclic electron flow and coupled phosphorylation in plant chloroplasts; for further details see refs. 23,183,226, and Chapter 4. Dotted lines indicate sites of inhibitor action.

processes described in section 12.4.1, has an action spectrum characteristic of photosystem I alone, i.e. with a "red rise" in quantum yield (e.g. ref. 221). It is commonly inhibited to a greater extent than is non-cyclic photophosphorylation in the same organism by inhibitors such as desaspidin, CCCP, antimycin A, DSPD and salicylaldoxime [108,176,208, contrast 92, 215] (see Fig.12.4).

Cyclic photophosphorylation is commonly light-saturated at much lower irradiances than is non-cyclic photophosphorylation in the same organism (e.g. refs. 116,209,230). This is true regardless of the absolute values of irradiance required: contrast the high irradiances needed for the "sun" microalgae used in the work cited above with the lower irradiances needed in the "shade" alga *Hydrodictyon africanum* [175,181,189,191].

The quantum yield of cyclic photophosphorylation in vivo has been estimated by Tanner et al. [221], using anaerobic glucose photoassimilation in *Chlorella* as a quantitatively important ATP-consuming reaction (see section 12.4.2). 1 quantum absorbed by photosystem I produces 1 ATP, implying an ATP/2e of 2 (see Table II).

The in vivo capacity for cyclic photophosphorylation has been estimated by the methods outlined in section 12.4.1. Bedell and Govindjee [11] report rates up to 200 μmoles ATP. mg chl^{-1}. h^{-1} in *Chlorella*. The experiments of Biggins [22] on the rates of cyclic and non-cyclic electron flow in *Porphyridium*, together with the photosynthetic rates reported by Leclerc [125] suggest a capacity of up to 200 μmoles ATP. mg chl^{-1} . h^{-1}, based on the *uncoupled* electron transfer rate and assuming an ATP/2e of 2. This latter estimate is not subject to the criticism that the rate observed may reflect the capacity for ATP utilisation rather than its synthesis, since it is based on an *uncoupled* rate of electron transport (cf. ref. 220). Other estimates for the capacity for cyclic photophosphorylation in sun-adapted eukaryotes are generally less than 100 μmoles. mg chl^{-1} . h^{-1} (see Table II and ref. 189). The observed quantum yield, low maximum capacity and early light saturation of in vivo cyclic photophosphorylation are related, internally consistent, parameters. The capacity for cyclic photophosphorylation, as for oxidative phosphorylation (section 12.4.2) is much lower (about 20 μmoles ATP. mg chl^{-1}. h^{-1}) in the shade-adapted alga *Hydrodictyon africanum* than in the sun-adapted eukaryotes mentioned above [189,191].

A wide range of ATP-requiring processes in green cells can be supported by ATP generated by cyclic photophosphorylation (see refs. 73,182,209). It is important to note that while some of these use ATP within the chloroplast (e.g. the synthesis of ADPG for starch synthesis in the chlorophyte line of evolution [21,42,182]), many are outside the chloroplast (e.g. all active transport processes at the plasmalemma and tonoplast, chloroplast movement in *Mougeottia*, and protein synthesis in 80s ribosomes; see ref. 182). This means that the export of ATP from its site of synthesis on the outer surface of the thylakoid [189] via the PGA-triosephosphate shuttle must occur (Chapters 5 and 6) under conditions in which cytoplasmic reoxidation of the reductant exported concurrently is very slow. A recycling of this reductant into the chloroplast, using the dicarboxylic acid porter, is a likely fate for it (see 211).

The capacity for cyclic photophosphorylation in vivo is similar to that of oxidative phosphorylation (Table II) [220], and the range of processes which they can support is similar.

However, the rates at which the two sources of ATP can support different reactions, even in the same organism, are very variable. The data in Table III, for various ATP-using processes in *Hydrodictyon africanum*, show a range of 0.85—9.0 for the ratio: rate of process under cyclic photophosphorylation conditions/rate of process under oxidative phosphorylation conditions. I have considered this problem elsewhere [185,187,189], and concluded that mitochondrial and/or chloroplast energy metabolism produces signals other than changes in ATP concentration (or the related energy charge or phosphorylation potential) which activate or inhibit different ATP-consuming reactions. This concept is widely accepted for the regulation of photosynthetic carbon metabolism [168], and in no way invalidates conclusions as to the *source* of ATP for these reactions based on experiments which isolate a single ATP-generating reaction. ATP can act as the energy source for a reaction without necessarily controlling the rate of the reaction.

The question of the extent to which cyclic photophosphorylation occurs under the stringent conditions specified earlier (i.e. the absence of a

TABLE III

Rates of various ATP-requiring reactions in *Hydrodictyon africanum* when ATP is supplied by either cyclic photophosphorylation or oxidative phosphorylation

Cyclic photophosphorylation is isolated by removal of CO_2 and O_2 from the gas phase in the presence of 1 μM DCMU in light (fluorescent); oxidative phosphorylation is isolated by aerobic conditions in the dark (see sections 12.4.2 and 12.4.3). Rates in pmole·cm^{-2}·s^{-1} (except 3-*O*-methylglucose influx, fmole·cm^{-2}·s^{-1}). (Data of lines 1—3 unpublished experiments of J.A. Raven, lines 4—6 from ref. 189.)

Process	Rate supported by cyclic ATP synthesis	Rate supported by oxidative phosphorylation	Ratio cyclic/ oxidative	References and comments
Ouabain-sensitive active K^+ influx	0.73 ± 0.11	0.08 ± 0.24	9	[172,173,181]
Ouabain-sensitive active Na^+ efflux	0.55 ± 0.11	0.23 ± 0.06	2.4	
Active $H_2PO_4^-$ influx at the plasmalemma	0.22 ± 0.03	0.15 ± 0.02	1.5	[183,185]
Active influx of 3-*O*-methyl glucose (10 μM) at plasmalemma	2.9 ± 0.3	3.4 ± 0.4	0.85	Flux in fmole. cm^{-2}·s^{-1}. [188,189]
Active influx and metabolism of glucose (10 mM)	3.2 ± 0.3	3.7 ± 0.4	0.86	Active glucose influx suggested by analogy with influx of analogue 3-*O*-methylglucose [188,189]
Conversion of sugar phosphates labelled with ^{14}C from $^{14}CO_2$ in photosynthesis into starch in a "pulse and chase" experiment	5.24 ± 0.64	3.92 ± 0.48	1.33	Method of Glagoleva and Zalenskii [65] and Glagoleva et al. [66]; [see 188,189]. Flux in pmole C·cm^{-2}·s^{-1}

functional photosystem II and of electron acceptors for photosystem I) is still a matter of dispute. When these conditions are imposed, the data on quantum yield (see above) and on quantitative involvement of cytochrome *f* [22] suggest that most of the photosystem I reaction centres can participate in cyclic electron flow. The quantum yield of non-cyclic electron flow (see section 12.3.3.1) suggests that most of the photosystem I units can also co-operate with photosystem II in non-cyclic electron flow (see ref. 22). The balance between cyclic and non-cyclic photophosphorylation, i.e. the switch between cyclic and non-cyclic operation of these photosystem I units, could be via the relative demand for ATP and reductant [8], and in vivo evidence for such regulation has been presented [100,174,181].

The unrestrained operation of such a regulation mechanism would involve substantial inhibition of cyclic electron transport when photosystem II was inhibited (e.g. by DCMU or in far-red light) and electron acceptors for photosystem I were present (e.g. O_2, CO_2). This would lead to a removal of electrons (see ref. 181) from the cyclic pathway. While some data support such an inhibition (e.g. refs. 63,79), there are also a large number of instances of the occurrence of cyclic photophosphorylation under these conditions (e.g. refs. 176,181,185,230). Oxidative phosphorylation has been ruled out in some instances by the presence of cyanide [176]. This suggests that there must be some mechanism which stops the "over-oxidation" of the cyclic electron transport pathway (see refs. 181,185). This could involve, in part, the isolation of some photosystem I units from photosystem II [181,185].

The data of Raven [181] and Biggins [22] are consistent with the occurrence of some cyclic photophosphorylation under conditions favourable for non-cyclic photophosphorylation. The data of Biggins show that about 10% of the photosystem I units are involved in cyclic electron flow under these conditions; since the rate constant for cyclic electron flow is less than that for non-cyclic, it only contributes some 2% to the total photosynthetic electron flow. In both the work of Raven [181] and of Biggins [22] no precautions were taken to eliminate nitrate as an electron acceptor for non-cyclic electron flow; these experiments may underestimate the contribution of cyclic ATP synthesis under these conditions (see section 12.5.2).

Conditions in which cyclic photophosphorylation may be of more importance may include maintenance ATP supply at or below the light compensation point of photosynthesis [22], and in photoheterotrophic growth of certain "acid flagellates". These organisms include obligate phototrophs which can grow photolithotrophically, or photoheterotrophically on acetate with a repressed PSCRC and photosystem II; under the latter conditions cyclic photophosphorylation may be a major ATP source [236].

Photoheterotrophic growth of eukaryotic algae under strictly anaerobic conditions (e.g. in the presence of DCMU to prevent photosynthetic O_2 evolution) using cyclic photophosphorylation as energy source did not occur in the experiments of Nuhrenberg et al. [152] or of Nakayami et al. [148]. The algae used could all grow heterotrophically in the dark using oxidative phosphorylation to supply ATP. Since the capacity for cyclic photophosphorylation is similar to that of oxidative phosphorylation (Table II) a likely explanation of the absence of anaerobic growth in the light is the O_2 requirement for biosynthesis (section 12.4.2).

Finally, it may be noted that while cyclic photophosphorylation has a wide taxonomic distribution [157,183,209], none of a wide range of ATP-requiring processes examined in *Chara corallina* could be supported by cyclic photophosphorylation in vivo. While explanations other than the complete absence of cyclic photophosphorylation are possible [211], this deserves further investigation, especially in view of the possibility that charophytes are obligate photolithotrophs [60], and hence may also be deficient in some aspect of respiratory energy transduction (cf. section 12.4.3).

12.4.5. Non-cyclic photophosphorylation

This process is characterised by the occurrence of ATP-requiring processes in vivo which require light absorbed by both photosystems I and II, and the presence of an electron acceptor such as CO_2, O_2 or NO_3^- (Fig. 12.2). In order to eliminate contributions from cyclic and oxidative phosphorylations, which are not ruled out by the conditions mentioned above, it is necessary to add one of the inhibitors (e.g. antimycin A or salicylaldoxime: section 12.4.4) which inhibit these two mechanisms of ATP synthesis more than the non-cyclic process. Thus it is easier to isolate non-cyclic photophosphorylation from alternative ATP-supplying processes than it is to isolate non-cyclic reductant generation from respiratory reductant generation (section 12.3.3; see also Chapter 5).

Dealing first with non-cyclic photophosphorylation with CO_2 as electron acceptor, evidence from experiments of the types discussed in section 12.4.1 suggests the involvement of this process in a number of ATP-requiring processes in vivo, of which the quantitatively most important is CO_2 fixation itself [183,209].

Much evidence supports the view that the ATP generated in non-cyclic electron flow (4 ATP per 2 NADPH if the ATP/2e is 2; see Chapter 4 and discussions in Chapter 5) is sufficient to supply the ATP required by the reactions of the photosynthetic carbon reduction cycle, i.e. 3 ATP per CO_2 fixed and 2 NADPH reoxidised [108,115,177,181,185,208,211,220]. More ATP and reductant are required per CO_2 fixed if the glycolate pathway is operative [74], but the ratio is still such that no additional ATP input is required (see Chapter 5).

However, some workers have argued for an involvement of cyclic [207] or O_2-linked non-cyclic ([79] and Chapter 5) ATP synthesis in steady-state photosynthesis by isolated chloroplasts. Some alternative explanations for these results are discussed by Klob et al. [115] and by Smith and Raven [211]. However, it is likely that photosynthesis by the C4 pathway does require an additional ATP input from cyclic photophosphorylation [24,74, 134].

These results (see also refs. 181,189,230) suggest that isolated non-cyclic photophosphorylation with CO_2 as electron acceptor has a relationship to irradiance more similar to that of photosynthesis, i.e. a much higher irradiance requirement for saturation, than does isolated cyclic photophosphorylation (section 12.4.4). This is true of both sun-adapted and shade-adapted plants [189].

A number of other ATP-requiring processes in vivo can occur under conditions in which non-cyclic photophosphorylation with CO_2 as terminal electron acceptor is the only source of ATP; these processes have characteristics (e.g. action spectrum, irradiance required for saturation, DCMU sensitivity) consistent with ATP supply by non-cyclic photophosphorylation. Most of the examples involve active ion transport at the plasmalemma (e.g. refs. 99,169,181,185,209,230,232). This availability of ATP synthesised in non-cyclic photophosphorylation with CO_2 as electron acceptor for processes other than CO_2 fixation is consistent with the ATP/NADPH ratios for consumption in CO_2 fixation and synthesis in non-cyclic photophosphorylation discussed above.

Evidence for ATP synthesis related to light-stimulated nitrate and nitrite reduction in *Ankistrodesmus* has been presented by Ullrich [227,228] and Ullrich-Eberius [229]. Under an N_2 atmosphere, in the presence of nitrate and nitrite, the characteristics of $^{32}P_i$ incorporation are those of non-cyclic photophosphorylation, i.e. insensitivity to antimycin, and inhibition by DCMU or the absence of light absorbed by photosystem II. The uncoupler CCCP inhibits both the $^{32}P_i$ incorporation and electron transport [228,229]; the absence of the classical uncoupler stimulation of electron flow suggests that, as with CO_2 fixation, the reduction of nitrate and nitrite consumes as well as synthesises ATP (see refs. 90,140,186). The significance of the ATP synthesis coupled to light-dependent nitrate reduction to the overall energetics of photolithotrophic growth is further discussed in section 12.5.2.

As regards ATP synthesis coupled to the use of O_2 as terminal electron acceptor for non-cyclic electron flow (pseudocyclic photophosphorylation) there is still some confusion [209]. Heber [79] and Gimmler [63] have presented evidence that this process can support chloroplast shrinkage in vivo (see Chapter 5). Raven [176,186,187] and Raven and Glidewell [192] found in *Hydrodictyon africanum* that active ATP-requiring K^+ and P_i influxes are light-stimulated by a process which needs both O_2 and photosystem II when oxidative phosphorylation, cyclic photophosphorylation and non-cyclic photophosphorylation with CO_2 or nitrate and electron acceptors have been eliminated by the addition of cyanide and antimycin A (see Figs. 12.2, 12.3, and 12.4).

Experiments with $^{18}O_2$ [67,68] show that there is a photosystem II-dependent uptake of O_2 in *H. africanum* which is greater at the CO_2 compensation point than at CO_2 saturation. The component which is stimulated by the absence of CO_2 is inhibited by both cyanide (at a concentration which inhibits both CO_2 fixation and dark O_2 uptake) and the uncoupler CCCP (at a concentration which abolishes CO_2 fixation and stimulates dark respiration) [67,68]. This component is probably due to O_2 uptake in glycolate synthesis by the ribulose diphosphate oxygenase reaction ([5,6,68] and also Chapter 8).

The remaining photosystem II dependent O_2 uptake is insensitive to both cyanide and CCCP, and is probably pseudocyclic O_2 uptake, with the associated electron transport coupled to ATP synthesis [67,68]. While O_2-requiring glycolate synthesis via the pathway discussed by Coombs and Whittingham [41] could contribute to this O_2 uptake, synthesis and metabolism of glycolate by this pathway consume ATP and NADPH in a lower ratio than that in which they are produced in non-cyclic photophosphorylation [74]. Thus glycolate synthesis by the Coombs and Whittingham pathway, as well as straightforward pseudocyclic electron transport, could produce a surplus of ATP which can be used for ATP-requiring processes in vivo [192]. Egneus et al. [52] have demonstrated a light-dependent $^{18}O_2$ uptake in intact, isolated *Spinacia* chloroplasts which exceeds that stoichiometric with the observed rate of glycolate synthesis; this extra O_2 uptake may well be related to O_2-linked non-cyclic electron flow coupled to ATP synthesis.

Thus there is evidence that non-cyclic electron flow with CO_2, nitrate, nitrite or (probably) O_2 as electron acceptor can be coupled to ATP synthesis. From the quantum requirement of 2 quanta per electron for non-cyclic electron flow (see Table II), and assuming that the ATP/2e ratio of 2 found in isolated chloroplasts (Chapter 4) applies at low irradiances in vivo, the in vivo quantum requirement of non-cyclic photophosphorylation is probably two quanta per ATP. It appears from earlier reports that the ATP/2e ratio of photophosphorylation was lower under light-limiting than under light-saturated conditions were due to damage to the thylakoids during chloroplast extraction [82,119].

Indirect evidence has been adduced as to the maximum rate of non-cyclic photophosphorylation in vivo by Ullrich [227] using the observed maximum rate of non-cyclic electron flow in vivo and the ATP/2e of 2 found in vitro for non-cyclic photophosphorylation, or of 1.5 for CO_2 assimilation. The data of Ullrich [228] with *Ankistrodesmus* suggest a capacity for non-cyclic photophosphorylation in excess of 2000 μmoles ATP. mg chl^{-1} .h^{-1}; similar values for *Chlorella* and *Scenedesmus* are suggested by the non-cyclic electron transport rates reviewed by Raven and Glidewell [191]. Even higher rates (up to 4000 μmoles ATP. mg chl^{-1} .h^{-1}) may be computed for sun-adapted angiosperms and non-chlorophyte microalgae [51,191]. These rates are rather higher than the maximum capacity of photophosphorylation thus far demonstrated in isolated chloroplasts [10]. More direct estimates of non-cyclic photophosphorylation rates in vivo yield lower values [183,209]. As with oxidative phosphorylation (section 12.4.3) and cyclic photophosphorylation (section 12.4.4), the capacity for non-cyclic photophosphorylation in vivo in the shade-adapted alga *Hydrodictyon africanum* is lower (about 150 μmoles ATP. mg chl^{-1} .h^{-1}) than the corresponding rate in sun-adapted plants.

The much greater capacity of non-cyclic photophosphorylation than of either cyclic or oxidative phosphorylations (Table II), together with the evidence for inhibition of both of the lower-capacity ATP-synthesising processes when non-cyclic photophosphorylation can occur (section 12.4.3, 12.4.4), suggests that non-cyclic photophosphorylation is the dominant source of ATP during photolithotrophic growth. This is further discussed in section 12.5.2.

12.4.6. Conclusions on ATP supply in the illuminated green cell

There is evidence that fermentation, oxidative phosphorylation, cyclic photophosphorylation and non-cyclic photophosphorylation can all occur in the illuminated green cell. The capacity of these processes, and their quantum requirements, in vivo are given in Table II. It is likely that the major ATP-supplying reaction under conditions of photolithotrophic growth is non-cyclic photophosphorylation with CO_2 (and, when present, nitrate) as electron acceptor. Cyclic and O_2-linked non-cyclic photophosphorylation are probably more important for maintenance and luxury accumulation processes in the light when growth cannot occur, while oxidative phosphorylation supplies these processes with ATP in the dark. Fermentation is important as a source of ATP for maintenance in dark-anaerobic conditions.

12.5. INTEGRATION OF ENERGY SUPPLY WITH GROWTH

12.5.1. Quantitative role of respiration in photolithotrophic growth

Evidence discussed in Sections 12.3 and 12.4 suggests that the supply of ATP and reductant for processes other than CO_2 reduction by the PSCRC during photolithotrophic growth is supplied directly from the photoreactions rather than by respiratory reactions. This is supported by an analysis of the quantitative role of "dark" respiration in photolithotrophic compared with heterotrophic growth [190].

Table IV shows the ratio of CO_2 produced by "dark" respiration to C assimilated into cell material for a number of *Chlorella* strains growing under various conditions.

The data show a spread of the CO_2/C ratio of 0.15—0.20 for photolithotrophic growth at light saturation and 0.75—1.20 for heterotrophic

TABLE IV

Ratio of CO_2 lost in dark respiratory processes to C assimilated (CO_2/C) during photolithotrophic (continuous illumination) and heterotrophic growth of various strains of *Chlorella*

For heterotrophic growth, the ratio was determined from cell weight gain and hexose consumed or CO_2 produced; for photolithotrophic growth, from dry weight gain and respiratory CO_2 production measured immediately after cessation of illumination; this is unlikely to overestimate the rate of dark respiratory processes in the light (sections 12.3.2 and 12.4.3).

Organism	Source of energy carbon nitrogen	Doubling time (h)	CO_2/C	References
Chorella sorokiniana (Tx 7-11-05) (high temperature)	Light (sat) CO_2 nitrate	2	0.25	[146,170,212]
	Glucose glucose nitrate	4	0.75	
Chlorella regularis (high temperature)	Light (sat) CO_2 nitrate	3	0.15	[53]
	Glucose glucose nitrate	4	0.75	
Chorella ellipsoidea	Light (sat) CO_2 nitrate	7	0.15-0.2	[91,145, 146, 200]
	Light (lim) CO_2 nitrate	33	0.30	
	Glucose glucose nitrate	14	0.90	
Chlorella pyrenoidosa (Emerson strain)	Light (sat) CO_2 nitrate	8	0.15	[43,147,200]
	Light (lim) CO_2 nitrate	17	0.65	
	Glucose glucose nitrate	16	1.20	
	Glucose glucose ammonium	16	0.95	

growth. The lower values are most interesting as they indicate the maximum efficiency of growth in terms of "dark" respiration (higher values may be due to various "idling" processes, discussed by Beevers [12]. The much lower values attained in photolithotrophic growth are real, and cannot be attributed to either the production of cells which need less energy per unit C assimilated in the light, or to a lower relative maintenance requirement due to faster growth in the light. Differences in cell composition between photolithotrophically and heterotrophically grown cells [53,142]; are in the wrong direction to account for the discrepancy, in that heterotrophically grown cells contain less protein (with a high energy requirement for synthesis [59]) and more carbohydrate.

As regards the maintenance requirement, the lower CO_2/C for autotrophic growth is still seen even when the growth rate is reduced to below that for heterotrophic growth by the use of low irradiances [117] (Table IV).

Thus the differences in CO_2/C found for heterotrophic compared with photolithotrophic growth are real. The minimum value of 0.75 for heterotrophic growth with nitrate as N source is close to that predicted in Table I for respiratory generation of ATP and reductant for heterotrophic growth, while that in the light (0.15) is close to the lower limit set by the CO_2 production involved in the synthesis of essential carbon skeletons, with all reductant and ATP supplied by photosynthetic partial processes (section 12.2.2.3).

These data contrast with the conclusion of Penning de Vries, et al. [165], arrived at on theoretical grounds, that an increase in the ATP/2e of oxidative phosphorylation should not lead to large increases in the respiratory efficiency of growth at high relative growth rates. Photophosphorylation may, in this context, be regarded as a means of increasing the ATP/2e ratio in respiration.

12.5.2. Contribution of various energy-supplying reactions to photolithotrophic growth

12.5.2.1. Growth at high irradiances

As was pointed out in section 12.2.2.1, the overall requirement for non-cyclic electron flow in photolithotrophic growth is the same regardless of whether processes such as nitrate reduction use photoproduced reductant directly or by regeneration from assimilated carbon. The evidence (sections 12.3 and 12.5.1) that the direct use of photoreductant is the common situation in the illuminated green cell has important repercussions for ATP supply. The direct use of photoreductant in reductive biosynthesis leaves much of the ATP generated in the coupled non-cyclic phosphorylation available for ATP-requiring processes. Indirect use via CO_2 fixation involves consumption of 0.75 of the ATP made in the non-cyclic electron flow in CO_2 fixation, leaving less for use by other processes.

Direct use of photoreductant in reductive biosyntheses other than CO_2

fixation thus gives a bonus in terms of ATP. This is quantitatively most important when the requirement for reductant is highest, i.e. during the ecologically predominant process of growth with nitrate as N source. The requirements summarised in Table I for growth of cells phototrophically with nitrate as N source show that 5.6 electrons are needed to assimilate 1C; if all of this is supplied directly as photoreductant, ATP/2e ratio in non-cyclic photophosphorylation of 2 supplies 5.6 ATP. This exceeds the minimal ATP requirement for assimilation of 1 CO_2 into cell material, i.e. no additional ATP synthesis is required. This may explain why the role of cyclic photophosphorylation is so small under conditions in which nitrate is available for growth (section 12.4.4).

With ammonium as N source, even with all reductant used directly from non-cyclic electron flow, only some 4.4 ATP are produced by non-cyclic photophosphorylation per C assimilated. This is less than the minimum estimate of 5 quoted in Table I, i.e. an additional ATP input of 0.6 ATP per net C assimilated from pseudocyclic, cyclic or oxidative phosphorylation is required for growth. For a growth rate of 300 μmoles C assimilated. mg chl^{-1} h^{-1}, this amounts to 180 μmoles ATP. mg chl^{-1} .h^{-1}, i.e. just within the capacity of either cyclic or oxidative (and probably pseudocyclic) ATP synthesis (Table II).

Since it has been assumed that all of the reductant used is made directly by non-cyclic photophosphorylation, the NADH generated in the synthesis of essential C skeletons (sections 12.2.2.3 and 12.5.1) would all be used in ATP synthesis by oxidative phosphorylation; this would exactly supply the 0.6 ATP per C assimilated. Alternatively, the use of some of this NADH in reductive biosynthesis would reduce the amount of ATP made in non-cyclic photophosphorylation as well as oxidative phosphorylation, and leave the production of extra ATP to cyclic or pseudocyclic photophosphorylation. Additional evidence is required before these possibilities can be distinguished.

12.5.2.2. Growth at low irradiances

The requirements for reductant and ATP per C assimilated are similar at high and low irradiances, save that the ATP requirement may be higher since the lower growth rate involves a relatively greater maintenance requirement. The major constraint on which energy transduction processes are used is, in this case, the observed quantum requirement for growth (Table II) rather than the capacity of the various pathways as was the case for growth at light saturation (section 12.5.2.1).

With nitrate as N source, no problems are encountered. If all the reductant is supplied to reductive biosynthesis directly from non-cyclic electron flow, the 12 quanta needed to assimilate 1 C can supply 6 ATP (5.6 via electron flow to CO_2 and nitrate, the rest from pseudocyclic photophosphorylation). Thus all the ATP needed for growth, and for an improbably large maintenance requirement, can be supplied by non-cyclic ATP synthesis.

Growth on ammonium as N source needs 10 quanta per C assimilated. If all the reductant for biosynthesis is supplied directly by non-cyclic photophosphorylation, these 10 quanta could supply 5 ATP if they were all used in non-cyclic photophosphorylation (4.4 via non-cyclic electron flow in reductive biosynthesis, the rest by pseudocyclic photophosphorylation). This just covers the ATP requirement for growth, but leaves no margin for maintenance. If cyclic photophosphorylation is used instead of pseudocyclic, the maximum ATP production from the 10 quanta is 5.6 (Table II); oxidative phosphorylation following CO_2 fixation could produce 5.4. Thus it is just possible to account for the minimum energy requirements for cell synthesis with ammonium as N source in terms of known mechanisms.

It must be pointed out that more pessimistic assumptions as to the ATP/2e ratio of non-cyclic photophosphorylation, the ATP required to assimilate 1 mole of C, and the flexibility of excitation energy distribution between the two photosystems, lead Kok [119] to calculate a larger quantum requirement for photolithotrophic growth by known mechanisms than is found experimentally (Table II). Patently, further work is required.

12.5.2.3. Economic advantages of direct use of photoproduced reductant and ATP

(i) Running costs. This topic has been discussed previously by Raven [179, 180,183,191]. When plant growth is limited by CO_2 supply [178], the direct use of photoproduced co-factors has the advantage that less reduced carbon is lost as CO_2 in dark respiratory processes (Table IV; section 12.5.1).

During growth at low irradiances the limiting resource is light; a combination of direct use of reductant and of ATP leads to a lower quantum requirement for growth than would be the case if they were regenerated by respiratory reactions. This is particularly marked in the case of growth on nitrate, the most common natural combined N source [133].

Wild et al. [237] have suggested that the direct use of photoproduced cofactors has a specific role in "shade" plants, which are able to grow at lower irradiances than "sun" plants. While the lower quantum requirement for growth which can result from the direct use of cofactors might be of adaptive significance to these plants, they are not characterised by a greater emphasis on the direct use of cofactors than are "sun" plants; some "sun" plants have taken this adaptation as far as is possible [191]. A reduced requirement for high-energy cofactors in producing and maintaining unit biomass seems to be a more reasonable explanation of the ability of "shade" plants to grow at very low irradiances [191].

(ii) Capital costs. For a green plant cell growing in continuous light (i.e. in a laboratory culture, in the polar summer or on the light side of a planet with the same side always exposed to its sun) there is a clear advantage in terms of the quantity of metabolic machinery required in the direct use of photoproduced cofactors. Thus for a given rate of ATP or reductant synthesis, the

direct pathway requires an investment of energy and materials in producing chloroplast thylakoid membranes. The indirect pathway requires a similar amount of thylakoid membrane, but in addition it requires stroma enzymes for CO_2 fixation, and respiratory enzymes for regeneration of reductant and ATP from reduced carbon.

For a green cell under more earth-natural conditions of a light—dark alternation every 24 h the advantages are less clear-cut. In order to carry out the same amount of energy-requiring biosynthesis during a 24-h period, the "direct" pathway only provides cofactors for half the time, and so it would be expected that it would involve a greater activity of biosynthetic enzymes than a corresponding heterotrophic cell which can use the cofactors over the entire 24 h.

The "direct" pathway thus involves a saving in capital costs in terms of the lower requirement for enzymes of carbon fixation and of the respiratory pathways for regeneration of cofactors, but involves a greater cost in producing extra biosynthetic enzymes. What is not clear is whether the balance of capital cost favours the "direct" pathway. If the balance is against the "direct" pathway, this must presumably be outweighed by the economy of running costs. This argument is based on the observed preponderance of "direct" use of cofactors in laboratory cultures (sections 12.3 and 12.4) and in nature (e.g. ref. 54).

It is observed (Table V) that, as might be expected from the observed predominance of direct use of photoproducts, the balance of capital investment in photolithotrophically as compared with heterotrophically grown algae is in favour of photosynthetic rather than respiratory machinery. The repression of chloroplast development during heterotrophic growth seems reasonable. The chloroplast can occupy up to half of the cytoplasmic volume of a green cell and contain up to half its total protein (e.g. refs. 9, 206; section 12.2.2.4), and with the exception of starch storage and nitrite reduction in the chlorophyte line of evolution (sections 12.2.2.4; 12.4.4) it is not needed for heterotrophic growth. Varying degrees of chloroplast repression, and increased capacity for respiration, are found in different algae under heterotrophic conditions (provided they can grow heterotrophically).

Since the respiratory capacity is expressed on a total cell weight or protein basis, and the chloroplast is a larger fraction of cell weight or protein in photolithotrophic cells, it might be objected that the decreased respiratory capacity is no greater than that of any non-chloroplast enzyme. However, the data reviewed in Table V suggest that there is a reduced respiratory capacity even on a cytoplasmic weight or protein basis, in agreement with the argument used earlier.

A further point of interest is the growth rates found under photolithotrophic compared to heterotrophic conditions (Tables IV and V). In algae with incomplete repression of the chloroplast under heterotrophic condi-

TABLE V

Growth rate, photosynthetic capacity and respiratory capacity of obligately photolithotrophic, obligately heterotrophic and facultative photolithotrophs-heterotrophs as a function of nutritional conditions

Nutritional type	Example	Heterotrophic growth rate as fraction of photolithotrophic	Effect of heterotrophic conditions on		References
			Photosynthetic capacity	Respiratory capacity	
Obligate photolithotroph	*Chlamydomonas eugametos*	0	No effect	No effect	[50,96]
Facultative photolithotroph-heterotroph	Many strains of *Chlorella, Scenedesmus* and *Chlamydomonas*	0.25—0.75	10—100% of rate in photolithotrophic cultures	Higher than in photolithotrophic cultures	[36,53,96,122, 154,170,212]
	Strains of *Euglena gracilis*	1	Essentially none in heterotrophic cultures	Higher than in photolithotrophic cultures	[38,96,203]
	Ochromonas malhamensis	5	Very low (low even in photolithotrophs)	?	[144]
Obligate heterotrophs	*Prototheca* *Polytoma* *Astasia*	∞	Not applicable: no photolithotrophic growth		[45,50]

tions, the photolithotrophic growth rate exceeds that found heterotrophically. In many strains of *Euglena gracilis*, which show complete repression of chloroplast synthesis under heterotrophic conditions, the heterotrophic growth rate does not exceed that found photolithotrophically. Only *Ochromonas malhamensis* shows the increase in growth rate under heterotrophic compared with photolithotrophic conditions expected when the heterotrophic cell has shed the burden of synthesising and maintaining a fully functional chloroplast [17]. Thus in facultative heterotrophs, photolithotrophic growth is faster than naive economic arguments might suggest.

12.5.3. Some special features of multicellular plants

Multicellular photolithotrophic algae and higher plants have cells without photosynthetically competent chloroplasts, even in illuminated parts of the plant (see refs. 48,56,182). Many meristematic and reproductive cells are heterotrophic; this may reflect a spatial separation of nuclear and cell division from photosynthesis, perhaps analogous to the temporal separation seen in many photolithotrophic unicells where cell division commonly occurs in the dark phase [38,219]. Despite this, the direct use of photoproducts occurs in the growth of *Cucurbita* and *Xanthium* leaves, as shown by the low CO_2/C ratio (Table IV; section 12.5.1 [93]). Even when the whole higher plant (e.g. *Zea*) is considered, the CO_2/C for its photolithotrophic growth (0.29—0.35) is lower than for heterotrophic growth of cultured roots (0.71) or embryos (0.55). These values were calculated from the data of Heichel [85] and Kandler [107] respectively. This topic is discussed in more detail by Raven [190].

A number of angiosperms forego the advantages of the direct use of photoproducts claimed for algae in section 12.5.2.3. Thus in a number of flowering plants, the major energy-requiring processes of nitrate reduction and symbiotic N_2 fixation [18,139] occur in underground parts of the plant [14,159,213], where they must perforce be powered by respiratory reactions. Nitrate reduction in leaves can lead to problems of pH regulation [194,195], while the oxygen sensitivity of nitrogenase makes for problems in the direct use of reductant from non-cyclic electron flow for N_2 reduction [214].

Even in leaves, the cells associated with quantitatively important active transport processes such as the loading of photosynthate and ions delivered in the transpiration stream into the phloem, salt excretion in halophytes via glands or into bladders, and stomatal guard cells have a higher ratio of mitochondria to chloroplasts than the mesophyll cells [2,30,31,123,130, 138,160,161]. However, it is found that photosynthetic partial processes can power these processes at rates equal to, or greater than those supported by respiration [71,86,95,130,167]. This may have considerable importance under natural conditions, in that the substrates for the phloem transport and salt excretion processes are supplied in the light by transpiration and photosynthesis, and stomatal opening frequently occurs in response to light. Thus these energy-requiring processes occur mainly in the light, when they have the option of using energy directly from the photoreactions.

The relative scarcity of chloroplasts in these cells specialised for active transport suggests that there may be symplastic transport of photoproducts from nearby mesophyll cells [28,130,182,cf.86]. However, despite the low CO_2 fixation rate reported for some of these cells [131], it is possible that these chloroplasts are specialised for ATP or reductant supply rather than CO_2 fixation. Certainly the photochemical capacity of guard cells is adequate to power the active transport involved in stomatal movement [158,161,166], and they may not be in symplastic connection with other leaf cells [2]. Nelson and Mayo [150] have reported the occurrence of functional stomata in the orchid *Paphidio pedillum* in which the guard cells lack chlorophyll detectable by fluorescence microscopy. Investigations of this plant with respect to energy sources for stomatal movements in the light, and of the occurrence of plasmaodesmata between the guard cells and other leaf cells would be of interest.

A final point is that chloroplasts occur in higher plants in locations separated by large diffusion barriers from external CO_2 [56]. It has been argued [182] that these chloroplasts cannot bring about net fixation of exogenous CO_2, and that their major function may be in generating ATP and reductant for other processes.

12.5.4. *Conclusions on relation of direct use of photoproducts to photolithotrophic growth*

Advantages of the direct use of photoproducts in photolithotrophic algal cells can be seen in terms of more efficient use of resources. This is particularly true of "running costs" under both CO_2-limiting and light-limiting growth conditions; the advantages in terms of use of "capital resources" are less immediately obvious. On balance, the direct use of photoproducts has economic advantages. The photolithotrophic growth rate of algae compared with the heterotrophic growth rate is even greater than would be expected in terms of the advantages of the direct use of photoproducts, and suggests that other factors must be involved.

The presence of heterotrophic cells in multicellular photolithotrophs and further specialisations, mean that interpretations are more complicated than in unicellular algae. The direct use of photoproduced cofactors has considerable advantages in these plants.

12.6. ENERGY SUPPLY FOR CHLOROPLAST SYNTHESIS AND MAINTENANCE IN THE DARK AND DURING GREENING

Kirk and Tilney-Bassett [114] review the distribution of dark repression of chloroplast synthesis, and the evidence (see also ref. 113) that the synthesis of chloroplast components during greening can occur even when photosynthesis by the developing chloroplast is inhibited by DCMU. Even when the development of chloroplasts is light-dependent, the light-requirement is informational rather than energetic. Regardless of the energy source for chloroplast synthesis, chloroplast maintenance must occur in the natural dark period.

The presence of the shuttles [81,188,234, and see Chapter 6 of this volume] for transfer of carbon skeletons, reductant and high-energy phosphate across the chloroplast outer membranes can account for the transport of mitochondrial ATP and cytoplasmic reductant into the chloroplast. In green plants in the restricted sense, the chloroplast contains the enzymes of the oxidative pentose phosphate pathway (section 12.3.2), so they can generate reductant by this pathway using imported carbohydrate, as well as using shuttles in the manner considered in Chapters 5 and 6.

12.7. ACKNOWLEDGEMENTS

The ideas presented here have evolved in large measure in discussion with my present and past colleagues, particularly Drs. T. ap Rees, F.L. Bendall, S.M. Glidewell, R.L. Lyne, E.A.C. MacRobbie and J.I. Sprent.

REFERENCES

1 Alexander, N.J., Gillham, N.W. and Boynton, J.E. (1974) Mol. Gen. Genet., [130], 275–290.
2 Allaway, N.G. and Setterfield, G. (1972) Can. J.Bot., [50], 1405–1413.
3 Anderson, L.E. and Advani, V.R. (1970) Plant Physiol., [45], 583–585.
4 Anderson, L.E., Lim Ng, T-C. and Park, K-E.Y. (1974) Plant Physiol., [53], 835–839.
5 Andrews, T.J., Lorimer, G.M. and Tolbert, N.E. (1971) Biochemistry, [10], 4777–4782.
6 Andrews, T.J., Lorimer, G.M. and Tolbert, N.E. (1973) Biochemistry, [12], 11–17.
7 Ap Rees, T. (1974) in Plant Biochemistry (Northcote, D.H.N., ed), Vol.II of MTP International Review of Science, Biochemistry Series One, pp. 89–128, Butterworth, London.
8 Arnon, D.I. (1963) in Photosynthetic Mechanisms in Green Plants (Kok, B. and Jagendorf, A.T., eds), pp. 195–212. N.R.C.- N.A.S. publication 1145, Washington, D.C.
9 Atkinson Jr., A.W., John, P.C.L. and Gunning, B.E.S. (1974) Protoplasma, [81], 77–110.
10 Avron, M. and Neumann, J. (1968) A. Rev. Plant Physiol., [19], 137–166.
11 Bedell, G.W. III and Govindjee (1973) Plant Cell Physiol., [14], 1081–1097.
12 Beevers, H. (1970) in Prediction and Measurement of Photosynthetic Productivity (Malik, I., ed.), pp. 209–214, Pudoc, Wageningen.
13 Beevers, H. (1974) Plant Physiol., [54], 437–442.
14 Beevers, L. and Hagemann, R.H. (1969) A. Rev. Plant Physiol., [20], 495–522.
15 Begum, F. and Syrett, P.J. (1970) Arch. Microbiol. [72], 344–352.
16 Ben-Amotz, A. and Avron, M. (1972) Plant Physiol., [49], 244–248.
17 Ben-Shaul, Y., Epstein, H.T. and Schiff, J.A. (1965) Can. J. Bot., [43], 129–136.

18 Bergerson, F.J. (1971) in Biological Nitrogen Fixation in Natural and Agricultural Habitats (Lie, T.A. and Mulder, E.G., eds), pp. 511—524, Nijhoff, The Hague.
19 Bernstein, E. (1964) J. Protozool., [11], 56—74.
20 Bernstein, E. (1968) in Methods in Cell Physiology (Prescott, D.H., ed), Vol.III, pp. 119—145, Academic Press, New York.
21 Bisulputra, T. (1974) in Algal Physiology and Biochemistry (Stewart, W.D.P., ed), pp. 124—160, Blackwell, Oxford.
22 Biggins, J. (1973) Biochemistry, [12], 1165—1169.
23 Biggins, J. (1974) FEBS Lett., [38], 311—314.
24 Black Jr., C.C. (1973) A. Rev. Plant Physiol., [24], 253—286.
25 Bongers, L.H.J., (1958) Neth. J. Agr. Sci., [6], 79—88.
26 Boulter, D., Ellis, R.J. and Yarwood, A. (1972) Biol. Revs., [47], 113—175.
27 Bourne, D.T. and Ranson, S.L. (1965) Plant Physiol., [40], 1178—1190.
28 Brinckmann, E. and Lüttge, U. (1974) Planta, [119], 47—57.
29 Burton, M.D. and Moore, J. (1974) J. Ultrastruct. Res., [48], 414—419.
30 Canny, M.J. (1973) Phloem Translocation. Cambridge University Press.
31 Cataldo, D.A. and Berlyn, G.P. (1974) Amer. J. Bot., [61], 957—963.
32 Canvin, D.T. and Atkins, C.A. (1974) Planta, [116], 207—224.
33 Chapman, E.A. and Graham, D. (1974a) Plant Physiol., [53], 879—885.
34 Chapman, E.A. and Graham, D. (1974b) Plant Physiol., [53], 886—892.
35 Chapman, E.A. and Osmond, C.B. (1974) Plant Physiol., [53], 893—898.
36 Cheniae, G.M. and Martin, I.F. (1973) Photochem. Photobiol., [17], 441—459.
37 Cook, J.R. (1963) J. Protozool., [10], 436—444.
38 Cook, J.R. (1968) in The Biology of *Euglena* (Buetow, D.E.,ed), Vol.I, pp. 243—314, Academic Press, New York.
39 Cooksey, K.E. (1972) Plant Physiol., [50], 1—6.
40 Cooksey, K.E. (1974) J. Phycol., [10], 253—257.
41 Coombs, J. and Whittingham, C.P. (1966) Proc. Roy. Soc. B., [164], 511—520.
42 Craigie, J.S. (1974) in Algal Physiology and Biochemistry (Stewart, W.D.P., ed), pp. 206—235, Blackwell, Oxford.
43 Cramer, M. and Myers, J. (1949) Plant Physiol., [24], 255—264.
44 Crawford, R.M.M. (1972) Flora, [161], 209—223.
45 Danforth, W.F. (1962) in Physiology and Biochemistry of Algae (Lewin, R.A., ed), pp. 99—123, Academic Press, New York.
46 Davies, D.D. (1973) S.E.B. Symposia, [28], 513—529.
47 Devlin, R. M. and Galloway, R.A. (1968) Physiol. Plant., [21], 11—25.
48 Dodge, J.D. (1973) The Fine Structure of Algal Cells, Academic Press, London.
49 Douce, R., Holtz, R.B. and Benson, A.A. (1973) J. Biol. Chem., [248], 7215—7222.
50 Droop, M.R. (1974) in Algal Physiology and Biochemistry (Stewart, W.D.P., ed), pp. 530—559. Blackwell, Oxford.
51 Dunstan, W.H. (1973) J. Exp. Mar. Biol. Ecol., [13], 181—187.
52 Egneus, H., Heber, U. and Matthieson, U. (1974) Z. Physiol. Chem., [355], 1190.
53 Endo, H., Nakajima, K., Chino, R. and Shirata, M. (1974) Agr. Biol. Chem., [38], 9—18.
54 Eppley, R.W., Packard, T.T. and MacIsaac, J.J. (1970) Mar. Biol., [6], 195—199.
55 Everson, R.G. and Gibbs, M. (1967) Plant Physiol., [42], 1153—1154.
56 Fahn, A. (1967) Plant Anatomy, Pergamon, Oxford.
57 Fergusson, A.W. (1972) Diss. Abstr., [32B], 597.
58 Fock, H., Bate, G.C. and Egle, K. (1974) Planta, [121], 9—16.
59 Forrest, W.W. and Walker, D.J. (1971) Adv. Microbiol. Physiol., [5], 213—274.
60 Forsberg, C. (1965) Physiol. Plant., [18], 275—290.
61 Gibbs, M. (1962a) in Physiology and Biochemistry of Algae (Lewin, R.A., ed), pp. 61—90. Academic Press, New York.

62 Gibbs, M. (1962b) in Physiology and Biochemistry of Algae (Lewin, R.A., ed), pp. 91—97, Academic Press, New York.
63 Gimmler, H. (1973) Z. Pflanzenphysiol., [68], 289—307.
64 Givan, C.V. and Leech, R.M. (1971) Biol. Revs., [46], 409—428.
65 Glagoleva, T. and Zalenskii, O.V. (1970) Photosynthetica, [4], 15—20.
66 Glagoleva, T., Chulanovskaya, M.V. and Zalenskii, O.V. (1972) Photosynthetica, [6], 354—363.
67 Glidewell, S.M. and Raven, J.A. (1975) J. Exp. Bot., [26], 479—488.
68 Glidewell, S.M. and Raven, J.A. (1976) J. Exp. Bot., [27], 200—204.
69 Goulding, K.M. and Merrett, M.J. (1966) J. Exp. Bot., [17], 678—689.
70 Green, W.G.E. and Israelstam, G.F. (1970) Physiol. Plant., [23], 217—231.
71 Habeshaw, D. (1969) J. Exp. Bot., [20], 64—71.
72 Hall, D.O. and Evans, M.C.W. (1972) Sub-cell. Biochem., [1], 197—206.
73 Halldall, P. (1970) Photobiology of Micro-organisms, Wiley Interscience, London.
74 Hatch, M.D. and Slack, C.R.. (1970) A. Rev. Plant Physiol., [21], 141—162.
75 Havenkamp-Obbema, R., Mooman, A. and Stegwee, D. (1974) Z. Pflanzenphysiol., [45], 277—286.
76 Hayaishi, O. (1962) in Oxygenases (Hayaishi, O., ed), pp. 1—31, Academic Press, New York.
77 Hayaishi, O. (1974) in Molecular Mechanisms of Oxygen Activation(Hayaishi, O., ed), pp. 1—28, Academic Press, New York.
78 Heber, U. (1963) Planta, [59], 600—616.
79 Heber, U. (1969) Biochim. Biophys. Acta, [180], 302—319.
80 Heber, U. (1973) Biochim. Biophys. Acta, [305], 140—152.
81 Heber, U. (1974) A. Rev. Plant Physiol., [25], 393—421.
82 Heber, U. and Kirk, M.R. (1975) Biochim. Biophys. Acta, [376], 136—150.
83 Heber, U. and Santarius, K.A. (1966) Biochim. Biophys. Acta, [109], 390—408.
84 Heber, U., Pon, N.G. and Heber, M. (1963) Plant Physiol., [38], 355—360.
85 Heichel, G.H. (1971) Photosynthetica, [5], 93—98.
86 Hill, B.S. and Hill, A.E. (1973) J. Membrane Biol., [12], 145—158.
87 Hirt, G., Tanner, W. and Kandler, O. (1971) Plant Physiol., [47], 841—843.
88 Hoch, G. and Owens, O. van H. (1963) in Photosynthetic Mechanisms of Green Plants (Kok, B. and Jagendorf, A.T., eds), pp. 409—420, N.R.C. - N.A.S. publication 1145, Washington, D.C.
89 Hoch, G., Owens, O.van H. and Kok, B. (1963) Arch. Biochem. Biophys., [101], 171—180.
90 Hoffmann, A. (1972) Planta, [102], 72—84.
91 Hoogenhout, M. and Amesz, J. (1965) Arch. Mikrobiol., [50], 10—25.
92 Hope, A.B., Lüttge, U. and Ball, E. (1974) Z. Pflanzenphysiol., [72], 1—10.
93 Hopkinson, J.M. (1964) J. Exp. Bot., [15], 125—137.
94 Hughes, A.P. (1971) Chem. and Ind., 1971. (No.32), 904—909.
95 Humble, G.D. and Hsaiao, T.C. (1970) Plant Physiol., [46], 483—487.
96 Huth, W. (1967) Flora, [158], 58—87.
97 Ikuma, H. (1972) A.Rev. Plant Physiol., [23], 419—436.
98 Jackson, W.A. and Volk, R.J. (1970) A. Rev. Plant Physiol., [21], 385—432.
99 Jeschke, W.D. (1972) Planta, [103], 164—180.
100 Jeschke, W.D. and Simonis, W. (1969) Planta, [88], 157—171.
101 Johansen, C. and Lüttge, U. (1974) Z. Pflanzenphysiol., [71], 189—199.
102 Johnson, H.S. (1972) Planta, [106], 273—277.
103 Kaiser, W. and Urbach, W. (1973) Ber. Deut. Bot. Ges., [86], 213—226.
104 Kaiser, W. and Urbach, W. (1974) Ber. Deut. Bot. Ges., [87], 145—153.
105 Kanazawa, T., Kanazawa, K., Kirk, M.R. and Bassham, J.A. (1970) Plant Cell Physiol., [11], 149—160.

106 Kandler, O. (1955a) Z. Naturforsch., [10B], 38—46.
107 Kandler, O. (1955b) Ann. Biol., [31], 373—383.
108 Kandler, O. and Tanner, W. (1966) Ber. Deut. Bot. Ges., [79], (48)—(57).
109 Kannangara, C.G. and Stumpf, P.K. (1972) Plant Physiol., [49], 497—501.
110 Kelly, D.P. (1971) A. Rev. Microbiol., [25], 177—210.
111 Kessler, E. (1964) A. Rev. Plant Physiol., [15], 57—72.
112 Kessler, E. (1974) in Algal Physiology and Biochemistry (Stewart, W.D.P., ed), pp. 456—473, Blackwell, Oxford.
113 Kirk, J.T.O. (1970) A. Rev. Biochem., [21], 11—42.
114 Kirk, J.T.O. and Tilney-Bassett, R.A.E. (1967) The Plastids, Freeman, London.
115 Klob, W. Kandler, O. and Tanner, W. (1973) Plant Physiol., [51], 825—827.
116 Klob, O. Tanner, W. and Kandler, O. (1971) in Proc. 2nd Int. Cong. Photosynthesis (Forti, G., Avron, M. and Melandri, A., eds), Vol. 3, pp.1998—2010, Junk, The Hague.
117 Kok, B. (1952) Acta, Bot. Neerl., [1], 445—467.
118 Kok, B. (1960) in Encyclopedia of Plant Physiology (Ruhland, W., ed), Vol. V/2, pp. 566—663, Springer, Berlin.
119 Kok, B. (1972) in Horizons in Bioenergetics (San Pietro, A. and Gest, H., eds), pp. 153—170, Academic Press, New York.
120 Komor, E. and Tanner, W. (1974) Z. Pflanzenphysiol., [71], 115—128.
121 Kylin, A. (1966) Physiol. Plant., [19], 883—887.
122 Latzko, E. and Gibbs, M. (1969) Plant Physiol., [44], 295—300.
123 Lauchli, A. (1972) A. Rev. Plant Physiol., [23], 197—218.
124 Lea, P.J. and Miflin, B.J. (1974) Nature, [251], 614—616.
125 Leclerc, J.C. (1967) Photosynthetica, [1], 179—191.
126 Leech, R.M. (1963) Biochim. Biophys. Acta, [71], 253—265.
127 Lewin, R.A. (1954) J. Gen. Microbiol., [11], 459—471.
128 Lloyd, D. (1974) in Algal Physiology and Biochemistry (Stewart, W.D.P., ed), pp. 505—529, Blackwell, Oxford.
129 Lorenzen, H. and Hesse, M. (1974) in Algal Physiology and Biochemistry (Stewart, W.D.P., ed), pp. 894—908., Blackwell, Oxford.
130 Lüttge, U. (1971) A. Rev. Plant Physiol., [22], 23—44.
131 Lüttge, U. and Osmond, C.B. (1970) Austr. J. Biol. Sci., [23], 17—26.
132 McGowan, R.E. and Gibbs, M. (1974) Plant Physiol., [54], 312—319.
133 Marshard, E. (1958) in Encyclopedia of Plant Physiology, Vol. VIII (Ruhland, W., ed), pp. 119—149, Springer, Berlin.
134 Mayne,B.C., Dee, A.M. and Edwards, D.G. (1975) Z. Pflanzenphysiol., [74], 275—291
135 Mazliak, P. (1973) A. Rev. Plant Physiol., [24], 287—310.
136 Miflin, B.J. (1972) Planta, [105], 225—233.
137 Miflin, B.J. and Beevers, H. (1974) Plant Physiol., [53], 870—874.
138 Milthorpe, F.L. and Moorby, J. (1969) A. Rev. Plant Physiol., [20], 117—138.
139 Minchin, F.R. and Pate, J.S. (1973) J. Exp. Bot., [24], 259—271.
140 Morris, I. (1974) in Algal Physiology and Biochemistry (Stewart, W.D.P., ed), pp. 583—609, Blackwell, Oxford.
141 Myers, J. (1947) J. Gen. Physiol., [30], 217—227.
142 Myers, J. (1951) A. Rev. Microbiol., [5], 157—180.
143 Myers, J. (1971) A. Rev. Plant Physiol., [22], 289—312.
144 Myers, J. and Graham, J-R. (1956) J. Cell. Comp. Physiol., [47], 397—414.
145 Myers, J. and Graham, J-R (1959) Plant Physiol., [34], 345—352.
146 Myers, J. and Graham, J-R. (1961) Plant Physiol., [36], 342—346.
147 Myers, J. and Johnston, J.A. (1949) Plant Physiol,, [24], 111—119.
148 Nakayama, O., Ueno, T. and Tsuchiya, F. (1974) Ferment. Technol., [52], 225—232.
149 Neilsen, A.H., Holm-Hansen, O. and Lewin, R.A. (1972) J. Gen. Microbiol., [71], 141—148.

150 Nelson, S.D. and Mayo, J.M. (1975) Can. J. Bot., [53], 1—7.
151 Neyra, C.A. and Hagemann, R.H. (1974) Plant Physiol., [54], 480—483.
152 Nuhrenberg, B., Lesemann, D. and Pirson, A. (1968) Planta, [79], 162—180.
153 Ogren, W.L. and Krogmann, D.W. (1965) J. Biol. Chem., [240], 4603—4608.
154 Ohad, I., Siekevitz, P. and Palade, G.E. (1967) J. Cell Biol., [35], 521—552.
155 Ohmann, E., Rindt, K.P. and Boriss, R. (1969) Z. Allg. Mikrobiol., [9], 557—564.
156 Opute, F.I. (1974a) J. Exp. Bot., [25], 798—809.
157 Opute, F.I. (1974b) J. Exp. Bot., [25], 810—822.
158 Pallas, Jr., J.E. and Dilley, R.A. (1972) Plant Physiol., [49], 649—650.
159 Pate, J.S. (1973) Soil Biol. Biochem., [5], 109—120.
160 Pate, J.S. and Gunning, B.E.S. (1972) A. Rev. Plant Physiol., [23], 173—196.
161 Pearson, C.J. and Milthorpe, F.L. (1974) Aust. J. Plant Physiol., [1], 221—236.
162 Pederson, T.A., Kirk, M. and Bassham, J.A. (1965) Physiol. Plant., [19], 219—231.
163 Penning de Vries, F.W.T. (1972) in Crop Processes in Controlled Environments (Rees, A.R., Cockshull, K.E., Hand, D.W. and Hurd, R.G., eds), pp. 327—347, Academic Press, London.
164 Penning de Vries, F.W.T. (1975) Ann. Bot., [39], 77—92.
165 Penning de Vries, F.W.T., Brunsting, A.H.M. and van Laar, M.M. (1974) J. Theoret. Biol., [45], 339—377.
166 Penny, M.G. and Bowling, D.J.F. (1974) Planta, [119], 17—25.
167 Plaut, Z. and Reinhold, L. (1969) Austr. J. Biol. Sci., [22], 1105—1113.
168 Preiss, J. and Kosuge, T. (1970) A. Rev. Plant Physiol., [21], 433—466.
169 Prins, H.B.A. (1974) Ph.D. Thesis, University of Groningen.
170 Pulich, W.H. and Ward, C.H. (1973) Plant Physiol., [51], 337—344.
171 Rambeck, W.A. and Bassham, J.A. (1973) Biochim. Biophys. Acta, [304], 725—735.
172 Raven, J. A. (1967a) J. Gen. Physiol., [50], 1607—1625.
173 Raven, J.A. (1967b) J. Gen. Physiol., [50], 1927—1940.
174 Raven, J.A. (1968) Abh. Deutsch. Akad. Wiss. Berlin, Nr. [4a], 145—151.
175 Raven, J.A. (1969a) New Phytol., [68], 45—62.
176 Raven, J.A. (1969b) New Phytol., [68], 1089—1113.
177 Raven, J.A. (1970a) J. Exp. Bot., [21], 1—16.
178 Raven, J.A. (1970b) Biol. Revs., [45], 167—221.
179 Raven, J.A. (1971a) Chem. and Ind., 1971, 859—865.
180 Raven, J.A. (1971b) Trans. Bot. Soc. Edinb., [41], 219—225.
181 Raven, J.A. (1971c) J. Exp. Bot., [22], 420—433.
182 Raven, J.A. (1972a) New Phytol., [71], 227—247.
183 Raven, J.A. (1974a) in Algal Physiology and Biochemistry (Stewart, W.D.P., ed), pp. 391—423, Blackwell, Oxford.
184 Raven, J.A. (1974b) in Algal Physiology and Biochemistry (Stewart, W.D.P., ed), pp. 434—455, Blackwell, Oxford.
185 Raven. J.A. (1974c) J. Exp. Bot., [25], 221—229.
186 Raven. J.A. (1975a) in Transport in Cells and Tissues (Lüttge, U. and Pitman, M.G., eds), Volume of Encyclopedia of Plant Physiology, New Series, in press, Springer, Berlin.
187 Raven. J.A. (1975b) in Ion Transport in Plant Cells and Tissues (Baker, D.A. and Hall, J.L., eds), Chapter 5, pp. 125—160, Elsevier, Amsterdam.
188 Raven, J.A. (1976a) New Phytol., [76], 195—204.
189 Raven, J.A. (1976b) New Phytol., [76], 205—212.
190 Raven, J.A. (1976c) Ann. Bot., [40], 581—602.
191 Raven, J.A. and Glidewell, S.M. (1975a) Photosynthetica, [9], 361—371.
192 Raven, J.A. and Glidewell, S.M. (1975b) New Phytol., [75], 197—204.
193 Raven, J.A. and Glidewell, S.M. (1975c) New Phytol., [75], 205—213.

194 Raven, J.A. and Smith, F.A. (1974) Can. J. Bot., [52], 1035—1048.
195 Raven, J.A. and Smith, F.A. (1976) New Phytol., [76], 415—431.
196 Ried, A. (1970) in Prediction and Measurement of Photosynthetic Productivity, (Malik, J., ed), pp. 231—246, Pudoc, Wageningen.
197 Ried, A., Muller, I. and Soeder, C.J. (1962) Vortr. Ges. Bot. Deut. Bot. Ges., [1], 187—194.
198 Reid, A., Setlik, I., Bossert, U. and Berkova, F. (1973) Photosynthetica, [7], 161—176.
199 Saha, S., Izawa, S. and Good, N.E. (1970). Biochim. Biophys. Acta, [223], 158—164.
200 Samejima, H. and Myers, J. (1958) J. Gen. Microbiol., [18], 107—117.
201 Sarkassian, G.S. and Fowler, M.W. (1974) Planta, [119], 335—350.
202 Sasa, T. (1961) Plant Cell Physiol., [2], 253—270.
203 Schiff, J.A. and Epstein, H.T. (1968) in The Biology of Euglena (Buetow, D.E., ed), Vol. 2, pp. 285—333, Academic Press, New York.
204 Schiff, J.A. and Hodson, R.C. (1973) A. Rev. Plant Physiol., [24], 381—414.
205 Schnarrenberger, C. and Oeser, A. (1974) Eur. J. Biochem., [45], 77—82.
206 Schotz, F., Bathelt, H., Arnold, C-G. and Schimmer, O. (1972) Protoplasma, [75], 229—254.
207 Schurmann, P., Buchanan, B.B. and Arnon, D.I. (1972) Biochim. Biophys. Acta, [267], 111—124.
208 Simonis, W. (1967) Ber. Deut. Bot. Ges., [80], 395—402.
209 Simonis, W. and Urbach, W. (1973) A. Rev. Plant Physiol., [24], 89—114.
210 Smillie, R.M. (1963) Can. J. Bot., [41], 123—154.
211 Smith, F.A. and Raven, J.A. (1974) New Phytol., [73], 1—12.
212 Sorokin, C. and Krauss, R.W. (1962) Plant Physiol., [37], 37—42.
213 Stewart, W.D.P. (1966) Nitrogen Fixation in Plants, Athlone, University of London.
214 Stewart, W.D.P. (1973) A. Rev. Microbiol., [27], 283—316.
215 Stuart, T.S. (1971) Planta, [96], 81—92.
216 Stumpf, P.K. and Boardman, N.K. (1970) J. Biol. Chem., [245], 2579—2599.
217 Syrett, P.J. (1962) in Physiology and Biochemistry of Algae (Lewin, R.A., ed), pp. 171—188, Academic Press, New York.
218 Syrett, P.J. and Wong, H-A. (1963) Biochem. J. [89], 308—315.
219 Tamiya, H. (1966) A. Rev. Plant Physiol., [27], 1—26.
220 Tanner, W., Löffler, M. and Kandler, O. (1969) Plant Physiol., [44], 422—429.
221 Tanner, W., Loos, E., Klob, W. and Kandler, O. (1968) Z. Pflanzenphysiol., [59], 301—303.
222 Tolbert, N.E. (1973) Current Topics in Cell Regulation, [7], 21—50.
223 Tolbert, N.E. (1974) in Algal Physiology and Biochemistry (Stewart, W.D.P., ed), pp. 474—504, Blackwell, Oxford.
224 Tolbert, N.E., Oeser, A., Yamazaki, R.K., Hagemann, R.H. and Kisaki, T. (1969) Plant Physiol., [44], 135—147.
225 Trebst, A. (1970) Ber. Deut. Bot. Ges. [83], 373—398.
226 Trebst, A. (1974) A. Rev. Plant Physiol., [25], 423—458.
227 Ullrich, W.R. (1972) in Proc. 2nd Int. Cong. Photosynthesis (Forti, G., Avron, H. and Melandri, A., eds), Vol. 3, 1329—1336, Junk, The Hague.
228 Ullrich, W.R. (1974) Planta, [116], 143—152.
229 Ullrich-Eberius, C.I. (1973) Planta, [115], 25—36.
230 Urbach, W. and Gimmler, H. (1970) Z. Pflanzenphysiol., [62], 276—286.
231 Van Oorschot, J.L.P. (1955) Meded. Landbouw, Wageningen, [55], 225—276.
232 Von Sumere, C.F. and Dedonder, A. (1971) Z. Pflanzenphysiol., [65], 159—175.
233 Walker, D.A. (1962) Biol. Revs., [37], 215—256.

234 Walker, D.A. (1974) in Plant Biochemistry (Northcote, D.H.N., ed), Vol. II of MTP International Review of Science, Biochemistry Series One, pp. 1—50, Butterworth, London.
235 Weismann, G.S. (1972) Plant Physiol., [49], 138—141.
236 Wiessner, W. (1970) in Photobiology of Microorganisms (Halldall, P., ed), pp. 135—164, Wiley Interscience, London.
237 Wild, A., Ke, B. and Shaw, E.R. (1973) Z. Pflanzenphysiol., [69], 344—350.
238 Willenbrink, J. (1964) Z. Naturforsch., [19B], 356—357.
239 Wintermans, J.F.G.M. (1955) Meded. Landbouw, Wageningen, [55], 69—126.
240 Yung, K.H. and Mudd. J.B. (1966) Plant Physiol., [41], 506—509.

Subject Index

Author Index